The Audubon Society
Field Guide to
North American Mushrooms

A Chanticleer Press Edition

The Audubon Society Field Guide to North American Mushrooms

Gary H. Lincoff,
President, North American Mycological
Association

Visual Key by
Carol Nehring

Alfred A. Knopf, New York

This is a Borzoi Book
Published by Alfred A. Knopf, Inc.

All rights reserved. Copyright 1981
under the International Union for the
protection of literary and artistic works
(Berne). Published in the United States
by Alfred A. Knopf, Inc., New York,
and simultaneously in Canada by
Random House of Canada Limited,
Toronto. Distributed by Random
House, Inc., New York.

Prepared and produced by
Chanticleer Press, Inc., New York.

Color reproductions by Nievergelt
Repro AG, Zurich, Switzerland.
Type set in Garamond by
Dix Type Inc., Syracuse, New York.
Printed and bound by Dai Nippon,
Tokyo, Japan.

Published December 1, 1981
Reprinted four times
Sixth Printing, September 1989

Library of Congress Catalog Number:
81-80827
ISBN: 0-394-51992-2

CONTENTS

11 Introduction
31 How to Use This Guide

Part I Color Plates
35 Key to the Color Plates
36 Thumb Tab Guide
52 *Color Plates*
Small, Fragile Gilled Mushrooms
1–108
Veiled Mushrooms with Free Gills
109–180
Veiled Mushrooms with Attached Gills
181–228
Mushrooms with Free Gills 229–234
Mushrooms with Attached Gills 235–
369
Boletes and Others 370–420
Chanterelles and Other Vase-shaped
Mushrooms 421–444
Stalked Polypores, Tooth Fungi, and
Others 445–480
Polypores and Other Shelflike
Mushrooms 481–549
Slimes, Jellies, and Crustlike Fungi
550–597
Cup-shaped Mushrooms and Bird's-nest
Fungi 598–633
Puffballs, Earthstars, Amanita Buttons,
and Others 634–681
Morels, Stinkhorns, and Other Club-
shaped Mushrooms 682–732
Coral-like Mushrooms 733–756

Contents

Part II Text

323 Ascomycetes
377 Basidiomycetes
843 Slime Molds

Part III Appendices

857 Spore Print Chart
871 Mushroom Poisoning
875 Cooking and Eating Wild Mushrooms
881 Glossary
887 Authors of Fungi
889 Picture Credits
893 Index

THE AUDUBON SOCIETY

The National Audubon Society is among the oldest and largest private conservation organizations in the world. With over 560,000 members and more than 500 local chapters across the country, the Society works in behalf of our natural heritage through environmental education and conservation action. It protects wildlife in more than eighty sanctuaries from coast to coast. It also operates outdoor education centers and ecology workshops and publishes the prizewinning AUDUBON magazine, AMERICAN BIRDS magazine, newsletters, films, and other educational materials. For further information regarding membership in the Society, write to the National Audubon Society, 950 Third Avenue, New York, New York 10022.

THE AUTHOR

Gary H. Lincoff is President of the
North American Mycological
Association and teaches field mycology
and botany at the New York Botanical
Garden. He is co-author of *Toxic and
Hallucinogenic Mushroom Poisoning* and
author of *A Guide to the Poisonous
Mushrooms in the Greater New York Area*.
Mr. Lincoff has traveled throughout
North America lecturing, conducting
workshops, and studying fungi and
wild plants.

ACKNOWLEDGMENTS

Many people have helped make this field guide what it is. Howard Bigelow, Peter Katsaros, Samuel Ristich, Clark Rogerson, Kit Scates, Walter Sturgeon, and Greg Wright read portions of the manuscript. Their field knowledge and critical skills helped in many ways to make this a better work. Bunji Tagawa, who illustrated the book, helped me work out many taxonomic problems, as did Sam Ristich, who aided me in almost every facet of the project. Any omissions or factual errors are, of course, my own.

The staff of the Library of the New York Botanical Garden facilitated my research, and the Library's resources made possible the writing of this book. The founders of the New York Mycological Society, John Cage and Guy Nearing, contributed much— John, whose writings first interested me in mushrooms, and Guy, who showed me that they can, indeed, be identified.

A field guide is a synthesis of the work of countless amateur and professional observers. I am especially indebted to such eminent mycologists as Richard P. Korf, Josiah L. Lowe, Orson K. Miller, Jr., René Pomerleau, Alexander H. Smith, Daniel E. Stuntz, and Harry D. Thiers, whose research on North

American mushrooms has made my task much easier. I also profited from the analytical acuity of Rolf Singer. I am also deeply grateful to all the regional mushroom clubs in the United States, and especially to the North American Mycological Association and its founder, Harry S. Knighton, for making it possible for people to get together in the pursuit of this most challenging of the natural sciences.

I owe special thanks to my wife, Irene Liberman, not only for her assistance and advice, but also for her patience. My father's encouragement and support have aided me in countless ways. I also want to thank the many good friends who stood by me during this project, especially Walter and Arline Deitch, who recommended me for this book; Adria Hillman, for her good counsel; Emanuel Salzman, for his perspective; and Sylvia Stein, whose enthusiasm kept my spirits up.

Finally I want to thank Paul Steiner and the staff of Chanticleer Press for their creative contributions and hard work: Gudrun Buettner, whose expertise helped solve many problems; Carol Nehring, who worked out the visual key and oversaw the layout; Susan Rayfield, who developed the plan for the guide; Helga Lose and John Holliday, who saw the book through production; and most of all, Mary Beth Brewer and Ann Whitman, who edited and coordinated the project.

INTRODUCTION

Mushrooms are among the most mysterious life forms. The ancient Greeks believed they came from Zeus's lightning because they appeared after rains and reproduced and grew inexplicably. In the Middle Ages, the circular patterns formed by some mushrooms were dubbed "fairy rings" and were thought to be the work of the "little people," who supposedly danced around them at midnight, performing magic rites. In the New World, some hallucinogenic mushrooms have been called "the food of the gods" and invested with supernatural powers. Mushrooms are mysterious, but not as they were once thought to be. They appear suddenly, and often in places where they have never been seen before. They have, in fact, been out of sight, growing underground or beneath bark. And much remains to be learned about fungi; some species contain dangerous toxins, many of which are not yet fully understood.

Some mushrooms are of course edible. Since Roman times, fungi have been famous as gourmet fare. Truffles, boletes, chanterelles, and morels, all of which grow in North America, often fetch fantastic prices. And with good reason: after tasting wild fungi, most people find the common cultivated mushroom bland and uninteresting. Mushrooms also play a vital role in the world's ecosystem. Many land plants could not thrive in their absence, since some establish a symbiotic relationship with fungi, exchanging essential nutrients. And were it not for mushrooms, which hasten decomposition, many dead plants and fallen trees would take far longer to decay.

This book is designed for anyone who wants to identify the common mushrooms of North America. The combination of vivid color photographs and nontechnical descriptions should make it relatively easy to do so.

Geographical Scope:

This guide covers North America north of Mexico. The ranges given in the descriptions indicate where the species have been found; they may occur elsewhere as well. Too little is known about the distribution of most North American mushrooms to be more specific.

What Is a Mushroom?

A mushroom is the fruiting body of a plant, the part of a fungus that typically appears above ground and contains its reproductive units, or spores.

Although mushrooms are usually considered members of the plant kingdom, they differ from most plants in that they lack chlorophyll and must rely on organic material for nutrition. They do this in 3 ways: as saprophytes, as parasites, and as mycorrhizae. Saprophytes live on dead organic matter, including dead wood, the dead tissue of living trees, dung, leaf litter, or conifer litter. Parasites attack living plants or animals. Mycorrhizal mushrooms have a symbiotic relationship with plants, usually either trees or shrubs. The underground, vegetative part of the fungus—the mycelium—extracts nutrients from the

substrate. The mycelium of a
mycorrhizal mushroom sheaths the root
ends of the flowering plant, expanding
the plant's root system; the mushroom
receives necessary carbohydrates from
the tree.

Parts of a Mushroom: A gilled mushroom is described in full here. (Complete details on the anatomy of other groups of mushrooms will be found in the text.) Each part is roughly analogous to parts of other mushrooms. For example, the gills in these fungi perform the same function as the tubes in boletes and polypores—the production of spores.

Cap: The cap is the most conspicuous part of a mushroom and its shape can be important in identifying different genera. At first, many mushroom caps are nearly round, conical, bell-shaped, or convex. As most species mature, the cap becomes broadly convex or flat, or develops uplifted edges. Some mushrooms are vase-shaped, or become vase-shaped with age. The center of the cap may be knobbed, either broadly or sharply, or it may be sunken. The margin of the cap may be inrolled, downcurved, straight, or upturned, and torn, wavy, hairy, or smooth. Some species have hanging veil remnants along the margin. Others are radially lined, deeply furrowed, wrinkled, or pitted. Caps may be dry, moist, sticky, or slimy when they are fresh. However, a sticky or slimy cap may appear dry when it is old or dried out; put a drop of water on it to see if it becomes sticky or slimy. The color of a cap that feels or looks moist often fades as moisture evaporates.

In addition to being wet or dry, cap surfaces may be smooth and hairless, powdery, granular, or adorned with radial lines, hairs, scales, or veil remnants. If hair is present, it may be silky, fibrous, or scalelike.

Bell-shaped

Conical

Knobbed

Flat

Sunken

Surfaces
Smooth

Velvety

Hairy to fibrous

With raised scales

With flat scales

With patches

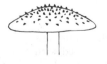 

The flesh of the cap lies beneath the surface layer, or cuticle. It may be th or thin, soft or tough, white or pigmented. When the flesh of *Lactariu* species is cut, it exudes a milky fluid called latex.

Mushrooms often have a distinctive odor, especially when they are crushed. Some smell like anise, fruit, cucumbers, green corn, creosote, fish, radishes, garlic, or raw potatoes.

Taste can be an important aid in differentiating between similar species. For example, different *Russula* species may taste mild, acrid, biting, or peppery. To determine the taste of a mushroom, chew a very small piece until the flavor becomes apparent and then spit the mushroom out. Study the deadly and poisonous species in advance and avoid testing them.

Gills: Gills are platelike structures on the underside of the cap; they radiate out from the stalk and produce the spores. They may be attached to the stalk directly, at a 90° angle (adnate), or obliquely, at a 45° angle (adnexed); some are notched just short of where they are attached to the stalk.

In many mushrooms, the gills run down the stalk slightly or deeply. In some, the gills pull away from the cap as it expands; these are referred to as seceding gills. In a few genera, the gills are free, or not attached to the stalk. The easiest way to observe gill attachment is to slice a mushroom in half longitudinally.

Gills are variously spaced. Some are crowded together so tightly that one cannot see any space between them; others may be close to one another or widely separated (distant). Most gills have thin edges that may be minutely fringed, toothed, or colored differently from the rest of the gills.

Stalk: Most gilled mushrooms have a stalk, usually located at the center of the cap; some are off-center or attached at one

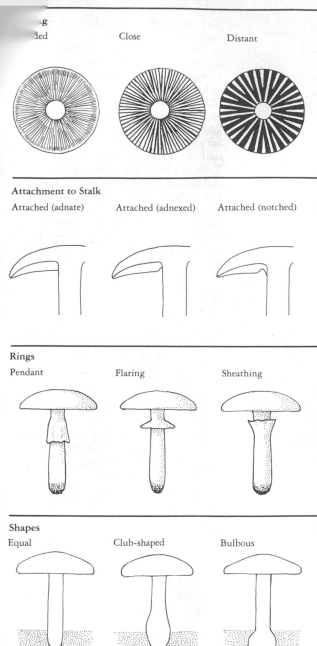

...g

...ed   Close   Distant

Attachment to Stalk

Attached (adnate)   Attached (adnexed)   Attached (notched)

Rings

Pendant   Flaring   Sheathing

Shapes

Equal   Club-shaped   Bulbous

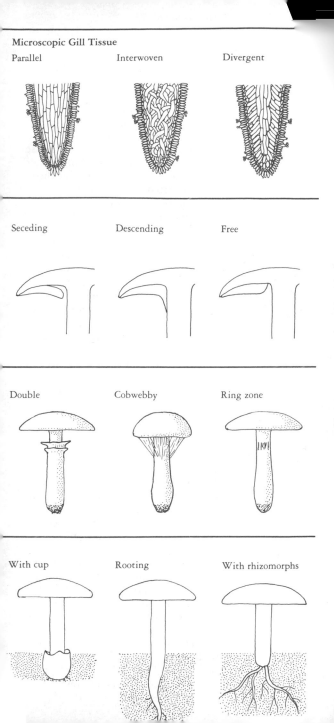

Microscopic Gill Tissue

Parallel

Interwoven

Divergent

Seceding

Descending

Free

Double

Cobwebby

Ring zone

With cup

Rooting

With rhizomorphs

central

side (lateral). In the few cases where a stalk is absent or very small and stublike, the mushrooms usually grow on wood, and are thus elevated sufficiently for spore dispersal. Stalks have various shapes. Some are even, or the same width from top to bottom. Others are club-shaped, flattened, or spindle-shaped. The base may be bulbous, abruptly bulbous, or tapering to a rootlike growth at the base, or it may have runners

off-center

(rhizomorphs) growing out of it. The stalk surface may be smooth, dotted, lined, netted, scaly, powdery, or hairy. In some species, the stalk snaps apart when bent; a few are rubbery. Stalks may be hollow, solid, or filled with cottony tissue. Many

stalkless

species have a ring, a skirt- or bandlike tissue, or a zone of fibers on the stalk left by the partial veil as the mushroom expands. Others have a cup (also called a volva) or bits of tissue about the stalk base; these are remnants of a universal veil.

Veils: Some mushrooms have veils, membranes that cover and protect either the entire immature mushroom or the immature gills. As the mushroom enlarges, the membrane ruptures, usually leaving traces on the cap or stalk.
If the membrane covers the entire mushroom, it is known as a universal veil; on rupturing, it may leave patches on the cap, or a saclike cup, patches, or bands of tissue at the stalk base.
A membrane that encloses the immature gills on the unexpanded cap is called a partial veil. When the cap expands, the partial veil breaks and may leave remnants along the cap margin or on the stalk. The remnants often appear as a ring or skirt around the stalk. Usually the ring left by a partial veil is single, but in some species it is 2-layered. In a few species, rings can be loosened and moved.

Classification of Mushrooms: Within the plant kingdom, fungi constitute the division Eumycota, which is split into subdivisions, 3 of which are included in this guide. Each subdivision is divided into classes, each class into orders, and each order into families. For example, the common commercial mushroom, *Agaricus bisporus,* belongs to the subdivision Basidiomycotina, which contains all fungi that produce spores on microscopic structures known as basidia; to the class Hymenomycetes, a grouping also based on microscopic spore characteristics, the location of the spores, and the way they are dispersed; to the order Agaricales, which contains all gilled mushrooms and boletes; and to the family Agaricaceae, which includes *Agaricus* species and their close relatives. Families are further divided into genera (singular, genus), and the genera into species. When people speak of a particular mushroom they usually have in mind a species, such as the Chanterelle (*Cantharellus cibarius*). Some species have named varieties. For example, the Cleft-foot Amanita (*Amanita brunnescens*) has a white variety, which is called *Amanita brunnescens* var. *pallida.* In some cases, the varieties are so unlike one another that some experts consider them different species or members of a species complex.

One of the problems facing the beginning mushroom hunter is the large number of names that have been given to the same species and the frequent changes in names. Until recently, mushrooms were classified according to their appearance in the field, and similar species were classified in the same families or genera. In the last few decades, mycologists working with microscopes have found that some mushrooms that resemble one another in the field are quite dissimilar microscopically. Other mycologists are experimenting with chemical analysis, and their findings have also altered the classification of mushrooms. Although some people still favor the groupings based solely on field characteristics, the new tendency is to group mushrooms together on the basis of microscopic and chemical properties as well. In the end this will give us a fuller understanding of the natural relationships that exist between groups of mushrooms.

How Mushrooms Reproduce and Grow:

All mushrooms produce millions of spores, microscopic reproductive units that are dispersed in various ways. Spores that germinate develop into hyphae (singular, hypha), threadlike strands that are collectively known as the mycelium (plural, mycelia). Each hyphal cell has one nucleus. If two genetically compatible mycelia come into contact, they join to form a mycelium containing cells with 2 nuclei. This mycelium with 2 nuclei per cell is the vegetative portion of the fungus; it takes in nourishment from various types of organic matter. Given the right conditions—not yet completely understood—the vegetative mycelium begins to develop small knots of hyphal tissue, which grow and become immature fruiting bodies. These expand and form mature mushrooms. Spores are dispersed, and the life cycle begins again.

The ascomycetes produce spores in saclike containers that are typically sausage-shaped. These are known as asci (singular, ascus). Most asci contain 8 spores, which are usually dispersed through an opening at the tip of the ascus. The asci line the exposed surface of cup fungi, and the pits and grooves of morels. In the flask fungi, the asci line the inside of minute, flask-shaped vessels known as perithecia. In truffles, the asci develop in the interior of the mushroom and the spores are dispersed when the mushroom decays or when

Where Spores Are Produced

Microscopic Spores

Ascus

spore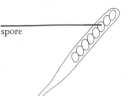

Basidium

spore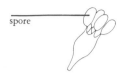

Morel

asci in pits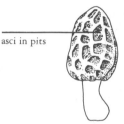

Cup Fungus

asci on inner surface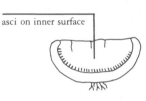

Chanterelle

basidia on outer surface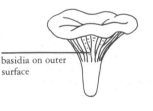

Puffball

basidia in spore mass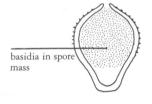

Gilled Mushroom

basidia on gills

Bolete

basidia in tubes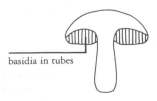

animals eat it and spread the spores in their droppings. Many ascomycetes can reproduce asexually, forming spores independent of asci.

The basidiomycetes produce their spores on short appendages known as sterigmata that project out of microscopic structures known as basidia (singular, basidium). The basidia are typically club-shaped. In the class Hymenomycetes, the basidia line the fertile surfaces of the mushrooms and the spores are actively discharged. In chanterelles, basidia line the mushroom's underside; in tooth fungi, they occur on the teeth; in pore fungi, in the tubes; and in gilled mushrooms, on the gills.

In the class Gasteromycetes, the spores are not self-propelled. In the bird's-nest fungi, they are dispersed as raindrops strike the "eggs"; in puffballs, they are shot (or "puffed") through an apical opening in the mushroom; in stinkhorns, the spores adhere to the legs of insects attracted to the foul-smelling slime the mushroom secretes. Other gasteromycetes decay, and thus release the enclosed spore mass.

Slime molds can reproduce sexually or asexually. Sexual reproduction takes place through the production of spores, which are either exposed on hairlike stalks or enclosed in spore cases. Asexual reproduction can occur if the plasmodium splits apart.

When to Hunt Mushrooms: Most mushrooms can be found for only a limited time, usually in either spring or fall. Some are present in the spring, then disappear during the hot, dry summer months, and reappear in the fall, when fungi are most abundant. In the Northeast and Midwest, the mushroom season begins in late April and continues until hard frost in the fall. Because the Southeast warms up earlier and stays warm longer, its season is correspondingly longer, and on the

Gulf Coast mushrooms often fruit abundantly through the winter. In the northern Rockies, mushrooms usually come up in spring and fall. In the southern Colorado Rockies, the best times to collect fungi are July and August, the warmest and wettest months. The Pacific Northwest has a short spring mushroom season, a long dry summer when few mushrooms are present, and then, if the rains come in time and the frost is late, a long fall mushroom season. In southern California and the Southwest, mushrooms occur during the rainy winter and spring, tapering off as it becomes warmer and drier.

Collecting Mushrooms: Collect mushrooms in a flat-bottomed basket. Take along a roll of wax paper and wrap each species you find; do not use plastic wrap since it hastens decay. This will keep species separate and fresh until you return home. A pocket knife or trowel is useful in extracting mushrooms from the ground; be very careful not to disturb the underground root system more than necessary. Bring note cards with you and jot down pertinent field data. In particular, note the habitat of the mushroom, including what type of tree it is growing on or near; whether it is growing singly, scattered, in groups, or in clusters; any distinctive odor or taste; the color of the cap, stalk, gills, pores, or teeth, and latex, which may change after the mushroom is picked. Note any color changes when it is bruised. You can also use the note cards to set up spore prints in the field; they will often be ready by the time you return home. If you are absolutely certain of the identification of an edible species, you can clean it in the field. Until you are experienced, however, it is best to take the mushroom home intact; the stalk base is often a crucial identification feature, and cleaning can remove diagnostic characteristics. The more characteristics you can observe, the better chance you have of identifying the mushroom. It helps to have fresh mushrooms rather than old ones, and to collect many specimens of one kind at various stages of growth.

How to Make a Spore Print: A spore print is essential for accurate identification of many mushrooms. To make one, cut off the mushroom's stalk close to the base. Place the cap, with the gills or pores facing down, on a piece of white paper. If you are in the field, enclose the cap and paper in wax paper and place them on the bottom of a basket. At home, cover them with a glass. Sometimes the spores fall more readily if you place a drop of water on the cap before you cover it. Some mushrooms produce spore prints in a few hours; others take much longer, sometimes overnight.

How to Identify Mushrooms: Because mushrooms are so varied and abundant, their identification may at first seem very difficult. This guide enables you to compare the mushrooms you encounter with color photographs of living mushrooms in their natural habitats. You can more easily find the mushroom you have seen because the photographs are arranged by shape and color. Distinguishing certain groups of look-alikes, such as LBM's ("little brown mushrooms"), may require microscopic examination or chemical analysis and can confound even an expert. Identifying even the genus of these mushrooms is a good accomplishment.
After you have located the picture of the species most like your specimen, check the field characteristics listed in the text description, and the spore print color. Also check the look-alikes section, and eliminate each, item by item. If you plan to eat the mushrooms you have found, make sure that the

edibility section and the comments describe them as safe. If you are absolutely sure of your identification, refer to the section on cooking mushrooms in the appendices of this guide for general safeguards and for recipes. Mushroom hunters who have microscopes can gain extra assurance that their identification is correct if they examine the spores microscopically. In short, before you eat any mushroom, check every possible source of error. If any doubt remains about the edibility of a species, do not eat it.

Spore Print Chart: The appendices include a chart based on spore print color that will help you to key down to genus many gilled mushrooms you find.

Edible and Poisonous Mushrooms: Many people become interested in wild mushrooms by the prospect of finding choice edible species. There are many safe edibles in North America, but some poisonous and deadly species as well. Learning to differentiate between them is, of course, essential. We cannot overemphasize the importance of gaining experience before you begin eating wild fungi. Join a mushroom club and benefit from the experience of other collectors. Never accept an identification unless all evidence confirms it. And always make a spore print. The appendices include hints on cooking wild mushrooms. There is also a section on the symptoms of mushroom poisoning. Be sure to read it before you eat any wild mushrooms.

Organization of the Color Plates: The photographs of mushrooms have been arranged according to the features you see in the field—shape and color. Thus, the group called "Boletes and Others" contains, in addition to boletes, a few stalked, fleshy polypores and a tooth fungus that somewhat resembles a bolete.

The color plates are arranged in the following order:

Small, Fragile Gilled Mushrooms
Veiled Mushrooms with Free Gills
Veiled Mushrooms with Attached Gills
Mushrooms with Free Gills
Mushrooms with Attached Gills
Boletes and Others
Chanterelles and Other Vase-shaped Mushrooms
Stalked Polypores, Tooth Fungi, and Others
Polypores and Other Shelflike Mushrooms
Slimes, Jellies, and Crustlike Fungi
Cup-shaped Mushrooms and Bird's-nest Fungi
Puffballs, Earthstars, Amanita Buttons, and Others
Morels, Stinkhorns, and Other Club-shaped Mushrooms
Coral-like Mushrooms

Thumb Tab Guide: The organization of the color plates is explained in a table preceding them. A silhouette of a typical member of each group appears on the left. Silhouettes of mushrooms within that group are shown on the right. For example, the silhouette of a cup fungus represents the group entitled "Cup-shaped Mushrooms and Bird's-nest Fungi"; the silhouette of a morel represents "Morels, Stinkhorns, and Other Club-shaped Mushrooms." This representative silhouette is also inset on a thumb tab at the left edge of each double page of color plates devoted to that group.

Captions: The caption under each photograph gives the plate number, common name of the mushroom, its size, and the page number of the text description. Those species known to be poisonous or hallucinogenic are indicated by a danger symbol (⊗). For most species, the width of the mushroom is given, but for fungi that are much taller than wide, the height is given. For the group called "Slimes, Jellies, and Crustlike Fungi" no measurements are provided since these mushrooms are so variable. The introduction to each color section indicates which measurements are given for the species in that section.

Do not base your decision to eat any mushroom on the absence of a poison symbol. The edibility of many North American mushrooms is unknown, and eating a species without sure knowledge that it is safe is a little like playing Russian roulette.

Organization of the Text: This guide covers members of 3 subdivisions of fungi, the Ascomycotina, the Basidiomycotina, and the Myxomycotina. Each subdivision is introduced by a discussion of its distinctive features, including physical appearance and how mushrooms in the group produce spores. Within each subdivision, important orders and families are described. These general descriptions are followed by an account of each species in the book. The families in the largest order of gilled mushrooms, the Agaricales, are arranged alphabetically, as are some of the other large groups. Elsewhere, related families are grouped together.

Plate Numbers: Each species account begins wth the number of the color plate or plates. Those species that are known to be poisonous or hallucinogenic are indicated by a danger symbol ($\otimes$).

Common Names: The common names of fungi have never been standardized; those given here reflect prevailing usage. When no widely accepted common name exists, one has been coined here, usually based on a translation of the scientific name. When alternate common names are widely used, we note them in the comments.

Scientific Names: Experienced mushroom hunters usually refer to fungi by their scientific names. These names are not hard to learn and are more accurate than their English counterparts. Common names may differ from region to region, but, with a few notable exceptions, a scientific name signifies the same fungus to mycologists everywhere. The scientific name of a species is always italicized and composed of two words, usually derived from Latin or Greek. The first, always capitalized, is the name of the genus; the second, always in lower case, is the species epithet. For example, *Amanita brunnescens* belongs to the genus *Amanita;* it is the only member of the genus *Amanita* that bears the name *brunnescens.*

Each scientific name is followed by one or more names in abbreviated form. For example, *Amanita brunnescens* is followed by "Atk.," an abbreviation of the name of G. F. Atkinson, who first named the mushroom and whose concept of the species is reflected in the description we provide. *Boletus chrysenteron* is followed by "Bull. ex St. Amans." This notation indicates that this mushroom was described by St. Amans, who, in doing so, used a concept of the species similar to that of J. B. F. Bulliard. *Alnicola melinoides* is followed by "(Fr.) Küh." This indicates that Fries originally described the species *melinoides* (he placed it in the genus *Naucoria*); R. Kühner transferred it to the genus *Alnicola.* Occasionally, the same scientific name has been unintentionally applied to 2 different species; knowing the name of a species' author will allow you to trace the history of any scientific name and to be sure that your species is identical to that discussed in other books.

The names of the mushroom's family and order appear beneath each species name.

Description: Each species description begins with a

general statement in italics about the appearance of the mushroom, including key features that are useful in immediate identification. Next, a mature mushroom is described feature by feature, including its surface, its spore-producing area (gills, tubes, spore mass, etc.); stalk (if present); veils (if present); and spores. Each section includes information about shape, color, texture, and size. Relevant measurements are included in both inches and millimeters or centimeters. If the species is different when it is young or very old, that information is included. The surface, flesh, or spores of some mushrooms change color when various chemical solutions are applied; that information is also provided when relevant. Marginal drawings illustrate features that are not visible in the color photographs. For mushroom hunters with microscopes, we describe the spores of every species. Spores are most easily seen at a magnification of about ×400. Their most important characteristics are given in the species descriptions, including size (measured in microns), shape, surface texture, and color. When relevant, we also note the presence or absence of sterile cells, called cystidia, and describe the gill tissue. Full information on spore characteristics can be found in *How to Identify Mushrooms to Genus III: Microscopic Features,* by D. Largent, D. Johnson, and R. Watling (Mad River Press, Inc.).

Edibility: Edibility status is included when known. The comments section often contains additional information, such as why caution must be exercised before eating a certain species. If there is no edibility category, there is insufficient information on whether the species is safe or poisonous; do not experiment.

Season: Each entry covers the period when the mushroom may be expected to appear. This time span can vary considerably, depending on climatic conditions and geography. If spring comes to your region later than usual, you can expect to see spring mushrooms later as well. And if summer is exceptionally dry, or if warm weather continues well into the fall, you may see fall mushrooms later.

Habitat: This section often begins with the mushroom's growth habit—whether it grows singly, in groups, or in clusters. We then tell where it occurs. If a species grows only near a particular tree or shrub, or on a specific substrate (cones, dung, conifer needles, leaf litter, etc.), that information is included.

Range: We indicate here where each species has been found, moving from north to south, and then from east to west. Remember that the distribution of most North American mushrooms is incompletely known; you may find a species outside the range indicated.

Look-alikes: Some mushroom species are so distinctive that they can be mistaken for no other. Most, however, can easily be confused with others. To help you distinguish between these very similar species, we describe the differences between them. By eliminating look-alikes, you make your identification more certain. A look-alike that is covered in full elsewhere in this guide is indicated by an asterisk.

Comments: Each species account concludes with additional information regarding the mushroom, including alternate names, facts on edible and poisonous species, uses, and folklore.

HOW TO USE THIS GUIDE

Example 1
Red mushroom growing from a tree

Walking in the woods you find a stalkless mushroom growing out of the base of a maple tree. The mushroom is shelflike, about 8" wide, flat, hard, and shiny red; its undersurface has tiny brownish pores.

1. Turn to the Thumb Tab Guide preceding the color plates and look for the silhouette that most closely resembles the mushroom you have seen. In the group called Polypores and Other Shelflike Mushrooms you find the silhouette for polypores, tooth fungi, Coral-pink Merulius, and False Turkey-tail, color plates 485, 489–491, 507–526, 536, 542, 543.
2. Check the color plates. Three mushrooms are shiny red, the Beefsteak Polypore, Ling Chih, and Hemlock Varnish Shelf, color plates 513–515. The captions indicate the sizes and refer you to text pages 455, 460, and 461.
3. Reading the text, you eliminate the Beefsteak Polypore because, unlike your mushroom, it is fleshy and its surface is somewhat gelatinous. You rule out the Hemlock Varnish Shelf since it grows on dead conifers. The description of Ling Chih confirms your identification.

Example 2
White mushroom on the forest floor

You find a tall white mushroom growing on the forest floor. It has white gills that are not attached to the stalk, a large ring toward the top of the stalk, and a smooth cap margin. With a knife, you carefully remove the mushroom from the ground and you see a saclike cup around the base.

1. In the Thumb Tab Guide you find the group called Veiled Mushrooms with Free Gills and you select the silhouette for the Death Cap and Destroying Angel, color plates 113, 123, and 124.

2. Turning to the color plates you quickly eliminate plate 113, the Death Cap, because it is greenish. Your mushroom very closely resembles the Destroying Angel, pictured in plates 123 and 124. The captions indicate that the mushroom is poisonous ⊗ and refer you to page 551.

3. The text describes the mushroom you have found, the deadly Destroying Angel. You take a spore print, which is white, confirming your identificaton.

Example 3
Orange-brown
mushroom on the
ground

You find a mushroom on the ground in a deciduous woods. The cap and stalk are orange-brown and the gills are attached to the stalk, close together, and white. As you handle the mushroom, you break the gills and they release a great deal of white fluid; the gills slowly turn brown where they were broken. The mushroom begins to smell fishlike.

1. In the Thumb Tab Guide, you find the group Mushrooms with Attached Gills and look for the silhouette that most closely resembles your mushroom. You find the silhouette for lactarii and turn to the color plates indicated.

2. Several of the plates resemble your mushroom, but one, plate 291, the Voluminous-latex Milky, looks most like it. You turn to the page indicated, 696.

3. The text description confirms that you have found the Voluminous-latex Milky, a choice edible mushroom.

Part I
Color Plates

The color plates on the following page are divided into 14 groups:

Small, Fragile Gilled Mushrooms
Veiled Mushrooms with Free Gills
Veiled Mushrooms with Attached Gills
Mushrooms with Free Gills
Mushrooms with Attached Gills
Boletes and Others
Chanterelles and Other Vase-shaped Mushrooms
Stalked Polypores, Tooth Fungi, and Others
Polypores and Other Shelflike Mushrooms
Slimes, Jellies, and Crustlike Fungi
Cup-shaped Mushrooms and Bird's-nest Fungi
Puffballs, Earthstars, Amanita Buttons, and Others
Morels, Stinkhorns, and Other Club-shaped Fungi
Coral-like Mushrooms

Thumb Tab Guide: To help you find the correct group, a table of silhouettes precedes the color plates. Each group is represented by a silhouette of a typical member of that group on the left side of the table. On the right, you will find the silhouettes of mushrooms found within that group. The representative silhouette for each group is repeated as a thumb tab at the left edge of each double page of color plates, providing a quick and convenient index to the color section.

Tab	Group	Plate Numbers
	Small, Fragile Gilled Mushrooms	1–108

Typical Shapes	Mushrooms	Plate
	marasmii and inky caps	1–4, 67
	conocybes, psilocybes, mycenas, entolomas, and Witch's Hat	5, 44, 45, 49, 50, 68, 70–72, 74–76, 85
	mycenas, collybias, naematolomas, psathyrellas, and the Velvet Foot	6, 22–24, 36, 61–64, 93, 98, 107
	collybias, Magnolia-cone Mushroom, and others	7–9, 11, 12, 21
	Lichen Agaric, Chanterelle Waxy Cap, Small Chanterelle, and others	10, 27, 28, 56–60, 97
	fiber heads, phaeocollybias, Fairy Ring Mushroom, and others	13–18, 43, 53, 73, 88, 89
	inky caps and others	19, 20, 41, 42, 51
	psathyrellas, marasmii, psilocybes, waxy caps, panaeoli, mycenas, entolomas, and others	25, 26, 29, 30, 32, 33, 35, 37–40, 46–48, 55, 65, 66, 69, 77, 78, 82–84, 86, 87, 90–92, 94–96, 99–104

Group	Plate Numbers
Small, Fragile Gilled Mushrooms (continued)	1–108
Veiled Mushrooms with Free Gills	109–180

Typical Shapes	Mushrooms	Plate Number
	collybias, Bluing Psilocybe, Deadly Conocybe, and others	31, 34, 52, 79–81, 105, 108
	volvariellas	109–111, 118
	amanitas	112, 115, 116, 148
	Death Cap and Destroying Angel	113, 123, 124
	amanitas	114, 138, 142
	amanitas	117, 128, 129, 132, 135, 136, 139, 145, 146
	amanitas	119, 120, 133, 140
	agarici, limacellas, and Smooth Lepiota	121, 122, 141, 153–157, 168
	amanitas	125–127, 130, 131, 134, 137, 143, 144, 147, 149
	agarici, amanitas, and Sharp-scaled Lepiota	150, 151, 163–

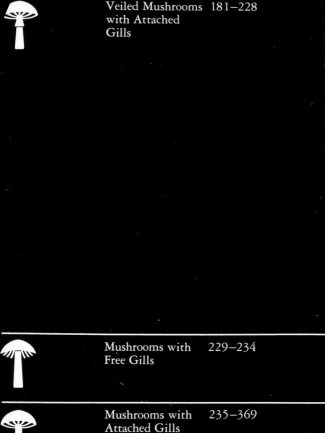

Veiled Mushrooms 181–228
with Attached
Gills

Mushrooms with 229–234
Free Gills

Mushrooms with 235–369
Attached Gills

Typical Shapes	Mushrooms	Plate N
	agarici and lepiotas	152, 158 169–180
	pholiotas and Scaly Fiber Head	181, 183–190
	stropharias, agrocybes, and others	182, 193, 198, 201–204, 206, 207, 214, 215, 222–226, 228
	cystodermas, armillarias, catathelasmas, and others	191, 192, 194–196, 205, 209–213
	stropharias, pholiotas, and Slimy-sheathed Waxy Cap	197, 199, 200, 208, 227
	gomphidii and chroogomphi	216–221
	plutei	229–234
	melanoleucas, entolomas, and Rooted Oudemansiella	235, 268, 274, 275, 282, 359, 367–369
	waxy caps, russulas, laccarias, trichs, mops, and others	236, 249, 254, 255, 257–261, 265, 278, 280, 296, 297, 301, 302, 306, 319, 321–323, 327–330, 339, 343, 345, 346, 348, 353, 355–358

Group	Plate Numbers
Mushrooms with Attached Gills (continued)	235–369
Boletes and Others	370–420

Typical Shapes	Mushrooms	Plate Num
	clitocybes, laccarias, and allies; leucopaxes and allies; waxy caps, russulas, entolomas, chanterelles, boletes, Jack O'Lantern, False Chanterelle, and others	237–239, 2᠁ 243, 245, 24᠁ 252, 253, 262 264, 266, 267 272, 273, 277, 279, 283, 286, 287, 303, 304, 307–311, 313, 320, 324–326, 334–336, 350, 352, 354, 360, 365, 366
	lactarii	240, 244, 247, 248, 250, 251, 256, 284, 285, 288–293, 295, 312, 337, 338, 349, 361–364
	fiber heads and tricholomas	269–271, 281, 314–318, 351
	cortinarii	276, 298–300, 305, 331–333, 340–342, 344, 347
	Lobster Mushroom	294
	boletes and Golden Hypomyces	370–372, 376–418
	Pinecone Tooth	373
	polypores	374, 375, 419, 420

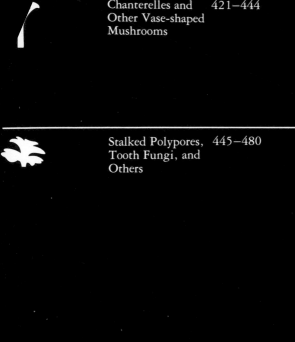

Chanterelles and   421–444
Other Vase-shaped
Mushrooms

Stalked Polypores,   445–480
Tooth Fungi, and
Others

Polypores and   481–549
Other Shelflike
Mushrooms

Typical Shapes	Mushrooms	Plate
	chanterelles, lentinelli, Jack O'Lantern, and others	421, 423, 427, 430–, 438, 440, 4
	chanterelles, thelephores, Apricot Jelly, and others	422, 424, 425, 428, 429, 436, 437, 439, 442, 443
	polypores and Yellow-green Hypomyces	445, 447–451, 453, 458, 460
	tooth fungi and Jelly Tooth	446, 455, 459, 466
	polypores, Ash-tree Bolete, and Black Tooth	452, 454, 456, 457, 461, 462, 464, 465, 468
	polypores, tooth fungi, and Crimped Gill	463, 469–480
	Thick-maze Oak Polypore	467
	polypores, patchment fungi, and Silver Leaf Fungus	481, 482, 534, 535, 537–541, 544, 545
	oyster mushrooms and Jack O'Lantern	483, 492, 495, 497
	oyster mushrooms, creps, panelli, lentinelli, and others	484, 486–488, 493, 494, 496, 498–506

Polypores and 481–549
Other Shelflike
Mushrooms
(continued)

Slimes, Jellies, and 550–597
Crustlike Fungi

Cup-shaped 598–633
Mushrooms and
Bird's-nest Fungi

Typical Shapes	Mushrooms	Plate Numbers
	polypores, tooth fungi, Coral-pink Merulius, and False Turkey-tail	485, 489–491, 507–526, 536, 542, 543
	polypores	527–531, 533
	tooth fungi, and others	532, 546–549
	slime molds and Cannon Fungus	550–557, 562
	slime molds and Spreading Yellow Tooth	558–561
	jelly fungi	563–567
	cup fungi, crust fungi, and Silky Parchment	568–571, 573–575
	parchment fungi, crust fungi, polypores, and others	572, 576–597
	earth tongues, jelly fungi, and cup fungi	598–605, 608–622, 625, 627–630
	cup fungi and Blueberry Cup	606, 607, 623, 624, 626

Thumb Tab	Group	Plate Numbers
	Cup-shaped Mushrooms and Bird's-nest Fungi (continued)	598–633
	Puffballs, Earthstars, Amanita Buttons, and Others	634–681
	Morels, Stinkhorns, and Other Club-shaped Mushrooms	682–732

Typical Shapes	Mushrooms	Plate Numbers
	bird's-nest fungi	631–633
	earthstars and Pink Crown	634–640, 642
	puffballs, flask fungi, and Wolf's-milk Slime	641, 645–656, 663–665, 667, 668
	gilled puffballs, Oregon White Truffle, and others	643, 644, 657–662, 669, 670
	Netted Rhodotus	666
	amanita buttons and Gem-studded Puffball	671–681
	earth tongues and Goldenthread Cordyceps	682–686, 693
	cordyceps, corals, stinkhorns, and others	687–692, 696–701
	stinkhorns	694, 695
	inky caps and stinkhorns, and Smooth Thimble-cap	702–709

Thumb Tab	Group	Plate Numbers
	Morels, Stinkhorns, and Other Club-shaped Mushrooms (continued)	682–732
	Coral-like Mushrooms	733–756

Typical Shapes	Mushrooms	Plate Numbers
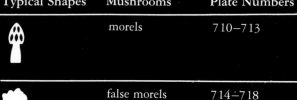	morels	710–713
	false morels	714–718

The color plates on the following pages are numbered to correspond with the numbers preceding the species descriptions.

Most mushrooms have gills—pla
structures on the undersurface of t.
cap that radiate from the stalk outw
Because there are many gilled
mushrooms in North America, they a
divided here into 5 groups based on
shape, the presence or absence of veils,
and gill attachment. Many choice
edibles are included, but so are species
that are poisonous or hallucinogenic.
Gill attachment and veil remnants are
critical features in differentiating gilled
mushrooms and should be examined
with great care. Cap width is given in
the captions.

Small, Fragile Gilled Mushrooms:
These mushrooms are rarely more than
2″ high. Their caps and stalks are
typically thin and delicate, and in a few
the stalks are rubbery to brittle.

Veiled Mushrooms with Free Gills:
These fungi have gills that are free of or
only slightly attached to the stalk.
Typically they have remnants of a
partial or universal veil, or both.

**Veiled Mushrooms with Attached
Gills:** Most of these mushrooms have
gills that are attached to the stalk.
They usually have remnants of a partial
veil; some also have remnants of a
universal veil.

Mushrooms with Free Gills: This
very small group is characterized by the
absence of a universal and partial veil
and by gills that are free from the stalk.

Mushrooms with Attached Gills:
These have gills that are attached to the
stalk; most lack a veil or veil remnants.
Also included here are some members
of the family Cortinariaceae; although
many have a veil, it is cobwebby and
usually disappears quickly.

1 **Black-footed Marasmius,** *w.* ⅜–¾″, *p. 770*

2 **Pinwheel Marasmius,** *w.* ¹⁄₁₆–⅝″, *p. 774*

3 **Japanese Umbrella Inky,** *w.* ⅜–1″, *p. 600*

4 Non-inky Coprinus, *w.* ¼–⅝″, *p. 598*

5 White Dunce Cap, *w.* ⅜–1″, *p. 560*

6 Common Mycena, *w.* 1–3″, *p. 780*

7 Magnolia-cone Mushroom, *w.* ¼–¾″, *p.* 797

8 Douglas-fir Collybia, *w.* ¼–¾″, *p.* 797

9 Conifer-cone Baeospora, *w.* ¼–¾″, *p.* 738

10　Lichen Agaric, *w.* ¼–1⅜″, *p.* 787

11　Tuberous Collybia, *w.* 1¼–4″, *p.* 757

'2　Powder Cap, *w.* ⅜–¾″, *p.* 738

⊗ 13   **Lilac Fiber Head,** *w. ⅜–1¼″, p. 631*

⊗ 14   **Blushing Fiber Head,** *w. 1–3″, p. 632*

⊗ 15   **White-disc Fiber Head,** *w. ⅝–1⅜″, p. 626*

16   White Fiber Head, *w.* ⅝–1¼″, *p. 629*

17   Fairy Ring Mushroom, *w.* ⅜–1⅝″, *p. 772*

18   Fairy Ring Mushroom, *w.* ⅜–1⅝″, *p. 772*

19 Alcohol Inky, *w.* 2–3″, *p.* 596

20 Semi-ovate Panaeolus, *w.* 1¼–3½″, *p.* 603

21 Parasitic Psathyrella, *w.* 1–2″, *p.* 606

22 Family Collybia, *w.* ⅜–1⅝″, *p.* 752

23 Clustered Collybia, *w.* 1–2″, *p.* 753

24 Tufted Collybia, *w.* 1–2″, *p.* 754

25 Common Psathyrella, *w. 1–4″, p. 604*

26 Lawn Mower's Mushroom, *w. ⅜–1¼″, p. 606*

27 Burn Site Mycena, *w. ⅜–2″, p. 786*

28   Fetid Marasmius, *w.* ⅜–1¼", *p. 778*

29   American Simocybe, *w.* ⅜–1⅜", *p. 638*

30   Brown Alder Mushroom, *w.* ⅜–1", *p. 611*

⊗ 31   Bluing Psilocybe, *w.* ¾–1⅝″, *p. 721*

32   Orange-yellow Marasmius, *w.* ¾–2½″, *p. 776*

⊗ 33   Blue-foot Psilocybe, *w.* ⅜–1⅜″, *p. 719*

34 Zoned-cap Collybia, *w.* ⅜–1″, *p.* 758

35 Spotted Collybia, *w.* 2–6″, *p.* 756

36 Brick Tops, *w.* 1⅝–4″, *p.* 710

37 Garlic Marasmius, *w.* ¼–1¼″, *p. 774*

38 Fused Marasmius, *w.* ⅜–1⅝″, *p. 771*

⊗ 39 Deadly Lawn Galerina, *w.* ⅜–1⅜″, *p. 622*

40   Velvet-cap Marasmius, *w.* ⅜–1¼″, *p. 773*

41   Orange-mat Coprinus, *w.* ¾–1¼″, *p. 602*

42   Mica Cap, *w.* 1–2″, *p. 600*

43 Sphagnum-bog Galerina, *w.* ⅜–1⅜″, *p. 621*

⊗ 44 Conifer Psilocybe, *w.* ¼–¾″, *p. 723*

45 Yellow-stalked Mycena, *w.* ⅜–¾″, *p. 779*

46 Round Stropharia, *w.* ⅜–2″, *p. 730*

47 Hemispheric Agrocybe, *w.* ⅜–1″, *p. 557*

48 Golden Waxy Cap, *w.* 1–2¾″, *p. 660*

49   Salmon Unicorn Entoloma, *w.* ⅜–1⅝″, *p. 644*

50   Yellow Unicorn Entoloma, *w.* ⅜–1¼″, *p. 643*

51   Yellow Bolbitius, *w.* ¾–2″, *p. 559*

52 Walnut Mycena, *w.* 3/8–5/8″, *p. 782*

53 Orange-gilled Waxy Cap, *w.* 3/8–2″, *p. 662*

54 Golden-scruffy Collybia, *w.* 1/4–3/4″, *p. 758*

55   Scaly-stalked Psilocybe, *w.* 1–3″, *p.* 724

56   Chanterelle Waxy Cap, *w.* ⅜–1⅜″, *p.* 656

57   Small Chanterelle, *w.* ¼–1¼″, *p.* 391

58   Fuzzy Foot, *w.* ⅛–1″, *p. 809*

59   Orange Moss Agaric, *w.* ⅛–⅝″, *p. 796*

60   Golden-gilled Gerronema, *w.* ¼–1⅝″, *p. 760*

⊗ 61   Sulfur Tuft, *w.* 1–3¼″, *p.* 709

62   Smoky-gilled Naematoloma, *w.* 1–3″, *p.* 708

63   Velvet Foot, *w.* 1–2″, *p.* 759

64 Orange Mycena, *w.* ⅜–2″, *p. 781*

65 Scarlet Waxy Cap, *w.* 1–2″, *p. 658*

66 Bitter Pholiota, *w.* 1–2″, *p. 712*

67 Orange Pinwheel Marasmius, *w.* ⅛–1¼″, *p.* 775

68 Coral Spring Mycena, *w.* ⅛–⅜″, *p.* 778

69 Fading Scarlet Waxy Cap, *w.* ¾–1⅝″, *p.* 663

70 Red-orange Mycena, *w.* ⅜–¾″, *p.* 785

71 Witch's Hat, *w.* ¾–3½″, *p.* 658

72 Witch's Hat, *w.* ¾–3½″, *p.* 658

73   Rooting Redwood Collybia, *w.* 1–6″, *p.* 743

⊗ 74   Bog Conocybe, *w.* ⅛–⅜″, *p.* 561

75   Brown Dunce Cap, *w.* ⅜–¾″, *p.* 561

76 Bleeding Mycena, *w.* ⅜–2″, *p. 781*

77 Dispersed Naematoloma, *w.* ⅜–1⅝″, *p. 709*

78 Fringed Tubaria, *w.* ⅜–1¼″, *p. 639*

79   Hairy-stalked Collybia, *w.* ⅜–1¼", *p. 756*

80   Oak-loving Collybia, *w.* ⅜–2", *p. 755*

⊗ **81   Deadly Conocybe,** *w.* ¼–1", *p. 559*

⊗ 82　Dung-loving Psilocybe, *w.* ⅜–1¼″, *p. 720*

83　Mountain Moss Psilocybe, *w.* ¼–1″, *p. 722*

⊗ 84　Stuntz's Blue Legs, *w.* ⅝–1⅝″, *p. 725*

⊗ 85   Liberty Cap, *w. ⅜–1″, p. 723*

⊗ 86   Potent Psilocybe, *w. ⅝–2¼″, p. 719*

87   Bell-cap Panaeolus, *w. ¾–2″, p. 602*

88 Pretty Phaeocollybia, *w.* 3/8–2″, *p. 634*

89 Kit's Phaeocollybia, *w.* 3/4–2 3/8″, *p. 634*

90 Cucumber-scented Mushroom, *w.* 3/8–2″, *p. 769*

⊗ 91   Girdled Panaeolus, *w. 1¼–2″, p. 604*

92   Corrugated-cap Psathyrella, *w. 2–4″, p. 608*

93   Clustered Psathyrella, *w. 1–2″, p. 607*

94 Large Mycena, *w.* 1–2″, *p.* 783

⊗ 95 Stuntz's Blue Legs, *w.* ⅝–1⅝″, *p.* 725

96 Ringed Psathyrella, *w.* 1–4″, *p.* 608

97 Grayling, *w. 1–2″, p. 741*

98 Spotted Mycena, *w. ¾–2″, p. 783*

99 Velvety Psathyrella, *w. 2–4″, p. 609*

100 Hairy-stalked Entoloma, *w.* ⅜–2″, *p. 648*

101 Blue-toothed Entoloma, *w.* ⅜–1⅝″, *p. 644*

102 Blue Mycena, *w.* ¼–¾″, *p. 785*

103   Parrot Mushroom, *w.* ⅜–1¼″, *p. 665*

104   Pink Mycena, *w.* 1–2″, *p. 784*

105   Lavender Baeospora, *w.* ⅜–1⅝″, *p. 739*

106   Violet Collybia, *w.* ⅜–1¼″, *p.* 755

107   Red-gilled Psathyrella, *w.* 1–2″, *p.* 605

108   Little Brown Collybia, *w.* ⅜–1¼″, *p.* 753

109   Tree Volvariella, *w. 2–8″, p. 677*

110   Tiny Volvariella, *w. ¼–1¼″, p. 678*

111   Parasitic Volvariella, *w. 1–3¼″, p. 679*

112 Grisette, *w.* 2–4″, *p.* 549

① 113 Death Cap, *w.* 2½–6″, *p.* 543

114 Hated Amanita, *w.* 2–5″, *p.* 548

115 Tawny Grisette, *w.* 2–4″, *p.* 536

116 Volvate Amanita, *w.* 2–3″, *p.* 552

117 Salmon Amanita, *w.* 1–5″, *p.* 553

118 Smooth Volvariella, *w.* 2–6″, *p.* 678

119 Powder-cap Amanita, *w.* 1–3″, *p.* 533

120 Powder-cap Amanita, *w.* 1–3″, *p.* 533

Veiled Mushrooms with Free Gills

121 Smooth Lepiota, *w. 2–4", p. 519*

122 Abruptly-bulbous Agaricus, *w. 3–6", p. 500*

⊗ 123 Destroying Angel, *w. 2–5", p. 551*

124 Destroying Angel, *w.* 2–5", *p. 551*

125 Citron Amanita, *w.* 2–4", *p. 531*

126 Cleft-foot Amanita, *w.* 1–6", *p. 527*

⊗ 127   Booted Amanita, *w.* 1–4″, *p.* 532

⊗ 128   Gemmed Amanita, *w.* 1–4″, *p.* 537

⊗ 129   Gemmed Amanita, *w.* 1–4″, *p.* 537

⊗ 130 **Panther,** *w.* 1–6″, *p. 541*

⊗ 131 **Panther,** *w.* 1–6″, *p. 541*

132 **Blusher,** *w.* 2–6″, *p. 545*

133   Strangulated Amanita, *w. 2–5″, p. 538*

34   Cleft-foot Amanita, *w. 1–6″, p. 527*

135   Yellow Blusher, *w. 2–4″, p. 535*

136 Yellow Patches, *w.* 1–3″, *p. 534*

⟨ 137 Yellow-orange Fly Agaric, *w.* 2–6″, *p. 540*

138 Coccora, *w.* 4–12″, *p. 529*

139   Yellow Patches, *w.* 1–3″, *p.* 534

140   False Caesar's Mushroom, *w.* 1–5″, *p.* 542

141   Slimy-veil Limacella, *w.* 1–3″, *p.* 554

142 American Caesar's Mushroom, *w.* 2–5¼", *p. 528*

⊗ 143 Fly Agaric, *w.* 2–10", *p. 539*

⊗ 144 Fly Agaric, *w.* 2–10", *p. 539*

145   Western Yellow Veil, *w.* 2–4″, *p. 526*

146   Stout-stalked Amanita, *w.* 2–6″, *p. 547*

⊗ 147   Cleft-foot Amanita, *w.* 1–6″, *p. 527*

148 Orange Spring Amanita, *w.* 2–6″, *p. 550*

⊗149 Gray-veil Amanita, *w.* 1–3″, *p. 544*

150 Prince, *w.* 4–10″, *p. 502*

⊗ 151   California Agaricus, *w.* 2–4″, *p.* 504

⊗ 152   Western Flat-topped Agaricus, *w.* 2–6″, *p.* 507

153   Meadow Mushroom, *w.* 1–4″, *p.* 505

154 Meadow Mushroom, *w.* 1–4″, *p. 505*

⊗ 155 Yellow-foot Agaricus, *w.* 2⅜–7″, *p. 509*

156 Horse Mushroom, *w.* 3–6″, *p. 501*

157 Spring Agaricus, *w.* 2–6″, *p.* 503

⊗ 158 Felt-ringed Agaricus, *w.* 3–6″, *p.* 506

159 Red-gilled Agaricus, *w.* ⅝–1⅝″, *p.* 523

160   Bleeding Agaricus, *w.* 1–4″, *p.* 505

161   Wine-colored Agaricus, *w.* 2–6″, *p.* 508

162   Deadly Lepiota, *w.* 1–2″, *p.* 517

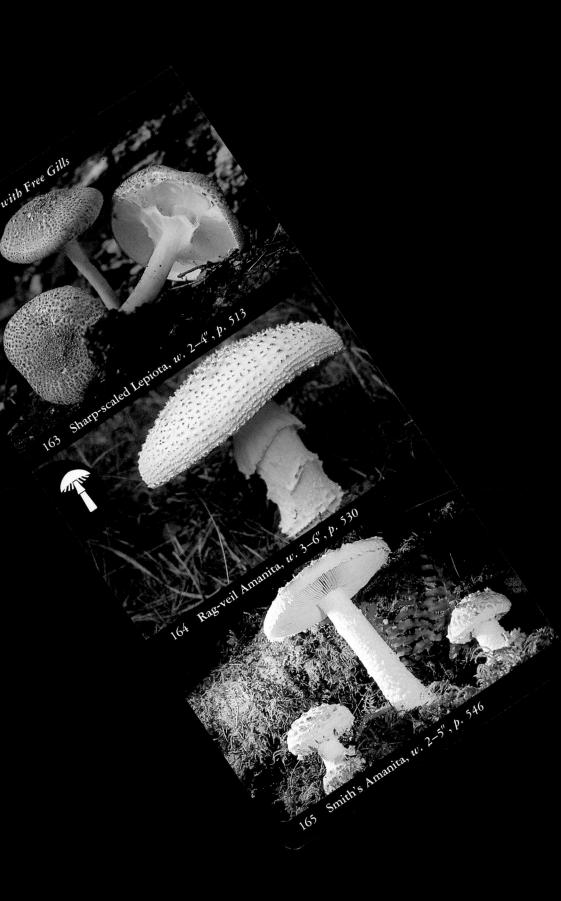

with Free Gills

163 Sharp-scaled Lepiota, *w. 2–4", p. 513*

164 Rag-veil Amanita, *w. 3–6", p. 530*

165 Smith's Amanita, *w. 2–5", p. 546*

166 Coker's Amanita, w. 3–6", p. 531

167 Western Woodland Amanita, w. 2–5", p. 546

168 Ringed Limacella, w. 1–3", p. 554

⊗ 169   Green-spored Lepiota, *w. 2–12″, p. 509*

⊗ 170   Green-spored Lepiota, *w. 2–12″, p. 509*

171   Shaggy Parasol, *w. 3–8″, p. 521*

172 Parasol, *w. 3–8″, p. 520*

173 Reddening Lepiota, *w. 1–6″, p. 513*

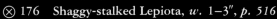

175 Red-tinged Lepiota, *w.* 1–3″, *p.* 522

⊗ 176 Shaggy-stalked Lepiota, *w.* 1–3″, *p.* 516

⊗ 177 Malodorous Lepiota, *w.* ⅝–2″, *p.* 517

178 Black-disc Lepiota, *w. ⅜–2", p. 514*

179 Onion-stalked Lepiota, *w. 1–3", p. 515*

180 Lemon-yellow Lepiota, *w. 1–2", p. 518*

181   Ground Pholiota, *w.* ³⁄₄–4", *p. 718*

182   Honey Mushroom, *w.* 1–4", *p. 736*

⊗ 183   Scaly Fiber Head, *w.* 1¼–2³⁄₄", *p. 633*

184 Sharp-scaly Pholiota, *w.* 1–4", *p.* 717

185 Yellow Pholiota, *w.* 1⅝–4", *p.* 715

'86 Golden Pholiota, *w.* 2–6", *p.* 712

⊗ 187   Scaly Pholiota, *w.* 1–4", *p.* 716

188   Sharp-scaly Pholiota, *w.* 1–4", *p.* 717

189   Burnt-ground Pholiota, *w.* 1–2", *p.* 713

190 **Powder-scale Pholiota,** *w.* ⅜–1⅝″, *p. 711*

191 **Common Conifer Cystoderma,** *w.* 1–2″, *p. 511*

192 **Golden False Pholiota,** *w.* 2–12″, *p. 523*

193   Knobbed Squamanita, *w.* 1¼–2⅜", *p. 524*

194   Pungent Cystoderma, *w.* 1–2", *p. 510*

195   Tuberous Cystoderma, *w.* ⅜–1⅜", *p. 512*

196   Shaggy-stalked Armillaria, *w. 2–5″, p. 731*

197   Questionable Stropharia, *w. 2–6″, p. 726*

198   Green Stropharia, *w. 1–3″, p. 725*

oms with Attached Gills

199 Slimy-sheathed Waxy Cap, *w.* 1–5″, *p.* 664

200 Lacerated Stropharia, *w.* 2–6″, *p.* 728

201 Gypsy, *w.* 2–6″, *p.* 635

202 Hard's Stropharia, *w.* 1–4″, *p. 728*

203 Wine-cap Stropharia, *w.* 2–6″, *p. 729*

204 Wine-cap Stropharia, *w.* 2–6″, *p. 729*

205 Scaly Yellow Armillaria, *w. 2–7", p. 734*

206 Honey Mushroom, *w. 1–4", p. 736*

207 Hard Agrocybe, *w. 1⅝–4", p. 557*

208   Destructive Pholiota, *w.* 3–8″, *p. 714*

209   Fragrant Armillaria, *w.* 2–5″, *p. 732*

210   Fetid Armillaria, *w.* 2–6″, *p. 735*

211   Imperial Cat, *w. 5–16″, p. 741*

212   Swollen-stalked Cat, *w. 3–6″, p. 742*

◊ 214 Big Laughing Gym, *w.* 3¼–7″, *p.* 623

◊ 215 Garland Stropharia, *w.* 1–2″, *p.* 727

'16 Clustered Gomphidius, *w.* 1–6″, *p.* 652

217 Rosy Gomphidius, *w.* 1⅝–2⅜″, *p. 653*

218 Slimy Gomphidius, *w.* 1–4″, *p. 652*

219 Wine-cap Chroogomphus, *w.* 1–3″, *p. 651*

220   Brownish Chroogomphus, *w.* 1–5″, *p.* 650

221   Wool Chroogomphus, *w.* 1–3″, *p.* 650

22   Scaly Lentinus, *w.* 2–5″, *p.* 766

223   Maple Agrocybe, *w.* 1¼–4″, *p. 556*

224   Ringed Tubaria, *w.* ⅜–2″, *p. 638*

225   Spring Agrocybe, *w.* 1¼–3½″, *p. 558*

⊗ 226   Common Large Psilocybe, *w.* ⅝–3½", *p. 721*

227   Changing Pholiota, *w.* ⅝–2⅜", *p. 715*

28   Deadly Galerina, *w.* 1–2½", *p. 620*

229   Yellow Pluteus, *w*. ⅜–1¼″, *p. 673*

230   Golden Granular Pluteus, *w*. 1–2¼″, *p. 675*

231   Fawn Mushroom, *w*. 1¼–4¾″, *p. 675*

232 Fawn Mushroom, *w.* 1¼–4¾", *p.* 675

233 Black-edged Pluteus, *w.* 1¼–4", *p.* 674

34 Pleated Pluteus, *w.* 1–2", *p.* 676

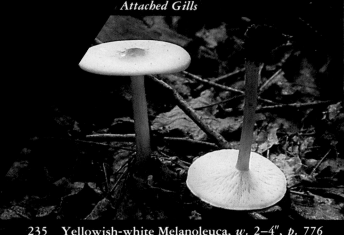

235   Yellowish-white Melanoleuca, *w.* 2–4″, *p.* 776

236   White Waxy Cap, *w.* 1–4″, *p.* 659

237   Fragrant Clitocybe, *w.* ⅜–2⅜″, *p.* 751

238 Crowded White Clitocybe, *w.* 2–6″, *p.* 746

239 White Leucopax, *w.* 1–4″, *p.* 767

'0 Peppery Milky, *w.* 2–6″, *p.* 690

⊗ 241   Sweating Mushroom, *w. ³⁄₈–1⁵⁄₈", p. 745*

242   Sweetbread Mushroom, *w. 2–4", p. 641*

243   Fringed Ripartites, *w. 1–2", p. 796*

244 Variegated Milky, *w.* 1¼–4″, *p.* 693

245 Sweating Mushroom, *w.* ⅜–1⅝″, *p.* 745

6 Turpentine Waxy Cap, *w.* 2–4″, *p.* 666

247 Willow Milky, *w.* 1⅝–9″, *p. 681*

248 Deceptive Milky, *w.* 2–10″, *p. 683*

249 Red-and-black Russula, *w.* 2½–7″, *p. 701*

250 Buff Fishy Milky, *w.* 1–3¼″, *p. 687*

⊗ 251 Pink-fringed Milky, *w.* 2–4¾″, *p. 694*

252 Short-stalked White Russula, *w.* 4–8″, *p. 698*

253 Aborted Entoloma, *w.* 2–4″, *p. 641*

254 Purple-gilled Laccaria, *w.* 2–8″, *p. 762*

255 Sandy Laccaria, *w.* 1–3″, *p. 763*

256 Burnt-sugar Milky, *w*. 1¼–6″, *p. 680*

257 Poplar Trich, *w*. 2–6″, *p. 802*

258 Amyloid Tricholoma, *w*. 2–4″, *p. 794*

⊗ 259 Poison Pie, *w.* 1¼–3½", *p. 624*

260 Golden-spotted Waxy Cap, *w.* 1–3", *p. 657*

261 Yellow-centered Waxy Cap, *w.* 1–3", *p. 660*

262 Wood Clitocybe, *w.* 1–2″, *p.* 747

263 Funnel Clitocybe, *w.* 2–3″, *p.* 747

264 Giant Clitocybe, *w.* 4–18″, *p.* 748

265 Platterful Mushroom, *w.* 2–5″, *p. 807*

266 Fried-chicken Mushroom, *w.* 1–5″, *p. 768*

267 Cloudy Clitocybe, *w.* 3–6″, *p. 749*

268   Rooted Oudemansiella, *w*. 1–4″, *p. 788*

) 269   Green-foot Fiber Head, *w*. ⅜–1⅝″, *p. 627*

70   Russet-scaly Trich, *w*. 1–3″, *p. 805*

⊗ 271 Woolly Fiber Head, *w.* ¾–1¼″, *p. 630*

272 Ringless Honey Mushroom, *w.* 1–4″, *p. 737*

274 **Straight-stalked Entoloma,** *w.* 1–2", *p.* 646

275 **Lead Poisoner,** *w.* 2¾–6", *p.* 645

276 **Slimy-banded Cort,** *w.* 1¼–4", *p.* 614

277   Bitter Leucopax, *w. 2–5″, p. 768*

278   Corpse Finder, *w. 1–2″, p. 625*

279   Smoky-brown Clitocybe, *w. 2–8″, p. 744*

280 Pink Calocybe, *w.* ⅝–1″, *p. 740*

281 Shingled Trich, *w.* 1–4″, *p. 800*

282 Early Spring Entoloma, *w.* 1–2″, *p. 646*

283 Cracked-cap Rhodocybe, *w. 1–2″, p. 648*

⊗ 284 Northern Bearded Milky, *w. 2⅜–8″, p. 690*

⊗ 285 Spotted-stalked Milky, *w. 3–10″, p. 692*

286   Firm Russula, *w.* 1⅜–7″, *p. 699*

287   Poison Paxillus, *w.* 1⅝–4¾″, *p. 671*

'88   Yellow-latex Milky, *w.* 1⅝–4¾″, *p. 695*

⊗ 289   Red-hot Milky, *w.* 1⅝–4¾″, *p. 691*

290   Hygrophorus Milky, *w.* 1¼–4″, *p. 685*

291   Voluminous-latex Milky, *w.* 2–5¼″, *p. 696*

292 Peck's Milky, *w*. 2–6″, *p. 689*

293 Corrugated-cap Milky, *w*. 2–8″, *p. 682*

294 Lobster Mushroom, *mold, p. 373*

295 **Aromatic Milky**, *w.* ⅝–2″, *p. 681*

⊗ 296 **Red-brown Trich**, *w.* 2–6″, *p. 802*

297 **Veiled Trich**, *w.* 1⅜–3½″, *p. 799*

298 Deadly Cort, *w*. 1–2″, *p. 615*

299 Saffron-colored Cort, *w*. 1–2″, *p. 614*

300 Red-gilled Cort, *w*. 1–2⅜″, *p. 618*

301   Little Gym, *w.* 1–2″, *p. 623*

302   Decorated Mop, *w.* 1¼–2⅜″, *p. 807*

303   Yellow Oyster Mop, *w.* 1–3″, *p. 809*

304 Late Fall Waxy Cap, *w. 1–3″, p. 661*

305 Variable Cort, *w. 2–4″, p. 617*

6 Almond-scented Russula, *w. 1–5″, p. 704*

307 Flame-colored Chanterelle, *w.* ⅜–2″, *p. 389*

308 Chanterelle, *w.* ⅜–6″, *p. 387*

309 Salmon Waxy Cap, *w.* 1–3″, *p. 664*

⊗ 310 Jack O'Lantern, *w.* 3–8″, *p.* 787

311 False Chanterelle, *w.* 1–2⅜″, *p.* 669

12 Orange-latex Milky, *w.* 2–5½″, *p.* 683

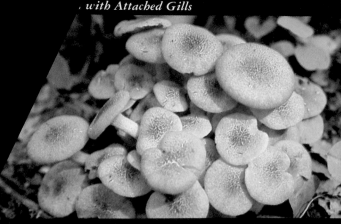

313   Ringless Honey Mushroom, *w.* 1–4″, *p.* 737

⊗ 314   Caesar's Fiber Head, *w.* 1–2″, *p.* 627

⊗ 315   Torn Fiber Head, *w.* ⅜–1⅝″, *p.* 630

, with Attached Gills

316 Black-nipple Fiber Head, *w.* ⅜–1″, *p. 629*

317 Pungent Fiber Head, *w.* 1–3″, *p. 632*

318 Straw-colored Fiber Head, *w.* 1–3″, *p. 628*

319 Separating Trich, *w.* 2–3", *p.* 804

320 Fat-footed Clitocybe, *w.* 1–3", *p.* 745

321 Graving Yellow Russula, *w.* 2–4", *p.* 698

322 **Velvet-footed Pax,** *w.* 1⅝–4¾″, *p.* 670

323 **Canary Trich,** *w.* 2–4″, *p.* 800

324 **Gilled Bolete,** *w.* 1–3″, *p.* 672

325 Larch Waxy Cap, *w.* 1–2″, *p. 667*

326 Gilled Bolete, *w.* 1–3″, *p. 672*

327 Variegated Mop, *w.* 2–4″, *p. 808*

⊗ 328　Emetic Russula, *w.* 1–3″, *p. 701*

329　Blackish-red Russula, *w.* 2⅜–4¾″, *p. 703*

330　Shellfish-scented Russula, *w.* 1–6″, *p. 707*

331 Bracelet Cort, *w. 2–5″, p. 612*

332 Cinnabar Cort, *w. 1¼–2⅜″, p. 613*

333 Blood-red Cort, *w. 1–2″, p. 618*

334 Rosy Russula, *w.* 1–4″, *p.* 705

335 Common Laccaria, *w.* ⅜–2″, *p.* 762

336 Purple-gilled Laccaria, *w.* 2–8″, *p.* 762

337 Indigo Milky, *w.* 2–6″, *p. 686*

⊗ 338 Common Violet-latex Milky, *w.* 1¼–4″, *p. 695*

339 Purple-bloom Russula, *w.* 1–2″, *p. 705*

340 Violet Cort, *w.* 2–4¾", *p. 620*

341 Pungent Cort, *w.* 2–5", *p. 619*

342 Silvery-violet Cort, *w.* 1¼–2⅜", *p. 611*

343 Fragile Russula, *w.* 1–3″, *p. 702*

344 Viscid Violet Cort, *w.* 1–2″, *p. 617*

345 Russulalike Waxy Cap, *w.* 2–5″, *p. 667*

346 Blewit, *w. 2–6", p. 749*

347 Bulbous Cort, *w. 2–5", p. 616*

348 Variable Russula, *w. 2–6", p. 706*

349 Silver-blue Milky, *w.* 2–3¼", *p. 688*

350 Anise-scented Clitocybe, *w.* 1–4", *p. 750*

351 Fibril Trich, *w.* 2–4", *p. 806*

352 **Black-and-white Clitocybula,** *w.* 1–4″, *p.* 751

353 Smoky Gray Trich, *w.* 2–5″, *p.* 803

354 Cloudy Clitocybe, *w.* 3–6″, *p.* 749

⊗ 355   Dirty Trich, *w.* 2–6″, *p.* 801

356   Green Quilt Russula, *w.* 2–6″, *p.* 700

357   Tacky Green Russula, *w.* 2–3⅜″, *p.* 697

358 Soapy Trich, *w.* 1⅝–3¼″, *p.* 804

359 Changeable Melanoleuca, *w.* 1–3″, *p.* 777

360 Gray Almond Waxy Cap, *w.* 1–4″, *p.* 654

361   Dirty Milky, *w.* 2–4″, *p.* 693

362   Gerard's Milky, *w.* 1⅜–4¾″, *p.* 684

363   Chocolate Milky, *w.* ¾–4″, *p.* 686

364 Slimy Lactarius, *w.* 1¼–3½″, *p.* 688

365 Dusky Waxy Cap, *w.* 2–5″, *p.* 656

366 Inocybelike Waxy Cap, *w.* 1–3″, *p.* 662

⊗ 367   Lead Poisoner, *w.* 2¾–6″, *p.* 645

368   Violet Entoloma, *w.* 1–2″, *p.* 647

369   Midnight-blue Entoloma, *w.* 2–5″, *p.* 642

Boletes and Others

Boletes are fleshy mushrooms that typically have a central stalk. All have a fertile surface composed of small, porelike openings, which are the mouths of tubes where the spores are produced. The bolete family contains a few toxic species; it also includes the largest number of safe edible mushrooms in North America. One, the King Bolete, is perhaps the most prized of all edible mushrooms. Also included here are a few stalked, fleshy polypores and the Pinecone Tooth. The cap width is given in the captions.

370   **Parasitic Bolete**, *w. ¾–3¼″, p. 571*

371   **Slippery Jill**, *w. 2–4″, p. 589*

372   **White Suillus**, *w. 1¼–4″, p. 587*

373   Pinecone Tooth, *w.* ⅜–1⅝", *p. 426*

374   Pale Beefsteak Polypore, *w.* 1–5", *p. 456*

375   Winter Polypore, *w.* ⅝–4", *p. 479*

376   Dotted-stalk Suillus, *w.* 2–6″, *p. 584*

377   Black Velvet Bolete, *w.* 1¼–6″, *p. 590*

378   Dark Bolete, *w.* 2–6″, *p. 595*

379 Old Man of the Woods, *w.* 1⅝–6″, *p. 580*

380 Lilac-brown Bolete, *w.* 2–4¾″, *p. 592*

381 Violet-gray Bolete, *w.* 1⅝–6″, *p. 594*

382   Bitter Bolete, *w.* 2–12″, *p.* 593

383   Spotted Bolete, *w.* 1⅜–4″, *p.* 565

384   Chestnut Bolete, *w.* 1¼–4″, *p.* 575

385 Graceful Bolete, *w.* 1¼–4″, *p.* 594

386 Pungent Suillus, *w.* 1⅝–5½″, *p.* 588

387 Short-stalked Suillus, *w.* 2–4″, *p.* 582

388   Red-capped Scaber Stalk, *w.* 2–8″, *p.* 577

389   Aspen Scaber Stalk, *w.* 1⅝–6″, *p.* 578

390   Common Scaber Stalk, *w.* 1⅝–4″, *p.* 578

391 Painted Suillus, *w. 1¼–4¾″, p. 587*

392 Western Painted Suillus, *w. 2¾–5½″, p. 585*

393 Hollow-stalked Larch Suillus, *w. 1¼–4″, p. 583*

394 Blue-staining Suillus, *w.* 2⅜–5½″, *p. 582*

395 Bay Bolete, *w.* 1¼–4″, *p. 565*

396 Admirable Bolete, *w.* 2¾–6″, *p. 569*

397 Burnt-orange Bolete, *w.* 2–4¾", *p. 591*

398 Peppery Bolete, *w.* ⅝–3½", *p. 571*

399 Yellow-cracked Bolete, *w.* 2–8", *p. 572*

400   Golden Hypomyces, *mold, p. 371*

401   Slippery Jack, *w.* 2–4¾", *p. 586*

402   Bluing Bolete, *w.* 1⅝–4¾", *p. 576*

403 Russell's Bolete, *w.* 1¼–5¼″, *p.* 563

404 Rosy Larch Bolete, *w.* 3–10″, *p.* 574

405 King Bolete, *w.* 3¼–10″, *p.* 568

406   Larch Suillus, *w.* 2–6″, *p. 584*

407   Shaggy-stalked Bolete, *w.* 1¼–3½″, *p. 563*

408   Frost's Bolete, *w.* 2–6″, *p. 568*

⊗ 409   Red-mouth Bolete, *w.* 2⅜–5¼", *p. 572*

410   Chrome-footed Bolete, *w.* 1¼–6", *p. 591*

411   Red Gyroporus, *w.* 1–3½", *p. 576*

412   Two-colored Bolete, *w. 2–6", p. 566*

413   Red-cracked Bolete, *w. 1¼–3¼", p. 567*

414   Zeller's Bolete, *w. 2–4", p. 573*

415 Tomentose Suillus, *w.* 2–4″, *p. 590*

416 Chicken-fat Suillus, *w.* 1¼–4″, *p. 581*

417 Powdery Sulfur Bolete, *w.* ⅜–4″, *p. 579*

418 Ornate-stalked Bolete, *w.* 1⅝–8″, *p.* 570

419 Rooting Polypore, *w.* 1⅜–10″, *p.* 480

420 Scaly Yellow Polypore, *w.* 4–6″, *p.* 442

Chanterelles and Other Vase-shaped Mushrooms

This group includes two of the most sought-after edibles in North America, the Chanterelle and the Black Trumpet, as well as other, less choice species, and the poisonous Jack O'Lantern. All are characterized by their vase- or funnel-like shape. Many, like the Chanterelle, have prominent gill-like ridges on their undersurface; others have true gills. The width of the cap is given in the captions.

421   Cinnabar-red Chanterelle, *w.* ⅜–1⅝", *p. 388*

422   Apricot Jelly, *w.* ⅜–2¾", *p. 383*

423   Cockle-shell Lentinus, *w.* 1–2", *p. 764*

424   Umbrella Polypore, *w.* ⅜–1⅝″, *p. 482*

425   Fragrant Chanterelle, *w.* 3–6″, *p. 395*

426   Jack O'Lantern, *w.* 3–8″, *p. 787*

427   Chanterelle, *w.* ⅜–6″, *p. 387*

428   Smooth Chanterelle, *w.* 1–4″, *p. 390*

429   Lobster Mushroom, *mold*, *p. 373*

430 Scaly Vase Chanterelle, *w.* 2–6″, *p. 396*

431 Scaly Vase Chanterelle, *w.* 2–6″, *p. 396*

432 White Chanterelle, *w.* 2–6″, *p. 392*

433   Leaflike Oyster, *w.* 1–3″, *p.* 760

434   Fragrant Hygrophoropsis, *w.* ⅜–1⅜″, *p.* 669

435   Yellow-footed Chanterelle, *w.* ⅜–2¾″, *p.* 393

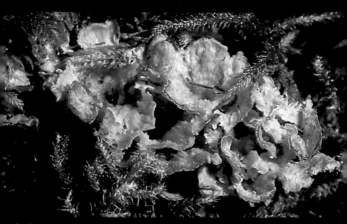

436 Stalked Stereum, *w.* ⅝–1¾", *p. 498*

437 Common Fiber Vase, *w.* 1–2", *p. 413*

438 Stalked Lentinellus, *w.* 1–2", *p. 764*

439   **Flat-topped Coral**, *w. 1–3″, p. 404*

440   **Pig's Ear Gomphus**, *w. 1–4″, p. 396*

441   **Trumpet Chanterelle**, *w. ⅜–3″, p. 392*

442 Vase Thelephore, *w.* 1–6″, *p. 413*

443 Black Trumpet, *w.* ⅜–3¼″, *p. 394*

444 Clustered Blue Chanterelle, *w.* 1–4″, *p. 397*

Stalked Polypores, Tooth Fungi, and Others

This very diverse group includes variously colored mushrooms of different habitats and seasons. Most are very large—including some of the largest mushrooms in North America—and many are edible. The majority are stalked, fleshy polypores, and many grow in overlapping, bouquetlike clusters. Also included here are a few stalked tooth fungi, as well as the Jelly Tooth. The width of the cap is given in the captions.

445   Shiny Cinnamon Polypore, *w.* ⅜–2″, *p. 450*

446   Spongy-footed Tooth, *w.* 1–4″, *p. 434*

447   Rooting Polypore, *w.* 1⅜–10″, *p. 480*

448 Woolly Velvet Polypore, *w.* 1¼–7¼″, *p. 471*

449 Montagne's Polypore, *w.* 1–4″, *p. 451*

450 Green's Polypore, *w.* 1–6″, *p. 452*

451   Yellow-green Hypomyces, *mold, p. 373*

452   Ash-tree Bolete, *w.* 2–4¾″, *p. 564*

453   Elegant Polypore, *w.* 2–4″, *p. 483*

454 Crested Polypore, *w. 2–8″, p. 442*

455 Sweet Tooth, *w. ⅝–6″, p. 428*

456 Scaly Yellow Polypore, *w. 4–6″, p. 442*

457   Blue-pored Polypore, *w. 1–6", p. 441*

458   Kurotake, *w. 2–5", p. 446*

459   Jelly Tooth, *w. 1–3", p. 383*

460 Sheep Polypore, *w.* 2–6″, *p. 444*

461 Bone Polypore, *w.* 1–4″, *p. 472*

462 Flett's Polypore, *w.* 4–8″, *p. 443*

463   Bondarzew's Polypore, *w.* 2⅜–10″, *p. 448*

464   Goat's Foot, *w.* 1–8″, *p. 445*

465   Bitter Iodine Polypore, *w.* 2–6″, *p. 480*

466   Scaly Tooth, *w. 2–8″, p. 434*

467   Thick-maze Oak Polypore, *w. 2–6″, p. 453*

468   Black Tooth, *w. 2–3″, p. 436*

469   Bluish Tooth, *w.* 1–4″, *p.* 432

470   Red-juice Tooth, *w.* 1–6″, *p.* 433

471   Black-staining Polypore, *w.* 2–8″, *p.* 470

472   Crimped Gill, *w.* ⅜–1″, *p. 493*

473   Umbrella Polypore, *w.* ⅜–1⅝″, *p. 482*

474   Hen of the Woods, *w.* ¾–2¾″, *p. 463*

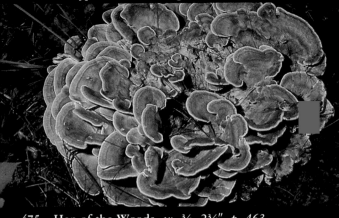

475   Hen of the Woods, *w.* ¾–2¾", *p. 463*

476   Berkeley's Polypore, *w.* 3–10", *p. 447*

477   Berkeley's Polypore, *w.* 3–10", *p. 447*

478 Chicken Mushroom, *w. 2–12″, p. 468*

479 Orange Rough-cap Tooth, *w. 1–4″, p. 431*

480 Dye Polypore, *w. 2–10″, p. 473*

Polypores and Other Shelflike Mushrooms

This is a group of large, woody mushrooms. All grow on wood and most either lack stalks or have lateral or off-center stalks. In this group are the woody polypores, large fleshy tooth fungi, small stereums, oyster mushrooms, and the poisonous Jack O'Lantern. The captions give the width of each.

481   Thin-maze Flat Polypore, *w.* 1–6″, *p. 454*

482   Turkey-tail, *w.* 1–4″, *p. 489*

⊗ 483   Jack O'Lantern, *w.* 3–8″, *p. 787*

484 Oyster Mushroom, *w. 2–8″, p. 793*

485 Little Nest Polypore, *w. ⅜–2″, p. 485*

486 Black Jelly Oyster, *w. ⅛–¼″, p. 795*

487   Common Split Gill, *w.* ⅜–1⅝″, *p. 493*

488   White Marasmius, *w.* ⅜–1¼″, *p. 770*

489   Pendulous-disc Polypore, *w.* 1/32–¼″, *p. 484*

490   White Cheese Polypore, *w.* ⅜–4″, *p. 491*

491   Birch Polypore, *w.* 1–10″, *p. 477*

492   Elm Oyster, *w.* 2–6″, *p. 761*

493   Angel's Wings, *w.* 1–4″, *p.* 792

494   Flat Crep, *w.* ⅜–1⅝″, *p.* 636

495   Veiled Oyster, *w.* 2–5″, *p.* 792

496 White Oysterette, *w.* ¹⁄₁₆–¾", *p.* 743

497 Oyster Mushroom, *w.* 2–8", *p.* 793

498 Late Fall Oyster, *w.* 1–4", *p.* 789

499   Stalkless Paxillus, *w.* 1–4″, *p.* 672

500   Orange Mock Oyster, *w.* 1–3″, *p.* 791

501   Luminescent Panellus, *w.* ⅜–1¼″, *p.* 790

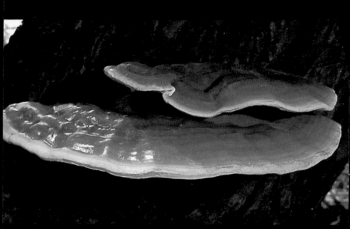

514 Ling Chih, *w.* 1–14", *p. 460*

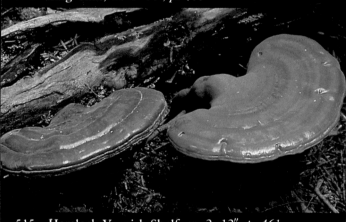

515 Hemlock Varnish Shelf, *w.* 2–12", *p. 461*

516 Yellow-red Gill Polypore, *w.* 1–4", *p. 463*

517 Mustard-yellow Polypore, *w.* 1–6″, *p.* 475

518 Artist's Conk, *w.* 2–20″, *p.* 460

519 Conifer-base Polypore, *w.* 1–10″, *p.* 465

520 Northern Tooth, *w.* 4–6″, *p.* 427

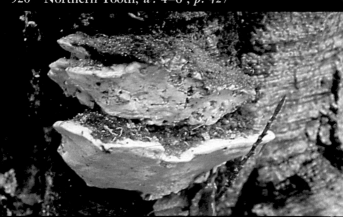

521 Mossy Maple Polypore, *w.* 1–8″, *p.* 472

522 Blue Cheese Polypore, *w.* ⅜–3¼″, *p.* 490

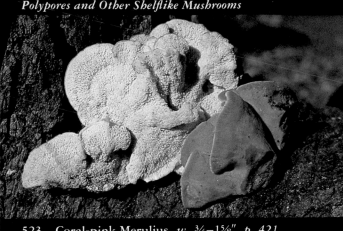

523   Coral-pink Merulius, *w.* ¾–1⅝″, *p. 421*

524   Rosy Polypore, *w.* 1–4″, *p. 458*

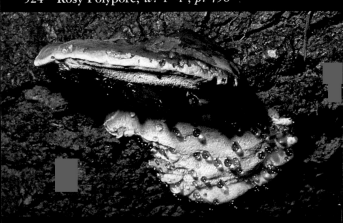

525   Resinous Polypore, *w.* 3–10″, *p. 468*

526 White Spongy Polypore, *w. 1–4″, p. 488*

527 Tinder Polypore, *w. 2–8″, p. 457*

528 Veiled Polypore, *w. ⅜–3⅜″, p. 452*

529   Larch Polypore, *w.* 2–8″, *p. 458*

530   Flecked-flesh Polypore, *w.* 2–10″, *p. 475*

531   Red-belted Polypore, *w.* 2–12″, *p. 459*

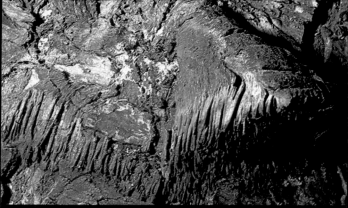

532 Indian Paint Fungus, *w.* 4–10″, *p.* 429

533 Cracked-cap Polypore, *w.* 2–12″, *p.* 476

534 Warted Oak Polypore, *w.* 3–16″, *p.* 466

535   Multicolor Gill Polypore, *w.* 1–4″, *p.* 469

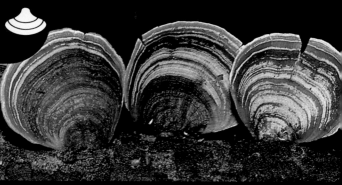

536   False Turkey-tail, *w.* ⅜–2¾″, *p.* 497

537   Hairy Parchment, *w.* ¼–¾″, *p.* 496

538   Violet Toothed Polypore, *w.* ⅜–3″, *p. 490*

539   Silver Leaf Fungus, *w.* ¼–¾″, *p. 418*

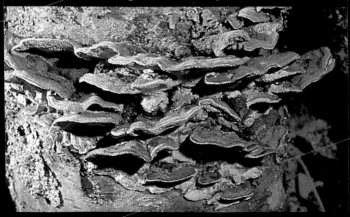

540   Mossy Maze Polypore, *w.* ¼–3″, *p. 450*

541 Bleeding Conifer Parchment, *w.* ⅛–⅜", *p.* 494

542 Kidney-shaped Tooth, *w.* 1–3", *p.* 435

543 Black-footed Polypore, *w.* 2–8", *p.* 478

544   Sweet Knot, *w.* ⅜–¾″, *p.* 462

545   Crowded Parchment, *w.* ⅛–⅝″, *p.* 496

546   Spongy Toothed Polypore, *w.* 1–2″, *p.* 488

547   Bearded Tooth, *w. 4–10″, p. 430*

548   Bear's Head Tooth, *w. 6–12″, p. 429*

549   Comb Tooth, *w. 4–10″, p. 431*

Slimes, Jellies, and Crustlike Fungi

 These mushrooms usually grow in a sheetlike or spreading form and may cover large areas. Two familiar pests, Dry Rot and Wet Rot, are included here, along with many parchmentlike fungi, some crust and cup fungi, a few jellies, and many slime molds.

550   Chocolate Tube Slime, *p. 853*

551   Cannon Fungus, *p. 830*

552   White-footed Slime, *p. 853*

553 Yellow-fuzz Cone Slime, *p. 850*

554 Insect-egg Slime, *p. 846*

555 Carnival Candy Slime, *p. 850*

556   Multigoblet Slime, *p. 852*

557   Yellow-fuzz Cone Slime, *p. 850*

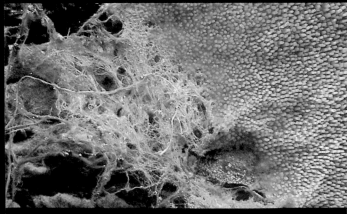

558   Spreading Yellow Tooth, *p. 423*

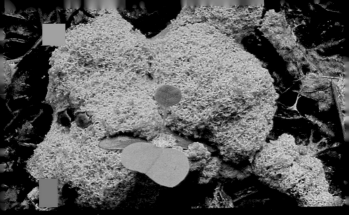

559   Scrambled-egg Slime, *p. 845*

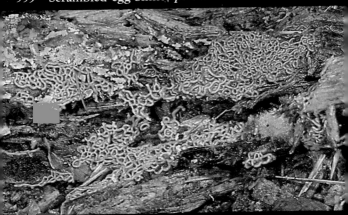

560   Pretzel Slime, *p. 851*

561   Many-headed Slime, *p. 847*

562   Many-headed Slime, *p. 847*

563   Black Jelly Roll, *p. 382*

564   Collybia Jelly, *parasite, p. 418*

565   Red Tree Brain, *p. 423*

566   Witches' Butter, *p. 385*

567   Orange Jelly, *p. 381*

568   Yellow Fairy Cups, *p. 362*

569   Burn Site Ochre Cup, *p. 352*

570   Burn Site Shield Cup, *p. 346*

571 Hophornbeam Disc, *p. 417*

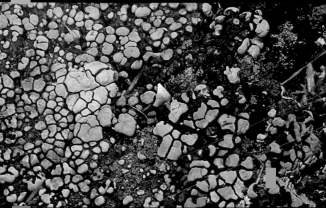

572 Ceramic Parchment, *p. 498*

573 Two-tone Parchment, *p. 420*

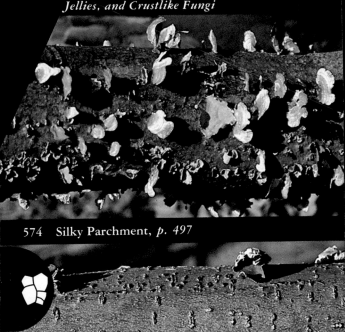

574   Silky Parchment, *p. 497*

575   Scurfy Alder Cup, *p. 365*

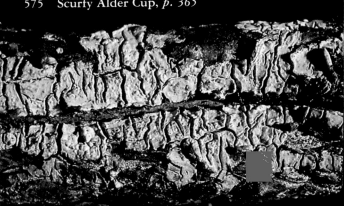

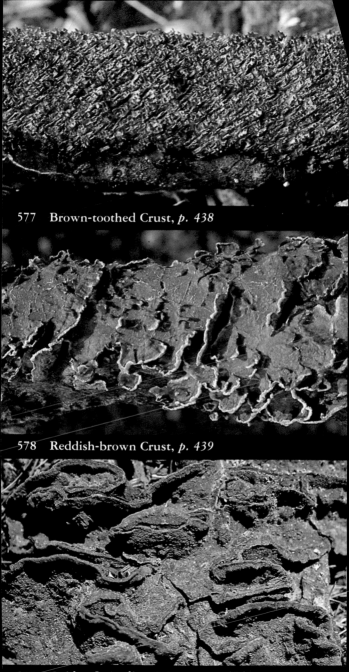

577 Brown-toothed Crust, *p. 438*

578 Reddish-brown Crust, *p. 439*

579 Golden Spreading Polypore, *p. 474*

580 Zoned Phlebia, *p. 425*

581 Orange Poria, *p. 484*

582 Smoky Polypore, *p. 445*

583 Conifer Parchment, *p. 422*

584 Milk-white Toothed Polypore, *p. 467*

586 Trembling Merulius, *p. 421*

587 Radiating Phlebia, *p. 424*

588 Ochre Spreading Tooth, *p. 437*

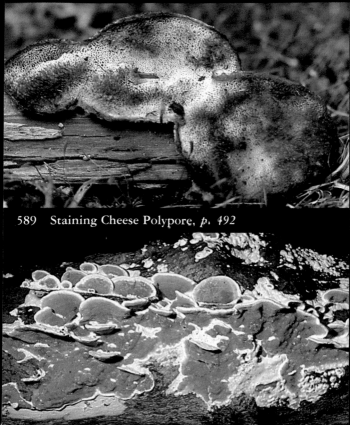

589 Staining Cheese Polypore, *p. 492*

590 Gelatinous-pored Polypore, *p. 449*

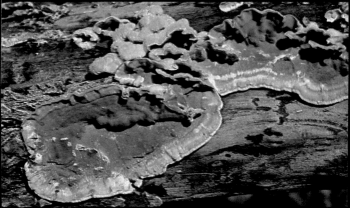

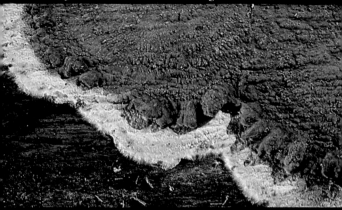

592   Wet Rot, *p. 415*

593   Dry Rot, *p. 415*

594   Tapioca Slime, *p. 852*

595 Velvet Blue Spread, *p. 420*

596 Tapioca Slime, *p. 852*

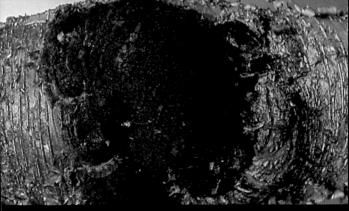

597 Clinker Polypore, *p. 467*

Cup-shaped Mushrooms and Bird's-nest Fungi

These cup- or saucer-shaped mushrooms grow on a variety of substrates, from soil to moss or rotting wood. Many are very small and brightly colored, and grow in large clusters, while others are larger and darkly colored, and grow singly and inconspicuously on the forest floor. Most lack stalks. True cup fungi are included here, along with bird's-nest mushrooms and some cup-shaped jellies. The captions give the width of each.

598   **Green Stain**, *w.* ¼″, *p. 361*

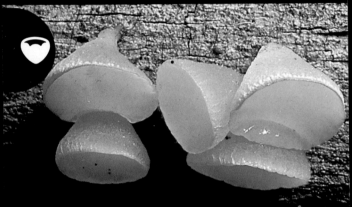

599   **Golden Jelly Cone**, *w.* ⅟₁₆–1″, *p. 382*

601   Bladder Cup, *w.* 2⅜–3¼", *p. 348*

602   Blue-staining Cup, *w.* ⅜–2", *p. 349*

603   Orange Peel, *w.* ¾–4", *p. 349*

604 Eyelash Cup, *w.* ¼–¾″, *p. 353*

605 Scarlet Cup, *w.* ¾–2⅜″, *p. 343*

606 Stalked Scarlet Cup, *w.* ¼–⅝″, *p. 344*

607 Shaggy Scarlet Cup, *w.* 1/4–3/8″, *p. 343*

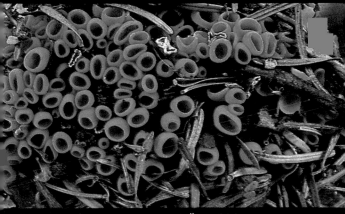

608 Pink Burn Cup, *w.* 1/4–5/8″, *p. 354*

609 Jellylike Black Urn, *w.* 3/8–3/4″, *p. 341*

610   Black Jelly Drops, *w.* ⅜–1⅝″, *p. 363*

611   Black Rubber Cup, *w.* 1–2″, *p. 342*

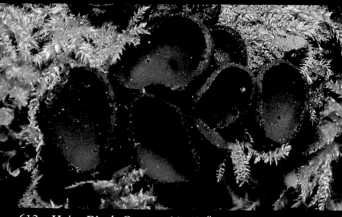

612   Hairy Black Cup, *w.* ¼–1¼″, *p. 341*

613   Devil's Urn, *w.* 1¼–3¼″, *p. 342*

614   Hairy Rubber Cup, *w.* ¾–1¼″, *p. 339*

615   Moose Antlers, *w.* 1–4″, *p. 345*

616 Purple Jelly Drops, *w.* ¼–⅜″, *p. 362*

617 Tree-Ear, *w.* 1–6″, *p. 380*

618 Pig's Ears, *w.* 1⅝–4″, *p. 331*

619 Crustlike Cup, *w.* ¾–2⅜″, *p. 331*

620 Recurved Cup, *w.* 2⅜–4¾″, *p. 347*

621 Veined Cup, *w.* 2–8″, *p. 330*

622   Ribbed-stalked Cup, *w.* ¾–3¼", *p. 332*

623   Long-stalked Gray Cup, *w.* 1¼–1⅝", *p. 335*

624   Blueberry Cup, *w.* ¼–⅜", *p. 360*

625 Common Brown Cup, *w.* 1¼–4″, *p. 347*

626 Elf Cup, *w.* ⅜–¾″, *p. 353*

627 Pyxie Cup, *w.* ¼–¾″, *p. 350*

628   Rose Goblet, *w.* ⅝–1⅜″, *p. 340*

629   Brown-haired White Cup, *w.* ⅜–1¼″, *p. 351*

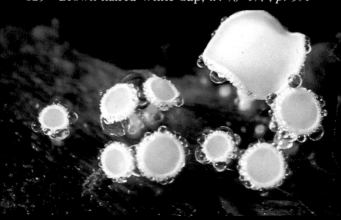

630   Stalked Hairy Fairy Cup, *w.* ¹⁄₃₂″, *p. 361*

631   Common Gel Bird's Nest, *w.* ⅛–⅜″, *p. 829*

632   Splash Cups, *w.* ¼–⅝″, *p. 828*

633   White-egg Bird's Nest, *w.* ¼–⅜″, *p. 828*

Puffballs, Earthstars, Amanita Buttons, and Others

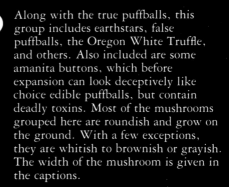

Along with the true puffballs, this group includes earthstars, false puffballs, the Oregon White Truffle, and others. Also included are some amanita buttons, which before expansion can look deceptively like choice edible puffballs, but contain deadly toxins. Most of the mushrooms grouped here are roundish and grow on the ground. With a few exceptions, they are whitish to brownish or grayish. The width of the mushroom is given in the captions.

634 Beaked Earthstar, *w.* 1⅝–2¼", *p. 818*

635 Collared Earthstar, *w.* 2¼–4½", *p. 819*

636 Rounded Earthstar, *w.* 1½–2¼", *p. 818*

637 Saltshaker Earthstar, *w.* 1⅜–4½″, *p.* 837

638 Barometer Earthstar, *w.* 2⅜–5¼″, *p.* 837

639 Arched Earthstar, *w.* 2¾–5″, *p.* 817

640   Pink Crown, *w.* 1¼–4", *p. 346*

64   Tumbling Puffball, *w.* 1¼–3½", *p. 820*

642   Earthstar Scleroderma, *w.* 5–8", *p. 839*

643 Fuzzy False Truffle, *w.* ¾–2¾″, *p. 350*

644 Puffball Agaric, *w.* ⅜–2¾″, *p. 811*

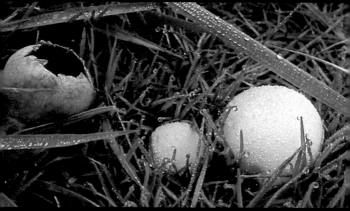

645 Western Lawn Puffbowl, *w.* ¾–1⅝″, *p. 827*

646 Granular Puffball, *w.* ⅜–¾", *p. 817*

647 Giant Puffball, *w.* 8–20", *p. 823*

648 Western Giant Puffball, *w.* 8–24", *p. 821*

649 Spiny Puffball, *w.* 1–2″, *p. 824*

650 Sculptured Puffball, *w.* 3¼–6″, *p. 820*

651 Tough Puffball, *w.* 2–8″, *p. 826*

652   Gem-studded Puffball, *w.* 1–2⅜″, *p. 825*

653   Purple-spored Puffball, *w.* 2¾–7″, *p. 822*

⊗ 654   Pigskin Poison Puffball, *w.* 1–4″, *p. 839*

655 Orange-staining Puffball, *w.* ¾–4″, *p. 824*

656 Skull-shaped Puffball, *w.* 3¼–8″, *p. 822*

657 Plated Puffball, *w.* ⅜–2″, *p. 815*

658   Pea-shaped Nidularia, *w.* ¹⁄₁₆–³⁄₈", *p. 830*

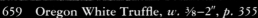

659   Oregon White Truffle, *w.* ³⁄₈–2", *p. 355*

660   Western Rhizopogon, *w.* ³⁄₈–2", *p. 813*

661 Stalked Yellow Trunc, *w. 1–3″, p. 813*

662 American False Russula, *w. ⅜–2″, p. 812*

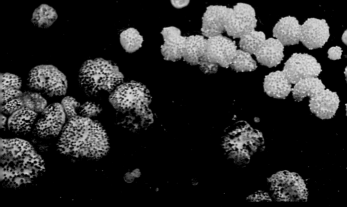

663 Yellow Cushion Hypocrea, *w. ½₂–⅛″, p. 371*

664   Pear-shaped Puffball, *w.* ⅝–1¾", *p. 826*

665   Wolf's-milk Slime, *w.* ⅛–⅝", *p. 848*

666   Netted Rhodotus, *w.* 1–2", *p. 795*

667 Red Cushion Hypoxylon, *w.* 1/16–5/8", *p.* 374

668 Carbon Balls, *w.* 3/4–15/8", *p.* 374

669 Carbon Cushion, *w.* 15/8–4", *p.* 375

670   Aborted Entoloma, *aborted stage,* p. 641

671   Coccora, *button,* p. 529

⊗ 672   Destroying Angel, *button,* p. 551

673 Death Cap, *button*, p. 543

674 Tawny Grisette, *button*, p. 536

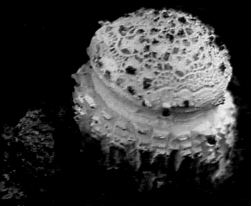

675 Rag-veil Amanita, *button*, p. 530

676   Gem-studded Puffball, *w.* 1–2⅜″, *p. 825*

⊗ 677   Yellow-orange Fly Agaric, *button, p. 540*

678   Yellow Blusher, *button, p. 535*

⊗ 679   **Panther,** *button, p. 541*

⊗ 680   **Fly Agaric,** *button, p. 539*

681   American Caesar's Mushroom, *button, p. 528*

Morels, Stinkhorns, and Other Club-shaped Mushrooms

 This group contains a wide variety of club-shaped and stalklike fungi, including the delectable morels and deadly false morels, as well as earth tongues, inky caps, and stinkhorns. These fungi grow upright from the ground or, in some instances, from wood. Some lack a distinguishable cap, while others have caps that are rounded, lobed, wrinkled, honeycombed, or saddle- to brain-shaped. Also included here are a few stalked puffballs and clublike coral fungi, and some of the parasitic flask fungi. The captions give the total height of the mushroom.

682   Velvety Earth Tongue, *h. 1⅝–3″, p. 357*

683   Goldenthread Cordyceps, *h. 3¾–4″, p. 370*

684   Swamp Beacon, *h. ⅞–2¼″, p. 358*

685 Orange Earth Tongue, *h.* ¾–3¼", *p. 358*

686 Irregular Earth Tongue, *h.* ¾–2¾", *p. 359*

687 Trooping Cordyceps, *h.* 1⅝–2⅜", *p. 369*

688   Pestle-shaped Coral, *h. 3–12"*, *p. 403*

689   Strap-shaped Coral, *h. 1–4"*, *p. 403*

690   White Green-algae Coral, *h. ¼–⅝"*, *p. 406*

691 Beetle Cordyceps, *h.* 2¾–4″, *p. 369*

692 Elegant Stinkhorn, *h.* 4–7″, *p. 835*

693 Velvety Fairy Fan, *h.* 1⅛–2″, *p. 359*

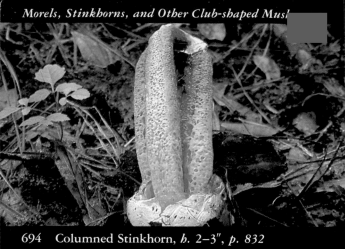

694   Columned Stinkhorn, *h.* 2–3″, *p. 832*

695   Stinky Squid, *h.* 1½–3″, *p. 833*

696   Fluted-stalked Fungus, *h.* 2–4″, *p. 332*

697 Dead Man's Fingers, *h. ¾–3¼", p. 376*

698 Dye-maker's False Puffball, *h. 2–8", p. 838*

699 Amanita Mold, *mold, p. 372*

700   Stalked Puffball-in-aspic, *h.* 1⅛–3¼″, *p.* 841

701   Club-shaped Stinkhorn, *h.* 1–2″, *p.* 834

702   Desert Inky Cap, *h.* 3⅛–12″, *p.* 814

703 Scaly Inky Cap, *h.* 4–8″, *p. 601*

704 Shaggy Mane, *h.* 4–14″, *p. 597*

705 Woolly-stalked Coprinus, *h.* 2¾–11″, *p. 599*

706   Ravenel's Stinkhorn, *h. 5¼–7¾", p. 835*

707   Netted Stinkhorn, *h. 6–7", p. 834*

708   Smooth Thimble-cap, *h. 2–3⅝", p. 329*

709 Lizard's Claw, *h.* 4–6″, *p. 832*

710 Yellow Morel, *h.* 3¾–5½″, *p. 327*

711 Half-free Morel, *h.* 3⅝–5⅝″, *p. 328*

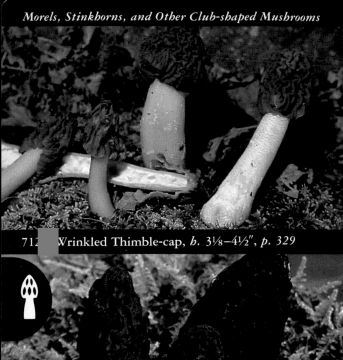

712 Wrinkled Thimble-cap, *h.* 3⅛–4½″, *p. 329*

713 Black Morel, *h.* 2¾–6″, *p. 326*

⊗ 714 Conifer False Morel, *h.* 2¾–6″, *p. 336*

715 Snowbank False Morel, *h.* 3⅝–8″, *p. 338*

☒ 716 Gabled False Morel, *h.* 7¼–9″, *p. 335*

717 Snowbank False Morel, *h.* 3⅝–8″, *p. 338*

718 **Thick-stalked False Morel,** *h.* 2–6″, *p.* 337

⊗ 719 **Saddle-shaped False Morel,** *h.* 1⅛–6″, *p.* 339

⊗ 720 **Saddle-shaped False Morel,** *h.* 1⅛–6″, *p.* 339

721 Fluted Black Helvella, *h. 2–6", p. 334*

722 Fluted White Helvella, *h. 1⅝–5⅛", p. 333*

723 Smooth-stalked Helvella, *h. ⅞–5⅝", p. 334*

724   Bladder Stalks, *h.* ½–1¾″, *p. 411*

725   Yellow Cudonia, *h.* 1⅛–2⅜″, *p. 363*

726   Ochre Jelly Club, *h.* 1⅛–3⅝″, *p. 364*

727   Green-headed Jelly Club, *h.* 1½–2⅞″, *p. 365*

728   Headlike Cordyceps, *h.* 1⅛–3⅞″, *p. 368*

729   Water Club, *h.* ¼–⅞″, *p. 366*

730   **Buried-stalk Puffball,** *h. 1–1⅞", p. 842*

731   **Desert Stalked Puffball,** *h. 7–17⅝", p. 841*

732   **Japanese-lantern Slime,** *h. ⅟₃₂–¼", p. 849*

Coral-like Mushrooms

Most of these mushrooms are coral fungi—club-shaped or many-branched forms that resemble underwater coral. Two of these, the Eastern and Rooting Cauliflower Mushrooms, are among the largest and most easily identified of choice edible mushrooms. Also included are a few coral-like jelly fungi and slime molds. The measurement given in the captions is usually the total height, including stalk or base.

733 Straight-branched Coral, *h. 2–4¾", p. 409*

734 Spindle-shaped Yellow Coral, *h. 2–6", p. 399*

735 Yellow Tuning Fork, *h. 1–4", p. 380*

736 Chocolate Tube Slime, *h.* ⅜–1″, *p. 853*

737 Purple Club Coral, *h.* 1–5″, *p. 400*

738 Carbon Antlers, *h.* 1⅝–3¼″, *p. 375*

739   White Worm Coral, *h. 2–6″, p. 400*

740   Coral Slime, *h. ¹⁄₃₂–³⁄₈″, p. 845*

741   Jellied False Coral, *h. 1–4″, p. 385*

742   White Coral, *h.* 1¼–5″, *p. 410*

743   Crested Coral, *h.* 1¼–4¼″, *p. 402*

744   Crown-tipped Coral, *h.* 2–5″, *p. 401*

⊗ 745   Jellied-base Coral, *h. 3–6″, p. 408*

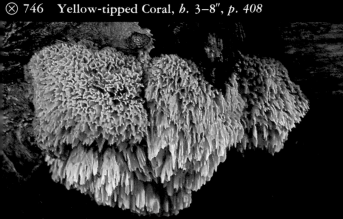

⊗ 746   Yellow-tipped Coral, *h. 3–8″, p. 408*

747   Orange Sponge Polypore, *w. 1–6″, p. 486*

748 Cotton-base Coral, *h.* ⅞–3″, *p.* 405

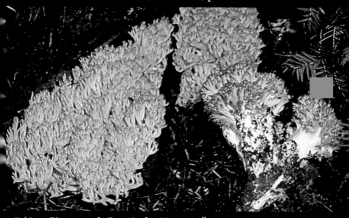

749 Clustered Coral, *h.* 3¾–8⅜″, *p.* 407

750 Light Red Coral, *h.* 2¾–6¼″, *p.* 406

751   Red Raspberry Slime, *h.* ⅛–¼″, *p. 848*

752   Violet-branched Coral, *h.* ¾–2⅜″, *p. 402*

753   Stalked Cauliflower Fungus, *h.* 10–15″, *p. 345*

754 Jelly Leaf, *h.* 2–4″, *p.* 384

755 Rooting Cauliflower Mushroom, *h.* 10–20″, *p.* 412

756 Eastern Cauliflower Mushroom, *h.* 6–10″, *p.* 411

The numbers preceding the species descriptions on the following pages correspond to the plate numbers in the color section.

ASCOMYCETES
(Subdivision Ascomycotina)

This is an extraordinarily large and diverse subdivision of the fungi. It includes the highly esteemed truffles and morels of haute cuisine, the brightly colored little cup fungi, and a broad range of minute fungi that cause plant and animal diseases—such as Dutch Elm Disease, Chestnut Blight, Leaf Curl, and Ringworm. Most of the larger fungi in this subdivision are restricted to 2 classes, Discomycetes and Pyrenomycetes, and are grouped together because of the way they produce spores. The spores develop within more or less cylindrical or round, saclike, microscopic structures known as asci. Further division in this subdivision into classes, orders, families, genera, and species is partially determined by the method of spore dispersal.

The size, shape, surface texture, and color of the asci and spores, as well as their reaction to certain chemicals, also play a part in the classification of these mushrooms. So while field examination is important, it may often prove insufficient for determining species with certainty.

DISC FUNGI
(Class Discomycetes)

The Discomycetes contain all of the large or brightly colored fungi that resemble cups or saucers, spoons, sponges, saddles, brains, tongues, urns, or fans, as well as those underground fungi called truffles. With the exception of truffles, all have an exposed fertile surface lined with microscopic saclike containers (asci) filled with spores, which are typically colorless.

CUP FUNGI AND ALLIES
(Order Pezizales)

The order Pezizales is made up of mushrooms whose spore-producing structures, or asci, line the inner or upper surface of a cup- or saucer-shaped fruiting body. Morels, false morels, saddle fungi, and many cup-shaped fungi are Pezizales.
The asci are usually cylindrical and generally contain 8 spores. The spores of most species are round or elliptical, large, and single-celled. The morels develop spores in the pits of the honeycombed head; the brain-shaped false morels, in the folds or wrinkles of the cap. The spores in saddle fungi are located around the outer surface of the saddlelike head, while the cup fungi bear asci on the upper or inner surface of the cup. The asci open at maturity and forcibly discharge their spores into the air. Sometimes a cloud of spores can be seen rising above the mushroom; occasionally, the "puff" of the discharging spores can even be heard. Members of this order are often common in the early spring, appearing before the season of most gilled mushrooms. Many grow on rotting wood, in wet areas, along roadsides, in dung, or on burned sites. Some

members of the order are small and
brightly colored: often, inspection of a
rotting log or a blackened area will
reveal several red, orange, or yellow cup
fungi. With luck, the mushroom
hunter may spot some flavorful morels.
Seven families within the order have
been described and are recognized here:
Morchellaceae (3 genera), Helvellaceae
(6 genera), Sarcosomataceae (11
genera), Sarcoscyphaceae (14 genera),
Ascobolaceae (6 genera), Pezizaceae (4
genera), and Pyronemataceae (48
genera). Most genera are small in
number or have few species represented
in North America. *Peziza* and *Helvella,*
however, are very large, with more than
100 described species.

713 Black Morel
Morchella elata Fr.
Morchellaceae, Pezizales

Description: *Black-ribbed, honeycombed cap on whitish
stalk.*
Cap: ¾–1⅝″ (2–4 cm) wide, ¾–2″
(2–5 cm) high; *elongate and narrowly
conical;* with dark gray to black
longitudinal and radial ribs (sometimes
irregular), and long, yellow-brown pits;
attached to stalk at base; hollow.
Stalk: 2–4″ (5–10 cm) long, ¾–1⅝″
(2–4 cm) thick; whitish, granular to
mealy; hollow.
Spores: 24–28 × 12–14 μ; elliptical,
smooth, located in pits.

Edibility: Choice, with caution.

Season: April–May.

Habitat: On the ground in coniferous woods,
especially spruce; on sandy soil in
mixed woods; associated with pines and
poplars in recently burned areas.

Range: Throughout North America.

Comments: The Black Morel may be a complex of
practically indistinguishable varieties,
including Peck's Morel (*M. angusticeps*)
and the Conical Morel (*M. conica*). The
Black Morel is often the first true morel

to appear in the spring, usually in poplar woods. Although it is an excellent edible, it may cause stomach upsets, especially if taken with alcoholic beverages. Most wild mushrooms are indigestible; do not eat them raw or in large quantities.

710 **Yellow Morel**
Morchella esculenta L. ex Fr.
Morchellaceae, Pezizales

Description: *Blond to yellow-brown, honeycombed cap on whitish stalk.*
Cap: 1⅝–2″ (4–5 cm) wide, 2¾–3½″ (7–9 cm) high; oval to elongate; with longitudinal, irregular, yellowish-brown ribs and round to elongate, yellowish pits; attached to stalk at base; hollow.
Stalk: 1–2″ (2.5–5 cm) long, ⅝–1″ (1.5–2.5 cm) thick, sometimes enlarged at base; whitish, with granular ribs; hollow.
Spores: 20–24 × 12–24 μ; elliptical, smooth, located in pits.

Edibility: Choice.
Season: Late April–early June; into August at high elevations; March–May on West Coast.
Habitat: Single to many, on the ground in old apple orchards and burned areas; associated with dead elms; also with tuliptrees, and ash, oak, and beech-maple woods.
Range: Throughout North America.
Look-alikes: *Verpa bohemica** has smaller, bell-shaped cap attached to stalk at top. False morels, such as the deadly *Gyromitra esculenta**, have wrinkled or brain-shaped, chambered caps.
Comments: Also called the "Honeycomb Morel." Sometimes it has white ribs and grayish pits, especially when immature, and it is then called the White Morel (*M. deliciosa*). Late in the season, the Yellow Morel is often huge and thick-footed, up to 12″ high and 6″ thick, and is

sometimes referred to as the Thick-
footed Morel (*M. crassipes*). The
appearance of the morel is the occasion
of an annual morel-hunting festival in
Boyne City, Michigan, where at the
start of a gun hundreds of people race
to find and collect as many morels as
possible in 90 minutes. The record for
one person is more than 900 morels. In
most places, however, finding a dozen
morels in an entire season is rewarding.

---

711 Half-free Morel
Morchella semilibera DC. ex Fr.
Morchellaceae, Pezizales

Description: *Yellow-brown, skirtlike, honeycombed cap
on whitish stalk.*
Cap: ⅜–1⅝" (1–4 cm) high and wide
at flaring base; conical; with
longitudinal and irregular brownish
ribs and round to elongate, yellow-
brown pits; attached to stalk midway
up cap; hollow.
Stalk: 3¼–4" (8–10 cm) long, ⅜–¾"
(1–2 cm) wide, thickening to 1⅜–1⅝"
(3.5–4 cm) at base; whitish, with faint
granular ribs; hollow.
Spores: 24–30 × 12–15 μ; elliptical,
smooth, located in pits.
Edibility: Good.
Season: April–early May.
Habitat: On the ground in damp, open woods;
near oak, beech, and tuliptrees.
Range: E. North America to Iowa; Pacific NW.
Look-alikes: *M. esculenta** and *M. elata** have caps
attached to stalk at base of cap. *Verpa
bohemica** has a skirtlike cap attached at
top of stalk.
Comments: The Half-free Morel appears about 7 to
10 days earlier than the Yellow Morel
(*M. esculenta*). Since most morels blend
well with their backgrounds, many
people have great difficulty learning to
spot them. It is good to practice
finding the Half-free Morel first, and so
develop some skill in time for the rest
of the Yellow Morel's season.

712 Wrinkled Thimble-cap
Verpa bohemica (Kromb.) Schroet.
Morchellaceae, Pezizales

Description: *Yellow-brown, wrinkled, thimblelike cap on whitish stalk.*
Cap: ⅜–1" (1–2.5 cm) wide, ¾–1¼" (2–3 cm) high; thimble- to bell-shaped; yellow-brown, with deeply wrinkled, skirtlike sides; *attached only at top of stalk;* hollow.
Stalk: 2⅜–3¼" (6–8 cm) long, ⅜–1" (1–2.5 cm) thick at base; cylindrical; whitish, with smooth to faintly granular ribs; stuffed or hollow.
Spores: 60–80 × 15–18 μ; huge, elliptical, smooth, in 2-spored sacs.
Edibility: Edible with caution.
Season: Late March–early May.
Habitat: On the ground, in damp, low-lying woods and wet areas.
Range: Throughout North America.
Look-alikes: *Morchella semilibera** is attached to stalk halfway down cap. Other *Morchellas** have stalk attached to base of cap.
Comments: Also called *Morchella bispora* and *Ptychoverpa bohemica.* This is the only morel-type mushroom with 2-spored sacs; all others have 8 spores. Although the Wrinkled Thimble-cap is sometimes eaten safely, it can cause lack of muscular coordination in susceptible individuals if consumed in large quantities or over several days.

708 Smooth Thimble-cap
Verpa conica Swartz ex Pers.
Morchellaceae, Pezizales

Description: *Brownish, smooth, thimblelike cap on whitish stalk.* 
Cap: ⅜–¾" (1–2 cm) wide, ⅜–1¼" (1–3 cm) high; thimble- to bell-shaped; brownish, with smooth, skirtlike sides; *attached only at top of stalk;* hollow.
Stalk: 1⅝–2⅜" (4–6 cm) long, ¼–⅝" (0.5–1.5 cm) thick; cylindrical;

whitish, with smooth to granular ribs;
stuffed or hollow.
Spores: 22–26 × 12–16 μ; elliptical,
smooth, located in pits.

Edibility: Edible.

Season: April–early May.

Habitat: On the ground in deciduous woods;
under old apple trees.

Range: Throughout the East; also Colorado,
California, and Pacific NW.

Look-alikes: *V. bohemica** has wrinkled, furrowed
cap. *Morchella semilibera** is attached to
stalk midway up cap.

Comments: The Smooth Thimble-cap is one of the
first mushrooms to appear in spring in
the North.

---

621 Veined Cup

Disciotis venosa (Pers. ex Fr.) Boud.
Morchellaceae, Pezizales

Description: *Conspicuously veined, reddish-brown cup on*
very short, stout stalk.
Cup: 2–8″ (5–20 cm) wide; cup-
shaped; outer surface dingy whitish,
inner surface reddish-brown, becoming
distinctly veined or wrinkled; often
irregularly shaped and splitting at
maturity.
Stalk: ¼–⅜″ (0.5–1 cm) long and
thick; short, stout, often buried in
soil.
Spores: 19–25 × 12–15 μ; broadly
elliptical, sometimes showing granules
at tips; pale yellow, brownish in mass
in spore sacs (asci).

Edibility: Edible with caution.

Season: March–May.

Habitat: On the ground in deciduous woods;
reported under conifers in Pacific NW.

Range: Throughout East; in Colorado, Utah,
New Mexico, and on West Coast.

Look-alikes: *Discina perlata** is less cuplike, less
veined, with warted spores.

Comments: Also called *Discina venosa* and *Peziza*
venosa. The Veined Cup is related to the
morels because its spores have granules
at the ends and its spore sacs (asci) are

not amyloid. Many people cook and eat this mushroom, but it is poisonous if consumed raw.

---

618 Pig's Ears
Discina perlata (Fr.) Fr.
Helvellaceae, Pezizales

Description: *Brown to tannish, wrinkled, cup- or ear-shaped fungus, sometimes with short, stout stalk.*
Cup: 1⅝–4" (4–10 cm) wide; disclike; exterior whitish, interior dark brown to tan; wrinkled to convoluted, edges turned down. Flesh brittle.
Stalk (when present): ¼–⅜" (0.5–1 cm) long and thick; brownish-tan.
Spores: 30–35 × 12–14 μ; spindle-shaped, minutely warted, with 3 oil drops and knobs at each end, located in cup.

Edibility: Edible with caution.
Season: May–July.
Habitat: Singly or in groups, on humus or rotten wood in coniferous areas; near melting snowbanks in western mountains.
Range: Northeast; Northwest; California.
Look-alikes: *Disciotis venosa** is more deeply veined, and has smooth spores; it is typically found in deciduous woods. Other similar species of *Discina* must be differentiated microscopically.
Comments: Also called *Peziza perlata* or *Acetabula ancilis*. The spores of *D. perlata* are quite similar to those of mushrooms in the genus *Gyromitra*, so that some mycologists place it there.

---

619 Crustlike Cup
Rhizina undulata Fr. ex Fr.
Helvellaceae, Pezizales

Description: *Dark brown, crustlike, upside-down cup; attached to burned ground by tangle of whitish, tough, rootlike hairs.*
Cup: ¾–2⅜" (2–6 cm) wide; shallowly cup-shaped to flat or crustlike; wavy

outer surface dark brown with paler
edge, lighter beneath; inner surface has
mass of thin cords attached to ground.
Spores: 22–40 × 8–11 μ; spindle-
shaped, smooth, with 2 or more oil
drops and knob at each end.

Season: July–October.

Habitat: On burned ground, especially under
conifers.

Range: Throughout N. North America.

Comments: Formerly known as *R. inflata.* This
fungus appears on burned ground a year
or so after a fire.

---

696 Fluted-stalked Fungus
Underwoodia columnaris Pk.
Helvellaceae, Pezizales

Description: *Tall, tan to brownish fluted clubs, fused
together at base.*
Club: 2–4″ (5–10 cm) high, ⅜–1¼″
(1–3 cm) thick; cylindrical to club-
shaped, cream to pale tan, becoming
brownish; deeply fluted. Interior with
large chambers.
Spores: 25–27 × 12–14 μ; elliptical,
smooth at first, becoming warted.

Season: June–July.

Habitat: On soil under hardwoods.

Range: E. Canada to New York and Michigan,
west to Kansas.

Comments: This species resembles a fluted-stalked
helvella, but does not have a saddle-
shaped cap.

---

622 Ribbed-stalked Cup
Helvella acetabulum (L. ex Fr.) Quél.
Helvellaceae, Pezizales

Description: *Brown cup with conspicuous white ribs
descending short, thick, white stalk.*
Cup: ¾–3¼″ (2–8 cm) wide, ⅜–1⅝″
(1–4 cm) high; cup-shaped; gray-
brown to dark brown above, sometimes
tinted violet; whitish, prominently
ribbed undersurface.
Stalk: ⅜–¾″ (1–2 cm) long, ¼–⅝″

(0.5—1.5 cm) thick; white, deeply ribbed, with chambers fused into underside of cup.

Spores: 18—22 × 12—14 μ; broadly elliptical.

Season: Early May—early July in East; December—April in Southwest.

Habitat: On the ground in woods and open places.

Range: Throughout North America.

Look-alikes: *H. griseoalba* has gray cup and shorter stalk.

Comments: Also known as the "Elfin Cap," *Paxina acetabulum,* and *Peziza acetabula.* This mushroom is easily identified and is sometimes found in large quantities, but there are no reliable reports about its edibility in North America.

722 Fluted White Helvella
Helvella crispa Scop. ex Fr.
Helvellaceae, Pezizales

Description: *Whitish, saddle-shaped cap on fluted stalk.* Cap: ⅜—2⅜" (1.5—6 cm) wide; ⅜—1⅜" (1—4 cm) high; edges rolled inward at first, then expanding; whitish to light buff above, hairy beneath. Stalk: 1¼—3½" (3—9 cm) long, ¼—1" (0.5—2.5 cm) thick, enlarged toward base; whitish, with deep fluting and ribs and minute hairs; attached to cap only at top. Interior chambered.

Spores: 18—21 × 10—13 μ; elliptical, with 1 large central oil drop.

Season: Late July to mid-October in East; December—April in Southwest; September—October in Northwest.

Habitat: On the ground in both deciduous and coniferous forests, and open grassy or wooded areas.

Range: East Coast; Colorado, Arizona, California, and Pacific NW.

Look-alikes: *H. lacunosa** has dark cap and is attached to stalk at several places.

Comments: One of the most common species of *Helvella,* this comes up in late-summer lawns near planted conifers.

723   Smooth-stalked Helvella
Helvella elastica Bull. ex Fr.
Helvellaceae, Pezizales

Description: *Brownish, saddle-shaped cap on slender stalk.*
Cap: ⅜–1⅜" (1–3.5 cm) wide, ¼–1⅝" (0.5–4 cm) high; saddle-shaped, with slightly incurved edge; pale grayish-yellow to gray, grayish-brown, or brown above; smooth, whitish to creamy-buff below.
Stalk: ⅝–4" (1.5–10 cm) long, ⅛–⅜" (0.3–1 cm) thick; *cylindrical;* same color as cap or paler, smooth.
Spores: $18–22 \times 11–13 \mu$; elliptical, smooth, with 1 large central oil drop.

Season: July–early November; December–April in Southwest.

Habitat: On the ground in both deciduous and coniferous woods.

Range: Throughout North America.

Look-alikes: *H. stevensii* has inrolled cap margin and hairy undersurface. Similar thin-stalked, saddle-shaped fungi can only be differentiated microscopically.

Comments: Also known as *Leptopodia elastica.* Most *Helvella* species are best identified microscopically; none are recommended as edibles.

---

721   Fluted Black Helvella
Helvella lacunosa Afz. ex Fr.
Helvellaceae, Pezizales

Description: *Dark, saddle-shaped cap on fluted stalk.*
Cap: ⅝–2" (1.5–5 cm) wide, ⅜–2" (1–5 cm) high; saddle-shaped to conical; dark brown to gray or gray-black; lobed to wrinkled above, smooth below.
Stalk: 1⅝–4" (4–10 cm) long, ⅝–¾" (1.5–2 cm) thick; grayish, deeply fluted or ribbed. Interior chambered.
Spores: $15–20 \times 9–12 \mu$; elliptical, smooth, with 1 large central oil drop.

Season: Mid-August to late October in East; August in Rocky Mountains; January–

February in Southwest.
Habitat: On the ground or decaying wood, in both deciduous and coniferous woods.
Range: Widely distributed, but especially common in Pacific NW.
Look-alikes: Similar dark, saddle-shaped fungi can only be differentiated microscopically.
Comments: Where this mushroom is common and well known, as in California, it is often eaten; however, it cannot be recommended because the related false morels (*Gyromitra*) are known to contain toxins.

623 Long-stalked Gray Cup
Helvella macropus (Pers. ex Fr.) Kar.
Helvellaceae, Pezizales

Description: *Gray cup on slender, hairy stalk.*
Cup: 1¼–1⅝″ (3–4 cm) wide; cuplike; gray, outer surface downy.
Stalk: 1¼–1⅝″ (3–4 cm) long, ¹⁄₁₆–⅛″ (1.5–3 mm) thick, enlarging slightly toward base; gray, covered with dense tufts of downy gray hairs.
Spores: 20–30 × 10–12 μ; elliptical to spindle-shaped, with 1 large central oil drop and 1 smaller drop at each end.
Season: Late June–early November in East; December–January along Pacific Coast.
Habitat: On the ground in both coniferous and deciduous woods.
Range: Throughout North America.
Look-alikes: Similar stalked cup fungi can only be reliably differentiated microscopically.
Comments: Also known as *Macroscyphus macropus* and *Peziza macropus*. Although this species looks like a cup fungus, it is related to the saddle fungi because of the oil drops in its spores.

⊗ 716 Gabled False Morel
Gyromitra brunnea Under.
Helvellaceae, Pezizales

Description: *Brownish, convoluted, saddle-shaped or brainlike cap on whitish, branched stalk.*

Cap: 4¾" (12 cm) wide, 2–4" (5–10 cm) high; convoluted; saddle-shaped to lobed, with folds arching up and out to compressed, unfused margin; reddish-brown to chocolate-brown above, underside pale buff to tan. Interior chambered; flesh brittle.

Stalk: 5¼" (13 cm) long, ¾–2" (2–5 cm) thick, enlarging somewhat toward base; ribbed to almost smooth, white. Interior nearly hollow to stuffed with cottony white tissue.

Spores: 28–30 × 12–15 μ; elliptical, finely warted to spiny, with 2 large oil drops.

Edibility: Poisonous.

Season: Late May–early June.

Habitat: On humus in hardwood forests.

Range: Maryland to Michigan, Ohio, Tennessee, Missouri, and Oklahoma.

Look-alikes: *G. caroliniana* is more convoluted and brainlike, with cap margin fused to stalk.

Comments: Also called *Helvella underwoodii*. The Gabled False Morel is reported to be poisonous and, although no deaths have been linked to it, its toxins can cause blood poisoning, diarrhea, severe headaches, and vomiting.

⊗ 714 **Conifer False Morel**
Gyromitra esculenta (Pers. ex Fr.) Fr.
Helvellaceae, Pezizales

Description: *Brownish, brainlike to saddle-shaped cap on short, whitish stalk.*

Cap: 1¼–4" (3–10 cm) wide, 2–4" (5–10 cm) high; brain-shaped; yellow, yellow-brown, or bay above, lighter below; wrinkled or folded but not pitted. Interior chambered; flesh brittle.

Stalk: ¾–2" (2–5 cm) long, ⅜–1" (1–2.5 cm) thick; whitish; unbranched but often fluted; scruffy to nearly smooth; hollow or with 1–2 chambers.

Spores: 18–22 (28) × 9–12 μ; elliptical, smooth, with 2 oil drops.

Edibility: Deadly.
Season: April—early June.
Habitat: On the ground under conifers.
Range: Throughout North America; most
common in North and in mountains.
Look-alikes: True morels have honeycombed caps
that are hollow, not chambered.
Comments: Also called *Helvella esculenta.* Other
common names include "Brain
Mushroom," "Beefsteak Morel,"
"Lorchel," and "Edible False Morel."
Scientists have discovered that the
Conifer False Morel develops a
compound similar to one used in the
manufacture of rocket fuel. It causes
acute illness and has been fatal in a few
instances; it also produces tumors in
laboratory animals. Its species name is
esculenta, or "edible," because the
toxins may be removed by drying and
rehydrating, or by a process of boiling,
rinsing, and boiling again. Even so, the
removal of toxins is sometimes
incomplete. In some locations where
the toxic concentration is apparently
minimal, the mushroom has been
cooked and eaten like a true morel with
no apparent ill effects. However, this
species cannot be recommended for
eating.

---

718 Thick-stalked False Morel
Gyromitra fastigiata (Kromb.) Rehm
Helvellaceae, Pezizales

Description: *Brownish, wrinkled to brainlike cap on
whitish, short, thick stalk.*
Cap: ¾–4″ (2–10 cm) wide, 1⅛–4″
(4–10 cm) high; brain-shaped to lobed,
with longitudinal folds; yellow-brown
to red-brown above, paler below.
Interior chambered; flesh brittle.
Stalk: ⅜–2″ (1–5 cm) long, 1–3″
(2.5–7.5 cm) thick; always thicker
than high, with long ribs and many
channels; whitish; appearing branched.
Spores: 22–32 × 10–14 μ; spindle-
shaped, with well-developed knobs at

each end; 1 large central oil drop and 2 smaller terminal ones.

Season: Late April–late May.

Habitat: On the ground in hardwood forests or mixed woods.

Range: NE. North America; Rocky Mountains; areas of Pacific NW.

Look-alikes: The deadly *G. esculenta** has unbranched stalk. *G. gigas** lacks knobs on ends of spores.

Comments: Because this species is closely related to the deadly Conifer False Morel (*G. esculenta*), it should not be eaten. Also known as *G. korfii.*

715, 717   Snowbank False Morel
Gyromitra gigas (Kromb.) Quél.
Helvellaceae, Pezizales

Description: *Brownish, brainlike cap on whitish, branched stalk.*
Cap: 1⅝–4″ (4–10 cm) wide and high; brainlike to deeply wrinkled; yellow-brown, becoming dingy chestnut. Interior chambered; often with cracked, brittle, whitish flesh.
Stalk: 2–4″ (5–10 cm) long and thick; somewhat short, whitish, and very broad, with multiple channels.
Spores: 24–36 × 10–15 μ; elliptical, smooth to finely warted, with short projection at each end.

Edibility: Choice, with caution.

Season: May–June.

Habitat: On humus under conifers, especially near melting snowbanks.

Range: On mountains in Colorado, Pacific NW., and California.

Look-alikes: *G. fastigiata** can only be differentiated by its spores.

Comments: Also known as the "Giant Helvella," the Snowbank False Morel is the most widely eaten false morel in North America. It often weighs up to 2 pounds. It is not known to occur east of the Rocky Mountains. All reported poisonings from false morels have been in the East and Midwest.

⊗ **719, 720 Saddle-shaped False Morel**
Gyromitra infula (Schaeff. ex Fr.) Quél.
Helvellaceae, Pezizales

Description: *Brownish, saddle-shaped cap on whitish or buff, unbranched stalk.*
Cap: 1¼–4" (3–10 cm) wide, ¾–4" (2–10 cm) high; saddle-shaped to 3-lobed, with incurved edge; reddish-brown to dark brown, wrinkled to convoluted. Interior chambered or hollow; flesh brittle.
Stalk: ⅜–2" (1–5 cm) long, ¾" (2 cm) thick; sometimes irregular; whitish to buff, unribbed; hollow.
Spores: 18–24 × 7–9 (12) μ; elliptical, smooth, with 2 large oil drops.

Edibility: Poisonous.

Season: July–October in East; November–April in West.

Habitat: On rotten wood or on ground in wood debris.

Range: Throughout North America.

Look-alikes: *G. ambigua* has lavender to violet tints on both head and stalk.

Comments: Also known as *Helvella infula* and the "Hooded Helvella." Although saddle-shaped and more similar to a *Helvella* than a *Gyromitra*, this mushroom is placed in the latter genus because of the number of oil drops in its spores.

---

614 Hairy Rubber Cup
Galiella rufa (Schw.) Nannf. & Korf
Sarcosomataceae, Pezizales

Description: *Cuplike fungus with tough, blackish-brown, hairy outer surface, reddish-brown, shallow center, and gelatinous flesh.*
Cup: ¾–1¼" (2–3 cm) wide; closed at first, opening to form shallow cup with incurved edges; outer surface blackish-brown, covered with clusters of hairs; inner surface reddish to reddish-brown, fading to tannish; skin tough. Internal layer of gelatinous flesh gives fungus a rubbery feel.

Stalk (when present): ⅜" (1 cm) long,
¼" (5 mm) thick; attached by dense
mass of black hairs.
Spores: 20 × 10 μ; elliptical, with
extremely narrow ends and fine warts.

Season: July–September.
Habitat: Clustered on deciduous wood.
Range: E. North America to Minnesota.
Look-alikes: *Bulgaria inquinans** lacks reddish-brown
color and ejects spores differently.
Comments: Although the Hairy Rubber Cup may
resemble a puffball before it opens, and
its gelatinous interior resembles a jelly
mushroom, it is cuplike at maturity
and bears no relationship to "puffs" or
"jellies."

---

628 Rose Goblet
Neournula pouchetii (Ber. & Rio.) Paden
Sarcosomataceae, Pezizales

Description: *Goblet-shaped cup with brownish outer*
surface and pink to rose inner surface, on
short, whitish stalk.
Cup: ⅝–1⅜" (1.5–3.5 cm) wide, ¾–
1¾" (2–4.5 cm) high; deep cup with
pale brown to purplish-brown outer
surface, pale pink to rose inner surface,
and toothed rim.
Stalk: ¾–1⅝" (2–4 cm) long, ¼–⅜"
(0.5–1 cm) thick; whitish; usually
buried.
Spores: 23–31.5 × 8–10.4 μ;
elliptical to oblong with rounded ends;
smooth, becoming covered with warts.
Season: April–July.
Habitat: In conifer litter, under western cedar
and hemlock.
Range: Pacific NW.; reported in E. Canada.
Look-alikes: *Galiella rufa** is thick-fleshed and
gelatinous, with shallower cup. *Urnula*
*craterium** has smooth spores and
gelatinous flesh that becomes tough and
leathery.
Comments: This unusual spring cup fungus is also
known as *N. nordmanensis.*

609 Jellylike Black Urn
Plectania melastoma (Sow. ex Fr.) Fkl.
Sarcosomataceae, Pezizales

Description: *Small, black, urn-shaped fungus with orange granules on outer surface.*
Cup: ⅜−¾" (1−2 cm) wide; closed at first, opening to form cup with incurved edges; *outer surface* black and downy, *bearing hairs encrusted with rusty orange granules,* usually near rim; inner surface black, smooth, gelatinous, glistening-shiny. Flesh gelatinous.
Stalk (when present): ¼−⅜" (0.5−1 cm) long and thick; tough; with gelatinous internal layer; attached by dense, wiry, black hairs.
Spores: 23−28 × 10−11 μ; spindle-shaped, smooth, containing many oil drops, which disappear at maturity.

Season: May−June.

Habitat: Single to clustered in conifer debris.

Range: Maine to British Columbia, south to Florida and Mexico.

Comments: This little black spring cup is differentiated from others by its bright, granule-encrusted outer surface.

612 Hairy Black Cup
Pseudoplectania nigrella (Pers. ex Fr.) Fkl.
Sarcosomataceae, Pezizales

Description: *Small, stalkless black cup with hairy outer surface.*
Cup: ¼−1¼" (0.5−3 cm) wide, ¼−⅝" (0.5−1.5 cm) high; cup-shaped to expanded; outer surface black, covered with black, twisted hairs; inner surface dull brown to blackish.
Spores: 12−14 μ; round, smooth, filled with many small oil drops; colorless to faintly brownish.

Season: May−July.

Habitat: Scattered or in groups on decaying coniferous wood; usually moss-covered.

Range: Common in North: Alaska; Pacific NW.; Wisconsin; New York.

Look-alikes: *Plectania** have elliptical spores.

342   Cup Fungi

Comments: Look for this mushroom in northern
mountains in the spring after the snow
melts.

---

611 Black Rubber Cup
Sarcosoma latahensis Paden & Tyl.
Sarcosomataceae, Pezizales

Description: *Black, rubbery cup with hairy, grayish-
black outer surface.*
Cup: 1–2″ (2.5–5 cm) wide, ⅜–1¼″
(1–3 cm) high; turban-shaped at first,
expanding to cup-shaped; outer surface
grayish or black, hairy; inner surface
dark purplish-black to black, shiny.
Flesh gelatinous at first, becoming
tough.
Stalk (when present): slight and stout.
Spores: 25–35 × 10–12 μ; elliptical,
smooth.
Season: May–June.
Habitat: On decayed wood and soil in coniferous
woods, often near melting snowbanks.
Range: Pacific NW.
Look-alikes: *S. globosa* is larger and urn-shaped, with
broad, swollen base and gelatinous
interior, and is found only in the East.
S. mexicana, a western species, is larger,
with stout, wrinkled stalk.
Comments: Identification often requires examining
a cross section of these cup fungi,
which reveals the amount of jelly in
their base, as well as studying spores
and other microscopic features.

---

613 Devil's Urn
Urnula craterium (Schw.) Fr.
Sarcosomataceae, Pezizales

Description: *Large, leathery, brown, urn-shaped cup.*
Cup: 1¼–3¼″ (3–8 cm) wide, 1⅝–
2⅜″ (4–6 cm) high; closed at first,
opening to reveal notched, incurved
edges; deeply cup-shaped; outer surface
dark brown, becoming black; inner
surface dark brown to black; scruffy;
texture gelatinous at first, becoming

tough and leathery.
Stalk: 1¼–1⅝" (3–4 cm) long, ¼"
(5 mm) thick; blackish, attached by
mass of black hairs.
Spores: 25–35 × 12–14 μ; elliptical,
smooth.

Season: Late March–May.
Habitat: Clustered on fallen deciduous wood,
especially oak.
Range: New York south to Alabama and
Mississippi, northwest to North
Dakota.
Comments: This is one of the first mushrooms
to appear in spring in the East.

---

607 **Shaggy Scarlet Cup**
Microstoma floccosa (Schw.) Rait.
Sarcoscyphaceae, Pezizales

Description: *Bright red cup covered on outer surface with*
white hairs; on short white stalk.
Cup: ¼–⅜" (0.5–1 cm) wide, ⅜" (1
cm) high; outer surface has long,
conspicuous, white hairs; inner surface
scarlet.
Stalk: 1¼–2" (3–5 cm) long, ¹⁄₁₆–¼"
(1.5–5 mm) thick; white, slender,
hairy.
Spores: 20–35 × 15–17 μ; elliptical,
with extremely narrow ends; smooth.
Season: June–August.
Habitat: Clustered on buried deciduous wood.
Range: E. North America, especially in
Southeast.
Comments: Also known as *Sarcoscypha floccosa* and
Anthopeziza floccosa. Although small, it
is typically clustered and conspicuous.

---

605 **Scarlet Cup**
Sarcoscypha coccinea (Scop. ex Fr.) Lamb.
Sarcoscyphaceae, Pezizales

Description: *Deep, bright red cup with white outer*
surface.
Cup: ¾–2⅜" (2–6 cm) wide, ¾–1¼"
(2–3 cm) high; cup-shaped, with
incurved edges; outer surface white,

with minute hairs; inner surface scarlet.
Stalk (when present): ⅜–1¼″ (1–3 cm)
long, ⅛–¼″ (3–5 mm) thick; white, at
times very short.
Spores: 26–40 × 10–12 μ; elliptical,
smooth, with a cluster of small oil
drops at each end.

Season: March–May; often into August in East;
November–April in California.

Habitat: On fallen hardwood branches in wet
places.

Range: East Coast; Pacific NW. and California.

Look-alikes: *S. occidentalis* * has smaller cup, well-
developed stalk, and appears later in
spring.

Comments: The Scarlet Cup is one of the first
mushrooms to look for in spring, as it
sometimes appears in March. It is easy
to see because of its size and bright
color.

---

606 Stalked Scarlet Cup
Sarcoscypha occidentalis (Sow.) Sacc.
Sarcoscyphaceae, Pezizales

Description: *Bright red cup on white stalk.*
Cup: ¼–⅝″ (0.5–1.5 cm) wide, ⅜″ (1
cm) high; shallow; outer surface whitish
and smooth; inner surface scarlet.
Stalk: ⅜–1¼″ (1–3 cm) long, ¹⁄₁₆″ (1.5
mm) thick; cylindrical, white.
Spores: 20–22 × 10–12 μ; elliptical,
usually with 1 oil drop at each end,
often surrounded by smaller drops.

Season: May–June; into September in cool
years.

Habitat: On fallen deciduous wood in wet
places.

Range: New England to Louisiana, west to
Kansas.

Look-alikes: *S. coccinea* * is taller and has a short
stalk or none.

Comments: The Stalked Scarlet Cup is readily
recognized by its small size, bright
color, white stalk, and habitat on fallen
branches. It is often found in the spring
when one is out gathering watercress
and wild leeks.

615 **Moose Antlers**
Wynnea americana Thax.
Sarcoscyphaceae, Pezizales

Description: *Clustered blackish-brown mushroom resembling moose antlers or rabbit ears.*
Mushroom: 1–4″ (2.5–10 cm) wide, 2⅜–5¼″ (6–13 cm) high; shaped like moose antlers or rabbit ears; outer surface blackish-brown; inner surface pinkish to reddish; arising from a tough, brown knot of underground tissue, 1⅝–2″ (4–5 cm) wide.
Spores: 32–40 × 15–16 μ; elliptical, pointed at each end, with longitudinal lines on surface and with several interior oil drops.
Season: July–September.
Habitat: On the ground under hardwoods.
Range: Throughout East.
Comments: The underground structure from which this species arises is the knotted, threadlike mycelium of the mushroom. The fungal threads, known as hyphae, convert elements of the soil into nutrients necessary for growth.

753 **Stalked Cauliflower Fungus**
Wynnea sparassoides Pfister
Sarcoscyphaceae, Pezizales

Description: *Long-stalked mushroom with large, yellow-brown, cauliflowerlike head.*
Mushroom: 2⅜–3¼″ (6–8 cm) wide and high; more or less round, cauliflower- to brain-shaped; contorted, beige.
Stalk: 8–12″ (20–30 cm) long, ¾–1¼″ (2–3 cm) thick; brown, solid.
Spores: 32–36 × 12–15 μ; with prominent longitudinal ridges on surface and with 3 interior oil drops.
Season: July–September.
Habitat: On soil under leaf litter.
Range: Massachusetts to New Jersey.
Comments: Although only recently described, the Stalked Cauliflower Fungus has probably been collected over the years.

570 Burn Site Shield Cup
Ascobolus carbonarius Kar.
Ascobolaceae, Pezizales

Description: *Small, blackish, shield-shaped cup; on
burned ground.*
Cup: ⅛–¼″ (3–5 mm) wide; often
clustered; rounded at first, becoming
cuplike, then flat; outer surface
greenish, becoming dark brown to
blackish, coarse, mealy; *inner surface
greenish at first, becoming blackish and
conspicuously dotted.*
Spores: 19–24 × 11–15 μ; elliptical,
warted; pale at first, becoming
purplish-brown.

Season: April–October.

Habitat: On burned ground.

Range: Throughout North America.

Comments: *Ascobolus* is a genus of mostly small,
dark mushrooms that typically grow on
dung. Like others of this genus, the
Burn Site Shield Cup is most easily
recognized at maturity, when its spore
sacs (asci) protrude far above the fertile
surface inside the cup.

---

640 Pink Crown
Sarcosphaera crassa (Sant. ex Steud.) Pouz.
Pezizaceae, Pezizales

Description: *Large, pinkish-violet, stalkless, crown-
shaped cup.*
Cup: 1¼–4″ (3–10 cm) wide and
high; globelike at first and partially
sunken in soil, later splitting into
7–10 rays above ground; outer surface
whitish; inner surface pinkish-violet,
grayish-pink, or grayish-lilac and
deeply cup-shaped, fleshy and thick-
walled. Flesh brittle.
Spores: 15–18 × 8–9 μ; elliptical with
blunt ends, smooth.

Edibility: Edible with caution.

Season: June–August.

Habitat: Solitary to clustered, immersed in soil
under coniferous or deciduous trees.

Range: Common in Rocky Mountains and

Pacific NW.; also reported in Michigan and New York.

Comments: Formerly known as *S. coronaria* and *S. eximia*. Because it usually grows just beneath the surface and only partially breaks through at maturity, it resembles a pinkish-violet hole in the ground. Some people become ill from eating this species, while others do not. It should always be thoroughly cooked.

625 Common Brown Cup
Peziza badio-confusa Korf
Pezizaceae, Pezizales

Description: *Very common large, stalkless, brown cup.*
Cup: 1¼–4″ (3–10 cm) wide; deeply cup-shaped to irregular; reddish-brown to dark brown, sometimes with olive-brown cast on inner surface. Flesh brittle.
Spores: 17–21 × 8–10 μ; elliptical, finely warted, with 2 oil drops. Spore sacs amyloid.
Edibility: Edible.
Season: May; occasionally throughout summer.
Habitat: On the ground in woods.
Range: Throughout North America.
Look-alikes: *P. badia,* uncommon and found in fall, is slightly darker and has wider spores. *Auricularia auricula** grows on wood and is a rubbery jelly fungus.
Comments: This is often irregularly shaped and resembles a discarded torn piece of rubber. It is edible when cooked.

620 Recurved Cup
Peziza repanda Pers.
Pezizaceae, Pezizales

Description: *Common large cup with white outer surface, brownish inner surface, and expanded or turned-down margin.*
Cup: 2⅜–4¾″ (6–12 cm) wide; saucerlike to shallowly cup-shaped, with margin even or splitting and bent backward or turned down; outer surface

whitish; inner surface buff-tan to brown; attached to wood by central knot of tissue.
Spores: 14–16 × 9–10 μ; smooth, without oil drops. Spore sacs amyloid.

Season: May–June; August–October.

Habitat: On rotten hardwood logs or soil.

Range: Throughout North America.

Look-alikes: *P. cerea, P. sylvestris,* and *P. varia* are so similar that they cannot be clearly differentiated on the basis of field characteristics, especially if *P. repanda* is found on the ground. *Pachyella clypeata* is attached to wood by whole lower surface.

Comments: Many distinctly different species are also called the Recurved Cup. As a result, although often listed as edible, mushrooms identified as this species should not be eaten.

---

601 Bladder Cup
Peziza vesciculosa Bull. ex St. Amans
Pezizaceae, Pezizales

Description: *Large, clustered, pale yellowish-brown cups; on dung.*
Cup: 2⅜–3¼" (6–8 cm) wide; deeply cup-shaped, often with strongly incurved margin; outer surface pale yellowish-brown and coarsely scruffy; inner surface light yellowish-brown. *Flesh forms blisters or bladders in center of cup.*
Spores: 20–24 × 11–14 μ; elliptical, smooth, without oil drops. Spore sacs amyloid.

Season: June–October; November–February on West Coast.

Habitat: Several to crowded, on manure heaps and rich manured soil.

Range: Throughout North America.

Comments: The Bladder Cup can be very common at times on manure and in manured gardens. It is a large, colorful cup that may appear quite regular in shape or densely packed and distorted.

603 Orange Peel
Aleuria aurantia (Fr.) Fkl.
Pyronemataceae, Pezizales

Description: *Common bright orange, shallow cup.*
Cup: ¾–4″ (2–10 cm) wide; saucer- to
shallowly cup-shaped; outer surface pale
whitish to orange-yellow, often with a
covering of minute hairs; inner surface
bright orange fading to orange-yellow.
Flesh brittle.
Spores: 18–22 × 9–10 μ; elliptical,
with conspicuous network of ridges.
Edibility: Edible.
Season: May–October.
Habitat: In groups or clusters, on hard soil along
roadsides, on disturbed soil of gardens,
newly planted lawns, and landslides.
Range: Throughout North America.
Comments: In color and shape, this species
resembles an orange peel—hence the
common name.

602 Blue-staining Cup
Caloscypha fulgens (Pers. ex Fr.) Boud.
Pyronemataceae, Pezizales

Description: *Yellowish-orange, stalkless cup fungus,
often stained bluish-green.*
Cup: ⅜–2″ (1–5 cm) wide; irregularly
cup-shaped, lopsided; outer surface blue
to greenish-blue; inner surface yellow to
orange with blue-green stains. Flesh
brittle.
Spores: 5–7 μ; round, smooth.
Season: May–July.
Habitat: Solitary to clustered, in boggy places
and in mountainous, coniferous areas.
Range: E. North America west to Montana;
Rocky Mountains and California.
Comments: This mushroom is the only blue- to
green-stained orange cup fungus found
in early spring in spruce bogs. In
mountainous, coniferous areas, where
it appears later than it does at lower
elevations, it grows near melting
snowbanks.

643 Fuzzy False Truffle
Geopora cooperi Hark.
Pyronemataceae, Pezizales

Description: *Trufflelike cup fungus with fuzzy, brownish outer surface and deeply convoluted, whitish inner surface.*
Mushroom: ¾–2¾" (2–7 cm) wide; more or less round; outer surface light to dark brown, somewhat furrowed, covered with long brown hairs; inner surface deeply convoluted, whitish.
Spores: 18–30 × 12–17 μ; broadly oval to elliptical, smooth, generally with 1 oil drop. Spore sacs cylindrical, tapering toward base, with lidlike cover; usually containing 8 spores; pore nearest tip of spore sac nonamyloid.
Season: Year-round.
Habitat: In soil or on surface of ground under Douglas-fir, noble fir, pine, spruce, and eucalyptus.
Range: Alaska to California.
Look-alikes: Other underground fungi can only be differentiated microscopically.
Comments: Also known as *G. harknessi*. Rodents, including squirrels, often transport this species to their food caches. Despite its appearance, the Fuzzy False Truffle is more closely related to the cup fungi than to truffles, with which it was originally classified.

627 Pyxie Cup
Geopyxis carbonaria (A. & S. ex Fr.) Sacc.
Pyronemataceae, Pezizales

Description: *Slender-stalked, ochre-brown to pinkish cup with toothed, whitish margin.*
Cup: ¼–¾" (0.5–2 cm) wide; cup-shaped; outer surface orange-brown or pink-tinged, with minute scales; margin toothed, whitish; inner surface ochre-brown or pink-tinged.
Stalk (when present): ¼–¾" (0.5–2 cm) long, ¹⁄₃₂–¹⁄₁₆" (1–1.5 mm) thick; slender, orange-brown to off-white.
Spores: 12–18 × 6–9 μ; narrowly

elliptical, more or less pointed at each
end, without oil drops.

Season: May–September.
Habitat: On charred wood or burned ground.
Range: Maine and New York, west to British
Columbia and California.
Look-alikes: *G. vulcanalis* is whitish to yellowish
and grows in moss in coniferous woods.
Tarzetta* have spores with 2 oil drops.
Comments: Other cup fungi, including species of
Aleuria, Anthracobia, Peziza, and
Tarzetta, may be found growing along
with the Pyxie Cup.

629 Brown-haired White Cup
Humaria hemisphaerica (Wigg. ex Fr.) Fkl.
Pyronemataceae, Pezizales

Description: *Whitish, stalkless cup covered with stiff,
brownish hairs.*
Cup: ⅜–1¼" (1–3 cm) wide, ¼–⅝"
(0.5–1.5 cm) high; cup-shaped; outer
surface fringed with stiff, brownish
hairs; inner surface off-white to grayish.
Spores: 25–27 × 12–15 μ; elliptical,
with small warts and 2 oil drops.
Season: July–August.
Habitat: Clustered to scattered on rich humus;
also near or on rotten wood.
Range: Throughout North America.
Comments: This species can be readily identified by
its deep cup shape, whitish interior,
and outer covering of stiff, brown hairs.

600 Yellow Rabbit Ears
Otidea leporina (Bat. ex Fr.) Fkl.
Pyronemataceae, Pezizales

Description: *Yellow fungus shaped like rabbit ears.*
Mushroom: ⅝–1" (1.5–2.5 cm) wide,
1–2" (2.5–5 cm) high; elongated, split
from top to base down shorter side;
inner and outer surfaces bright yellow
to yellowish-brown.
Spores: 12–14 × 6–8 μ; elliptical,
smooth, with 2 small oil drops.
Season: August–September.

Habitat: Single to several, on soil or debris in coniferous woods.

Range: New Hampshire, New York, and Maryland, west to Colorado, Washington, and California.

Look-alikes: *O. onotica* has a pinkish cast on its inner surface. *O. smithii* has a grayish-brown outer surface. *O. alutacea* has a grayish-brown inner surface and usually grows in dense clusters.

Comments: Rabbit-ear fungi (*Otidea*) look like cup fungi that are standing on end; often they are irregularly split and not quite erect.

---

569   Burn Site Ochre Cup

Anthracobia melaloma (A. & S.) Boud.
Pyronemataceae, Pezizales

Description: *Small, stalkless, yellow-ochre disc with dark-hairy outer surface; on burned ground.*
Cup: $\frac{1}{16}-\frac{3}{8}''$ (0.15–1 cm) wide; often in broad masses, clustered or crowded; cup-shaped, expanding to disclike, with elevated margin; outer surface with closely flattened, dark brown, downy hairs; inner surface yellow-brown to ochre-orange.
Spores: $14-22 \times 7-11$ μ; elliptical, smooth, with 2 oil drops.

Season: July–January.

Habitat: On burned ground and burned wood.

Range: Throughout North America.

Look-alikes: *Pyronema omphalodes* is smaller, has reddish tones and hairless outer surface; it forms dense clusters and has conspicuous, white, threadlike mat at base.

Comments: Within months of the disastrous 1980 eruption of Mount St. Helens, this species appeared in the central blast zone. In any burned site, a succession of mushrooms will appear as soon as conditions become favorable for fruiting. For some species, heated soil or sterilized soil is required, while others need the release of certain mineral nutrients into the soil. For several years after a burn, a variety of mushrooms, mostly cup fungi, will appear, often in great numbers.

604 Eyelash Cup
Scutellinia scutellata (L. ex Fr.) Lamb.
Pyronemataceae, Pezizales

Description: *Small, clustered, red to orange cup with dark, eyelashlike hairs.*
Cup: ¼–¾″ (0.5–2 cm) wide; closed at first, opening to form a shallow cup; outer surface and margin conspicuously fringed with eyelashlike, brown hairs; inner surface bright red to orange.
Spores: 18–19 × 10–12 μ; elliptical, roughened, filled with oil drops.
Season: June–November.
Habitat: Usually densely grouped on rotten wood and adjacent soil.
Range: Throughout North America.
Look-alikes: *Cheilymenia* has a paler center, transparent or yellowish hairs, and smooth spores without oil drops. Other species of *Scutellinia* grow on the ground and differ microscopically.
Comments: This small, brightly colored cup is distinctive because it is ringed by long, dark hairs.

626 Elf Cup
Tarzetta cupularis (L. ex Fr.) Lamb.
Pyronemataceae, Pezizales

Description: *Small, grayish-tan cup with toothed margin and short stalk.*
Cup: ⅜–¾″ (1–2 cm) wide and high; cup-shaped, with rounded-toothed margin; grayish-tan, covered with minute granules.
Stalk: slight.
Spores: 18–23 × 10–15 μ; broadly elliptical, smooth, with 2 large oil drops.
Season: June–September.
Habitat: On burned ground, soil, and moss in coniferous woods.

Range: Quebec to West Virginia, west to
Colorado and Pacific NW.

Look-alikes: *Geopyxis vulcanalis* is pale yellow.

Comments: Formerly called *Geopyxis cupularis,* this
was transferred to *Tarzetta* because
Geopyxis spores lack oil drops. Such
features, especially in the cup fungi, are
now often important in establishing
clearly defined genera.

---

608 Pink Burn Cup
Tarzetta rosea (Rea) Dennis
Pyronemataceae, Pezizales

Description: *Pink to red cup with short stalk; growing in
masses on burned ground.*
Cup: ¼–⅝" (0.5–1.5 cm) wide and
high; cup-shaped, with margin often
splitting; bright pink to red.
Stalk (when present): short, whitish.
Spores: $17-20 \times 9-11 \ \mu$; elliptical,
smooth, with 2 large oil drops.

Season: June–September.

Habitat: Several to many on burned ground.

Range: Widely distributed in North America.

Comments: Formerly known as *Pustularia rosea.*
Several cup fungi inhabit burned
sites; many, like this mushroom, are
found nowhere else. Mushroom hunters
who are interested in these fungi, and
even more interested in morels, listen
for reports of forest fires and hunt the
burned area the following spring.

TRUFFLES
(Order Tuberales)

Truffles are the most expensive edible
mushrooms in the world. They grow
underground and somewhat resemble
puffballs. As with all ascomycetes, their
spores are formed in sacs (asci). Truffles
are generally large, fleshy to waxy,
roundish to lumpy fungi that develop a
marbled flesh as their spores mature,
and produce a pungent aroma. They
grow near particular trees, especially

oaks; some are mycorrhizal. They can be found from spring through fall. Their spores are not forcibly discharged; spore dispersal is achieved by the decay of the fungus or because its aroma attracts a burrowing rodent that eats it and then excretes it some distance away.

Not all underground fungi are truffles, and not all truffles are edible, nor is any common European truffle known to grow in North America. Because North American truffles have not been systematically studied (except on the West Coast, where they are relatively abundant and quite varied), the eating of American truffles cannot be recommended. Based on animal studies, however, no truffle is known to be poisonous.

There are 3 common families of truffles, and more than 2 dozen genera, which are classified according to the exterior and interior structures of the mushroom and the shape and color of the spores. One unrelated, similar-looking family, the Elaphomycetaceae, which is best known as the host for several species of *Cordyceps* (and as an aphrodisiac in England), has mushrooms that disintegrate into a partially powdery mass.

---

659 Oregon White Truffle
Tuber gibbosum Hark.
Tuberaceae, Tuberales

Description: *Small, brownish, nearly round or convoluted fungus, marbled within; growing underground under Douglas-fir.*
Mushroom: ⅜–2″ (1–5 cm) wide; round to somewhat irregular and convoluted; buff to light brown or brown, minutely scaled and matted. Flesh conspicuously marbled with white veins and canals.
Spores: 36–52 × 28–38 μ; long, elliptical, ornamented, dark yellowish-

brown, not forcibly discharged.
Edibility: Choice, with caution.
Season: September—May.
Habitat: Underground under Douglas-fir;
reported under oak.
Range: British Columbia to N. California.
Comments: In Europe several species of truffles are
gathered commercially. Nowadays these
are found with the help of a trained dog
or muzzled pig, although until
recently young children were taught
to spot the cloud of yellow truffle flies
hovering above a site. Truffles have
been studied for more than a century in
North America, but only recently have
people taken to "truffling" as a
pastime. A clawlike garden fork called
a truffle rake is used to dig about under
suspected trees, especially along the
Oregon coast. Truffles and their look-
alikes are also found stashed in rodent
nests. Identification is extremely
difficult for most species, but the effort
may eventually prove to be highly
rewarding. The Oregon White Truffle,
for example, is rated by some as highly
as the European White Truffle (*T.
magnatum*).

EARTH TONGUES
(Order Helotiales)

The order Helotiales contains the earth
tongues, fairy fans, and swamp beacons
of the woods, as well as the hairy fairy
cups, jelly drops, and jelly clubs. They
are mostly small, variously shaped
mushrooms that grow on plant stems,
wood, and wet leaves; a few grow on
the ground. Although often brightly
colored and usually abundant, they are
easily overlooked by mushroom
hunters. Turning over fallen leaves or
examining plant stems, especially
composites like goldenrod, will
sometimes uncover a wealth of fairy
cups. Although none can be
recommended as edible, at least

1 species is associated with food—the Green Stain has been used in the making of Tunbridgeware, the dark green-stained veneer of English dinner tables and wooden plates.

The spore sacs (asci) are usually club-shaped and have a pore at the tip through which spores are forcibly discharged into the air. The spores are usually small, smooth, narrow, and very long, and often have 2 or more cells.

Eight families and more than 170 genera have been described. One member of the order Ostrapales is included here with the earth tongues.

682 Velvety Earth Tongue
Trichoglossum hirsutum (Fr.) Boud.
Geoglossaceae, Helotiales

Description: *Black club with velvety stalk and lance-shaped, flattened head.*
Head: ¼" (5 mm) wide at top, ¹⁄₁₆–⅛" (1.5–3 mm) wide near base, ⅜–⅝" (1–1.5 cm) high; club- to lance-shaped or elliptical; upper part somewhat compressed, with long, black spines.
Stalk: 1¼–2⅜" (3–6 cm) long, ⅛" (3 mm) thick; densely velvety, composed of parallel clusters of brown hairs and long, black spines.
Spores: 100–150 × 6–7 μ; cylindrical, brown, layered in spore sac; 16-celled.

Season: July–October.

Habitat: On rotting wood or soil, or in sphagnum moss.

Range: Ontario south to Louisiana, west to California.

Look-alikes: *T. velutipes* and other closely related species differ only microscopically. *Geoglossum* species are smooth and lack minute spines.

Comments: To distinguish *Trichoglossum* from *Geoglossum,* the mushrooms should be examined with a hand lens to check for minute spines.

685 Orange Earth Tongue

Microglossum rufum (Schw.) Under.
Geoglossaceae, Helotiales

Description: *Bright orange to yellowish, somewhat spoonlike club.*
Head: ⅛–⅝″ (0.3–1.5 cm) wide, ⅜–1⅝″ (1–4 cm) high; compressed, orange to yellow.
Stalk: ⅜–1⅝″ (1–4 cm) long, 1/16–⅛″ (1.5–3 mm) thick; cylindrical, orange to yellow.
Spores: 18–40 × 4–6 μ; sausage-shaped, smooth.

Season: July–September.

Habitat: Scattered to clustered, in sphagnum moss, on rotten logs, and on leaf litter.

Range: Throughout North America.

Look-alikes: *Neolecta irregularis**, usually found in groups of several, shows great irregularity in shape.

Comments: This mushroom can be easily distinguished from other species of *Microglossum,* which are green to brown. The Orange Earth Tongue is similar to the various dark earth tongues, species of *Geoglossum* and *Trichoglossum.*

---

684 Swamp Beacon

Mitrula paludosa Fr.
Geoglossaceae, Helotiales

Description: *White-stalked club with yellowish head; in wet places in spring.*
Head: ⅛–⅜″ (0.3–1 cm) wide, ¼–⅝″ (0.5–1.5 cm) high; elliptical to pear-shaped; clear yellow to pale orange, smooth.
Stalk: ⅝–1⅝″ (1.5–4 cm) long, 1/16″ (1.5 mm) thick; pure satiny white to pink-tinted, smooth, translucent.
Spores: 10–18 × 2.5–3 μ; cylindrical, with rounded ends.

Season: May–July.

Habitat: On decaying leaves in wet soil, swamps, sphagnum bogs, or temporary pools in woods.

Range: Nova Scotia south to Tennessee, west

across N. North America to Montana
and British Columbia.

Look-alikes: Other species of *Mitrula* can only be
differentiated microscopically.

Comments: In late-spring woods, the Swamp
Beacon often appears to glow in shallow
pools. Formerly known as *M. phalloides.*

686 Irregular Earth Tongue
Neolecta irregularis (Pk.) Korf & Rog.
Geoglossaceae, Helotiales

Description: *Irregularly contorted club with yellowish
head and white stalk.*
Head: ¼–⅝" (0.5–1.5 cm) wide, ⅜–
2" (1–5 cm) high; irregular, contorted,
compressed; yellow.
Stalk (when present): ⅜–¾" (1–2 cm)
long, ¹⁄₁₆–¼" (1.5–5 mm) thick; satiny
white, dusted with fine powder.
Spores: 6–10 × 4–5 μ; elliptical,
smooth.

Season: June–October.

Habitat: Scattered or in groups, on the ground
among pine needles on mosses.

Range: Throughout coniferous areas of North
America.

Look-alikes: *Clavaria fusiformis,* a basidiomycete, is a
clustered, usually unbranched coral
mushroom, typically spindle-shaped.

Comments: Also known as *Mitrula irregularis* and
Spragueola irregularis. The Irregular
Earth Tongue is usually found in large
numbers in summer woods.

693 Velvety Fairy Fan
Spathularia velutipes Cke. & Farlow
Geoglossaceae, Helotiales

Description: *Velvety brown, stalked club fungus with
yellowish-brown, fan-shaped head.*
Head: ⅜–1¼" (1–3 cm) wide, ⅜"
(1 cm) high; fan- to spoon-shaped,
much compressed; yellowish to
brownish-yellow.
Stalk: ¾–1⅝" (2–4 cm) long, ⅝" (1.5
cm) thick, narrowing at base to ⅛–¼"

360   *Earth Tongues*

(3–5 mm); bay-brown and minutely velvety; attached to substrate by orange fungal hairs.
Spores: 33–43 × 2 μ; smooth, needle-shaped, becoming many-celled.

Season: August–September.

Habitat: In groups or clusters, on rotting hardwood logs; on ground under pines.

Range: New Hampshire to North Carolina, west to Minnesota.

Look-alikes: *S. flavida* has pale yellow to buff head and whitish to yellowish, nonvelvety stalk. *S. spathulata* grows only in California, has yellow-brown to reddish-brown head and stalk.

Comments: Also known as *Spathulariopsis velutipes*. Fairy fans are readily recognized by their broadly compressed heads.

---

624 Blueberry Cup
Monilinia vaccinii-corymbosi (Rde.) Hon.
Sclerotiniaceae, Helotiales

Description: *Small, brown, stalked cup fungus growing on fallen overwintered highbush blueberries.*
Cup: ¼–⅜″ (0.5–1 cm) wide and high; cup-shaped, somewhat closed at first but expanding, with flared margin; inner surface fawn-colored; outer surface brownish, satiny.
Stalk: ⅜–1¼″ (1–3 cm) long, 1/32–1/16″ (1–1.5 mm) thick; clove-brown, smooth.
Spores: 14–18 × 9–10 μ; elliptical. Tip of spore sac (ascus) amyloid.

Season: March–May.

Habitat: Single to several, in wet soil, swamps, and bogs on mummified fallen fruits or twigs of highbush blueberry.

Range: Massachusetts to North Carolina.

Comments: Similar mushrooms in several genera in the family Sclerotiniaceae grow on other fallen, rotting fruits, such as peaches and cherries; sometimes these mushrooms grow on twigs and roots or out of buried tuberlike structures (sclerotia). Many were formerly known as species of *Sclerotinia,* and most appear in the spring. One notable exception is the Common Wood Ciboria (*Ciboria peckiana*), which grows on wood in the fall.

598 Green Stain
Chlorociboria aeruginascens (Nyl.) Kan.
Dermatiaceae, Helotiales

Description: *Very small, stalked, blue-green cup; on blue-green-stained dead wood.*
Cup: ¼″ (5 mm) wide; cup-shaped, becoming flattened; blue-green, sometimes with yellowish tints; smooth, thin-fleshed.
Stalk: ⅟₃₂–⅟₁₆″ (1–1.5 mm) long; short, slender.
Spores: $6-10 \times 1.5-2$ μ; spindle-shaped, with small oil drop at each end.
Season: June–November.
Habitat: On rotting logs and barkless wood, especially oak.
Range: Throughout North America.
Look-alikes: Other greenish cup fungi can only be differentiated microscopically.
Comments: The fungal threads of this mushroom grow in wood; the pigment that they contain turns the wood green.

630 Stalked Hairy Fairy Cup
Dasyscyphus virgineus S.F.G.
Hyaloscyphaceae, Helotiales

Description: *Tiny, long-stalked, cream-colored cup covered with white hairs; on dead twigs and plant stems.*
Cup: ⅟₃₂″ (1 mm) wide; cup-shaped; cream, covered with white hairs.
Stalk: ⅟₁₆–⅛″ (1.5–3 mm) long; densely covered with long, white hairs.
Spores: $6-10 \times 1.5-2.5$ μ; spindle-shaped.
Season: Spring and fall.
Habitat: Often in large groups, on dead twigs, stems, beech burs, and birch catkins.
Range: Throughout North America.

Look-alikes: Similar species can only be
differentiated microscopically.
Comments: The Stalked Hairy Fairy Cup is one of a
number of very small, beautiful cup
fungi that are covered with hairs.

---

616 Purple Jelly Drops
Ascocoryne sarcoides (Jacq. ex S.F.G.)
Groves & Wilson
Leotiaceae, Helotiales

Description: *Purplish, jellylike, turban- to cup-shaped*
mushroom; clustered on rotten wood.
Cup: ¼–⅜″ (0.5–1 cm) wide;
clustered; turban-shaped or shallow,
with curved to flat center; reddish-
purple. Flesh jellylike.
Stalk (when present): very short.
Spores: $10-19 \times 3-5$ μ; elliptical to
asymmetrical; 1-celled at first,
becoming 2- to 4-celled; 2 oil drops.
Season: September–October.
Habitat: On stumps and fallen logs.
Range: Throughout North America.
Comments: This species may resemble jelly fungi,
but it has spore sacs (asci) rather than
club-shaped spore-bearing structures
(basidia).

---

568 Yellow Fairy Cups
Bisporella citrina (Batsch ex Fr.) Korf & Carp.
Leotiaceae, Helotiales

Description: *Very small, bright lemon-yellow cup on*
short, thick, stalklike base; in clusters on
decaying wood.
Cup: ⅛″ (3 mm) wide; shallowly cup-
shaped, narrowing to small base;
lemon-yellow, smooth.
Spores: $9-14 \times 3-5$ μ; elliptical, often
becoming 2-celled; oil drops at each
end.
Season: July–November.
Habitat: On decaying wood.
Range: Throughout North America.
Look-alikes: *Phaeohelotium flavum* has brown spores
and brownish tint at maturity.

Comments: Also known as *Helotium citrinum* and *Calycella citrina.* Somewhat similar mushrooms grow on herbaceous plants and mosses, and the Acorn Cup (*Hymenoscyphus fructigenus*) grows on fallen acorns and hickory nuts.

610 Black Jelly Drops
Bulgaria inquinans Fr.
Leotiaceae, Helotiales

Description: *Blackish, gelatinous, turban-shaped, stalkless mushroom with flat top.*
Mushroom: ⅜–1⅝" (1–4 cm) wide, ¾–1⅝" (2–4 cm) high; rounded to turban-shaped, becoming flat-topped; outer surface dull brown and rough; inner surface blackish and shiny. Flesh tough and gelatinous.
Spores: 11–14 × 6–7 μ; kidney-shaped; upper 4 in every sac (ascus) very dark brown, lower 4 colorless.
Season: August–September.
Habitat: Clustered on bark of dead hardwoods, especially oak.
Range: New York to Alabama, west to Pacific NW.
Comments: As the common name suggests, this mushroom looks like jelly drops or licorice drops.

725 Yellow Cudonia
Cudonia lutea (Pk.) Sacc.
Leotiaceae, Helotiales

Description: *Dry, yellowish to buff, stalked club with convoluted head.*
Head: ⅜–⅝" (1–1.5 cm) wide and high; convoluted, brainlike; yellowish to salmon-buff, smooth. Flesh firm to leathery.
Stalk: ¾–1¾" (2–4.5 cm) long, ¼" (5 mm) thick; pale yellow, somewhat mealy.
Spores: 45–78 × 2 μ; needle-shaped, smooth, becoming many-celled.
Season: July–September.

Habitat: On leaf litter under beeches and other hardwoods.
Range: E. Canada to Tennessee.
Look-alikes: *C. circinans* has brownish stalk, cream-buff to pale brownish head, and is usually found in coniferous woods. *C. monticola* is pinkish-brown and occurs in spring, often near melting snow in mountainous areas of the Pacific NW. *Leotia lubrica** is larger and distinctly gelatinous.
Comments: Although its members are usually small and inconspicuous, *Cudonia* is a distinctive genus of club fungi; these mushrooms are fertile only on the upper surface, and generally have needle-shaped spores. *Cudonia* is related to *Leotia, Mitrula,* and *Spathularia.* The Yellow Cudonia is often misidentified as a "dry" *Leotia.*

726 Ochre Jelly Club
Leotia lubrica Pers. ex Fr.
Leotiaceae, Helotiales

Description: *Gelatinous, ochre-yellow, stalked club with convoluted head.*
Head: ⅜–1⅝″ (1–4 cm) wide and high; smooth or furrowed, convex and more or less convoluted; margin inrolled; buff, ochre, or cinnamon, at times tinged with olive.
Stalk: ¾–2″ (2–5 cm) long, ¼–⅜″ (0.5–1 cm) thick; pale buff to yellow, minutely scaly.
Spores: $20–25 \times 5–6 \mu$; slightly spindle-shaped with rounded ends, often slightly curved, becoming 6- to 8-celled.
Season: July–October.
Habitat: Several to abundant, on soil or litter in coniferous, mixed, or scrub oak woods.
Range: Throughout North America.
Look-alikes: *L. atrovirens* is completely green. *L. viscosa** has green head. *Cudonia** species are smaller, not jellylike.
Comments: The Ochre Jelly Club is often found in large clusters.

727 Green-headed Jelly Club
Leotia viscosa Fr.
Leotiaceae, Helotiales

Description: *Gelatinous, ochre-yellow, stalked club with green, convoluted head.*
Head: ¼–⅜" (0.5–1 cm) wide, ¾–1¼" (2–3 cm) high; convex, with inrolled, convoluted margin; olive-green to dark green.
Stalk: ¾–1⅛" (2–4 cm) long, ¼–⅜" (0.5–1 cm) thick; white, yellow, or orange, with green dots or particles.
Spores: 17–26 × 4–6 μ; somewhat spindle-shaped with rounded ends, often slightly curved.
Season: July–September.
Habitat: Clustered or in groups, on soil or rotten wood.
Range: Throughout North America.
Look-alikes: *L. lubrica** is sometimes tinged greenish overall. *L. atrovirens* has green head and green or pale green stalk.
Comments: The jelly clubs (*Leotia*) are gelatinous fungi, but like the jelly drops they have spore sacs (asci) rather than the clublike spore structures (basidia) of true jelly fungi.

575 Scurfy Alder Cup
Encoelia furfuracea (Roth ex Pers.) Kar.
Leotiaceae, Helotiales

Description: *Clustered, scurfy, leathery, cinnamon cups growing out of living alder or hazel branches.*
Cup: ⅜–⅝" (1–1.5 cm) wide; rounded, becoming irregularly cup-shaped, with incurved margin; outer surface brownish, covered with whitish to rust, branlike scruff, leathery; inner surface cinnamon.
Spores: 6–11 × 2–4 μ; narrowly elliptical or sausage-shaped; 1-celled.
Season: November–June.
Habitat: On living alder and hazel branches.
Range: Across N. North America, south at least as far as Iowa and Pennsylvania.

Comments: Also known as *Cenangium furfuraceum.*
Because of its leathery texture and
spreading, irregularly shaped, clustered
cups, the Scurfy Alder Cup might at
first glance be confused with some kind
of parchment fungus (*Stereum*). But this
species is one of a number of cup fungi
that are so restricted in habitat that it is
often possible to identify the species
simply on the basis of the host.

729 Water Club
Vibrissea truncorum A. & S. ex Fr.
Stictidaceae, Ostropales

Description: *Small, white-stalked club with yellow-
orange, convex, inrolled, rumpled head; on
branches in mountain streams.*
Head: $\frac{1}{16}-\frac{1}{4}''$ (1.5–5 mm) wide and
high; convex, inrolled; yellow-orange to
reddish-orange, rumpled.
Stalk: $\frac{1}{4}-\frac{5}{8}''$ (0.5–1.5 cm) long, $\frac{1}{32}-$
$\frac{1}{16}''$ (1–1.5 mm) thick; white, with
minute hairs.
Spores: $200 \times 1.5\ \mu$; needle-shaped,
many-celled.
Season: June–August.
Habitat: In groups or clustered, on partly or
wholly submerged branches in cold
mountain streams.
Range: Throughout N. North America; also
higher elevations in South.
Look-alikes: *Mitrula paludosa** has an elliptical head.
Leotia* and *Cudonia** species grow on
the ground or on rotting wood, not in
water.
Comments: Like the various earth tongues and jelly
clubs of the Helotiales, the Water Club
differs from the cup fungi of the
Pezizales by having spore sacs (asci)
with a pore at the apex, rather than a
lid that flips open to discharge the
spores. It differs from the Helotiales in
that it has cylindrical rather than club-
shaped asci.

FLASK FUNGI
(Class Pyrenomycetes)

The class Pyrenomycetes contains fungi that are mostly minute and typically grow on plants and other fungi. Unlike Discomycetes, they develop enclosed fertile surfaces, which are composed of microscopic, flask-shaped bodies (perithecia) lined within by spore sacs (asci). Some familiar members of this class are the Chestnut Blight Fungus, Dutch Elm Disease Fungus, Coral Spot, and Ergot, as well as a variety of parasitic fungi and oddities like Dead Man's Fingers.

OSTIOLE FLASKS
(Order Sphaeriales)

This is a broadly defined order that contains many species that are superficially extremely dissimilar. Dead Man's Fingers is a blackened cluster that sticks up out of the ground like a hand from the grave. Ergot, a fungus that infects rye and other grasses, can cause hallucinations and gangrene if eaten, but is used in medicines to relieve migraine and reduce postnatal contractions. Other Sphaeriales include a large array of brightly colored, minute plant parasites and a number of moldlike fungi that parasitize some common gilled mushrooms and boletes. All of the species placed in this order have minute, flasklike vessels called perithecia that contain the spore sacs (asci) of the fungi. The perithecia are often densely packed and numerous, and create a broad, pustular fertile surface. In those species that produce noticeable mushrooms, the perithecia occur either on a compact, mattresslike structure called a stroma or on a loose, cottony mat known as a subiculum. A stroma (pl. stromata) may be cushion-shaped, hemispherical, club-shaped, or irregularly elongated. Subicula appear as molds on other mushrooms. The order contains several different groups of fungi that can be distinguished on the basis of the shape, color, ornamentation, and segmentation of the spores. Most species discharge their spores through a canal-like opening, or ostiole, at the tip of the perithecium.

728 Headlike Cordyceps
Cordyceps capitata (Holmsk. ex Fr.) Link
Clavicipitaceae, Sphaeriales

Description: *Small, dark, round-headed mushroom on yellow stalk, arising from underground, trufflelike fungus.*
Head: ⅜–⅝" (1–1.5 cm) wide and high; somewhat rounded; dark brown to black, surface finely roughened by protruding necks of immersed, flasklike vessels (perithecia).
Stalk: ¾–3¼" (2–8 cm) long, ⅜–⅝" (1–1.5 cm) thick; yellow, stout, smooth to somewhat furrowed.
Spores: threadlike spores break into 7 short, rodlike part-spores while still in the sac (ascus). Part-spores 14–20 × 2–3 μ; cylindrical, smooth.
Season: September–November.
Habitat: Single to several, in forests, on underground species of *Elaphomyces*, especially *E. granulatus*.
Range: Widely distributed in North America.
Look-alikes: *C. canadensis* has part-spores that are spindle-shaped and larger (20–50 × 3–7.5 μ).

Comments: Unless completely unearthed, so that it can be seen growing out of a trufflelike fungus, the Headlike Cordyceps is likely to be misidentified. Neither this mushroom nor the *Elaphomyces* on which it grows is known to be edible, but both are used—the *Cordyceps* by shamans in Mexico for divining the future and the *Elaphomyces* by some Europeans and Orientals as an aphrodisiac.

691 Beetle Cordyceps
Cordyceps melolonthae (Tul.) Sacc.
Clavicipitaceae, Sphaeriales

Description: *Yellow-orange, stalked mushroom with oval, cream-colored head arising from underground scarab beetle larvae.*
Head: ⅜–⅝″ (1–1.5 cm) wide, ¾–1¼″ (2–3 cm) high; oval; white to yellowish, surface finely roughened by protruding necks of immersed, flasklike vessels (perithecia).
Stalk: 2–2¾″ (5–7 cm) long, ⅛–⅜″ (0.3–1 cm) thick; slender to stout, smooth, yellowish.
Spores: part-spores 4–8 × 1–1.5 μ; elliptical, smooth.
Season: June–September.
Habitat: Single to several, on beetle larvae in soil or rotten wood.
Range: SE. North America.
Comments: This is one of several species of *Cordyceps* that parasitize insects. The mushroom usually attaches itself to the larval state of the insect, but it can attach itself to the mature insect as well. The mushroom's hyphae—a mat or cottony stuffing composed of whitish threads—penetrates the body of the insect and mummifies it. When the insect dies, the fungus matures and discharges its spores.

687 Trooping Cordyceps
Cordyceps militaris (L. ex St. Amans) Link
Clavicipitaceae, Sphaeriales

Description: *Reddish-orange, cylindrical mushroom arising from underground insect pupae or larvae.*
Head: ¼″ (5 mm) wide, ⅜–¾″ (1–2 cm) high; cylindrical to spindle-shaped; reddish-orange, finely roughened by protruding necks of immersed, flasklike vessels (perithecia).
Stalk: 1¼–1⅝″ (3–4 cm) long, ⅛–¼″ (3–5 mm) thick, slender and tapering to base; red-orange or paler; smooth.

Spores: 300–500 × 1–1.5 μ;
threadlike and many-celled; breaking
into part-spores, 3.5–6 × 1–1.5 μ;
barrel-shaped, smooth.

Season: September–November.

Habitat: Single to several or many, on pupae and
larvae of moths and butterflies.

Range: New England to North Carolina, west
to California.

Comments: This is the most commonly seen species
of *Cordyceps*. It is especially abundant in
the Southeast and is typically found in
groups or clusters, although it
occasionally grows singly.

---

683   Goldenthread Cordyceps
Cordyceps ophioglossoides (Ehr. ex Fr.) Link
Clavicipitaceae, Sphaeriales

Description: *Cylindrical, reddish-brown mushroom with*
yellow stalk, attached by tangled yellow
threads to underground, trufflelike fungus.
Head: ⅜–⅝″ (1–1.5 cm) wide, ¾–1″
(2–2.5 cm) high; cylindrical to oval;
reddish-brown, smooth, becoming
finely roughened by protruding necks of
immersed, flasklike vessels (perithecia).
Stalk: 3″ (7.5 cm) long, ⅛–⅜″ (0.3–1
cm) thick; yellow, smooth, with thick,
yellowish basal threads attached to
underground, trufflelike fungus.
Spores: part-spores 2.5–5 × 2 μ;
elliptical, smooth.

Season: August–October.

Habitat: Single to several, in forests on
underground false truffles (*Elaphomyces*).

Range: E. North America.

Comments: Accurate identification of this
mushroom requires unearthing it
completely to see the bright yellow
threads at the base that attach to the
false truffle. The growth of *Cordyceps*
interferes with the production of spores
in the false truffle.

663 Yellow Cushion Hypocrea
Hypocrea gelatinosa (Tode ex Fr.) Fr.
Hypocreaceae, Sphaeriales

Description: *Small, fleshy to gelatinous, cushion-shaped stalkless, yellowish mushroom, becoming green and dotted.*
Mushroom: $\frac{1}{32}$–$\frac{1}{8}$" (1–3 mm) wide; circular, cushion-shaped, fleshy, soft, and translucent to gelatinous; light yellow to yellowish-white, becoming greenish to greenish-black at maturity; surface roughened by protruding necks of mostly immersed, rounded, flasklike vessels (perithecia).
Spores: at first elliptical, greenish, finely warted, and 2-celled; later separating at cross-wall into 2 part-spores. Upper part-spore 4 μ, round; lower part-spore 5–6 × 3–4 μ, elliptical.

Season: September–October.
Habitat: On decaying wood.
Range: Maine to North Carolina, west to Ohio and Iowa.
Look-alikes: *H. citrina* and *H. sulphurea* form irregularly shaped mushrooms over substrate.
Comments: Also known as *Chromocrea gelatinosa* and *Creopus gelatinosa*.

400 Golden Hypomyces
Hypomyces chrysospermus Tul.
Hypocreaceae, Sphaeriales

Description: *Three-stage mold that grows on various boletes: first white, then powdery lemon-yellow, finally reddish-brown.*
Mold: white mold stage (*Verticillium*) produces elliptical, 1-celled, thin-walled, smooth, asexual spores (10–30 × 5–12 μ); powdery yellow stage (*Sepedonium chrysospermum*) produces round, 1-celled, thick-walled, warted, yellow, asexual spores (10–25 μ in diameter); reddish-brown stage (*Hypomyces*) produces orange-yellow or reddish-brown, flasklike vessels

(perithecia).
Spores: 25–30 × 5–6 μ; spindle-shaped, transparent; unequally 2-celled.

Season: June–September.

Habitat: On species of boletes, such as *Boletus chrysenteron, B. parasiticus,* and *Gyroporus castaneus;* reported on *Paxillus involutus* and species of *Rhizopogon.*

Range: Throughout North America.

Comments: Also called *Apiocrea chrysosperma.* This fungus has 3 distinct stages of growth. The white mold that forms on boletes and related mushrooms lasts but a short time, and is replaced by a yellow powder that lasts until the host has begun to disintegrate. Both the white and yellow stages produce asexual spores. The reddish-brown sexual stage, which develops when the host is in advanced decay, is not commonly seen or collected.

699 Amanita Mold
Hypomyces hyalinus (Schw. ex Fr.) Tul.
Hypocreaceae, Sphaeriales

Description: *White to pink-tinged mold covering various amanitas, distorting them into thick, club-shaped forms.*
Mold: whitish to pink-tinged; produces reddish or amber, flasklike vessels (perithecia).
Spores: 13–22 × 4.5–6.5 μ; spindle-shaped, strongly warted, transparent; unequally 2-celled.

Season: June–October.

Habitat: Grows on *Amanita rubescens, A. flavorubescens, A. frostiana,* and possibly *A. bisporigera.*

Range: Maine to North Carolina and Colorado.

Comments: This parasite forms such a dense, thick covering on its *Amanita* hosts that it is usually impossible to identify them. Therefore, although the Amanita Mold would be edible when found growing on the Blusher (*A. rubescens*), it would be deadly on the Two-spored Destroying Angel (*A. bisporigera*).

294, 429 Lobster Mushroom
Hypomyces lactifluorum (Schw. ex Fr.) Tul.
Hypocreaceae, Sphaeriales

Description: *Bumpy, bright orange to orange-red mold growing on various white species of* Lactarius *and* Russula *mushrooms.*
Mold: orange, orange-red, or reddish-purple, with fine bumps; produces orange to reddish, flasklike vessels (perithecia).
Spores: 35–50 × 4.5 μ; spindle-shaped, strongly warted, transparent; equally 2-celled.

Edibility: Choice, with caution.

Season: July–October.

Habitat: In woods on white *Lactarius* and *Russula* mushrooms.

Range: Throughout North America.

Comments: The parasitic Lobster Mushroom transforms its ordinarily unpalatable *Lactarius* and *Russula* hosts into excellent edibles. The mushrooms should be well cleaned and cooked. If the host cannot be confirmed, avoid this; it could potentially parasitize a poisonous species.

451 Yellow-green Hypomyces
Hypomyces luteovirens (Fr.) Tul.
Hypocreaceae, Sphaeriales

Description: *Yellowish-green, bumpy mold covering gills and upper stalk of various* Lactarius *and* Russula *mushrooms.*
Mold: yellowish-green to dark green, with fine bumps; produces dark green, flasklike vessels (perithecia).
Spores: 28–35 × 4.5–5.5 μ; spindle-shaped, finely warted, transparent; 1-celled.

Season: July–November.

Habitat: In woods on various *Russula* and *Lactarius* mushrooms.

Range: New England to North Carolina, west to Texas, Colorado, and California.

Comments: Also called *Peckiella viridis*. Since the mold covers the gills of its host, it can only be detected when the host mushroom is turned over. It should not be eaten.

---

668   Carbon Balls
Daldinia concentrica (Bolt. ex Fr.) Ces. & DeNot.
Xylariaceae, Sphaeriales

Description: *Small, hard, black ball; on wood.*
Mushroom: ¾–1⅛" (2–4 cm) wide; round; grayish-white at first, then reddish-brown, finally black and somewhat shiny; surface becoming dotted with minute, bumplike pores, powdery black at maturity. Flesh dark purplish-brown, fibrous and carbonlike, with 6–40 darker concentric zones. Spores: 12–17 × 6–9 μ; elliptical, with 1 flattened side; smooth, black. Spore print brown to black.

Season: June–September; overwinters.

Habitat: Usually in clusters, on dead branches and stumps of deciduous wood; also on wounds of living deciduous trees.

Range: Maine to Florida, west to North Dakota and Pacific NW.

Comments: When a Carbon Ball is sliced in half, its striking concentric zones are revealed. Also known as "Cramp Balls."

---

667   Red Cushion Hypoxylon
Hypoxylon fragiforme (Pers. ex Fr.) Kickx
Xylariaceae, Sphaeriales

Description: *Round; salmon-pink, becoming brick red, then blackening.*
Mushroom: ⅟₁₆–⅝" (0.15–1.5 cm) wide, ⅟₁₆–⅜" (0.15–1 cm) high; round, grayish-white at first, then salmon-pink, becoming brick red at maturity, blackening with age; surface minutely bumpy; single layer of flasklike vessels (perithecia) just beneath surface. Flesh hard, black. Spores: 11–15 × 5–7 μ; more or less

spindle-shaped, with 1 flattened side;
dark brown; 1–3 oil drops. Spore print
dark brown to blackish.

Season: July–November; overwinters.
Habitat: Usually in swarms, on bark of dead or
dying beeches.
Range: E. North America.
Look-alikes: *H. argillaceum* is found on ash trees.
H. fuscum grows on alders and birches.
Comments: All species in the family Xylariaceae are
grayish-white at first, changing color as
their spores mature.

669 Carbon Cushion
Ustulina deusta (Fr.) Pet.
Xylariaceae, Sphaeriales

Description: *Brittle, blackish, crustlike cushion.*
Mushroom: 1⅛–4″ (4–10 cm) wide,
⅛–¼″ (3–5 mm) thick, sometimes
forming sheets up to 20″ (50 cm) long;
irregular cushion or thick, soft crust;
grayish-white when young, soon black,
then brittle and easily detached from
wood; flasklike vessels (perithecia)
large, with bumplike pores immersed
in whitish flesh.
Spores: 28–34 × 7–10 μ; spindle-
shaped, with 1 side more or less
flattened; black. Spore print black.
Season: Summer–early winter; overwinters.
Habitat: On stumps and dead roots of deciduous
trees, such as beech, maple, or ash.
Range: Widely distributed in North America.
Comments: Most amateur mushroom hunters
overlook this species, mistaking it for
burned wood.

738 Carbon Antlers
Xylaria hypoxylon (L. ex Hook.) Grev.
Xylariaceae, Sphaeriales

Description: *Thin-stalked, forked, woody club; white
near top, black near base.*
Club: 1⅛–3¼″ (4–8 cm) high, ¼–⅜″
(0.5–1 cm) thick; thin-stalked,
branched, and more or less cylindrical;

upper portion at first powdery white because of asexual spores, then becoming black; lower portion black, hairy. Flesh white, woody, tough. Spores: $11-14 \times 5-6 \mu$; bean-shaped, black. Spore print blackish.

Season: June—August in East; October in Pacific NW.; December—February in California.

Habitat: On dead wood.

Range: Throughout North America.

Look-alikes: Some coral mushrooms (*Clavaria** and *Pterula*) may appear similar but are fleshy.

Comments: Like Dead Man's Fingers (*X. polymorpha*) and Carbon Balls (*Daldinia concentrica*), this mushroom produces its sexual spores in sacs (asci) lining the insides of flasklike vessels (perithecia). The open ends of these vessels perforate the surface of the mushroom and appear as pores or minute bumps.

---

697 Dead Man's Fingers
Xylaria polymorpha (Pers. ex Mér.) Grev.
Xylariaceae, Sphaeriales

Description: *Short-stalked, thick, distorted, fingerlike clubs; white to buff, becoming black.*
Club: $\frac{3}{4}-3\frac{1}{4}"$ (2–8 cm) high, $\frac{3}{8}-1\frac{1}{4}"$ (1–3 cm) thick; short-stalked, powdery white to buff at first when covered with asexual spores, becoming black at maturity; carbonlike, crustlike surface minutely wrinkled and finely cracked. Flesh white, tough.
Spores: $20-30 \times 5-10 \mu$; narrow, spindle-shaped, with 1 flattened side. Spore print dark brown to black.

Season: June—October.

Habitat: Usually in clusters, on rotting deciduous wood, at bases of stumps, especially maple and beech.

Range: Throughout North America.

Comments: The whitish asexual stage occurs in early spring; the carbonlike sexual stage follows in summer.

BASIDIOMYCETES
(Subdivision Basidiomycotina)

The Basidiomycetes form the subdivision that contains all the woody fungi and nearly all the large fleshy fungi. It includes all but a few of the known edible and poisonous mushrooms. The gilled mushrooms, boletes, and puffballs belong here, as do the jelly fungi, chanterelles, corals, tooth fungi, polypores, and crust fungi. All these fungi produce their spores on appendages protruding from variously designed (usually club-shaped) microscopic structures known as basidia. Classification of classes, orders, families, genera, and species within this subdivision is determined partly by the method of spore dispersal, partly by the size, shape, surface, and color of the spores, and partly by the response of the basidia and spores to chemical reagents. Field examination of the mushrooms is important but often insufficient to determine identification with certainty.

EXPOSED HYMENIUM FUNGI
(Class Hymenomycetes)

The class Hymenomycetes contains the
gilled mushrooms, boletes, polypores,
crust fungi, tooth fungi, corals,
chanterelles, and jelly fungi. All of the
species in this class have a fruiting body
with an exposed fertile surface lined
with microscopic club-shaped or
irregular structures known as basidia.
The spores develop on and are
discharged from short appendages
called sterigmata, which in most
species grow from the top of the
basidia.

JELLY FUNGI
(Order Tremellales)

The order Tremellales is broadly
defined here to include all gelatinous
mushrooms that possess microscopic,
club-shaped spore-bearing structures
called basidia, which are either
segmented or resemble a tuning fork.
Jelly fungi can be soft and gelatinous or
stiff and rubbery. They vary
considerably in shape and can appear
bloblike, brainlike, earlike, or leaflike,
or may even resemble other kinds of
mushrooms. Some are transparent or
colored yellow, orange, and red; others
are dark brown to black. None is
known to be poisonous, and a few are
popular edibles. Microscopic
examination is sometimes necessary for
accurate identification, since jelly fungi
resemble some cup fungi, which are
unrelated and produce spores in saclike
structures called asci.
For the convenience of those people
without access to a microscope, some
jelly fungi usually distinguished from
the Tremellales on the basis of
microscopic differences are grouped
together here. The Orange Jelly and
other related jelly fungi, for example,

may look similar to *Tremella* and its
allies, but they have differently
designed basidia and are usually
classified separately to call attention to
this basic reproductive difference.

617 Tree-Ear
Auricularia auricula (Hook.) Under.
Auriculariaceae, Auriculariales

Description: *Large, brownish, rubbery, ear-shaped.*
Mushroom: 1–6″ (2.5–15 cm) wide;
cuplike to ear-shaped; upper surface
smooth, wavy, reddish-brown to
blackish; undersurface with dense, silky
covering or minutely hairy, deeply and
irregularly ribbed and veined, reddish-
to yellowish-brown. Flesh thin,
rubbery.
Spores: 12–15 × 4–6 μ; sausage-
shaped, smooth, colorless. Basidia
cylindrical, with cross-walls; spores
form on upper surface. Spore print
white.

Edibility: Edible.

Season: May–June; September–December.

Habitat: Several to many, on coniferous wood;
sometimes on deciduous wood.

Range: Throughout North America.

Look-alikes: *Peziza badio-confusa** and other similarly
colored cup fungi are very brittle, break
easily, and grow on the ground.

Comments: Also known as the "Wood-Ear." A
closely related species, Mo-Ehr
(*A. polytricha*), is widely cultivated and
sold commercially at Chinese markets.
It was recently reported to affect blood
coagulation, and may contribute to the
low incidence of coronary artery disease
in China.

735 Yellow Tuning Fork
Calocera viscosa (Pers. ex Fr.) Fr.
Dacrymycetaceae, Dacrymycetales

Description: *Yellowish, tough, gelatinous, coral-like
mushroom, forked at tip of branches.*

Mushroom: ⅜–1″ (1–2.5 cm) wide, 1–4″ (2.5–10 cm) high; branching from slender, deeply rooted, white base; upper branches usually forked; golden- to orange-yellow, slimy. Flesh tough, gelatinous.
Spores: 9–14 × 3–5 μ; sausage-shaped, smooth; becoming 2-celled. Basidia shaped like a tuning fork; spores form on all sides of mushroom. Spore print ochre-yellow.

Season: August–November.

Habitat: Several to many, on coniferous wood.

Range: Throughout N. North America; south to North Carolina in coniferous woods.

Look-alikes: Similar coral fungi are brittle and easily broken.

Comments: Also known as the "Yellow False Coral." A very small related species, the Clublike Tuning Fork (*C. cornea*), is usually unbranched and club-shaped, with a pointed tip; like the Yellow Tuning Fork, it grows in groups on pine logs and other woods.

---

567 **Orange Jelly**
Dacrymyces palmatus (Schw.) Bres.
Dacrymycetaceae, Dacrymycetales

Description: *Irregularly lobed mass of bright orange jelly with white point of attachment.*
Mushroom: ⅜–2⅜″ (1–6 cm) wide, ⅜–1″ (1–2.5 cm) high; brainlike to lobed mass; yellowish-orange, with white basal attachment; reddish, horny when dry. Flesh gelatinous, tough.
Spores: 17–25 × 6–8 μ; cylindrical to sausage-shaped, smooth; becoming 8- to 10-celled. Basidia shaped like a tuning fork; spores form on upper surface of mushroom. Spore print yellowish.

Edibility: Edible.

Season: May–November.

Habitat: On coniferous logs.

Range: Throughout North America.

Look-alikes: *Tremella mesenterica** lacks white basal attachment, grows on deciduous wood.

Comments: The Orange Jelly is edible, but should
be boiled or steamed, rather than
sautéed.

599 Golden Jelly Cone
Guepiniopsis alpinus (Tr. & Ear.) Bres.
Dacrymycetaceae, Dacrymycetales

Description: *Small, golden, gelatinous, conelike cups.*
Mushroom: $\frac{1}{16}-1''$ (0.15–2.5 cm)
wide, $\frac{1}{16}-\frac{3}{8}''$ (0.15–1 cm) high; cone-
to cup-shaped; golden-yellow to
orange, drying reddish-orange. Flesh
gelatinous.
Spores: $12-18 \times 4-6 \mu$; sausage-
shaped, smooth; becoming 4- to 5-
celled. Basidia shaped like a tuning
fork; spores form in center of cone.
Spore print yellowish.
Season: May–June.
Habitat: On coniferous wood debris, logs,
stumps, and on living coniferous twigs.
Range: Rocky Mountains west to Washington
and California.
Look-alikes: Cup fungi are brittle, with saclike
spore-bearing structures (asci).
Comments: This species is often abundant during
spring snow melt. It grows out of
cracks in dead coniferous wood and is
attached by a point.

563 Black Jelly Roll
Exidia glandulosa Bull. ex Fr.
Tremellaceae, Tremellales

Description: *Small, fused, blackish, cuplike to brain-
shaped jelly, with smooth or warted surface.*
Mushroom: $\frac{3}{8}-\frac{3}{4}''$ (1–2 cm) wide;
blisterlike, fusing with others to form a
long, narrow row up to 6'' (15 cm)
long; individuals brainlike to
irregularly contorted; translucent at
first, becoming reddish-black to olive-
black to black; smooth to warty. Flesh
gelatinous, thin.
Spores: $10-16 \times 4-5 \mu$; sausage-
shaped, smooth. Basidia have

longitudinal walls; spores form on warty surface. Spore print white.

Season: April—May; November—January.

Habitat: In rows, on deciduous wood such as oak, willow, and alder.

Range: Throughout North America.

Comments: There are several species of *Exidia*, all beginning as more or less translucent blisters. The Granular Jelly Roll (*E. nucleata*) turns a reddish-brown and contains granular material. The Pale Jelly Roll (*E. alba*) is white. The Amber Jelly Roll (*E. recisa*) has a small stalk and becomes yellowish-brown.

422 Apricot Jelly
Phlogiotis helvelloides (Fr.) Mar.
Tremellaceae, Tremellales

Description: *Apricot-colored, funnel-shaped jelly with off-center stalk; on the ground.*
Mushroom: ⅜—2¾" (1—7 cm) wide, 1—4" (2.5—10 cm) high; somewhat funnel- to spoon-shaped, with wavy-lobed margin; apricot, pink, or rose, smooth to slightly veined. Flesh rubbery.
Stalk: short, off-center.
Spores: $9–12 \times 4–6 \mu$; elliptical, smooth. Basidia with longitudinal walls; spores form on undersurface. Spore print white.

Edibility: Edible.

Season: May—July; August—October.

Habitat: On the ground under conifers or on rotten wood.

Range: Widely distributed; most common in Pacific NW.

Comments: This attractive mushroom is edible, but more for color and texture than taste.

459 Jelly Tooth
Pseudohydnum gelatinosum (Scop. ex Fr.) Kar.
Tremellaceae, Tremellales

Description: *Tonguelike, translucent whitish jelly, covered with teeth on undersurface.*

Mushroom: 1–3″ (2.5–7.5 cm) wide, 1–2″ (2.5–5 cm) high; tonguelike or spoon-shaped; upper surface dull, translucent, whitish to grayish, sometimes with brown tones; undersurface covered with short, toothlike, white spines. Flesh gelatinous.

Stalk (when present): lateral.

Spores: 5–7 μ; nearly round, smooth. Basidia have longitudinal walls; spores form on teeth. Spore print white.

Edibility: Edible.

Season: September–November.

Habitat: On rotting wood in areas with conifers.

Range: Throughout North America.

Comments: Also called "Jelly False Tooth," this is the only toothed jelly fungus. True tooth fungi (*Hydnum* and allies) are fleshy to brittle, not gelatinous. Like some other jelly fungi, the Jelly Tooth is edible, although it has little or no flavor.

---

754 **Jelly Leaf**
Tremella foliacea Pers. ex Fr.
Tremellaceae, Tremellales

Description: *Large, brownish, lettucelike jelly.*

Mushroom: 1–8″ (2.5–20 cm) wide, 2–4″ (5–10 cm) high; broadly leafy to wavy; reddish-brown to dark brown. Flesh gelatinous, thin to thick.

Spores: 7–12 × 7–9 μ; round to nearly round, smooth. Basidia have longitudinal walls; spores form on outer surfaces. Spore print yellowish.

Edibility: Edible.

Season: July–November.

Habitat: On stumps and logs of beeches and other deciduous trees; also reported on conifers.

Range: Throughout North America.

Comments: This mushroom can be quite large, and it usually appears late in the season. The Yellow Jelly Leaf (*T. frondosa*), a yellowish-buff but otherwise similar mushroom, is probably only a variant.

566 Witches' Butter
Tremella mesenterica Ret. ex Fr.
Tremellaceae, Tremellales

Description: *Irregularly lobed mass of golden jelly.*
Mushroom: 1–4″ (2.5–10 cm) wide,
1¼–1⅝″ (3–4 cm) high; globular to
brainlike or lobed; orange-yellow or
clear yellow, orange-red on drying.
Flesh tough, horny when dry.
Spores: 7–15 × 6–10 μ; oval to
elliptical, smooth. Basidia have
longitudinal walls; spores form on outer
surfaces. Spore print yellowish.
Edibility: Edible.
Season: Year-round.
Habitat: On deciduous wood, especially oak,
beech, and alder.
Range: Throughout North America.
Look-alikes: *Dacrymyces palmatus** has a white point
of attachment, grows on coniferous wood.
Comments: This fungus can be found during
warming winter thaws, throughout
cool, wet springs, and in summer in
the mountains. The Yellow Witches'
Butter (*T. lutescens*), probably just a
variant, is smaller, pale yellow, and
more lobed than contorted. Witches'
Butter is sometimes added to soups.

741 Jellied False Coral
Tremellodendron pallidum (Schw.) Burt
Tremellaceae, Tremellales

Description: *Small, whitish, leathery, coral-like jelly.*
Mushroom: 2–6″ (5–15 cm) wide, 1–
4″ (2.5–10 cm) high; coral-like with
upright branches, but flattened and
fused in parts; white to off-white, buff
when dry. Flesh tough, somewhat
gelatinous.
Spores: 7–11 × 4–6 μ; sausage-
shaped, smooth. Basidia have
longitudinal walls. Spore print white.
Edibility: Edible.
Season: July–November.
Habitat: On the ground in deciduous or mixed
woods.

Range: Quebec to South Carolina, west to
Missouri and Minnesota.
Look-alikes: Coral fungi are brittle and break easily.
Comments: Also known as _T. schweinitzii_. Although
edible, its flavor is reminiscent of
rancid buttered popcorn. The White
False Coral (_T. candidum_), which is
possibly just a variant, is smaller and
more spreading; it is the most common
species in the Quebec area.

CORAL AND PORE FUNGI AND ALLIES
(Order Aphyllophorales)

The Aphyllophorales is an artificial
order of mushrooms that includes the
chanterelles, coral fungi, tooth fungi,
crust and parchment fungi, and
polypores. Like all agarics and boletes,
most species in this group produce
spores on four small stalks (sterigmata),
projecting from the top of simple,
clublike microscopic structures
(basidia). In contrast to agarics and
boletes, these fungi usually lack any
kind of veil covering, as well as true
gills or readily detachable tubes.
The Aphyllophorales include a few of
our best edible mushrooms; however, a
great many—including the polypores
and the tooth fungi—are too woody or
too bitter to be used as food. A few
coral mushrooms are reported to
function as laxatives or to cause gastric
illness. None of the species in this order
is known to be fatally poisonous. While
this order includes some delightfully
fragrant and beautiful mushrooms, it
also contains some of the most
destructive—for example, some
polypores and crust fungi—that
parasitize healthy trees, and cause
extensive wood decay.
Although more than 20 families have
been described in the order, this guide
divides the order into 9 families, based
on their appearance in the field.

CHANTERELLE FAMILY
(Cantharellaceae)

The Chanterelle family includes some of our best known edible wild mushrooms, such as the spicy Chanterelle of open oak woods and western conifer forests, and the fragrant Black Trumpet of eastern beech woods. There are 4 genera in the family: *Cantharellus, Craterellus, Gomphus,* and *Polyozellus.* Most are orangish or yellow; a few are gray to brownish-black, one is white, one is blue, and a few have albino color forms. Most are either convex or vase-shaped. All lack true gills. Instead, as the mushroom develops, it produces spores on ridges, folds, or on a nearly smooth surface. Most chanterelles grow on the ground; they are found in summer in the East, and in fall and winter in the West. Only a few mushrooms in this family, such as the Scaly Vase Chanterelle, are known to cause some discomfort on being eaten. However, poisonings occur through misidentification, such as when the toxic Jack O'Lantern is mistaken for the delectable Chanterelle.

308, 427 **Chanterelle**
Cantharellus cibarius Fr.
Cantharellaceae, Aphyllophorales

Description: *Bright yellow to orange cap with wavy margin and yellow-orange, forked, thick-edged ridges descending stalk; fragrant.*

Cap: ⅜–6" (1–15 cm) wide; convex, becoming flat with inrolled wavy margin, sunken in center; somewhat finely hairy or fibrous to smooth; yellow to orange-yellow. Odorless or with fragrance like apricots; taste mild to spicy-peppery.
Fertile Surface: narrow, thick-edged ridges, forked and crossveined, nearly distant, descending stalk; pale yellow to orange.

Stalk: 1–3″ (2.5–7.5 cm) long, ¼–1″ (0.5–2.5 cm) thick, sometimes enlarged at either end; smooth or with small, flattened fibers; yellowish to whitish, sometimes bruising orange. Flesh solid, white.

Spores: 8–11 × 4–6 μ; elliptical, smooth, colorless. Spore print pale buff to pale yellow.

Edibility: Choice, with caution.

Season: June–September in Southeast; July–August in Northeast; September–November in Northwest; November–February in California.

Habitat: Single to many, on the ground under oaks or conifers.

Range: Throughout North America.

Look-alikes: The poisonous *Omphalotus olearius** grows clustered on stumps or buried wood, has close, unforked, sharp-edged gills, and sometimes has disagreeable odor. *Hygrophoropsis aurantiaca** has crowded, forked gills, soft flesh, and is odorless. *C. lateritius** does not have well-formed gill-like ridges. *C. subalbidus** is more massive, whitish but bruising yellow to orange, odorless, and has white spore print.

Comments: The Chanterelle, also known as "Girolle" and "Pfifferling," is probably the most popular and prized edible wild mushroom in the world. Beware of confusing it with toxic look-alikes, however, particularly the poisonous Jack O'Lantern (*Omphalotus olearius*).

---

421 Cinnabar-red Chanterelle
Cantharellus cinnabarinus Schw.
Cantharellaceae, Aphyllophorales

Description: *Small, reddish-orange cap with pinkish, forked, thick-edged ridges descending dull red stalk.*

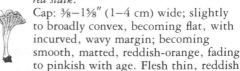

Cap: ⅜–1⅝″ (1–4 cm) wide; slightly to broadly convex, becoming flat, with incurved, wavy margin; becoming smooth, matted, reddish-orange, fading to pinkish with age. Flesh thin, reddish

near top, whitish near base.
Fertile Surface: well-formed, thick-edged ridges, forked and crossveined, distant, descending stalk, pinkish.
Stalk: ⅝–1⅝" (1.5–4 cm) long, ⅛–⅜" (0.3–1 cm) thick, sometimes tapering downward, sometimes curved; fibrous, dull red. Flesh solid, white.
Spores: 6–11 × 4–6 μ; elliptical to oblong, smooth, colorless. Spore print pinkish-cream.

Edibility: Good.

Season: Late June–October.

Habitat: Single to many, on the ground in moss and along pathsides, especially in open oak woods.

Range: E. Canada to Florida, west to Indiana.

Look-alikes: Hygrophorus* species have sharp-edged, unforked gills.

Comments: Though not as choice as the Chanterelle (C. cibarius), this species is attractive and often abundant.

307 Flame-colored Chanterelle
Cantharellus ignicolor Pet.
Cantharellaceae, Aphyllophorales

Description: Small, yellow-orange cap with orange to brownish, forked, thick-edged ridges descending stalk.
Cap: ⅜–2" (1–5 cm) wide; convex, with inrolled margin and sunken center, becoming flat or with decurved to wavy margin; smooth to somewhat rough, watery-looking in rain, orange to yellow-orange. Flesh thin, dull orange. Odorless or sometimes faintly fragrant.
Fertile Surface: narrow, blunt-edged, forked ridges with crossveins, distant, descending stalk; orange-yellow, wine-buff, or brownish.
Stalk: ¾–2⅜" (2–6 cm) long, ¹⁄₁₆–⅝" (0.15–1.5 cm) thick, sometimes slightly enlarged at base; compressed and becoming hollow with age; smooth, yellow to orange.
Spores: 9–13 × 6–9 μ; broadly

elliptical, smooth, colorless. Spore
print pale ochre-salmon.

Season: July—September.
Habitat: Scattered or clustered, on the ground
under deciduous and coniferous trees.
Range: E. Canada to Georgia and Michigan.
Look-alikes: *C. tubaeformis** has darker cap with
wavy margin, and whitish to pale
yellowish spore print.
Comments: Although some have speculated that
this chanterelle is edible, there has been
no evidence to support this view.

---

428 Smooth Chanterelle
Cantharellus lateritius (Berk.) Sing.
Cantharellaceae, Aphyllophorales

Description: *Yellow cap with wavy margin and orange
to yellow, smooth to slightly ridged or veined
undersurface descending stalk; fragrant.*
Cap: 1–4″ (2.5–10 cm) wide; flat to
decurved, with lobed to wavy, arched
margin, becoming funnel-shaped;
smooth, orange to yellow. Flesh thick
at center, orange near cap surface,
whitish near stalk. Odor fruity.
Fertile Surface: smooth to shallowly
ridged or veined and crossveined;
orange-yellow, often tinted pinkish.
Stalk: 1–4″ (2.5–10 cm) long, ¼–1″
(0.5–2.5 cm) thick, sometimes
enlarged at either end; often slightly
curved, off-center, orange-yellow. Flesh
white, solid, becoming hollow.
Spores: 7.5–12.5 × 4.5–6.5 μ;
elliptical, smooth, colorless. Spore
print pinkish-yellow.
Edibility: Choice.
Season: July—September.
Habitat: On the ground under oaks, especially
along paths.
Range: NE. North America.
Look-alikes: *C. cibarius** has well-formed, gill-like
ridges. *Craterellus odoratus** is typically
clustered and bouquetlike and grows in
the South. *Clavariadelphus truncatus** is
odorless, club-shaped, and has wrinkled
outer surface.

Comments: This chanterelle is abundant during hot summers in urban areas of the Northeast, and is also common in southern woods. Many say that it equals the Chanterelle (*Cantharellus cibarius*) in fragrance and flavor. Formerly known as *Craterellus cantharellus*.

57 **Small Chanterelle**
Cantharellus minor Pk.
Cantharellaceae, Aphyllophorales

Description: *Small, yellow-orange cap with forked, yellow-orange, thick-edged ridges descending stalk.*
Cap: ¼–1¼" (0.5–3 cm) wide; at first convex with inrolled margin, becoming flat to sunken with decurved margin, finally funnel-shaped with arched, wavy margin; waxy-fragile, smooth, dull yellow to orange. Flesh thin, yellow to orange.
Fertile Surface: very narrow, forked ridges with blunt edges and crossveins; ridges close to almost distant, descending stalk, yellow to orange.
Stalk: ⅝–2" (1.5–5 cm) long, ⅛–⅜" (0.3–1 cm) thick; central, often somewhat curved, smooth, yellow to orange. Flesh solid, becoming hollow.
Spores: 6–11.5 × 4–6.5 μ; broadly elliptical to oblong, smooth, colorless. Spore print pale yellowish-orange.
Edibility: Good.
Season: July–August.
Habitat: Scattered to many, on ground in moss in deciduous woods, especially beech-maple.
Range: E. North America to Gulf Coast.
Look-alikes: *C. cinnabarinus** is fleshy-brittle and reddish to pinkish, fading with weathering. *C. luteocomus* has wrinkled to poorly formed, gill-like ridges. *Hygrophorus** has unforked gills.
Comments: The Small Chanterelle is identified by its size, mossy habitat, and waxy-fragile structure.

432 White Chanterelle
Cantharellus subalbidus A.H.S. & Morse
Cantharellaceae, Aphyllophorales

Description: *Whitish cap with whitish, forked, thick-edged ridges descending stalk; bruising orange to orange-brown.*
Cap: 2–6″ (5–15 cm) wide; flat to somewhat sunken; smooth, becoming slightly scaly with age, whitish overall, bruising rusty-yellow to orange or orange-brown. Flesh thick, firm, white.
Fertile Surface: shallow, blunt-edged, forked, crossveined, distant ridges descending stalk; whitish.
Stalk: ¾–2⅜″ (2–6 cm) long, ⅜–1¼″ (1–3 cm) thick; dry, smooth, stout; white, discoloring brownish. Flesh solid, firm.
Spores: 7–9 × 5–5.5 μ; elliptical, smooth, colorless. Spore print white.

Edibility: Choice.

Season: September–November.

Habitat: Scattered to numerous, on the ground under Douglas-fir and lodgepole pine or scrub pine; also with mixed oaks.

Range: Coastal areas of Pacific NW.

Look-alikes: *C. cibarius** is yellow to orange. *Clitocybe** and *Hygrophorus** have unforked gills.

Comments: Meaty and flavorful, this is one of the best edible mushrooms. Sometimes the Chanterelle (*Cantharellus cibarius*) will be found growing with the White Chanterelle, but the former is more common at lower elevations.

441 Trumpet Chanterelle
Cantharellus tubaeformis Fr.
Cantharellaceae, Aphyllophorales

Description: *Yellow-brown to dark brown cap, with yellow to gray, violet, or brownish, forked, thick-edged ridges descending stalk.*
Cap: ⅜–3″ (1–7.5 cm) wide; convex, becoming flat, with decurved, inrolled margin, becoming arched or wavy; smooth to wrinkled or rough to slightly

scaly; dark yellow-brown to dark brown, with gray radial streaks, fading with age. Flesh thin, grayish-yellow.
Fertile Surface: narrow, thick-edged ridges, forked and crossveined, distant, descending stalk; yellowish to gray, violet, or brownish.
Stalk: 1–2″ (2.5–5 cm) long, ⅛–⅜″ (0.3–1 cm) thick; cylindrical to compressed, smooth to furrowed; yellow or yellow-orange to grayish-orange. Flesh solid, becoming hollow.
Spores: 8–12 × 6–10 μ; broadly elliptical, smooth, colorless. Spore print cream to yellowish.

Edibility: Good.

Season: July–October.

Habitat: On sphagnum moss in bogs, wet mossy ground, or moss-covered, decayed coniferous logs.

Range: N. North America.

Look-alikes: *C. infundibuliformis* has blackish-brown, smooth cap, lemon yellow, hollow stalk, and buff spore print.

Comments: The Trumpet Chanterelle is common in fall; unlike most chanterelles, it can be found growing on rotting wood.

435 Yellow-footed Chanterelle
Cantharellus xanthopus (Pers.) Duby
Cantharellaceae, Aphyllophorales

Description: *Orange-brown to brownish cap, with ochre-brown to buff, wrinkled undersurface descending stalk.*
Cap: ⅜–2¾″ (1–7 cm) wide; at first convex, becoming flat with decurved, crimped, cut, or wavy margin and sunken, soon torn center; later vase-shaped with raised margin; orange-yellow, with firmly attached, coarse, small, brownish fibers or scales, rendering surface ochre to orange-brown to brownish. Flesh very thin, pale buff to orangish.
Fertile Surface: smooth to slightly veined or wrinkled, descending stalk, ochre-brown to buff.

Stalk: ¾–2" (2–5 cm) long, ⅟₁₆–⅝"
(0.15–1.5 cm) thick, sometimes
narrowing below; smooth, orange,
often curved. Flesh solid, becoming
hollow.

Spores: 9–11 × 6–7.5 μ; elliptical,
smooth, colorless. Spore print pale
orange-buff.

Edibility: Edible.

Season: July–September.

Habitat: Numerous, on moss or mossy logs in
low, wet woods of mixed hardwoods,
hemlock, and pine.

Range: E. North America.

Look-alikes: *C. luteocomus* has smooth, orange-yellow
cap. *C. tubaeformis** has well-formed
gill-like ridges, often with violet tint.

Comments: The Yellow-footed Chanterelle is one of
several species formerly known as *C.
lutescens.*

---

443 Black Trumpet
Craterellus fallax A.H.S.
Cantharellaceae, Aphyllophorales

Description: *Vase-shaped mushroom with small, flat,
brownish scales on top and within; outer
surface smooth to wrinkled, gray to dark
brown or black; fragrant.*

Mushroom: ⅜–3¼" (1–8 cm) wide,
1¼–5½" (3–14 cm) high; funnel-
shaped and hollow, with decurved to
elevated, wavy margin; dry, scaly, gray
to dark brown or blackish, fading to
gray-brown. Flesh thin, brittle, gray-
brown or darker. Odor and taste
fragrant, fruity.

Fertile Surface: smooth to slightly
veined, brown to gray with a bloom
when immature, ochre to salmon-tinted
when mature, becoming blackish.

Spores: 10–20 × 7–11.5 μ; broadly
elliptical, smooth, colorless. Spore
print ochre-buff to ochre-orange.

Edibility: Choice.

Season: July–November.

Habitat: Numerous, under beech, oak, and other
deciduous trees.

Range: Throughout North America.
Look-alikes: *C. cornucopioides* has white spore print. *C. cinereus* has well-formed, forked, gill-like ridges. *Thelephora** species are leathery-tough, have spiny spores.
Comments: This fragrant mushroom is often smelled before it is seen. Look for it along paths in oak and beech woods. It is excellent in flavor and texture when added to other foods and is preserved best by drying.

425 Fragrant Chanterelle
Craterellus odoratus (Schw.) Fr.
Cantharellaceae, Aphyllophorales

Description: *Large, clustered, bouquet-shaped, orange fungus; very fragrant.*
Mushroom: 3–6" (7.5–15 cm) wide and high; composed of several funnel-shaped, hollow extensions, with wavy, lobed margins, arising from single stalk; smooth, bright orange. Flesh thin, orange. Odor strongly fragrant and fruity.
Fertile Surface: smooth to vaguely wrinkled, orange-yellow.
Stalk: short, not clearly separated from fertile outer surface, hollow.
Spores: 8.9–11.8 × 4.4–6.3 μ; elliptical to narrowly oval, smooth, colorless. Spore print pale apricot-colored.
Edibility: Edible.
Season: August–September.
Habitat: On the ground in open woods.
Range: North Carolina to Florida.
Look-alikes: *Cantharellus lateritius** is typically single-capped and not as deeply funnel-shaped; *Craterellus confluens*, a rare southeastern species, is multicapped, with stuffed or solid, fused stalk that is clearly differentiated from caps.
Comments: This species is one of the most fragrant mushrooms. Although some people eat very fresh, young specimens, the mature Fragrant Chanterelle develops an unpleasant odor and taste.

440 Pig's Ear Gomphus
Gomphus clavatus (Fr.) S.F.G.
Cantharellaceae, Aphyllophorales

Description: *Violet to buff, smooth to scaly mushroom with wavy margin; undersurface buff, violet, or ochre, wrinkled to ridged.*
Mushroom: 1–4″ (2.5–10 cm) wide; often overlapping in clusters to 6″ (15 cm) wide; compressed and somewhat fused, cylindrical to blunt, center sunken, margin becoming elevated and wavy; moist to dry, hairy to somewhat scaly in center, smooth at margin, violet, becoming yellow-buff. Flesh thick, brittle, whitish to buff; odor and taste not distinctive.
Fertile Surface: shallow, wrinkled, blunt-edged ridges or veins descending stalk, with crossveins. Buff to violet or violet-brown, becoming ochre.
Stalk: ⅜–2″ (1–5 cm) long, ⅜–¾″ (1–2 cm) thick; usually very short, often curved; sometimes fused with one or more adjacent stalks; smooth to finely hairy, buff to pale lilac. Flesh solid, white.
Spores: 10–13 × 4–6.5 μ; elliptical, warted, colorless. Spore print ochre.
Edibility: Choice.
Season: August–October.
Habitat: Scattered to clustered, on the ground under conifers, especially spruce and fir.
Range: Across N. United States; Pacific NW. to California.
Comments: Common in fall in coniferous woods of the Pacific Northwest. The wavy cap sometimes resembles a pig's ear, but this common name is also used for entirely unrelated mushrooms, such as *Discina perlata*.

430, 431 Scaly Vase Chanterelle
Gomphus floccosus (Schw.) Sing.
Cantharellaceae, Aphyllophorales

Description: *Vase-shaped, yellow, orange, or reddish mushroom, coarsely scaly on top and within;*

outer surface wrinkled to blunt-ridged, yellowish to ochre.
Mushroom: 2–6" (5–15 cm) wide at top, 3–8" (7.5–20 cm) high; cylindrical and blunt at first, broadening and becoming deeply sunken to funnel-shaped, with wavy to lobed margin; moist to sticky, with scales flattened near margin, coarse and down-curled near center, yellow-orange, orange, reddish-orange. Flesh thin to fairly thick, whitish.
Fertile Surface: shallowly wrinkled, veined or with low, blunt ridges; yellow to cream or ochre.
Stalk: 2–4" (5–10 cm) long, ⅝–2" (1.5–5 cm) thick; tapering downward, smooth or covered with minute fibers, orange to yellowish, solid, becoming hollow. Flesh whitish.
Spores: 11.5–14 × 7–8 μ; elliptical, wrinkled to warted, colorless. Spore print ochre.

Edibility: Edible, but not recommended.
Season: Late June–September.
Habitat: Under conifers or in mixed woods.
Range: Throughout North America.
Look-alikes: *G. kauffmanii* is very large, coarsely scaly, and tannish to brown.
Comments: Some recognize the bright red, coarse cap with a milk-white fertile surface as a distinct species, *G. bonari*. Although eaten and enjoyed by some, the Scaly Vase Chanterelle and its related look-alikes should be avoided. They contain an indigestible acid and are sometimes sour and unpalatable.

444 **Clustered Blue Chanterelle**
Polyozellus multiplex (Under.) Murr.
Cantharellaceae, Aphyllophorales

Description: *Blue to purple mushroom with many fan- to spoon-shaped caps with veined undersides.*
Mushroom: 1–4" (2.5–10 cm) wide, in compact masses up to 39" (1 m) wide and 2⅜–6" (6–15 cm) high; fan- to spoon-shaped caps with irregularly

wavy and lobed margin, center often sunken; dull purple to purplish-gray, blackening with age. Flesh thick, soft, brittle, purplish or grayish-violet; dark olive with KOH. Odor aromatic; taste mild.

Fertile Surface: veined to netted or porelike in part; purplish, with a bloom.

Stalk: 1–2″ (2.5–5 cm) long, ⅜–1″ (1–2.5 cm) thick; short, irregularly compound or fused; purplish-black; solid or sometimes tubular, brittle.

Spores: 6–8.5 × 5.5–8 μ; round to broadly elliptical, angular and warted, colorless. Spore print white.

Edibility: Edible.

Season: Late June–October.

Habitat: Clustered on the ground under spruce and fir.

Range: N. North America; Rocky Mountains.

Look-alikes: *Gomphus clavatus** is larger, grows in smaller clusters, soon becomes brownish, does not stain greenish with KOH, and has an ochre spore print.

Comments: Although the Clustered Blue Chanterelle is rarely seen in the eastern United States, its large size and unusual color make it easy to identify.

CORAL FUNGUS FAMILY
(Clavariaceae)

Coral fungi resemble pieces of underwater coral, and are club-shaped to many-branched, white to yellow, ochre, orange, pink, red, or purple. Most grow on the ground, some on logs and stumps, and all seem to be most plentiful in late summer and fall in coniferous woods. Many coral fungi are eaten and, although a few are known to cause a laxative effect or gastric upset, no serious poisonings have been reported.

Simple, unbranched coral fungi may appear similar to some brightly colored ascomycete club fungi or even a few

kinds of chanterelles, although the
latter 2 may be rapidly distinguished
with the aid of a microscope. More than
half a dozen distinct genera are now
recognized in what was formerly known
as the genus *Clavaria.* These genera
have been distinguished in part on the
basis of microscopic and chemical
characteristics. Although a small
number of them are sometimes placed
in different families, all coral fungi
have been kept here in 1 family. In
addition, 3 other genera are placed in
it: *Physalacria, Sparassis,* and *Thelphora,*
all of which, although macroscopically
dissimilar, are somewhat related to the
coral fungi through the characteristics
of their spore-bearing surfaces.

734 Spindle-shaped Yellow Coral
Clavulinopsis fusiformis (Fr.) Cor.
Clavariaceae, Aphyllophorales

Description: *Clusters of spindle-shaped, yellow clubs
fused together at base.*
Mushroom: ⅟₁₆–⅜" (0.15–1 cm) wide,
2–6" (5–15 cm) high; spindle-shaped,
round to flattened; fused and somewhat
branched at base, sometimes branched
once or twice at top; bright yellow,
smooth to longitudinally wrinkled or
grooved. Flesh yellow, firm to brittle;
interior becoming hollow.
Stalk: indistinct; white-hairy at base.
Spores: 5–9 × 4.5–9 μ; round to
broadly oval, smooth, colorless. Spore
print white to yellowish.
Edibility: Edible.
Season: July–October.
Habitat: Densely clustered, in grass in woods
and fields.
Range: E. North America; reported in Oregon
and California.
Look-alikes: *C. aurantiocinnabarina* is reddish-orange
and hollow. *C. laeticolor,* also known as
C. pulchra, is golden-orange and
smaller. *C. helvola* is unbranched,
smaller, and thinner. Jelly fungi, such

as *Calocera viscosa**, are rubbery. Club fungi, such as *Microglossum rufum**, are unbranched and unfused, and produce their spores in asci, not on basidia.

Comments: Also known as *Clavaria fusiformis,* this very common coral is edible.

737 Purple Club Coral
Clavaria purpurea Fr.
Clavariaceae, Aphyllophorales

Description: *Clusters of spindle-shaped, purplish clubs.*
Mushroom: ⅟₁₆–¼" (1.5–5 mm) wide, 1–5" (2.5–12.5 cm) high; cylindrical to spindle-shaped, slender; purple to grayish-purple or brownish smoky-purple. Flesh white to purplish, brittle. Odor unpleasant or indistinct.
Stalk: indistinct; white-hairy at base.
Spores: 5.5–9 × 3–5 μ; elliptical to oblong, smooth, colorless. Spore print white.

Edibility: Edible.

Season: September–October.

Habitat: Clustered on the ground in mountains, under spruce and fir.

Range: Across N. North America; south in mountains.

Comments: This coral is most abundant in the Pacific Northwest and the Rockies of Colorado. Other purplish corals are much taller and thinner or branched.

739 White Worm Coral
Clavaria vermicularis Fr.
Clavariaceae, Aphyllophorales

Description: *White, curved, wormlike clubs.*
Mushroom: ⅛–¼" (3–5 mm) wide, 2–6" (5–15 cm) high; cylindrical to spindle-shaped, becoming flattened, often curved or wormlike; white, yellowing with age. Flesh white, very brittle.
Stalk: indistinct; cluster branched only at base.
Spores: 5–7 × 3–4 μ; elliptical,

smooth, colorless. Spore print white.

Edibility: Edible.

Season: July—September.

Habitat: Densely clustered, in grass, woods, and fields.

Range: Quebec to North Carolina, west to Minnesota; Pacific NW. to California.

Comments: Resembling a cluster of rising white worms, this common, brittle coral is easily seen against a green background. When wet, the White Worm Coral becomes almost translucent.

---

744 Crown-tipped Coral
Clavicorona pyxidata (Fr.) Doty
Clavariaceae, Aphyllophorales

Description: *Many-branched, yellowish coral with crownlike tips; on wood.*

Mushroom: ¾–2⅜" (2—6 cm) wide, 2—5" (5—12.5 cm) high; many-branched, usually in equal 3-sided groupings, with tips cup-shaped and crownlike; yellow, becoming dull ochre, tannish-white, or pink-tinged. Flesh whitish, tough. Taste usually peppery.
Stalk: 1/16–1/8" (1.5—3 mm) thick, very short; whitish to brownish-pink.
Spores: 4–5 × 2–3 μ; elliptical, smooth, colorless, amyloid. Spore print white.

Edibility: Edible.

Season: June—September.

Habitat: On dead wood, especially willow, poplar, and aspen.

Range: Throughout North America.

Look-alikes: *C. avellanea* is found only on coniferous wood in Pacific NW.

Comments: Formerly known as *Clavaria pyxidata*. This species is often common in aspen woods in late spring and early summer. One of few corals that grow directly on wood, it is easy to recognize because it has boxlike (pyxidate) branches, a somewhat peppery taste, and amyloid spores.

752 Violet-branched Coral
Clavulina amethystina (Fr.) Donk
Clavariaceae, Aphyllophorales

Description: *Small, many-branched, lilac-violet coral.*
Mushroom: ¾–3¼″ (2–8 cm) wide,
¾–2⅜″ (2–6 cm) high; many-
branched; tips blunt to narrowed; lilac-
violet, smooth to wrinkled. Flesh lilac-
violet, slightly brittle.
Stalk: short and stout or indistinct.
Spores: 7–12 × 6–8 μ; elliptical to
nearly round, colorless; basidia 2-
spored. Spore print white.

Edibility: Edible.

Season: July–September.

Habitat: On the ground in deciduous woods.

Range: E. North America.

Look-alikes: *Clavaria zollingeri* is only slightly
branched, has 4-spored basidia with
smaller spores, and is more brittle.

Comments: Formerly known as *Clavaria
amethystina,* the Violet-branched Coral
differs from members of the genus
Clavaria by its cylindrical, 2-spored
basidia with strongly incurved spore
stalks. *Clavaria* species have clublike,
4-spored basidia with straight stalks.

743 Crested Coral
Clavulina cristata (Fr.) Schroet.
Clavariaceae, Aphyllophorales

Description: *Branched, whitish coral with toothed tips.*
Mushroom: 1″ (2.5 cm) wide, 1–3″
(2.5–7.5 cm) high; many-branched,
crested, with toothlike, jagged tips;
whitish, becoming yellowish or ochre.
Stalk (when present): ¼–1¼″ (0.5–3
cm) long; branching below and above.
Spores: 7–11 × 6.5–10 μ; nearly
round, smooth, colorless; basidia 2-
spored. Spore print white.

Edibility: Edible.

Season: June–October.

Habitat: On the ground in woods.

Range: Throughout North America.

Look-alikes: *C. cinerea* is ash-gray, sometimes tinged

with purple. *C. rugosa* has few branches.

Comments: Formerly known as *Clavaria cristata.*
Both the Crested Coral and the similar
Gray Coral (*Clavulina cinerea*) are often
parasitized by another fungus that
makes the species difficult to
distinguish.

---

689 Strap-shaped Coral
Clavariadelphus ligula (Fr.) Donk
Clavariaceae, Aphyllophorales

Description: *Small, slender, yellowish, clublike coral.*
Mushroom: ¼–⅝″ (0.5–1.5 cm) wide,
1–4″ (2.5–10 cm) high; elongate to
club-shaped, rarely forked, often
flattened or spoonlike; yellowish to
ochre, sometimes reddish. Flesh white.
Stalk: indistinct; white-hairy at base.
Spores: 8–18 × 3–6 μ; narrowly
elliptical, smooth, colorless. Spore
print white to pale yellowish.

Season: July–November.

Habitat: On conifer needles; reported on humus
in deciduous woods.

Range: Throughout N. North America; also in
California.

Look-alikes: *C. sachalinensis* has larger, buff-colored
spores. *C. pistillaris** is much larger,
more club-shaped, and ochre to orange
or reddish-brown. *Microglossum rufum**
has saclike spore-bearing structures
(asci) rather than basidia.

Comments: Formerly known as *Clavaria ligula,* this
yellowish coral resembles the petal-like
rays of the dandelion.

---

688 Pestle-shaped Coral
Clavariadelphus pistillaris (Fr.) Donk
Clavariaceae, Aphyllophorales

Description: *Yellow to ochre, clublike coral with bitter
taste.*
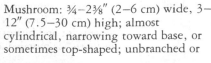
Mushroom: ¾–2⅜″ (2–6 cm) wide, 3–
12″ (7.5–30 cm) high; almost
cylindrical, narrowing toward base, or
sometimes top-shaped; unbranched or

rarely forked; yellowish, ochre, or
reddish-brown, darker on bruising,
often longitudinally wrinkled. Flesh
firm to spongy, white, darkening on
bruising. Taste typically bitter.
Stalk: indistinct; white-hairy at base.
Spores: $11-16 \times 6-10 \mu$; elliptical,
smooth, colorless. Spore print white to
yellow-tinged.

Edibility: Edible.
Season: July–October.
Habitat: On the ground in woods.
Range: E. North America; also West Coast.
Look-alikes: *C. truncatus** is shorter, much broader
across top, and not bitter. *C. mucronatus*
has sharp, spinelike central tip.
Macrotyphula fistulosa is very long and
slender.
Comments: Formerly known as *Clavaria pistillaris,*
this coral is often very large and
clublike. Though edible, it is
unpalatable. It is one of the few coral
fungi that turn brown when bruised.

---

439 Flat-topped Coral
Clavariadelphus truncatus (Quél.) Donk
Clavariaceae, Aphyllophorales

Description: *Yellow to ochre, clublike coral, often broad
and flattened at top.*

Mushroom: 1–3″ (2.5–7.5 cm) wide or
wider at top, 2–6″ (5–15 cm) high;
flat-topped and narrowing toward base;
yellowish-ochre to orange, wrinkled.
Flesh whitish to ochre, darker on
bruising, firm to spongy. Taste sweet
to bland.
Stalk: indistinct; white-hairy at base.
Spores: $9-13 \times 5-8 \mu$; elliptical,
smooth, colorless. Spore print ochre.
Edibility: Good.
Season: August–October.
Habitat: On the ground in coniferous woods.
Range: Throughout North America.
Look-alikes: *C. borealis,* common in coniferous
woods of the Pacific NW., has a white
spore print and bittersweet taste. *C.
pistillaris** is rounded and fertile on top,

occurs in deciduous woods, and tastes bitter. *Cantharellus lateritius** has fragrant, fruity odor, fertile surface with slight wrinkles, and shallow, gill-like ridges.

Comments: Formerly known as *Clavaria truncata*, this edible coral is especially common in the cold, wet northern coniferous woods of the Rocky Mountains, the Smokies, and California. Unlike similar club-shaped corals, it is truncated and sterile on the top.

748 Cotton-base Coral
Lentaria byssiseda (Fr.) Cor.
Clavariaceae, Aphyllophorales

Description: *Creamy to pinkish-tan, short-stalked, branched, pliant-tough coral.*
Mushroom: ⅜–2″ (1–5 cm) wide, ¾–2⅜″ (2–6 cm) high; usually clustered, with few to many bunched branches; tips of branches short, slender; cream, yellowish-white, or tan, often pink-tinged, becoming brownish with age. Flesh pliant, tough. Taste somewhat bitter.
Stalk (when present): ⅛–⅝″ (0.3–1.5 cm) long, ¹⁄₃₂–⅛″ (1–3 mm) thick; short, whitish, scaly, *rising out of creamy white, feltlike, hairy mycelial patch;* runners (rhizomorphs) often present.
Spores: 10–18 × 3–6 μ; cylindrical to oblong, smooth, colorless to buff. Spore print white to pale ochre.
Season: August–November.
Habitat: On twigs, leaves, cones, and rotten wood of deciduous and coniferous trees.
Range: Throughout North America.
Look-alikes: *Ramaria stricta** has yellowish branch tips and ornamented spores. *Ramaria** species occur mostly on the ground and are more colorful; most other differences are microscopic.
Comments: Other *Lentaria* species have been described in North America, but these may prove to be the same as, or variants of, the Cotton-base Coral.

690 **White Green-algae Coral**
Multiclavula mucida (Fr.) Pet.
Clavariaceae, Aphyllophorales

Description: *Many small, spindle-shaped, white clubs on green, algae-covered, rotten logs.*
Mushroom: $1/32-1/16''$ (1–1.5 mm) wide, $1/4-5/8''$ (0.5–1.5 cm) high; simple or rarely forked, slender, cylindrical to spindle-shaped, narrowing toward base; white to sometimes yellowish or cream. Flesh tough, waxy. Taste woody.
Stalk: indistinct.
Spores: $4.5-7.5 \times 1.8-3$ μ; narrowly elliptical, smooth, colorless. Spore print white.
Season: August–September.
Habitat: On rotting wood, usually covered with green algae; sometimes also on algae-covered soil.
Range: N. North America, especially in Northeast.
Comments: Formerly known as *Clavaria mucida*, this small, white coral is usually found in great numbers on algae-covered logs in wet woods. Too small to be of interest as an edible, this species blends in well with its background and is frequently overlooked by mushroom hunters.

750 **Light Red Coral**
Ramaria araiospora Marr & Stuntz
Clavariaceae, Aphyllophorales

Description: *Branching, red coral with yellowing tips.*
Mushroom: $1-3''$ (2.5–7.5 cm) wide, $2-5''$ (5–12.5 cm) high; many-branched, often from base; red to light red tips becoming yellow to orange, rising from white to yellowish-white base. Flesh fleshy-fibrous.
Stalk (when present): $3/4-1\frac{1}{4}''$ (2–3 cm) long, $3/8-5/8''$ (1–1.5 cm) thick; single.
Spores: $8-13 \times 3-4.5$ μ; cylindrical, finely warted, colorless. Spore print yellowish.
Edibility: Good.

Season: September–November.
Habitat: On the ground under western hemlock.
Range: Pacific NW. to California.
Look-alikes: *R. subbotrytis* is coral-pink, fading to creamy ochre. The poisonous *R. formosa** has peach to salmon-pink branches with yellowish tips.
Comments: This species is gathered and eaten in great quantities. The variety *rubella* is magenta and lacks the yellow branch tips.

749 Clustered Coral
Ramaria botrytis (Fr.) Rick.
Clavariaceae, Aphyllophorales

Description: *Large, pinkish, cauliflowerlike, with many branches.*
Mushroom: 2–5″ (5–12.5 cm) wide, 3–6″ (7.5–15 cm) high; densely and compactly branched; branches white, with reddish or purplish tips. Flesh white.
Stalk: ¾–2⅜″ (2–6 cm) long, ¾–1¼″ (2–3 cm) thick; single; white; sometimes cone-shaped.
Spores: 11–17 × 4–6 μ; almost cylindrical, with longitudinal lines, ochre to buff. Spore print ochre to pale orange.
Edibility: Choice.
Season: August–October.
Habitat: On the ground under conifers, especially spruce and fir.
Range: New England to North Carolina, west to S. California and British Columbia.
Look-alikes: *R. strasseri* is a northern coral with a spicy, sweet odor and is usually white, bruising darker, sometimes with pinkish tips.
Comments: Formerly known as *Clavaria botrytis*, the Clustered Coral is now the type species of *Ramaria*, a genus of many-branched corals that are nonamyloid and have a colored spore print, ornamented spores, and a fertile layer (upper branches) that turns green in 10 percent ferric sulfate. Species of *Ramaria* are best

differentiated by chemical and
microscopic characteristics. Many
mushroom lovers consider this species
to be the most flavorful large coral.

⊗ 746 **Yellow-tipped Coral**
Ramaria formosa (Fr.) Quél.
Clavariaceae, Aphyllophorales

Description: *Large, many-branched, orange-red coral
with yellowish tips; bruising brownish on
handling.*
Mushroom: 2–6″ (5–15 cm) wide, 3–
8″ (7.5–20 cm) high; many-branched
from single, massive base; pinkish-
orange to light red with light yellow
tips, bruising brownish on handling.
Flesh pinkish-orange to light red,
fibrous above, spongy below; brittle
when dry.
Stalk: indistinct; white-hairy at base.
Spores: 9–12 × 4.5–6 μ; elliptical,
warted, colorless. Spore print golden-
yellow.
Edibility: Poisonous.
Season: July–November.
Habitat: On the ground under conifers,
especially Douglas-fir and western
hemlock.
Range: Throughout N. North America.
Look-alikes: Species in the *R. aurea* complex do not
bruise on handling nor turn brown with
age.
Comments: Formerly known as *Clavaria formosa,*
this is one of the few coral mushrooms
reported to cause diarrhea or gastric
upset. It is best distinguished from
similar corals by the brown staining of
the mature flesh.

⊗ 745 **Jellied-base Coral**
Ramaria gelatinosa (Cok.) Cor. var.
oregonensis Marr & Stuntz
Clavariaceae, Aphyllophorales

Description: *Branching, orange to orange-brown,
stalkless coral with gelatinous base flesh.*

Mushroom: 2–5″ (5–12.5 cm) wide,
3–6″ (7.5–15 cm) high; many-
branched from compound gelatinous
base; orange to orange-brown,
sometimes aging with violet-gray
tones. Flesh translucent, gelatinous,
pale grayish-orange.
Spores: 7–10 × 4.5–6 μ; elliptical,
coarsely warted and ornamented,
colorless. Spore print golden-yellow.

Edibility: Poisonous.

Season: September–November.

Habitat: On the ground under western
hemlock.

Range: Pacific NW.

Comments: There are at least 2 varieties of this
species: var. *gelatinosa* has creamy
branches that become buff-pink to
pinkish-brown with age. Few other
species of *Ramaria* have a gelatinous
consistency; none should be confused
with the elastic-tough jelly fungi,
which also have very different basidia
and spores. None of the gelatinous
corals should be eaten, especially the
Jellied-base Coral, because of reported
gastric upsets.

733 Straight-branched Coral
Ramaria stricta (Fr.) Quél.
Clavariaceae, Aphyllophorales

Description: *Yellowish coral with compact, upright
branches.*
Mushroom: 1–3″ (2.5–7.5 cm) wide,
2–4″ (5–10 cm) high; many compact,
straight, parallel branches; light yellow
above, bruising light brown, grayish-
orange below. Flesh brownish-white,
darkening on bruising. Odor sometimes
unpleasant; taste sometimes bitter.
Stalk: $\frac{1}{16}$–¾″ (0.15–2 cm) long, $\frac{1}{8}$–
⅝″ (0.3–1.5 cm) thick; sometimes
indistinct; velvety white at base.
Spores: 7–10 × 3.5–5.5 μ; elliptical,
minutely warted, colorless. Spore print
golden-yellow.

Season: July–October.

Habitat: On deciduous or coniferous wood.
Range: Throughout North America.
Look-alikes: The few species of coral that grow on wood must be differentiated microscopically.
Comments: Formerly known as *Clavaria stricta,* the Straight-branched Coral has 2 recognizable varieties: var. *stricta* is yellowish, very compact, and grows on coniferous wood; var. *concolor* is grayish-orange, more openly branched, and grows on deciduous wood.

---

742 White Coral
Ramariopsis kunzei (Fr.) Donk
Clavariaceae, Aphyllophorales

Description: *White to pinkish, branched coral with scruffy-hairy stalk base.*
Mushroom: 1–3″ (2.5–7.5 cm) wide, 1–4″ (2.5–10 cm) high; usually many-branched; white to pinkish. Flesh pliant to fragile.
Stalk: ¼–1″ (0.5–2.5 cm) long, ⅛–¼″ (3–5 mm) thick; indistinct or short; scruffy-hairy.
Spores: 3–5.5 × 2.3–4.5 μ; broadly elliptical to nearly round, with minute spines; colorless. Spore print white.
Edibility: Edible.
Season: July–October.
Habitat: On the ground or, rarely, on decaying wood in forests.
Range: N. North America; New England to North Carolina; Washington to S. California.
Look-alikes: *Ramaria** have colored spores and a fertile layer that stains greenish in ferric sulfate. *Clavaria** have smooth spores. *Clavulina** have 2-spored, smooth, cylindrical basidia. Species of *Clavulinopsis** are brightly colored, with smooth spores. Other common species of *Ramariopsis* are brightly colored.
Comments: Formerly known as *Clavaria kunzei.* This species is best identified in the field by its branched structure, white to pinkish-tinged color, and fragile flesh.

724 Bladder Stalks
Physalacria inflata (Schw.) Fr.
Clavariaceae, Aphyllophorales

Description: *Small, whitish bladders with stalks.*
Mushroom: ¹⁄₃₂–³⁄₈" (0.1–1 cm) wide,
³⁄₈–1" (1–2.5 cm) high; rounded to
flattened bladderlike head, often
somewhat wavy or folded; whitish or
yellowish, smooth; bladder hollow,
waxy-firm. Flesh membranous.
Stalk: ⅛–¾" (0.3–2 cm) long, ¹⁄₆₄"
(0.5 mm) thick; very thin, firm, finely
hairy.
Spores: 4–6 × 2–3.5 μ; elliptical,
smooth, colorless. Spore print white.
Season: August–September.
Habitat: Clustered and usually numerous on
rotten wood and dead leaves.
Range: E. North America west to Wisconsin
and Nebraska.
Comments: This small, sometimes common late-
summer mushroom of the northern
woods is distinctive and easily
recognized.

756 Eastern Cauliflower Mushroom
Sparassis crispa Wulf. ex Fr.
Clavariaceae, Aphyllophorales

Description: *Large, stalkless, rounded mushroom with
finely wavy, white to pale yellowish, flat-
edged, leaflike branches.*
Mushroom: 6–12" (15–30 cm) wide,
6–10" (15–25 cm) high; large,
somewhat rounded, rosettelike,
composed of variously curled, folded,
lobed, and flattened branches, usually
wavy; white to creamy yellow, cordlike
root at base. Flesh white, fleshy-tough.
Spores: 4–7 × 3–4 μ; oval, smooth,
colorless. Spore print white.
Edibility: Choice.
Season: July–October.
Habitat: Open oak and sandy oak-pine woods.
Range: E. North America.
Look-alikes: *S. radicata** has very long rooting stalk
and thinner branches.

Comments: The Eastern Cauliflower Mushroom can
be common in wet summers, but is rare
in dry weather. When fresh, it is white
or whitish, has a crisp texture, and is
usually large enough to provide several
meals. Nothing known to be poisonous
even remotely resembles this
Elizabethan ruff of a mushroom. Also
known as *S. herbstii.*

---

755 **Rooting Cauliflower Mushroom**
Sparassis radicata Weir
Clavariaceae, Aphyllophorales

Description: *Large, rounded cauliflower- or lettucelike
mass with crimped, white to pale yellowish,
leaflike branches and long, rooted stalk.*
Mushroom: 6–14" (15–35.5 cm) wide,
10–20" (25–50 cm) high; large,
somewhat rounded, composed of series
of flattened, crimped, leaflike branches;
white to yellowish. Flesh white, fleshy-
tough.
Stalk: 3–5" (7.5–12.5 cm) long, 1–2"
(2.5–5) thick, narrowing toward base;
deeply rooted; perennial; black.
Spores: 5–6.5 × 3–3.5 μ; oval,
smooth, colorless. Spore print white.
Edibility: Choice.
Season: September–November.
Habitat: On the ground under conifers and on
conifer roots and stumps.
Range: West of Rocky Mountains from British
Columbia to N. California.
Look-alikes: *S. crispa** is an eastern species with
thicker, shorter, much more flattened
branches, and cordlike root at base.
Comments: This is an excellent edible mushroom
that sometimes looks more like a head
of lettuce than a cauliflower. While it
normally weighs 2–3 pounds, it can
weigh as much as 50 pounds. It should
be cooked slowly or stewed to make it
tender and more digestible.

437 Common Fiber Vase
Thelephora terrestris Fr.
Clavariaceae, Aphyllophorales

Description: *Circular to vase-shaped, single to overlapping, brown, fibrous caps with hairy, torn margin and wrinkled undersurface.*
Mushroom: 1–2″ (2.5–5 cm) wide and high; solitary or overlapping caps, often in rosettes up to 4¾″ (12 cm) wide; upper surface hairy, fibrous to scaly, light to dark brown, turning blackish with age, with whitish, even margin becoming torn and hairy; undersurface brownish, smooth, with radiating wrinkles, or with minute, nipplelike projections. Flesh leathery, thin. Odor like moldy earth.
Stalk (when present): very short.
Spores: 8–12 × 6–9 μ; angularly elliptical and lobed, sparsely spiny to nearly smooth, purplish. Spore print purplish to purplish-brown.

Season: Year-round; mostly July–December.

Habitat: On the ground, especially in sandy soil, under pines; also on roots, stumps, and seedlings.

Range: Throughout N. North America; south along the Appalachians to Georgia.

Look-alikes: *T. griseozonata,* possibly just a variant, has concentrically banded cap. *T. intybacea* has white, hairy margin. Species of *Craterellus** are fleshy. Stalked stereums, such as *Cotylidia diaphana**, have smooth spores and white spore print.

Comments: This fungus can be quite destructive, strangling tree seedlings.

---

442 Vase Thelephore
Thelephora vialis Schw.
Clavariaceae, Aphyllophorales

Description: *Large, leathery, dirty white to grayish-purple, vase-shaped mushroom.*
Mushroom: 1–6″ (2.5–15 cm) wide, 1–4″ (2.5–10 cm) high; clustered,

spoon- to fan-shaped, or fused and vase- to cup-shaped, arising from common base; radially lined to slightly scaly, hairy toward base, becoming smooth; dirty white or yellowish to dull brown or grayish-purple. Flesh leathery, thick. Odor sharp to fetid on drying.

Fertile Surface: smooth, becoming slightly wrinkled, with minute, nipplelike projections; pale yellow, becoming gray-brown.

Stalk: ¼–2" (0.5–5 cm) long and thick; whitish to gray, somewhat hairy.

Spores: 4.5–8 × 4.5–6.5 μ; angular and lobed, slightly spiny, olive-buff.

Season: August–October.

Habitat: On the ground in deciduous woods, especially oak.

Range: Vermont south to South Carolina, west to Illinois.

Comments: *Stereum* species are typically shelflike. *Thelephora* species may be erect and somewhat vaselike, or may encrust twigs, appearing shelflike with a torn, fringed margin.

DRY ROT FAMILY
(Coniophoraceae)

This is a small family of fleshy to crustlike, spreading mushrooms that grow on wood. Species in this family have brown, smooth, thick-walled spores; the inner wall is cyanophilous, or has a strong blue reaction to the chemical reagent cotton-blue. Dry Rot and Wet Rot, the 2 species included here, are among the best-known members of this family because they are severely destructive to building wood. They produce rhizomorphs—runners of fungal tissue from the mushroom base —that can grow to a length of several yards; they act as conduits for moisture.

592 Wet Rot
Coniophora puteana (Schum. ex Fr.) Kar.
Coniophoraceae, Aphyllophorales

Description: *Spreading, tawny to olive-brownish, smooth to bumpy crust with cottony-white margin and projecting runners; on logs.*
Mushroom: 1⅛–16″ (4–40 cm) wide, spreading sheetlike on wood; circular to elongated, with sterile margin and fertile central zone; brownish, becoming olive-yellow, and yellowish to cottony-white along margin; smooth to uneven, wavy or bumpy, with low, broad, dome-shaped warts and projecting runners (rhizomorphs). Flesh 0.3–1 mm thick.
Fertile Surface: irregular, wrinkled, and wavy to bumpy; yellowish, becoming olive to bronze.
Spores: 10–14 × 6–7 μ; elliptical, smooth, olive-brown; thick-walled, with cyanophilous inner wall.

Season: July–February; year-round where winters are mild.
Habitat: On coniferous and deciduous logs; also on standing timbers in buildings, especially in cellars and wet areas.
Range: Quebec to Washington, D.C., west to British Columbia and California.
Look-alikes: Other species of *Coniophora* can be distinguished only with a microscope.
Comments: Also known as *C. cerebella*, Wet Rot causes discoloration and cracking of wood. It requires dampness to grow.

593 Dry Rot
Serpula lacrimans (Fr.) Schroet.
Coniophoraceae, Aphyllophorales

Description: *Large, spongy-fleshy, spreading crust with ochre, honeycombed folds, hairy, white margin, and projecting runners.*
Mushroom: 3¼–6″ (8–15 cm) wide, spreading up to 20″ (50 cm) or more long; flat, fanlike, sometimes projecting and becoming bracketlike and somewhat stub-stalked on vertical

surfaces; with inflated, hairy, white, sterile margin; grayish-white runners (rhizomorphs) up to 4″ (10 cm) long; pitted, honeycombed folds yellowish- or reddish-brown, drying rusty-brown. Flesh 0.2–1.5 cm thick; yellowish, composed mostly of fertile surface. Odor becoming musty, unpleasant. Fertile Surface: covering most of mushroom with large, wavy folds; porelike or elongated and somewhat toothed on vertical surfaces. Pores 1–3 mm wide, up to 1 mm deep; irregular, shallow, rust-yellow to brownish. Stalk: absent, but stublike 'stalks' present on some bracketlike projections. Spores: $8–12.5 \times 4.5–6 \mu$; smooth, elliptical to somewhat flattened on 1 side, yellowish; thick-walled, with cyanophilous inner wall. Spore print brownish.

Season: June–January; year-round indoors.

Habitat: On coniferous logs in woods; under floorboards in buildings.

Range: Canada south to Connecticut, west to Arizona.

Look-alikes: *S. himantioides* is leathery and entirely crustlike. *Merulius** species have colorless spores. *Phlebia** do not develop a porelike surface.

Comments: Formerly known as *Merulius lacrimans,* the infamous Dry Rot is much more frequently encountered indoors than out. It is more destructive than Wet Rot because it can attack dry wood: its runners may extend several meters in search of moisture and serve the fungal body as a water-transport mechanism. It thrives in poorly ventilated areas.

CRUST FUNGUS FAMILY
(Corticiaceae)

Most of the crust fungi, which look like thin, spreading, leathery sheets—or even paint smears on wood—belong here; in addition, a few tumorlike fungi are included. Their fertile surfaces may be smooth or warted, toothed or waxy, or even somewhat gelatinous and translucent. More than 60 genera are now accepted in this broadly defined family. Most of the species described here are parasites or they are saprophytes, which disintegrate plants already dead; none is known to be edible. Those crust fungi with pores or some kind of tubular surface are included in the family Polyporaceae.

---

571 Hophornbeam Disc

Aleurodiscus oakesii (Berk. & Curt.) Hoeh. & Litsch.
Corticiaceae, Aphyllophorales

Description: *Disc-shaped, fleshy to leathery, attached at center; white-hairy underneath, concave and brownish above.*
Mushroom: $\frac{1}{32}$–$\frac{1}{16}$" (1–1.5 mm) wide; disc-shaped; scattered, becoming fused into masses 1 × 2 cm, with free, elevated, incurved margin; fleshy to leathery, with concave, brownish upper surface; undersurface hairy, white.
Fertile Surface: concave; buff-pink, drying brownish and somewhat powdery.
Spores: 18–21 × 12–13 μ; elliptical to ovate, slightly warted or roughened, colorless, amyloid. Cystidia cylindrical, with short, lateral branches at tip and sides. Spore print white.

Season: July–December; may overwinter in mild climates.

Habitat: On bark of dead hophornbeams and other deciduous wood.

Range: Canada to Alabama, west to Missouri.

Look-alikes: *A. amorphus* grows on coniferous wood. *Dendrothele candida* has nonamyloid spores and typically grows on the bark of living oaks. Cup fungi in genus *Cenagium* are usually not leathery and produce their spores in asci.

Comments: This species is readily identified by its shape, texture, and host tree.

539 Silver Leaf Fungus
Chondrostereum purpureum (Fr.) Pouz.
Corticiaceae, Aphyllophorales

Description: *Leathery and sheetlike, or with projecting overlapping caps, ochre to buff and wrinkled above; underside smooth, purplish to brownish, and waxy.*
Cap: ¼–¾" (0.5–2 cm) wide, sometimes in masses 4–20" (10–50 cm) long; sometimes overlapping, sometimes crimped or lobed; light ochre-buff to cinnamon-buff; hairy to hairless and wrinkled.
Fertile Surface: waxy and smooth to minutely powdery; light purple-drab to dark wine-buff.
Spores: 5–6.5 × 2–3 μ; broadly cylindrical, smooth, colorless. Spore print whitish.

Season: June–April; may persist for several years.

Habitat: On apple, plum, and other deciduous trees; also on poplar stumps and logs.

Range: Canada to Delaware, west to British Columbia and Oregon.

Comments: Also known as *Stereum purpureum*, this fungus causes the silver leaf disease of apple and plum trees. It enters the bark through wounds, producing the silvery appearance of the leaves; the disease kills branches, and then the whole tree.

564 Collybia Jelly
Christiansenia mycetophila (Pk.) Ginns & Sunh.
Corticiaceae, Aphyllophorales

Description: *Gelatinous and waxy, pale cream-yellow, brainlike patches, often in massed, tumorlike clusters; on Oak-loving Collybia (Collybia dryophila).*
Mushroom: to ¹⁄₆₄" (0.1–0.5 mm) thick; glassy, gelatinous, cream-yellow film covering round, wartlike, or brain-shaped to somewhat compressed gelatinous growths ¼–2" (0.5–5 cm) wide or larger. Growths, as many as

35, sometimes fused, are composed of tissue of host mushroom; they can occur on its cap, gills, and stalk.
Spores: $6-9 \times 1.5-2.3$ μ; narrowly elliptical to cylindrical, colorless; produces 2 types of conidiospores; basidial tips with oily contents.

Season: Summer–fall.

Habitat: On Oak-loving Collybia.

Range: Throughout North America.

Look-alikes: Other species of *Christiansenia* can be distinguished only with a microscope.

Comments: Until recently, this was thought to be a jelly fungus and was known as *Tremella mycetophila*. It is now believed that this fungus induces the tumorlike tissue growth of the Oak-loving Collybia and uses it as the base on which to develop.

---

576 Buff Crust
Corticium bombycinum (Som.) Bres.
Corticiaceae, Aphyllophorales

Description: *Irregular and sheetlike, but split into pieces; spongy-soft, white to pinkish-buff.*
Mushroom: ¾–1¼" (2–3 cm) wide, 1¼–4" (3–10 cm) long; spreading and irregular; white, becoming pinkish- or cream-buff, margin cottony-white to hairy; smooth or cracking into pieces, becoming uneven to rough. Flesh spongy-soft, thick, white.
Fertile Surface: smooth to rough, buff.
Spores: $6-10 \times 5-6$ μ; elliptical, smooth, colorless.

Season: July–March; sometimes overwinters.

Habitat: On bark of living and dead willow and alder; also reported on birch, maple, linden, poplar, and pine.

Range: Canada to New York, west to Texas, Arizona, and Washington.

Comments: This is one of the more common species of *Corticium*, now a residual genus of crust fungi best differentiated from similar genera by microscopic features that it lacks: colored spores (cystidia) and bristles (setae).

595   Velvet Blue Spread
Pulcherricium caeruleum (Pers.) Parm.
Corticiaceae, Aphyllophorales

Description: *Roundish, spreading, velvety blue patches.*
Mushroom: 1¼–4" (3–10 cm) wide;
crustlike, round or nearly so; usually
many massed together, separable from
wood when fresh; indigo or darker,
velvety, often somewhat faded toward
margin. Flesh membranous, soft, thick.
Fertile Surface: velvety, blue.
Spores: $6-10 \times 4-5 \mu$; elliptical,
smooth, colorless.

Season: August–November; sometimes
overwinters.

Habitat: On underside of decaying oak branches;
also on other deciduous wood.

Range: North Carolina to Florida, west to
Arkansas and Illinois.

Comments: Formerly known as *Corticium caeruleum*,
this species is common in the southern
Appalachians. It resembles blue velvet
upholstery on an oak branch.

---

573   Two-tone Parchment
Laxitextum bicolor (Pers. ex Fr.) Lentz
Corticiaceae, Aphyllophorales

Description: *Spongy, pliant, shell-like, and overlapping
or sheetlike with projecting caps; coffee-
brown and feltlike above, white below.*
Mushroom: ¾–2" (2–5 cm) wide, or
spreading in sheets up to 7 × 15 cm;
coffee-brown; feltlike to wrinkled
above, membranous, and indistinctly
hairy; becoming smooth or
concentrically furrowed; undersurface
white. Flesh spongy.
Fertile Surface: smooth to powdery;
white, but drying light buff to pale
pinkish buff.
Spores: $3.5-4.5 \times 2-3 \mu$; elliptical to
oval, smooth, colorless. Gloeocystidia
$20-60 \times 5-7 \mu$. Spore print whitish.

Season: April–December; may overwinter.

Habitat: On rotting deciduous wood; common
on alder.

Range: Canada to Florida; Washington and
Oregon; Indiana, Missouri, and Texas.
Comments: Formerly known as *Stereum fuscum.*

---

523 Coral-pink Merulius
Merulius incarnatus Schw.
Corticiaceae, Aphyllophorales

Description: *Vivid coral-pink, leathery, stalkless,*
semicircular, overlapping caps with pinkish-
buff, porelike surface below.
Mushroom: ¾–1⅝" (2–4 cm) wide,
1⅝–3¼" (4–8 cm) long; semicircular,
with wavy margin, mostly overlapping;
coral-pink, drying to salmon-buff;
finely hairy. Flesh 2–4 mm thick, off-
white to buff, spongy to leathery.
Fertile Surface: porelike, formed by
network of radiating, branched folds;
pinkish-ochre to salmon-buff.
Spores: 4–4.5 × 2–2.5 μ; elliptical,
smooth, colorless. Spore print white.
Season: September–October.
Habitat: On logs and stumps of white oak,
beech, birch, and maple.
Range: New York to North Carolina and
Louisiana, west to Ohio and Missouri.
Comments: The Coral-pink Merulius is common in
the Mississippi Valley and widely
distributed through eastern North
America.

---

586 Trembling Merulius
Merulius tremellosus Schrad. ex Fr.
Corticiaceae, Aphyllophorales

Description: *Thick and sheetlike, with projecting,*
woolly, white cap; gelatinous below, with
orange to pinkish, porelike surface.
Mushroom: 1–3" (2.5–7.5 cm) wide;
spreading sheetlike, often laterally
fused, margins becoming free and
upturned, projecting up to ⅝" (1.5
cm), sometimes overlapping; white,
hairy-woolly. Flesh 0.5–2 mm thick,
white, fleshy to waxy and gelatinous,
drying horny.

Fertile Surface: porelike, formed by network of radiating folds with crossveins. Pores 1−1.5 mm wide; angular and becoming elongated, about 0.5 mm deep; orange-buff to pinkish, somewhat gelatinous-translucent.
Spores: 3−3.5 × 0.5−1 μ; sausage-shaped, smooth, colorless. Spore print white.

Season: July−January.

Habitat: On birch, maple, and other deciduous trees; rarely on coniferous wood.

Range: Throughout North America.

Look-alikes: *M. corium* has smaller pores, about 3 per mm. *M. ambiguus* grows on conifers. *Serpula** species have dark spores. *Phlebia** species lack porelike surface.

Comments: Although the fertile surface is always porelike, *Merulius* is not a polypore genus: its pores are not the mouths of tubes, and its hymenium, or fertile surface, is continuous, covering both ridges and pits, and not a tubal lining.

---

583 Conifer Parchment
Peniophora gigantea (Fr.) Mass.
Corticiaceae, Aphyllophorales

Description: *Large, white to pinkish-buff, waxy, parchmentlike growth, curling when dry.*
Mushroom: 1¼−12″ (3−30 cm) wide; broadly attached and spreading; colorless to white to pinkish-buff, waxy, swollen when moist, separable from host; when dry, horny and parchmentlike, with fibrous, radiating, white margin sometimes becoming free and curling away from wood.
Fertile Surface: pale pinkish- to olive-buff.
Spores: 4.5−5 × 2.5−3 μ; elliptical, smooth, colorless Cystidia encrusted, 40−50 × 8−12 μ.

Season: June−January.

Habitat: On bark and wood of dead conifers, especially pine, fir, and hemlock.

Range: Canada to Texas, Mexico, west to Pacific States.

Comments: Also known as *Phlebia gigantea*, this species is used to control the Conifer-base Polypore (*Heterobasidion annosum*), a disease-carrying fungus that infects conifers. In pine plantations, saw blades are painted with a solution of spores of the Conifer Parchment, or the spores are inoculated into pine stumps, so that the stump will be colonized by this species and not by the Conifer-base Polypore.

---

565 Red Tree Brain
Peniophora rufa (Fr.) Boid.
Corticiaceae, Aphyllophorales

Description: *Small, red, waxy, cartilaginous, stalkless, wartlike, and coarsely wrinkled.*
Mushroom: ¼–⅜" (0.5–1 cm) wide; single or laterally fused; smooth, with free margin; broadly centrally attached. Flesh 0.7–2 mm thick, waxy.
Fertile Surface: convex to flat, becoming coarsely wrinkled; red to purple-brown, often with grayish bloom.
Spores: 6–8.5 × 1.5–2 μ; cylindrical, curved, smooth, colorless, with encrusted, colorless hyphae. Gloeocystidia numerous. Spore print white.
Season: March–December; sometimes overwinters.
Habitat: On dead twigs and branches of trembling aspen and other poplars.
Range: Canada to Massachusetts, west to North Dakota and Colorado.
Comments: Also known as *Stereum rufum* and *Cryptochaete rufa*, this species bursts out from the bark of fallen limbs and logs.

---

558 Spreading Yellow Tooth
Phanerochaete chrysorhiza (Torr.) Gilb.
Corticiaceae, Aphyllophorales

Description: *Sheetlike, bright orange-yellow, toothed surface with conspicuously spreading,*

branched, cordlike, orange strands.
Mushroom: ¾–2⅜" (2–6 cm) wide or larger; sheetlike and flattened against log, sometimes covering log's entire undersurface, ochre to salmon or bright orange, with small fibers; margin white, becoming yellowish; runners (rhizomorphs) 10–20 cm long, cordlike, bright orange, branching, growing from underside and penetrating log. Flesh membranous to leathery.
Fertile Surface: composed of teeth 1–4 mm long, slender, roundish, and crowded to fused; yellow to bright orange.
Spores: 3.5–4.5 × 2–2.5 μ; elliptical, smooth, colorless.

Season: June–January.

Habitat: On undersides of decayed deciduous logs.

Range: Maine south to North Carolina, west to Minnesota.

Comments: Also known as *Hydnum chryscomum, Mycoacia fragilissima,* and *Oxydontia fragilissima.* The long, bright orange runners, or rhizomorphs, make this species easy to recognize.

587 Radiating Phlebia
Phlebia radiata Fr.
Corticiaceae, Aphyllophorales

Description: *Gelatinous, pinkish-orange to red mushroom growing in crowded, radiating folds or wrinkles, with hairy margin.*
Mushroom: ⅜–1⅝" (1–4 cm) wide; fused in large sheets, but individual fungi discrete with observable limits; circular, raised with ridges and warts, sometimes fusing in rows (not becoming porelike) radiating from point of attachment; pinkish to orange-red or coral-red, with hairy margin. Flesh 0.5–1 mm thick, waxy-soft, somewhat gelatinous.
Fertile Surface: covering entire fungus.
Spores: 3.5–4.5 × 1.5–2 μ; sausage-

shaped to elliptical, with sharp tip; smooth, colorless. Spore print whitish.

Season: August–November; may overwinter.

Habitat: On rotting wood of both deciduous and coniferous trees.

Range: Throughout North America.

Look-alikes: *Merulius** species have porelike fertile surface.

Comments: The color, form, and texture of this species make it a distinctive mushroom.

580 Zoned Phlebia
Punctularia strigoso-zonata (Schw.) Tal.
Corticiaceae, Aphyllophorales

Description: *Leathery-waxy, hairy, almost circular, stalkless, flat, reddish-brown caps, often fused, with multicolored concentric zones.* Mushroom: ¼–1¼" (0.5–3 cm) wide, often laterally fused and up to 6" (15 cm) long; spreading, flat to upturned, almost circular; chestnut, becoming grayish; coarsely hairy, developing concentric furrows and multicolored zones. Flesh 0.2–0.5 mm thick, glistening, sometimes gelatinous; reddish-brown to purplish-black.
Fertile Surface: smooth at first, soon developing radially elongated, crowded wrinkles and knobs; chestnut, becoming black and polished.
Spores: 6–8 × 3.5–4 μ; elliptical, smooth, colorless. Spore print white.

Season: Year-round.

Habitat: On branches and dead trees of poplar, beech, and oak.

Range: Throughout N. North America.

Look-alikes: *Phlebia radiata** has raised radial ridges, not concentrically furrowed zones. Most *Stereum** species have smooth undersurface.

Comments: Formerly placed in the genera *Stereum, Phlebia,* and *Phaeophlebia,* this species is now even believed by some to belong to a distinct family, the Punctulariaceae. Although its fertile surface, as in *Phlebia,* covers a layer of folds, it also covers wartlike knobs. Unlike *Phlebia,* these folds and knobs are cushions separated by narrow, sterile fissures filled with an amorphous mineral matter.

TOOTH FUNGUS FAMILY
(Hydnaceae)

Tooth fungi produce their spores on teeth instead of gills or in tubes. They differ from coral fungi in having spinelike teeth that are geotropic, hanging down toward the ground, and from toothlike polypores in having conical teeth instead of elongated, flattened, torn tubes. Some tooth fungi grow on the ground, some on trees; many species seem to prefer late-summer and fall coniferous woods. A few tooth fungi, such as the Sweet Tooth, are highly prized edibles; while others, like the Bear's Head Tooth, are edible though occasionally bitter. Although some family members are too bitter or tough to be palatable, none is known to be poisonous. At one time, all tooth fungi were included in the genus *Hydnum,* but now more than a dozen genera are recognized, based on a combination of field characteristics and microscopic features. Some tooth fungi that have been included here are often placed in different families.

373 Pinecone Tooth
Auriscalpium vulgare S.F.G.
Hydnaceae, Aphyllophorales

Description: *Small, dark, long-stalked; on fallen cones.*
Cap: ⅜–1⅝" (1–4 cm) wide; kidney-shaped, laterally attached to stalk; dark brown, with dense, dark brown hairs Flesh 1 mm thick; firm, light brown.
Spines: 1–3 mm long; needlelike, dense, very fine; whitish, darkening to brown.
Stalk: 1–3" (2.5–7.5 cm) long, ¹⁄₆₄–

$\frac{1}{16}''$ (0.5–1.5 mm) thick; off-center, slender, somewhat rigid; dark brown, densely hairy.

Spores: 4.8–5.6 × 4–5.2 μ; almost round to short and oval, minutely spiny at maturity, colorless, amyloid. Spore print white.

Season: August–November.

Habitat: Single to several, on fallen rotting pinecones; also reported on Douglas-fir cones.

Range: Across N. North America.

Comments: Formerly known as *Hydnum auriscalpium*. This mushroom is easily identified because it is found only on conifer cones. Sometimes placed in the family Auriscalpiaceae.

520 Northern Tooth
Climacodon septentrionale (Fr.) Kar.
Hydnaceae, Aphyllophorales

Description: *Large, overlapping, yellowish-white caps with toothed undersides; stalkless.*

Cap: 4–6″ (10–15 cm) wide, in clusters 6–12″ (15–30 cm) high, 1–2″ (2.5–5 cm) thick near base, thinning toward margin; shelflike, growing in overlapping, horizontal clusters from solid base, with caps progressively smaller toward top and bottom of cluster; off-white to buff, turning yellow-brown with age, densely hairy to rough. Flesh 2–4 cm thick; white, zoned, fibrous, tough, and elastic. Odor mild when fresh, becoming rank and hamlike on drying; taste mild when fresh, bitter with age.

Spines: 0.5–2 cm long; narrow, crowded, pliant; dull white, drying to yellowish; tips ragged.

Spores: 4–5.5 × 2.5–3 μ; elliptical, smooth, colorless; thick-walled, with encrusted tip. Spore print white.

Season: July–October.

Habitat: High up on living sugar maples; also reported on beeches.

Range: NE. North America.

Comments: Formerly known as *Hydnum septentrionale* and *Steccherinum septentrionale*. This large tooth mushroom looks like a polypore but for its teeth. It grows in wounds of living deciduous trees and rots the heartwood.

---

455   Sweet Tooth
Dentinum repandum (Fr.) S.F.G.
Hydnaceae, Aphyllophorales

Description: *Fleshy, orange to buff, irregularly shaped cap with pale teeth descending slightly off-center stalk.*
Cap: ⅝–6″ (1.5–15 cm) wide; convex to flat or slightly depressed, with inrolled margin becoming wavy; buff to orange, turning dark orange on bruising, smooth to slightly scruffy.
Flesh 1–2 cm thick; white, bruising ochre.
Spines: 0.3–1 cm long; fleshy, yellowish-white to yellowish, becoming ochre on bruising; irregularly descending stalk.
Stalk: 1–4″ (2.5–10 cm) long, ⅜–1¼″ (1–3 cm) thick; central to slightly off-center, dry, smooth, solid; whitish-yellow to pale yellow, bruising ochre.
Spores: 6.5–9 × 6.5–8 μ; nearly round, smooth, colorless. Spore print white.
Edibility: Choice.
Season: July–November.
Habitat: On the ground under deciduous and coniferous trees.
Range: Throughout North America.
Look-alikes: *D. umbilicatum,* found in wet coniferous woods, is smaller and thinner, with sunken cap center.
Comments: Formerly known as *Hydnum repandum,* this species has color variants ranging from white to orange. The typical form is an excellent edible mushroom with a mild flavor said to be like that of oysters.

532 Indian Paint Fungus
Echinodontium tinctorium Ell. & Ev.
Hydnaceae, Aphyllophorales

Description: *Large, woody, hoof-shaped, olive-black,
stalkless conk with teeth and bright orange
flesh; on western conifers.*
Mushroom: 4–10″ (10–25 cm) wide,
3–6″ (7.5–15 cm) deep and high; hoof-
shaped; often moss-covered, olive-black
with olive-brown margin; cracked and
rough, with soft, fine hairs. Flesh
woody-tough and zoned, bright orange,
cinnamon or rust-red.
Spines: 1–3 cm long; stout, gray, with
blunt ends that are brittle, dry, and
flattened.
Spores: 5.5–8 × 3.5–6 μ; elliptical,
minutely spiny, colorless. Spore print
white.

Season: Year-round.

Habitat: On coniferous wood, usually western
grand fir and western hemlock.

Range: Rocky Mountains, Pacific NW., and
NW. Canada.

Comments: Some mycologists place this in its own
family, the Echinodontiaceae. Its
common name refers to the fact that
Indians used this mushroom to make
red war paint; it is still used by
hobbyists as a yarn dye.

548 Bear's Head Tooth
Hericium coralloides (Fr.) S.F.G.
Hydnaceae, Aphyllophorales

Description: *Large, whitish mass, toothed in many small
tufts; on eastern deciduous trees.*

Mushroom: 6–12″ (15–30 cm) wide,
8–20″ (20–50 cm) high; with stout,
tufted stems branching repeatedly;
white, becoming cream. Flesh brittle to
somewhat tough, white.
Spines: 0.5–1.5 cm long; stout,
tapering to point, in clusters on ends of
branches.
Stalk (when present): indistinct.
Spores: 5.2–7 × 4.5–6 μ; nearly

round and smooth to finely roughened
or dotted; colorless. Spore print white.

Edibility: Choice.

Season: August–November.

Habitat: On old logs and stumps; on wounds of
living trees, especially maple, beech,
oak, and hickory.

Range: NE. North America to North Carolina
and Minnesota.

Look-alikes: *H. abietis,* found on conifers in the
Pacific NW., is larger and salmon-buff
to whitish with age.

Comments: Also called "Waterfall Hydnum." Both
H. coralloides and *H. abietis* species are
edible and very good when cooked
slowly.

547 Bearded Tooth
Hericium erinaceus (Fr.) Pers.
Hydnaceae, Aphyllophorales

Description: *Large, whitish, beardlike mass, with long
teeth; on deciduous trees.*
Mushroom: 4–10″ (10–25 cm) wide
and high; oval to roundish solid mass of
spines; white, becoming yellowish;
long, hanging, beardlike spines. Base
stout, tough, rooting.
Spines: 1–4 cm long; hanging,
covering sides; formed in lines rather
than in tufts or on branches.
Spores: 5–6.5 × 4–5.5 μ; almost
round, smooth to minutely roughened
with dots, colorless. Spore print white.

Edibility: Choice.

Season: August–November.

Habitat: On living trees, especially oak, maple,
and beech.

Range: S. United States from Florida to
California; reported in New York,
Michigan, and Pacific NW.

Comments: Also called the "Unbranched
Hericium," the "Satyr's Beard," and
the "Hedgehog Mushroom." It is only
a choice edible when young and very
fresh, because it turns sour with age.

549 Comb Tooth
Hericium ramosum (Bull. ex Mér.) Let.
Hydnaceae, Aphyllophorales

Description: *Large, whitish, toothed mass, along sides of branches; on decaying deciduous wood.*
Mushroom: 4–10″ (10–25 cm) wide, 3–6″ (7.5–15 cm) high; many-branched and coarsely toothed on both sides of branches; white to creamy. Flesh soft, brittle, white.
Spines: 0.5–1 cm long, to 2.5 cm long in tufts; white to creamy; branches usually with numerous spines on sides or undersurface.
Stalk: indistinct, laterally attached, stublike, hairy.
Spores: 3–5 × 3–4 μ; almost round; finely roughened to almost smooth, colorless. Spore print white.
Edibility: Choice.
Season: August–October.
Habitat: On decaying deciduous logs, such as maple, beech, and birch.
Range: Throughout North America.
Look-alikes: *H. coralloides** has tufts of teeth at ends of branches. *H. erinaceus** has long, beardlike, unbranched teeth.
Comments: The Comb Tooth is reportedly the most widespread and common species of *Hericium* in North America. Like other *Hericium* species, it is a very good edible when young.

479 Orange Rough-cap Tooth
Hydnellum aurantiacum (Fr.) Kar.
Hydnaceae, Aphyllophorales

Description: *Bright orange to rust cap with numerous rounded projections, and brownish spines descending orange stalk; under conifers.*
Cap: 1–4″ (2.5–10 cm) wide; becoming somewhat sunken, with many short, rounded projections; finely hairy to matted, bright orange-salmon to rusty. Flesh 2-layered: upper layer thin, soft, corky; lower layer zoned, tough, reddish-brown.

Spines: 1.5–2.5 mm long; short, descending stalk, grayish-tan to reddish-brown, becoming dark brown. Stalk: ⅜–¾″ (1–2 cm) long and thick; short, orange, becoming rusty to dark brown. Flesh 2-layered: outer layer thin; inner core tough, zoned, reddish. Spores: 5.5–7.5 × 5–6 μ; roundish, strongly warted, brownish. Spore print brown.

Season: August–October.

Habitat: On the ground under conifers.

Range: Nova Scotia to Florida, west to Colorado and Pacific NW.

Comments: This mushroom is usually found in late summer and fall under conifers in mountains.

---

469 Bluish Tooth

Hydnellum caeruleum (Horn. ex Pers.) Kar.
Hydnaceae, Aphyllophorales

Description: *Blue-tinged, whitish cap with spines descending buff-colored stalk; under conifers.*
Cap: 1–4″ (2.5–10 cm) wide; single to sometimes fused, convex to flat or somewhat sunken; smooth or with small, round projections; velvety-soft to pitted; whitish, tinged blue, darkening to brown, bruising rusty. Flesh 2-layered: upper layer spongy; lower layer tough, fibrous, buff, zoned with bluish bands and brownish tints.
Spines: 3–5 mm long; close, descending stalk, off-white to blue, becoming dark brown.
Stalk: 1–2″ (2.5–5 cm) long, ⅜–¾″ (1–2 cm) thick; bulbous, buff-colored. Flesh 2-layered: outer layer thin; inner core tough-fibrous, orange-red with bands of blue. Mycelium straw-colored.
Spores: 4.5–7 × 4.5–5 μ; rounded, strongly warted, brownish. Spore print brown.

Season: August–October.

Habitat: Several to fused, on the ground under conifers.

Range: E. North America; Great Lakes region; Pacific NW. to California.

Look-alikes: *H. suaveolens* has violet-blue mycelium, dirty white to brownish cap, and stalk with dark violet tints, violet-zoned flesh, and very fragrant odor.

Comments: The Bluish Tooth can be identified readily when it is cut, revealing the blue-zoned cap flesh and reddish-orange stalk flesh. Its unusual color and beauty compensate for its lack of edibility.

470 Red-juice Tooth
Hydnellum peckii Bank.
Hydnaceae, Aphyllophorales

Description: *Whitish to pinkish cap exuding drops of red juice, with brownish spines descending stalk; under conifers.*
Cap: 1–6″ (2.5–15 cm) wide; flat to sunken; smooth, becoming uneven with many jagged, matted projections; white, becoming pinkish, exuding drops of reddish juice. Flesh leathery, zoned, brownish. Taste peppery; odor slightly disagreeable.
Spines: 2–6 mm long; crowded, descending stalk; pinkish, becoming brown.
Stalk: 1–3″ (2.5–7.5 cm) long, ⅜–¾″ (1–2 cm) thick; bulbous, whitish to pinkish. Flesh solid, dark reddish-brown.
Spores: 4.5–5.5 × 3.5–4.5 μ; roundish, strongly warted, brownish. Spore print brown.

Season: September–October.

Habitat: On the ground under conifers.

Range: Nova Scotia to North Carolina in mountains; Pacific NW.

Look-alikes: *H. diabolus* has strong fragrant-pungent odor and hairy surface.

Comments: The vivid red drops on the surface of this mushroom make it easily recognizable. It is especially common in the Pacific Northwest.

446 Spongy-footed Tooth
Hydnellum spongiosipes (Pk.) Pouz.
Hydnaceae, Aphyllophorales

Description: *Brownish cap with brown spines descending*
spongy, bulbous stalk; under oak.
Cap: 1–4″ (2.5–10 cm) wide; convex
to flat or sunken if single, often
misshapen if fused; finely hairy, uneven
or with slight, rounded projections;
cinnamon-brown to grayish-brown,
darkening on bruising. Flesh 2-layered:
upper layer thick, spongy; lower layer
thin, hard, cinnamon-brown.
Spines: 4–6 mm long; close,
descending stalk, brownish, bruising
dark brown.
Stalk: 1–4″ (2.5–10 cm) long, ¼–1″
(0.5–2.5 cm) thick, up to twice as
thick at bulbous base; dark brown.
Flesh 2-layered: outer layer thick, felty-
spongy; inner core hard, zoned, dark
brown.
Spores: 5.5–7 × 5–6 μ; rounded,
warted, brownish. Spore print brown.
Season: August–October.
Habitat: On the ground in oak woods.
Range: E. Canada to Florida, west to Michigan
and Iowa.
Look-alikes: *H. pineticola* grows under pines.
Comments: Most species of *Hydnellum* occur under
conifers, and often enclose debris or
parts of other plants as they grow.

---

466 Scaly Tooth
Hydnum imbricatum Fr.
Hydnaceae, Aphyllophorales

Description: *Coarsely scaled brown cap, brown teeth,*
and stalk.
Cap: 2–8″ (5–20 cm) wide; convex to
flat, becoming sunken and torn with
age; dry, covered with raised, brownish
scales; light brown when fresh,
becoming darker overall. Flesh 1–1.5
cm thick; white to brownish, soft,
fragile. Odor not distinctive; taste mild
to somewhat sharp.

Spines: 0.5–1.5 cm long; pale brown
when fresh, becoming darker; slightly
descending stalk.
Stalk: 1⅝–4″ (4–10 cm) long, ⅝–1⅜″
(1.5–3.5 cm) thick; dry, smooth, pale
brown; hollow with age.
Spores: 6–8 × 5–7.2 μ; nearly round,
with large, irregular warts. Spore print
brownish.
Edibility: Edible.
Season: June–October; common in late spring
and early summer.
Habitat: On the ground in coniferous,
deciduous, and mixed woods.
Range: Throughout North America.
Look-alikes: *H. scabrosum* has blackish stalk base,
flesh that darkens with KOH, and very
bitter taste.
Comments: This is a common mushroom of late
spring, and seems to vary in taste from
mild to unpleasant. The mild
specimens are edible, though not
particularly delectable.

542 **Kidney-shaped Tooth**
Mycorrhaphium adustum (Schw.) M. Gees.
Hydnaceae, Aphyllophorales

Description: *Kidney-shaped, tough, thin, tan cap with
short, brownish spines.*
Cap: 1–3″ (2.5–7.5 cm) wide; kidney-
to fan-shaped or circular, often fused to
somewhat overlapping; flattened;
roughened, minutely velvety, whitish
to tan, bruising smoky gray; margin
faintly zoned, very thin and wavy, often
aging black. Flesh thin, tough, white.
Spines: 1–3 mm long; more or less
flattened and generally fused in groups
of 2 or more; appearing forked at tips;
white, becoming pinkish- to cinnamon-
brown, margin becoming blackish.
Stalk (when present): ¾–1¼″ (2–3 cm)
long, ⅜–¾″ (1–2 cm) thick; lateral,
velvety, tough, whitish.
Spores: 2.5–4 × 1–1.25 μ;
cylindrical, smooth, colorless. Spore
print white.

Season: July—November.
Habitat: On logs and fallen branches of oak and other deciduous trees.
Range: E. Canada to North Carolina, west to Michigan.
Comments: Formerly placed in the genera *Hydnum* and *Steccherinum,* this species is often found in small clusters with fused caps, and is readily recognized by its pale color and tough, thin caps.

468 Black Tooth
Phellodon niger (Fr.) Kar.
Hydnaceae, Aphyllophorales

Description: *Tough, blackish cap with grayish spines.*

Cap: 2—3" (5—7.5 cm) wide, often fused; convex, becoming flat to sunken; roughened, velvety to hairy, gray, becoming brownish-gray to violet-black, grayish about margin. Flesh 2-layered: upper layer soft; lower layer firm-tough, black.
Spines: 2.5—4 mm long; light gray, darkening on bruising.
Stalk: 1—2" (2.5—5 cm) long, ¼—1" (0.5—2.5 cm) thick; grayish.
Spores: 4—5 μ; roundish, spiny, colorless. Spore print white.
Season: August—October.
Habitat: Single to several or clustered or fused, in mixed woods, especially under pine, spruce, and hemlock.
Range: E. Canada to North Carolina, west to Michigan and Wyoming.
Look-alikes: *P. atratus* has slender stalk and bluish-black cap. *P. tomentosus* has strongly zoned, brownish cap and fragrant odor.
Comments: A variety (*alboniger*) of this species has at first a whitish cap, which ages brownish with a blue-gray margin. The genus *Phellodon* differs from *Hydnellum* by its white spore print, from *Hydnum* by its spore print color and tough flesh, and from *Dentinum* by its dark colors and tough flesh.

588 Ochre Spreading Tooth
Steccherinum ochraceum (Pers. ex Fr.) S.F.G.
Hydnaceae, Aphyllophorales

Description: *Spreading, ochre-colored fungus with projecting, stalkless caps and short spines; on dead wood.*
Cap: ⅛–1⅝" (0.3–4 cm) wide, ⅛–⅝" (0.3–1.5 cm) high; projecting from broad, spreading base fungus; cap circular, often overlapping, sometimes absent; zoned, hairy, ochre; margin whitish. Flesh thin, tough, whitish.
Spines: 1–1.5 mm long; very short, compressed, often forked, ochre.
Spores: 3–4 × 2–2.5 μ; oval, smooth, colorless. Cystidia large and projecting. Spore print white.

Season: June–October; sometimes year-round.

Habitat: On dead deciduous wood, such as beech, maple, and viburnum.

Range: Throughout North America, especially in E. and C. United States.

Comments: This mushroom is often found as a spreading mass adhering to dead wood; it is one of several stalkless tooth or spine fungi. It superficially resembles stalkless polypores and crust fungi and can be found growing along with them.

HYMENOCHAETE FAMILY
(Hymenochaetaceae)

This is a small family of leathery, crusted fungi. They are typically stalkless, and may be flat and smooth, shelflike, or toothlike and ragged. On the upper surface, they have minute bristles, or setae, which can be seen with a hand lens; these mushrooms grow on wood where they cause a white rot. The flesh is brownish and turns a darker brown or black when touched with a solution of potassium hydroxide (KOH). At one time, many of these fungi were included with the leather fungi in the family Thelephoraceae.

Some mycologists now consider the genera *Inonotus, Onnia, Phellinus,* and *Coltricia* to be part of the Hymenochaete family, while others place these genera in a family of their own, the Mucronoporaceae. In this guide, they are kept together with the other pore fungi in the Polyporaceae. None of the Hymenochaetaceae is known to be edible.

577 Brown-toothed Crust
Hydnochaete olivaceum (Schw.) Bank.
Hymenochaetaceae, Aphyllophorales

Description: *Broadly spreading, leathery, olive-brown to cinnamon-brown crust with jagged, toothlike surface.*
Mushroom: 2–4" (5–10 cm) wide or wider; 4–8" (10–20 cm) long; broadly spreading against bark; olive-brown to brown, warted, toothlike, or nearly porelike when teeth are fused. Flesh 1–3 mm thick; leathery, brown; dark brown or black with KOH.
Fertile Surface: teeth 3–6 mm long; conical near base, becoming flat, blunt and ragged near tip. Bristles (setae) 35–150 × 9–13 μ; sharp, tapered at both ends, thick-walled, reddish-brown; *present on teeth.*
Spores: 4.5–6.5 × 1.2–1.5 μ; sausage-shaped, smooth, colorless. Spore print white.
Season: June–October; may persist for years.
Habitat: On underside of dead branches of deciduous wood, especially oak.
Range: Vermont to Alabama and Louisiana, west to Wisconsin and Texas.
Look-alikes: *Irpex** species are usually lighter in color and lack setae on flattened teeth. Crustlike tooth fungi are fleshy, not leathery, and lack setae.
Comments: Also known as *Irpex cinnamomeus* and *Hydnoporia fuscescens.* This species spreads like a sheet on dead oak branches. The genera *Hydnochaete* and *Hymenochaete* may be most readily

distinguished by the placement of their bristles: those in *Hydnochaete* are found only on the teeth, while those in *Hymenochaete* grow from the hymenium, or fertile surface.

578 Reddish-brown Crust
Hymenochaete badio-ferruginea (Mont.) Lév.
Hymenochaetaceae, Aphyllophorales

Description: *Leathery crust formed by overlapping shell-like caps; silky-brown with reddish-brown bands, brown underneath.*
Mushroom: ⅛–⅜″ (0.3–1 cm) wide, ⅛–¼″ (3–5 mm) long; sometimes merging with others; crustlike portion (when present): 1–3″ (2.5–7.5 cm) long, ¼–1″ (0.5–2.5 cm) wide; caps thin, shell-like, overlapping, with margins merging or somewhat free and extended; brown, with concentric, smooth, reddish-brown zones. Flesh golden-brown; blackish with KOH.
Fertile Surface: brown, smooth to minutely cracked, *with conical setae.*
Spores: 4–6 × 1–2 μ; cylindrical and somewhat curved, smooth, colorless. Spore print white.
Season: June–October; may persist for years.
Habitat: On rotting stumps of deciduous trees.
Range: Canada to North Carolina, west to Iowa.
Look-alikes: *H. tabacina,* the most common northern species, grows horizontally along logs without bark, and becomes deeply cracked in radial systems.
Comments: *Hymenochaete* is recognized as a genus of crustlike mushrooms, with flesh turning blackish with KOH and possessing bristlelike structures called setae growing from the fertile surface.

POLYPORE FAMILY
(Polyporaceae)

The family Polyporaceae, as defined here, includes all those mushrooms, except boletes, that have their microscopic, spore-producing basidia located on the inside walls of tubes. Polypore means "many-pored," and the pores are the mouths of the tubes. In most cases, they can be seen by turning the mushroom over and examining the undersurface of the cap.

Most polypores are fleshy-tough to woody; many can survive frost and are perennial. They usually grow on wood and are shelflike or stalkless. Some grow on the ground or on buried wood; most of these are stalked. Still others grow crustlike on wood.

Some polypores are choice edibles; others have medicinal value and are used in research; a few have found use as ornaments, dyes, or tinder. None is known to be fatally poisonous, although a few that are tender enough to eat can cause indigestion, and some people do have an allergic reaction to well-known and popular edibles. Some polypores are virulent pathogens of living coniferous and deciduous trees, but many play an essential role in overall woodland ecology as deadwood decomposers.

Before the turn of the century, most polypores were included in the genus *Polyporus,* although all that united most of them was a pore surface. As emphasis came to be placed on differentiating microscopic characteristics, the genus *Polyporus* was restricted to include only those polypores that are annual and stalked, with pale-colored flesh and cylindrical, colorless spores. Now 50 or so different genera are recognized. Some researchers place certain genera in new families. These include *Boletopsis* (family Boletopsidaceae); *Bondarzewia* (Bondarzewiaceae); *Coltricia, Inonotus, Onnia,* and *Phellinus* (Mucronoporaceae); *Fistulina* and *Pseudofistulina* (Fistulinaceae); and *Ganoderma* (Ganodermataceae). In this guide all remain in the family Polyporaceae.

Although some fleshy, stalked polypores that grow on the ground resemble boletes, they differ in a number of ways, the most important of which is that all of the tubes on a bolete are formed before the cap expands; polypores can overwinter under favorable conditions and continue to produce new tubes. Also, bolete tube cells are divergent; polypore cells are not. The best field characteristics for separating boletes from stalked polypores are that boletes decay readily and their tubes, with rare exception, can be easily detached from the flesh of the cap.

457 Blue-pored Polypore
Albatrellus caeruleoporus (Pk.) Pouz.
Polyporaceae, Aphyllophorales

Description: *Fleshy, blue cap with blue pores descending stalk; on ground.*
Cap: 1–6″ (2.5–15 cm) wide; single to several fused together; convex, becoming flat and nearly circular; indigo to blue-gray, aging brownish; dry, smooth to slightly hairy and rough. Flesh 0.1–1 cm thick; white, aging reddish; red with KOH.
Tubes: 2–5 mm long; descending stalk. Pores (1–5 per mm) angular, indigo.
Stalk: 1–3″ (2.5–7.5 cm) long, ¼–1″ (0.5–2.5 cm) thick; central or off-center; indigo, discoloring with age; smooth to somewhat pitted.
Spores: 4–6 μ or 4–6 × 3–5 μ; roundish or oval, smooth, colorless. Spore print white.
Season: September–October.
Habitat: On the ground in mixed hemlock and deciduous woods; usually moist areas.
Range: E. Canada to North Carolina.
Comments: Also known as *Polyporus caeruleoporus*. This is one of very few blue mushrooms. It grows abundantly each fall for several years.

454 Crested Polypore
Albatrellus cristatus (Fr.) Kotl. & Pouz.
Polyporaceae, Aphyllophorales

Description: *Fleshy, yellowish to yellowish-green cap*
with white to greenish-yellow pores
descending stalk; on ground.
Cap: 2–8″ (5–20 cm) wide; single to
several fused together; circular or
irregular, convex, becoming flat;
yellowish-green to yellowish-brown;
dry, smooth or finely hairy, becoming
cracked or crested. Flesh 0.2–2 cm
thick; white, bruising yellowish-green;
slowly turning reddish with KOH.
Tubes: 1–5 mm long; descending
stalk. Pores (1–3 per mm) angular,
white to greenish-yellow.
Stalk: 1¼–2⅜″ (3–6 cm) long, ⅜–1″
(1–2.5 cm) thick; central or off-center;
yellowish, smooth.
Spores: 5–7 × 4–5 μ; oval to almost
round, smooth, colorless. Spore print
white.
Season: September–October.
Habitat: On the ground in deciduous and
coniferous woods.
Range: E. Canada to Alabama, west to Ohio,
Wisconsin, and Missouri; especially
common in SE. United States.
Look-alikes: *A. ellisii** has sulfur-yellow, coarsely
scaly cap. *A. sylvestris* has ochre-olive to
yellow-green, slightly scaly cap and
roughened spores. *A. peckianus* has
yellow to cinnamon-buff, smooth cap.
Boletes have detachable tubes.
Comments: Also known as *Polyporus cristatus*.

420, 456 Scaly Yellow Polypore
Albatrellus ellisii (Berk.) Pouz.
Polyporaceae, Aphyllophorales

Description: *Fleshy, sulfur-yellow, coarsely scaly cap*
with white to yellowish pores, bruising
greenish, descending stalk; on ground.
Cap: 4–6″ (10–15 cm) wide; single;
convex, becoming flat, sometimes
circular; dry, with overlapping scales;

sulfur-yellow with greenish tinge. Flesh 20 mm thick; white, bruising yellowish-green.

Tubes: 2–5 mm long; descending stalk. Pores (1–2 per mm) angular, becoming torn with age; white, turning yellow, bruising greenish.

Stalk: 2¾–3¼″ (7–8 cm) long, 1⅝–2″ (4–5 cm) thick; off-center, with light to dark yellow surface network.

Spores: 8–9 × 5–7 μ; oval to elliptical, smooth, colorless. Spore print white.

Season: September–October.

Habitat: On the ground in mixed woods.

Range: Reported from New Jersey, Alabama, California, and Pacific NW.

Look-alikes: *A. cristatus** is yellowish-green, smooth, becoming cracked. *A. peckianus* has smooth cap. *A. sylvestris* is olive-ochre and smooth to slightly scaly.

Comments: Also called *Polyporus ellisii.* The Scaly Yellow Polypore is probably more common than reports indicate, as it is often mistaken for several of the more widely known species.

---

462 Flett's Polypore
Albatrellus flettii (Morse) Pouz.
Polyporaceae, Aphyllophorales

Description: *Fleshy, greenish-blue to ochre cap, with white to salmon pores descending stalk; on ground.*

Cap: 4–8″ (10–20 cm) wide; single to several fused together; convex to flat or concave, circular or irregular; greenish-blue, becoming dingy ochre; dry, smooth. Flesh 1–1.5 cm thick; white. Odor and taste mealy.

Tubes: 1–7 mm long; descending stalk. Pores (1–4 per mm) angular, becoming torn with age; white, becoming salmon.

Stalk: 2⅜–6″ (6–15 cm) long, ¾–1⅜″ (2–3.5 cm) thick; usually off-center; white, turning dingy ochre; smooth near base.

Spores: $3.5-4 \times 2.5-3$ μ; elliptical to almost round, smooth, colorless, amyloid. Spore print white.

Season: September–October; into December in California.

Habitat: On the ground in mixed woods.

Range: Pacific NW. and California.

Look-alikes: *A. caeruleoporus**, an eastern species, has blue pores.

Comments: Also known as *Polyporus flettii.*

---

460   Sheep Polypore
Albatrellus ovinus (Fr.) Kotl. & Pouz.
Polyporaceae, Aphyllophorales

Description: *Fleshy, white to pinkish-buff cap with white to yellow pores descending stalk; on ground.*

Cap: 2–6" (5–15 cm) wide; single to several fused together; convex and circular or fused and irregular; white, becoming pale tan to pinkish-buff; dry, smooth, becoming slightly scaly with age. Flesh 6–20 mm thick; white, drying yellow to greenish. Odor and taste mild, sometimes bitterish.
Tubes: 1–2 mm long; descending stalk. Pores (2–4 per mm) angular; white to yellow, with greenish or reddish tinge when dry.
Stalk: 1–3" (2.5–7.5 cm) long, ⅜–1¼" (1–3 cm) thick, enlarging downward, pointed at base; usually central; smooth; white, bruising pink.
Spores: $3-4.5 \times 3-3.5$ μ; elliptical to almost round, smooth, colorless. Spore print white.

Edibility: Edible.

Season: July–December.

Habitat: On the ground near conifers.

Range: Throughout N. North America; Maine to Tennessee; Rocky Mountains; Washington into California.

Look-alikes: *A. confluens* is typically clustered, with orange-tinted caps and bitter flavor.
A. subrubescens bruises orange and has amyloid spores. Boletes have detachable tubes.

Comments: Also known as *Polyporus ovinus*. The Sheep Polypore is the most widespread and abundant *Albatrellus* species.

---

464 Goat's Foot
Albatrellus pescaprae (Pers. ex Fr.) Pouz.
Polyporaceae, Aphyllophorales

Description: *Fleshy, scaly, gray to brown cap with white to greenish-yellow pores descending stalk; on ground.*
Cap: 1–8″ (2.5–20 cm) wide; single to somewhat fused; convex, almost circular; gray, pinkish-brown, brown, or blackish-brown; dry, somewhat scaly. Flesh 0.5–2 cm thick; white, bruising pinkish.
Tubes: 2–5 mm long; descending stalk. Pores 1–2 mm wide; large, angular; white or yellowish to greenish-yellow, sometimes becoming pinkish.
Stalk: 1–3″ (2.5–7.5 cm) long, ⅜–1⅝″ (1–4 cm) thick, bulbous near base; simple or branched toward top, lateral or off-center; white or yellowish at base, pale near cap; smooth.
Spores: 8–11 × 5–6 μ; oval to broadly elliptical, smooth, colorless. Spore print white.
Season: September–January.
Habitat: On the ground in mixed woods.
Range: New York to Alabama; Washington to California.
Look-alikes: *A. avellaneus* is smooth to slightly scaly, bruises yellow. *A. hirtus* has short, stiff hairs, often grows on wood, and is bitter. *A. sublividus* is purplish-gray, and *A. sylvestris* is ochre-olive; both have ornamented spores.
Comments: Also called *Polyporus pescaprae*.

---

582 Smoky Polypore
Bjerkandera adusta (Fr.) Kar.
Polyporaceae, Aphyllophorales

Description: *Tough, overlapping, projecting, smoky-gray, stalkless caps with grayish pores.*

Cap: ⅜–3″ (1–7.5 cm) wide;
overlapping, convex to flat, with thin
margin; white or tan to smoky-gray,
drying black along margin; dry, hairy.
Flesh 1–6 mm thick; white, usually
separated from tubes by thin, dark line.
Tubes: 2 mm long. Pores (5–7 per
mm) minute, almost circular,
becoming angular; grayish, bruising or
aging black.
Spores: 4–5.5 × 2.5–3 μ; oblong or
elliptical to almost cylindrical, smooth,
colorless. Spore print white.

Season: July–November; usually overwinters.

Habitat: On dead deciduous and coniferous
wood.

Range: Throughout North America.

Look-alikes: *B. fumosa* is larger, with thicker flesh
and an aniselike odor when fresh.

Comments: Also known as *Polyporus adustus*.

---

458 Kurotake
Boletopsis subsquamosa (L. ex Fr.) Kotl. &
Pouz.
Polyporaceae, Aphyllophorales

Description: *Fleshy, gray to brownish or blackish,*
smooth cap with whitish pores somewhat
descending stalk; on ground.
Cap: 2–5″ (5–12.5 cm) wide; flat or
convex to somewhat concave, nearly
circular, with slightly inrolled, wavy,
furrowed margin; olive-, bluish-, or
brownish-gray to brownish or black;
smooth, developing small radial fibers
or slight scales at center. Flesh 0.5–2.5
cm thick; white to gray, tinged pink-
violet; greenish-gray when dry. Taste
mild to bitter when moist, sweetish to
spicy when dry.
Tubes: 2–5 mm long; slightly
descending stalk. Pores (1–2 per mm)
circular, becoming irregular; white to
grayish, turning gray-brown to brown
when dry.
Stalk: 1⅜–3″ (4–7.5 cm) long, ⅜–1″
(1–2.5 cm) thick; short, stocky, central
to off-center; gray to olive-brown,

smooth or with fine scales.
Spores: 4.5–7 × 4–5 μ; almost round
to oval, with warty projections,
colorless to slightly brownish. Spore
print white or light brown.

Edibility: Edible.

Season: September–October.

Habitat: On the ground in deciduous and
coniferous woods, especially under pine
in northern mountains and
northwestern coastal sand dunes.

Range: New Hampshire to Alabama; Midwest;
Pacific NW.

Look-alikes: *Albatrellus pescaprae** is brownish and
scaly, with smooth spores. Boletes have
longer, detachable tubes.

Comments: Kurotake is esteemed by the Japanese,
who reportedly soak this sometimes
bitter mushroom in brine before
preparing it as food. The Gray
Boletopsis (*Polyporus griseus,* also known
as *Boletopsis grisea*) and the White-black
Boletopsis (*B. leucomelaena*), once
considered separate species, are now
believed to be forms of the highly
variable Kurotake.

476, 477 Berkeley's Polypore
Bondarzewia berkeleyi (Fr.) Sing.
Polyporaceae, Aphyllophorales

Description: *One to several fleshy, tough, cream caps with
whitish pores descending stalk.*
Cap: 3–10″ (7.5–25 cm) wide; convex,
becoming flat and depressed in center;
creamy white to yellowish; dry, hairy or
smooth, often rough or pitted. Flesh
0.3–2 cm thick; firm, becoming
tough, white. Odor becoming rank;
taste mild, bitter with age.
Tubes: 3–5 mm long; descending
stalk. Pores 0.5–2 mm wide; angular,
often torn; white, becoming dingy.
Stalk: 2–4″ (5–10 cm) long, 1¼–2″
(3–5 cm) thick; yellowish, knobby,
rooting.
Spores: 6–8 μ; round, strongly warted,
colorless, amyloid. Spore print white.

Edibility: Edible.

Season: July–October.

Habitat: On the ground near base or buried roots of deciduous trees, especially oak.

Range: Canada and Massachusetts to Louisiana; Midwest; Texas; Pacific NW.

Look-alikes: *Grifola frondosa** has many smaller, thinner, gray-brownish caps. *Laetiporus sulphureus** is orange to bright yellow, less tough-fleshed, and tart but not bitter. *Meripilus giganteus** blackens when bruised along cap edge and on pore surface.

Comments: Also known as *Polyporus berkeleyi.* Berkeley's Polypore becomes tough and intensely bitter when mature. It grows on tree roots, and when it first appears often looks like a huge hand with chunky fingers reaching out of the ground. As it matures, the "fingers" thicken and fan out to become overlapping shelves, often attaining a span of more than 3' and weighing more than 50 pounds.

---

463 Bondarzew's Polypore
Bondarzewia montana (Quél.) Sing.
Polyporaceae, Aphyllophorales

Description: *One to several fleshy, tough, ochre to brownish caps with whitish pores descending single short, thick stalk.*
Cap: 2⅜–10″ (6–25 cm) wide; convex, becoming flat and sunken; ochre to dark brown, dry, velvety. Flesh 0.3–2 cm thick; tough, becoming hard, white.
Tubes: 1–6 mm long; descending stalk. Pores (1–2 per mm) angular, whitish.
Stalk: 2–5″ (5–12.5 cm) long, ¾–1⅝″ (2–4 cm) thick; short and thick above ground; brown, velvety, rooting.
Spores: 5–7 μ; round, warted or ridged, colorless, amyloid. Spore print white.

Edibility: Edible.

Season: September–November.

Habitat: On the ground or on buried wood, under conifers: pine, fir, spruce, Douglas-fir.

Range: Pacific NW.

Comments: Also known as *Polyporus montanus*. *Bondarzewia* is a genus of fleshy, tough polypores that grow on wood and have spores with ornamentations that turn blue in Melzer's solution. Because no other polypores have this spore characteristic, this genus is placed by some in its own family, the Bondarzewiaceae. Because the spores of this genus resemble those of the gilled mushroom genera *Lactarius* and *Russula*, some mycologists have placed the 3 together in their own order, the Russulales.

---

590 Gelatinous-pored Polypore
Caloporus dichrous (Fr.) Ryv.
Polyporaceae, Aphyllophorales

Description: *Leathery and spreading, with projecting, stalkless, small, white, densely hairy caps, often overlapping, with waxy, separable, pinkish pores.*
Cap: ¼–2" (0.5–5 cm) wide; overlapping, flat, with narrow margin; white, dry, hairy to velvety. Flesh 1–4 mm thick; white.
Tubes: 0.5–1 mm long. Pores (5–8 per mm) minute, circular to angular; forming a separable elastic layer; buff, pinkish, or reddish-purple.
Spores: 3–5 × 0.5–1.5 μ; sausage-shaped to cylindrical, smooth, colorless. Spore print white.

Season: August–October; overwinters.

Habitat: On dead wood, usually deciduous logs.

Range: Throughout North America.

Look-alikes: When very young and in spreading form, it resembles *Poria** species.

Comments: Also known as *Polyporus dichrous*. The scientific name means "2-colored mushroom with beautiful pores." It can spread over a broad area.

540 Mossy Maze Polypore
Cerrena unicolor (Fr.) Murr.
Polyporaceae, Aphyllophorales

Description: *Leathery, spreading, stalkless, with
overlapping, projecting, thin, grayish,
zoned, hairy, often algae-covered caps with
mazelike tubes and smoky pores.*
Cap: ¼–3″ (0.5–7.5 cm) wide;
overlapping, flat; white to grayish,
often algae-covered; dry, with dense,
stiff hairs; strongly zoned. Flesh 0.5–1
mm thick; white, with thin, dark line
separating it from hairy cap surface.
Tubes: 0.5–4 mm long; mazelike,
becoming toothlike or rarely porelike.
Pores (2–3 per mm) white to smoky.
Spores: 4.5–5.5 × 2.5–3.5 μ; oblong
to elliptical, smooth, colorless. Spore
print white.
Season: Year-round.
Habitat: On wood, usually deciduous.
Range: Throughout N. North America;
scattered in Southeast and Midwest.
Look-alikes: *Daedalea quercina** and *Daedaleopsis
confragosa** have fewer hairs. *Irpex
lacteus** has long, flattened, toothlike
tubes. Tooth fungi (*Hydnum** and
related genera) have conical teeth.
Comments: Also known as *Daedalea unicolor,* the
Mossy Maze is usually covered with
green algae.

445 Shiny Cinnamon Polypore
Coltricia cinnamomea (Pers.) Murr.
Polyporaceae, Aphyllophorales

Description: *Small, thin, reddish-brown, silky, shiny,
zoned cap with brownish pores and velvety,
reddish-brown stalk; on ground.*
Cap: ⅜–2″ (1–5 cm) wide; single,
circular to laterally compressed, flat to
sunken, with thin margin; bright
cinnamon to amber-brown or darker;
dry, silky, shiny, fibrous, zoned;
margin lined or finely fringed. Flesh
0.5–1 mm thick; pliant when fresh,
rust-brown; black with KOH.

Tubes: 0.5–3 mm long. Pores (2–3 per mm) angular, yellow-brown to cinnamon.
Stalk: ⅜–1⅝" (1–4 cm) long, ¹⁄₃₂–⅛" (1–3 mm) thick; central, tough; reddish-brown, velvety or hairy.
Spores: 6–10 × 4.5–7 μ; elliptical, smooth, colorless to pale brown.

Season: June–November.

Habitat: Single, in hard-packed soil, along paths and in dense moss.

Range: E. Canada and Maine to Louisiana; Midwest; California.

Look-alikes: *C. perennis* has larger, thicker, dull yellowish to golden-brown cap, and is usually found on burned sites.

Comments: This is also known as *Polyporus cinnamomeus,* but is placed in *Coltricia* because of its rust-brown flesh that blackens with KOH.

449 Montagne's Polypore
Coltricia montagnei var. *montagnei* Fr.
Polyporaceae, Aphyllophorales

Description: *Fleshy, tough, yellow- to rust-brown, hairy, zoned cap with white to brownish pores and hairy, brown stalk.*
Cap: 1–4" (2.5–10 cm) wide; usually single and circular, occasionally laterally fused; convex to flat or sunken; yellow- to rust-brown or darker; silky-hairy, somewhat uneven and zoned. Flesh 1–5 mm thick; brown; black with KOH.
Tubes: 1–5 mm long; descending stalk slightly. Pores 0.5–2 mm wide; large, angular, brownish; pore walls near margin or stalk can appear mazelike or concentrically furrowed.
Stalk: 1–2" (2.5–5 cm) long, ⅛–⅝" (0.3–1.5 cm) thick; central, often poorly developed; brown, hairy.
Spores: 9–15 × 6–7.5 μ; oblong or elliptical, smooth, pale brown.

Season: July–October.

Habitat: On the ground or on buried coniferous wood.

Range: N. North America; Maine to North Carolina.

Look-alikes: *C. perennis* has thin-fleshed, zoned cap and long, thin stalk. *Onnia tomentosa** and *O. circinata* are golden- to cinnamon-brown, have a more regular pore surface, and possess setae.

Comments: Also known as *Polyporus montagnei*.

---

450 Green's Polypore
Coltricia montagnei var. *greenei* Fr.
Polyporaceae, Aphyllophorales

Description: *Fleshy, tough, yellow- to rust-brown, hairy cap with brownish pores or concentric plates, and velvety brown stalk.*
Cap: 1–6″ (2.5–15 cm) wide; usually single and circular, occasionally laterally fused; convex to flat or sunken; yellow- to rust-brown or darker; densely hairy, becoming smooth, sometimes zoned. Flesh 5–10 mm thick; brown; black with KOH.
Tubes: 5–8 mm long; descending stalk slightly; polygonal when young, but soon breaking up into continuous, concentric, gill-like furrows; often irregularly porelike near margin or stalk; brownish.
Stalk: 1–3″ (2.5–7.5 cm) long, ¼–1″ (0.5–2.5 cm) thick; central, brown to rust-brown, velvety.
Spores: 9–15 × 6–7.5 μ; oblong or elliptical, smooth, pale brown.

Season: July–October.

Habitat: On the ground in mixed woods.

Range: Massachusetts to North Carolina, west to Ohio, Iowa, and Michigan.

Comments: Formerly this species was called *Cyclomyces greenei*.

---

528 Veiled Polypore
Cryptoporus volvatus (Pk.) Hub.
Polyporaceae, Aphyllophorales

Description: *Tough to woody, whitish to yellowish, stalkless spore with white to brownish*

pores covered by thick, veil-like membrane
that breaks open near base.

Cap: ⅝–3⅜" (1.5–8.5 cm) wide;
round, with margin descending down
over pores forming thick, *veil-like
covering;* whitish or yellow, turning
ochre to brownish when dry; smooth,
with thin crust. Flesh 0.2–1 cm thick;
white.

Tubes: 2–5 mm long. Pores (3–4 per
mm) circular, white to brownish.

Spores: 8–12 × 3–5 μ; elongate or
elliptical to cylindrical, smooth,
colorless. Spore print pinkish-buff.

Season: May–August; overwinters in California.

Habitat: On living or dead conifers.

Range: Throughout N. North America; Maine
to West Virginia; Pacific NW. to
California; Rocky Mountains in
Colorado.

Comments: Also known as *Polyporus volvatus.* This is
the only "veiled" polypore and looks
somewhat like a puffball. There is a
small, circular perforation at the base,
which opens slightly to reveal a hollow
interior with an upper wall of tubes and
pores.

467 **Thick-maze Oak Polypore**
Daedalea quercina Fr.
Polyporaceae, Aphyllophorales

Description: *Tough, grayish to ochre, stalkless cap with
mazelike tubes and large, white to brownish
pores.*

Cap: 2–6" (5–15 cm) wide; convex to
flat, with thick margin; white, aging to
ochre-brown or black; dry, with fine
hairs; smooth, becoming cracked to
furrowed. Flesh 0.2–1.5 cm thick; off-
white to light brown.

Tubes: 0.5–3 cm long, 1 mm or more
wide, with thick walls; mazelike,
sometimes gill-like, rarely porelike;
white to brownish.

Spores: 5.5–7 × 2.5–3.5 μ; elliptical
to cylindrical, smooth, colorless. Spore
print white.

Season: Year-round.
Habitat: On dead oak.
Range: Maine to North Carolina, west to Ohio
and Iowa.
Look-alikes: *Daedaleopsis confragosa** has thin cap
with thin-walled tubes, narrow pores.
Comments: The mazelike pore surface of this
species is the source of its genus name,
which refers to the legendary Greek
Daedalus, who designed the labyrinth
of the Minotaur in ancient Crete.

481 Thin-maze Flat Polypore
Daedaleopsis confragosa (Fr.) Schroet.
Polyporaceae, Aphyllophorales

Description: *Tough, grayish to brown, zoned or
furrowed, stalkless cap with mazelike tubes
and white to pale brown pores.*
Cap: 1–6″ (2.5–15 cm) wide; convex to
flat, with thin, sharp margin; grayish
to brownish, sometimes bruising
pinkish; finely hairy or smooth, zoned,
coarsely wrinkled to bumpy. Flesh 0.2–
2 cm thick; white to brownish.
Tubes: 0.1–1.5 cm long, 0.5–1.5 mm
wide, with thin walls; mazelike, or
somewhat porelike or gill-like. Pores
white to brownish, sometimes bruising
pink.
Spores: 7–11 × 2–3 μ; cylindrical,
curved, smooth, colorless. Spore print
white.
Season: June–December; usually persists for
several years.
Habitat: On dead wood or wounds in living
trees.
Range: Throughout Canada; Maine to Florida;
Texas; Midwest; Pacific NW.
Look-alikes: *Daedalea quercina**, which grows on
oak, is thicker and wider, with thick-
walled tubes, large, broad pores that do
not discolor on bruising, and much
smaller spores.
Comments: Also known as *Daedalea confragosa*. The
Thin-maze Flat Polypore grows on
willow, birch, and some other
deciduous trees, but usually not on oak.

508 Hexagonal-pored Polypore
Favolus alveolaris (DC. ex Fr.) Quél.
Polyporaceae, Aphyllophorales

Description: *Fleshy, tough, scaly, reddish-yellow to cream cap with radial rows of 6-sided, white pores.*
Cap: ⅜–4″ (1–10 cm) wide; semicircular to kidney- or fan-shaped; reddish-yellow to pale red, weathering to whitish; somewhat scaly, becoming smooth. Flesh 0.5–2 mm thick; white.
Tubes: 1–5 mm long; slightly descending stalk. Pores 0.5–3 × 0.5–2 mm; 6-sided to diamond-shaped or honeycombed, usually in radial rows; white, drying yellowish.
Stalk (when present): ¼–⅜″ (0.5–1 cm) long, ¹⁄₁₆–¼″ (1.5–5 mm) thick; lateral, stubby.
Spores: 9–11 × 3–3.5 μ; cylindrical, smooth, colorless. Spore print white.

Edibility: Edible.

Season: May–November; conspicuous in May and June.

Habitat: On dead branches of deciduous trees, especially hickory, oak, willow, elm, poplar, and beech.

Range: Quebec to Manitoba; Maine to Alabama; Midwest; Rocky Mountains.

Look-alikes: *Polyporus arcularius** has circular cap and central stalk.

Comments: Also known as *Favolus canadensis, F. europaeus,* and *Polyporus mori.* This species is edible but usually too tough to be palatable.

513 Beefsteak Polypore
Fistulina hepatica Schaeff. ex Fr.
Polyporaceae, Aphyllophorales

Description: *Fleshy, somewhat gelatinous, juicy, flat, reddish cap with separate tubes and pinkish-yellow pores.*
Cap: 3–10″ (7.5–25 cm) wide; spoon-shaped to semicircular, flat; blood-red, roughened, slimy, gelatinous, becoming smooth. *Flesh* 2–6 cm thick;

off-white to pinkish, streaked with red, *zoned, soft and juicy when fresh.*

Tubes: 10–15 mm long; *free.* Pores 1 mm wide; circular; whitish to yellowish-buff, becoming reddish-brown.

Stalk (when present): 2–4″ (5–10 cm) long, ⅜–1¼″ (1–3 cm) thick; lateral, very short, thick; blood-red.

Spores: 4.5–6 × 3–4 μ; oval, smooth, colorless to pale yellow. Spore print pinkish-salmon.

Edibility: Good.

Season: July–October.

Habitat: On dead oak trunks and stumps, or at base of living oak.

Range: Widely distributed; most common in E. North America; rare in Canada.

Comments: The genus name, *Fistulina,* means "small pipes" and refers to the unique arrangement of the tubes, which are closely packed but free of one another. *Hepatica* derives from the liverlike appearance of this mushroom, which is acidic and tart but a good edible.

---

374 Pale Beefsteak Polypore
Pseudofistulina radicata (Schw.) Burd.
Polyporaceae, Aphyllophorales

Description: *Fleshy, tough, powdery gray-brown, flat cap with separate tubes, cream pores, and lateral, rooting stalk.*

Cap: 1–5″ (2.5–12.5 cm) wide; kidney- to fan-shaped, with incurved margin; gray-brown to pale brown or pale red; velvety, powdery and dry; *covered with long, cylindrical, spiny cells.* Flesh 0.4–7 mm thick; white to cream.

Tubes: 0.4–3 mm long; *free.* Pores (5–8 per mm) minute, circular, white to cream.

Stalk: ⅜–6″ (1–15 cm) long, ¼–¾″ (0.5–2 cm) thick; lateral and rooting; underground part of variable length; brownish, finely velvety.

Spores: 3 μ; nearly round, smooth, colorless. Spore print white.

Season: August—October.
Habitat: On roots of dead oak.
Range: Common only in SE. North America; reported from New Hampshire to South Carolina.
Comments: Formerly known as *Fistulina pallida*. Like the Beefsteak Polypore (*Fistulina hepatica*), this species has closely packed tubes that are not attached to one another.

527 Tinder Polypore
Fomes fomentarius (Fr.) Kickx
Polyporaceae, Aphyllophorales

Description: *Woody, dark grayish-brown, generally crusted, hoof-shaped, stalkless cap with brown pores.*
Cap: 2—8″ (5—20 cm) wide; hoof-shaped; gray to gray-brown or gray-black, with dark brown-black zones; dry, velvety to smooth, with hard, horny, thick crust. Flesh 0.3—3 cm thick; dark tan to brown.
Tubes: 0.5—6 cm long; very thick, not distinctly layered. Pores (3—4 per mm) circular, gray to brown.
Spores: 15—20 × 4.5—7 μ; cylindrical, smooth, colorless. Spore print white.
Season: Year-round.
Habitat: On dead deciduous trees or wounds in living trees, including maple, birch, beech, hickory, poplar, and cherry.
Range: Throughout N. North America, south to North Carolina and Texas.
Look-alikes: *Fomitopsis** species have whitish flesh. *Ganoderma applanatum** is generally flat-shelved, has readily bruising pore surface, and brown, blunt-edged spores with ornamented walls. *Phellinus** species are more like cork than wood and usually have setae.
Comments: This common, hoof-shaped conk is known in Europe as "Amadou," and has been used there for centuries as a kind of punkwood for the quick ignition of fires. It has also been used in the cauterization of wounds.

524 Rosy Polypore
Fomitopsis cajanderi (Kar.) Kotl. & Pouz.
Polyporaceae, Aphyllophorales

Description: *Tough, pinkish-brown, stalkless cap with pinkish-red to brownish pores.*
Cap: 1–4″ (2.5–10 cm) wide; flat to convex, with sharp margin; pinkish-red to pinkish gray-brown, finely hairy, becoming smooth or roughened and zoned, but not encrusted or cracked. Flesh 2–15 mm thick; rosy-pink to pinkish-brown.
Tubes: 1–3 mm long per season. Pores (4–5 per mm) circular, pinkish-red, becoming reddish-brown.
Spores: 4–7 × 1.5–2 μ; narrow, cylindrical, curved, smooth, colorless. Spore print whitish.
Season: Year-round.
Habitat: On dead conifers, such as pine, hemlock, fir, Douglas-fir, spruce, larch, and juniper; rarely on deciduous trees.
Range: Throughout Canada, south to Florida, Midwest, Rocky Mountains, and Pacific NW.
Look-alikes: *F. rosea* has wider spores.
Comments: Also known as *Fomes subroseus.*

529 Larch Polypore
Fomitopsis officinalis (Fr.) Bond. & Sing.
Polyporaceae, Aphyllophorales

Description: *Tough, white to yellowish, hoof-shaped, stalkless cap with whitish pores.*
Cap: 2–8″ (5–20 cm) wide; knoblike, becoming hoof-shaped; white to yellowish, dry, slightly encrusted to furrowed and cracked with age. Flesh 2–5 cm thick; cheesy when fresh, white. Odor mealy; taste bitter.
Tubes: 3–20 mm long per season. Pores (3–4 per mm) angular, whitish.
Spores: 4–5.5 × 3–4 μ; broadly elliptical, smooth, colorless. Spore print whitish.
Season: Year-round.
Habitat: On conifers, especially larch and pine.

Range: Michigan west to Washington, south to California and Colorado.

Comments: Also known as *Laricifomes officinalis,* *Fomes laricis,* and *Fomes officinalis.* The species name, *officinalis,* refers to this species' use as an official drug plant in pharmacopoeia. It was once believed to be a panacea for all ills, is said to be an effective laxative, and has been used as a substitute for quinine.

---

531 Red-belted Polypore
Fomitopsis pinicola (Fr.) Kar.
Polyporaceae, Aphyllophorales

Description: *Hard, brownish, stalkless cap with reddish, resinous crust and white to yellowish pores.*
Cap: 2–12" (5–30 cm) wide; flat to convex or hoof-shaped, with thickened, rounded margin; ruby-red to brown or black resinous crust in concentric zones, usually red near margin, furrowed with age. Flesh 0.5–2 cm thick; cream, light yellow, or brown, bruising pinkish; reddish with KOH.
Tubes: 3–5 mm long per season; distinctly layered. Pores (3–5 per mm) angular, whitish, bruising yellow.
Spores: 6–8 × 3.5–4 μ; cylindrical to oblong-elliptical, smooth, colorless. Spore print whitish.

Season: Year-round.

Habitat: On dead trees, stumps, or logs; occasionally on living trees.

Range: Throughout Canada, south to North Carolina, Ohio, California, and Arizona.

Look-alikes: Other species of *Fomitopsis** lack glossy, resinous crust or red marginal belt. *Fomes fomentarius** lacks resinous crust and red belt, has brown flesh staining brown with KOH. *Ganoderma applanatum** has thick, dark, nonresinous crust, is typically flat-shelved.

Comments: Also known as *Fomes pinicola.* The Red-belted Polypore is found on more than 100 host trees.

518 Artist's Conk
Ganoderma applanatum (Pers. ex Wall.) Pat.
Polyporaceae, Aphyllophorales

Description: *Woody, horny, flat to hoof-shaped, gray to brownish, stalkless cap with readily bruising white pores.*
Cap: 2–20″ (5–50 cm) wide; convex to steeply hoof-shaped; gray to brown to grayish-black; often warty and zoned with hard, horny crust that becomes furrowed or cracked. Flesh 0.5–5 cm thick; brown.
Tubes: 4–12 mm long per season; distinctly layered. Pores (4–6 per mm) circular, white, readily bruising darker, often aging yellowish to pale brownish.
Spores: 6.5–9.5 × 5–7 μ; broadly elliptical and blunt at 1 end, with thick double wall (outer perforated, inner chambered); brown. Spore print brown.

Season: Year-round.

Habitat: On dead wood, particularly of deciduous trees, but also reported on conifers and wounds in living trees.

Range: Throughout North America.

Look-alikes: *Fomitopsis pinicola** has reddish, resinous band near cap margin, cream flesh, and smooth spores.

Comments: This is the famous Artist's Conk: the white pore surface bruises easily, and a detailed drawing can be etched into the surface that on drying will remain permanently.

514 Ling Chih
Ganoderma lucidum (Ley. ex Fr.) Kar.
Polyporaceae, Aphyllophorales

Description: *Soft, corky, flat, zoned, red-varnished cap with white to dull brown pores.*
Cap: 1–14″ (2.5–35.5 cm) wide; often overlapping, circular to semicircular or kidney-shaped; with shiny, dark red to reddish-black varnish, often ochre toward margin; smooth, somewhat zoned. Flesh 0.2–3 cm thick; whitish to deep brown.

Tubes: 2—18 mm long. Pores (4—6 per mm) circular to angular, whitish to dull brown.
Stalk (when present): 1—4″ (2.5—10 cm) long, ¼—1⅝″ (0.5—4 cm) thick; lateral, varnished red, encrusted.
Spores: 7—12 × 6—8 μ; elliptical, with double wall (outer perforated, inner chambered); brown. Spore print brown.

Season: May—November; may overwinter.

Habitat: At base of many living deciduous trees, especially maple; also on stumps.

Range: New England to Florida; Midwest to Texas; Pacific NW. and California.

Look-alikes: *G. curtisii* is a southern species, has bright ochre cap, darker in age, and is always stalked. *G. oregonense* is northwestern and usually very large. *G. tsugae** is bright to dark red-orange, usually stalked, and grows on conifers.

Comments: In its stalked form this species is believed to be the Ling Chih of the ancient Chinese, the "mushroom of immortality" or the "herb of spiritual potency." A candy made of the essence of Ling Chih is sold in Chinese markets in New York City.

515 **Hemlock Varnish Shelf**
Ganoderma tsugae Murr.
Polyporaceae, Aphyllophorales

Description: *Soft, corky, flat, zoned, shiny, varnished, reddish cap with white to brownish pores.*
Cap: 2—12″ (5—30 cm) wide; kidney- to fan-shaped; shiny reddish to brownish-orange or black varnish, often with white to orange margin; smooth to powdery, somewhat zoned or furrowed. Flesh 0.5—3 cm thick; white.
Tubes: 0.3—1 cm long. Pores (4—6 per mm) circular to angular, white to brown, discoloring on bruising.
Stalk (when present): 1—6″ (2.5—15 cm) long, ⅜—1⅝″ (1—4 cm) thick; lateral or sometimes central; varnished and encrusted like cap.
Spores: 9—11 × 6—8 μ; elliptical and

blunt at 1 end, with thick double wall
(outer perforated, inner chambered);
pale brown. Spore print brown.

Season: May–November; may overwinter.

Habitat: On dead coniferous wood, especially
hemlock, spruce, and pine.

Range: E. Canada to Maine, North Carolina,
and Midwest.

Look-alikes: *G. oregonense* is much larger, thicker,
darker, and duller in color. Both *G.
curtisii* and *G. lucidum* * grow on
deciduous trees.

Comments: When first forming in May and early
June, this polypore resembles a
protruding knob on hemlock trees and
stumps. At this stage, very little
reddish crust has formed, and the flesh
is tender.

544 Sweet Knot
Globifomes graveolens (Schw.) Murr.
Polyporaceae, Aphyllophorales

Description: *Large mass of many overlapping, small,
thin, leathery, brownish caps with gray to
gray-brown pores; stalkless, attached to
wood by solid central core.*
Cap: ⅜–¾″ (1–2 cm) wide; densely
overlapping, round to cylindrical
clusters, fused at sides, with downcurved margin; steel-gray to pale brown
with whitish margin, becoming rust- to
cinnamon-brown or darker, aging
grayish-black to black; leathery to
rigid, slightly encrusted, powdery to
smooth. Flesh 1–4 mm thick; brown.
Odor (when present) fragrant.
Tubes: 1–4 mm long. Pores (3–4 per
mm) circular, concealed by caps; gray,
becoming gray-brown to brown.
Spores: 9–12.5 × 3–4.5 μ;
cylindrical, smooth, colorless to light
brown. Spore print brown.

Season: July–October; overwinters.

Habitat: On logs and trunks of deciduous trees,
especially maple, hickory, beech, sweet
gum, and oak.

Range: New York to Alabama; Midwest.

Comments: Also known as *Polyporus graveolens*. The Sweet Knot is often odorless, but can still be recognized by its massed, overlapping caps.

---

516 **Yellow-red Gill Polypore**
Gloeophyllum sepiarium (Fr.) Kar.
Polyporaceae, Aphyllophorales

Description: *Tough, spreading, with projecting hairy, zoned, yellowish-red, stalkless cap and leathery, gill-like pores.*
Cap: 1–4″ (2.5–10 cm) wide; often overlapping, flat, semicircular; bright yellowish-red to brown, with bright white, yellow, or orange margin; hairy to nearly smooth, zoned. Flesh 1–5 mm thick; yellowish- or rust-brown; black with KOH.
Tubes: 5–7 mm long; gill-like, crowded, sometimes mixed with pores. Pores (1.5–2 per mm) ochre to golden-brown.
Spores: 9–13 × 3–5 μ; cylindrical, smooth, colorless. Spore print white.
Season: June–November; may overwinter.
Habitat: On dead conifers; reported on hardwoods.
Range: Throughout North America.
Look-alikes: *G. trabeum* is gray to dull brown, becoming blackish, with pore surface partly poroid, partly mazelike, partly gill-like. *Lenzites betulina** is grayish-white, with zoned cap and white flesh.
Comments: Also known as *Lenzites saepiaria*. *Gloeophyllum* is differentiated in part from *Lenzites* by the colored flesh.

474, 475 **Hen of the Woods**
Grifola frondosa (Fr.) S.F.G.
Polyporaceae, Aphyllophorales

Description: *Large, clustered mass of grayish-brown, fleshy, spoon-shaped caps with whitish pores and lateral, white stalks branching from compound base.*
Cap: ¾–2¾″ (2–7 cm) wide;

overlapping, flat, fan- to spoon-shaped; grayish to gray-brown; dry, smooth or finely fibrous to roughened. Flesh 3–5 mm thick; white.
Tubes: 2–3 mm long; descending stalk. Pores (1–3 per mm) angular, white to yellowish.
Stalk: rudimentary or very short and thick; many-branched; white, smooth.
Spores: 5–7 × 3.5–5 μ; broadly elliptical, smooth, colorless. Spore print white.

Edibility: Choice.

Season: September–November.

Habitat: On ground at base of oak and other deciduous trees, and some conifers; also on stumps.

Range: Canada to Louisiana; Midwest; Idaho.

Look-alikes: *Meripilus giganteus** has large, thick-fleshed shelves; edges and pores bruise black. *Polyporus umbellatus** has centrally stalked circular caps. *P. illudens,* only in Idaho, is yellowish and grows on the ground under conifers.

Comments: Also called *Polyporus frondosus.* This choice edible is often hard to differentiate from the fallen leaves that are usually around it. It can be found for many years at the same spot or in the same area, often in clusters 10–20″ wide, weighing 5 to 100 pounds.

510 Tender Nesting Polypore
Hapalopilus nidulans (Fr.) Kar.
Polyporaceae, Aphyllophorales

Description: *Fleshy, orange-brown, stalkless cap with brownish pores.*
Cap: 1–5″ (2.5–12.5 cm) wide; convex; ochre to tawny-brown, finely hairy or smooth; spongy at first, drying hard. Flesh 2–40 mm thick; soft and watery when fresh, tawny-brown; purple with KOH.
Tubes: 2–10 mm long. Pores (2–4 per mm) angular or sinuous; yellowish- to cinnamon-brown.
Spores: 3.5–5 × 2–3 μ; elliptical to

cylindrical, smooth, colorless. Spore
print white.

Season: June–November; overwinters.

Habitat: On dead deciduous wood, such as
maple, beech, hickory, oak, willow.

Range: Canada to Louisiana, Ohio, and Kansas.

Look-alikes: *Aurantioporus croceus* is larger, waxy to
greasy on handling, and red with KOH.

Comments: Also known as *Polyporus nidulans.* This
eastern mushroom is similar in form
and texture to the rather colorless genus
Tyromyces; species identification can be
confirmed by the reaction to KOH.

519 Conifer-base Polypore
Heterobasidion annosum (Fr.) Bref.
Polyporaceae, Aphyllophorales

Description: *Hard, brownish, furrowed, stalkless cap, or
irregular projection from spreading crust,
with white to yellowish pores.*
Cap: 1–10″ (2.5–25 cm) wide; flat,
with wavy margin; single to several,
sometimes fused in long rows, often
projecting from spreading, crustlike
mass; off-white, gray-brown, or dark
brown, tinged red to blackish with age;
compactly hairy, becoming smooth,
with thin, black, crustlike skin, usually
concentrically zoned or furrowed. Flesh
0.2–1 cm thick; white to pinkish-
cinnamon.
Tubes: 2–10 mm long, often growing
only a single layer (3–5 mm) per
season. Pores (2–5 per mm) indistinctly
layered, circular to angular; white to
yellowish; often deformed, receding
from old areas.
Spores: 3.5–5 × 3–4 μ; broadly
elliptical to almost round, minutely
spiny, colorless. Spore print white.

Season: Year-round.

Habitat: At base of stumps, dead trunks, or on
roots of conifers; also on some
hardwoods and timbers.

Range: Maine to Florida; Missouri; Colorado;
New Mexico; Pacific NW.

Look-alikes: *Fomitopsis pinicola** is more regular in

shape, has a prominent, colored crust, distinctly stratified tube layers, and smooth spores. *Phellinus** species have orange to mustard-yellow to rust-brown flesh and grow near deciduous trees.

Comments: Also known as *Fomes annosus,* this is one of the fungi most destructive to conifers.

---

534   **Warted Oak Polypore**
Inonotus dryadeus (Fr.) Murr.
Polyporaceae, Aphyllophorales

Description: *Tough, stalkless, white to grayish-brown, warted to bumpy, crusted cap with grayish-brown pores.*
Cap: 3–16″ (7.5–40 cm) wide; single or sometimes fused, convex to irregular, with thick, rounded margin, exuding water drops when young; whitish to gray, becoming gray-yellow, red-brown, or blackish; flat-hairy, becoming smooth; uneven to warty or bumpy. Flesh 1.5–5 cm thick; soft and spongy when fresh, yellow- to rust-brown; black with KOH.
Tubes: 0.5–3 cm long. Pores (3–5 per mm) round to angular; whitish to gray-brown, often with silvery sheen, becoming dark.
Stalk (when present): very short.
Spores: 7–8.5 × 6.5–8 μ; round, smooth, whitish to pale yellow. Setae present.
Season: September–October; overwinters.
Habitat: At base of oaks, or on recent stumps; reported on conifers in Pacific NW.
Range: New York to Georgia, west to Missouri and Texas; Pacific NW.
Comments: Also known as *Polyporus dryadeus.*
Inonotus refers to the fibrous, hairy surface of the cap of most species, and *dryadeus* to this species' association with oak.

597 Clinker Polypore
Inonotus obliquus (Fr.) Pil.
Polyporaceae, Aphyllophorales

Description: *Large, black, stalkless, cracked canker.*
Mushroom: sterile core 10–16″ (25–40 cm) wide; black, very hard, deeply cracked; fertile portion up to 5 mm thick, crustlike, thin, dark brown.
Tubes: 5–10 mm long; oblique, usually split in front. Pores (3–5 per mm) angular to elongate; whitish, becoming dark brown.
Spores: 7.5–10 × 5–7.5 μ; broadly elliptical; smooth; colorless, becoming light yellow. Setae present.

Season: Year-round.

Habitat: On birch; reported on elm and alder.

Range: Throughout N. North America.

Comments: Also called the "Birch Canker Polypore" and *Poria obliqua*. This species is called the Clinker because the canker resembles something that has been burned and fused together.

584 Milk-white Toothed Polypore
Irpex lacteus (Fr.) Fr.
Polyporaceae, Aphyllophorales

Description: *Thin, leathery, spreading, with projecting white, hairy, stalkless caps; white to yellowish tubes breaking into teeth.*
Cap (when present): ⅜–1⅝″ (1–4 cm) wide; commonly overlapping or laterally fused, projecting from spreading, elongated, crustlike mass; white, drying yellowish; hairy to woolly. Flesh 0.5–2 mm thick, white.
Tubes: 1–5 mm long; soon breaking into flattened teeth. Pores (2 per mm) angular to curving to irregular; white to cream-yellow.
Spores: 5–6 × 2–3 μ; elliptical to cylindrical, smooth, colorless. Cystidia long, encrusted. Spore print white.

Season: Annual; overwinters.

Habitat: On dead deciduous wood; also reported on some conifers.

Range: Throughout North America.
Comments: Also known as *Polyporus tulipiferae*, referring to one of its many hosts, the tuliptree.

---

525 Resinous Polypore
Ischnoderma resinosum (Fr.) Kar.
Polyporaceae, Aphyllophorales

Description: *Large, flattened, hairy, brownish, stalkless cap with thick margin and whitish pores.*
Cap: 3–10″ (7.5–25 cm) wide; single to overlapping; semicircular, with thick, rounded margin, exuding water drops when young; ochre to dark brown, becoming blackish; velvety, hairy, becoming smooth at maturity, sometimes concentrically furrowed; often with dark, shiny, crusty, resinous zones, often radially wrinkled. Flesh 1–2.5 cm thick; tough and watery when young, drying hard, whitish to tan; darker with KOH.
Tubes: 1–10 mm long. Pores (4–6 per mm) circular to angular; white to pale brown, bruising darker.
Spores: 4.5–6.5 × 1.5–2.5 μ; cylindrical to sausage-shaped, smooth, colorless. Spore print whitish.
Season: September–October.
Habitat: On logs and stumps of deciduous trees.
Range: Widely distributed in North America.
Look-alikes: *Ganoderma** species have varnished crust. *Fistulina hepatica** has gelatinous layer and free tubes.
Comments: Also known as *Polyporus resinosus.* Its coniferous-wood look-alike, the Resinous Conifer Polypore (*I. benzoinum*), has thinner, ochre to brown flesh and tubes.

478, 511 Chicken Mushroom
Laetiporus sulphureus (Fr.) Murr.
Polyporaceae, Aphyllophorales

Description: *Single to overlapping clusters of fleshy, smooth, orange-red to orange-yellow caps* with sulfur-yellow pores.

Cap: 2–12″ (5–30 cm) wide; usually
overlapping, flat, semicircular to fan-
shaped; salmon to sulfur-yellow or
bright orange, weathering to white;
smooth. Flesh 0.4–2 cm thick, white,
light yellow or pale salmon.
Tubes: 1–4 mm long. Pores (2–4 per
mm) angular, bright sulfur-yellow.
Stalk (when present): rudimentary.
Spores: 5–7 × 3.5–5 μ; broadly
elliptical to almost round, smooth,
colorless. Spore print white.

Edibility: Choice.

Season: May–November.

Habitat: On stumps, trunks, and logs of
deciduous and coniferous trees; also on
living trees and buried roots.

Range: E. Canada to Florida; Midwest; Pacific
NW south to California.

Comments: Also called the "Sulfur Shelf" and
Polyporus sulphureus. This popular
mushroom can grow in large
overlapping clusters of 5 to 50 or more,
with shelves weighing up to a pound
each, or in rosettes 10–30″ wide. A
choice edible, it tastes like chicken. It
becomes somewhat indigestible as it
ages and, in some, causes an allergic
reaction, such as swollen lips.
Specimens from a few tree hosts, such
as eucalyptus, can cause digestive
upset. A variety, *semialbinus,* has a
salmon-colored cap and white spores.

---

535 Multicolor Gill Polypore
Lenzites betulina (Fr.) Fr.
Polyporaceae, Aphyllophorales

Description: *Tough, spreading, with projecting, hairy,
gray to brownish, multizoned, stalkless cap
with leathery, whitish, gill-like pores.*

Cap: 1–4″ (2.5–10 cm) wide; flat,
semicircular; whitish to grayish,
sometimes brownish, with narrow,
concentric, multicolored orange and
brown zones; hairy. Flesh 1–2 mm
thick; white.

Tubes: 8–12 mm long; gill-like or sometimes porelike near margin. Pores (1–1.5 per mm) white.
Spores: 4.5–6 × 2–3 μ; cylindrical, curved, smooth, colorless. Spore print white.

Season: July–November; overwinters.

Habitat: On dead deciduous wood; also reported on conifers.

Range: Throughout Canada, south to Florida; Midwest; Texas; Pacific NW. and California.

Comments: Despite this mushroom's scientific name, birch is only 1 of more than 2 dozen host trees. When turned over it resembles a gilled mushroom; the "gills" are actually a radially arranged variation of tubes and pores. If there is any confusion in the field, it can be resolved by noting the tough, leathery texture of the "gills."

471 Black-staining Polypore
Meripilus giganteus (Fr.) Kar.
Polyporaceae, Aphyllophorales

Description: *Several to many large, fleshy, grayish-yellow caps, with whitish pores bruising blackish, and very short, thick, ochre stalk.*
Cap: 2–8″ (5–20 cm) wide; in clusters; semicircular to fan- or spoon-shaped, with thin, sharp margin; grayish to yellowish-drab, becoming smoky and dark, blackening along margin; finely hairy or fibrous, radially wrinkled.
Flesh 1–10 mm thick; white.
Tubes: 1–3 mm long. Pores (4–7 per mm) angular, becoming torn; white, bruising blackish.
Stalk: rudimentary or very short and thick; ochre, smooth to fibrous.
Spores: 6–7 × 4.5–6 μ; broadly elliptical to almost round, smooth, colorless. Spore print white.

Edibility: Good.

Season: July–September.

Habitat: On ground around stumps or living

deciduous trees, especially oak and beech.

Range: Massachusetts to Louisiana and Missouri.

Look-alikes: *Grifola frondosa** and *Polyporus umbellatus** have many small, thin-fleshed caps. *Laetiporus sulphureus** is yellowish to orange-red, with yellow pores. *Bondarzewia berkeleyi** is clustered, thick fleshed, cream-yellow. None bruises black.

Comments: Also known as *Polyporus giganteus*. This relatively common mushroom can often be found in clusters 10−16″ wide, weighing up to 30 pounds, although clumps weighing 3−10 pounds are more typical. When very young it tastes and feels like beef liver.

448 **Woolly Velvet Polypore**
Onnia tomentosa (Fr.) Kar.
Polyporaceae, Aphyllophorales

Description: *Velvety, ochre to rust-brown cap and stalk, with brownish pores.*
Cap: 1¼−7¼″ (3−18 cm) wide; circular to semicircular or fan-shaped; ochre to rust-brown, velvety, hairy. Flesh 3−10 mm thick; ochre to rust-brown; black with KOH.
Tubes: 1.5−7 mm long. Pores (2−4 per mm) angular to somewhat irregular; brownish.
Stalk (when present): 1−2″ (2.5−5 cm) long, ¼−¾″ (0.5−2 cm) thick; lateral, off-center or central, often rudimentary; ochre to dark rust-brown.
Spores: 4.5−7 × 2.5−4 μ; elliptical, smooth, golden-yellow. Setae straight and pointed.

Season: August−October.

Habitat: On ground attached to buried wood, or on coniferous trunks and stumps.

Range: Throughout N. North America, south to Alabama; Pacific NW. to California.

Look-alikes: *O. circinata* is stalkless, grows on wood or spruce roots, and has thicker cap and hooked setae.

Comments: Also known by other generic names:
Polyporus, Coltricia, and *Mucronoporus.*
This relatively common mushroom can
be used to produce a variety of dyes.

461 Bone Polypore
Osteina obducta (Berk.) Donk
Polyporaceae, Aphyllophorales

Description: *One to several tough to hard, smooth, white*
to grayish caps with whitish pores
descending short, white stalk.
Cap: 1–4″ (2.5–10 cm) wide; circular
to spoon-shaped, fanning out from
single stalk; white to gray, or light
brown, smooth. Flesh 3–10 mm thick;
white.
Tubes: 1–3 mm long; slightly
descending stalk. Pores (3–5 per mm)
minute, angular; white, drying
yellowish.
Stalk: ⅜–1¼″ (1–3 cm) long, ⅛–¼″
(3–5 mm) thick; short to nearly absent,
off-center to lateral; white to cream;
smooth.
Spores: 4.5–6.5 × 2–3 μ; cylindrical,
smooth, colorless. Spore print white.
Season: July–November.
Habitat: On dead coniferous wood and on birch.
Range: E. Canada; N. United States; Pacific
NW.; California; Colorado.
Look-alikes: *Tyromyces chioneus** and related species
are stalkless and much softer.
Comments: Formerly called *Polyporus osseus.* Both
Osteina and *Osseus* mean "bonelike,"
referring to the color and hardness of
the dried mushroom.

521 Mossy Maple Polypore
Oxyporus populinus (Fr.) Donk
Polyporaceae, Aphyllophorales

Description: *Tough, white, overlapping, stalkless caps*
with white to yellowish-orange pores;
typically covered with moss.
Cap: 1–8″ (2.5–20 cm) wide; usually
overlapping, in large vertical masses;

semicircular to elongated; white to gray
or somewhat ochre when fresh, drying
dark cream; finely and densely hairy,
becoming smooth to slightly warty,
typically covered with moss. Flesh 0.3–
1 cm thick; watery, firm, white to
ochre.
Tubes: 2–5 mm long, 4 cm deep in old
specimens; distinct new layer each
season (up to 20 layers). Pores (4–6 per
mm) round to angular; white to cream
or yellowish-orange.
Spores: 4–5.5 μ; round, smooth,
colorless. Spore print white.

Season: Year-round.

Habitat: On trunk in wounds near base of living
maple; also on other deciduous trees.

Range: E. Canada to North Carolina and
Georgia, west to Minnesota and
Missouri.

Comments: Also known as *Fomes connatus*. *Populinus*
is a misnomer, at least in North
America, where the mushroom is most
frequently found in wounds at the base
of living maples, not poplars.

---

480 Dye Polypore
Phaeolus schweinitzii (Fr.) Pat.
Polyporaceae, Aphyllophorales

Description: *One to several fleshy, tough, ochre to rust-
brown, velvety to woolly caps with greenish-
yellow pores.*
Cap: 2–10″ (5–25 cm) wide; circular to
semicircular or fan-shaped, with
tapering base and sharp, wavy margin;
ochre to orange or rust-brown, lighter
at margin; hairy to woolly, somewhat
rough, weathering to smooth. Flesh
0.2–3 cm thick; yellowish to rust,
black with KOH.
Tubes: 1–10 mm long. Pores 0.5–1
mm wide along margin, 2–3 mm wide
near stalk; angular to somewhat fused;
greenish-yellow, darkening on
bruising, becoming rust to dark brown.
Stalk (when present): ⅜–2⅜″ (1–6 cm)
long, ⅜–1⅝″ (1–4 cm) thick; single or

several fused together and expanding
upward; central to off-center; ochre to
dark rust-brown, hairy, soft.
Spores: 5–8 × 3.5–4.5 μ; elliptical,
smooth, colorless. Cystidia long, thin.
Spore print whitish.

Season: June–November.
Habitat: On roots, stumps, or trunks of conifers;
also reported on some deciduous trees.
Range: N. North America south to Florida,
west to Colorado and California.
Comments: Also known as *Polyporus schweinitzii*.
The young mushrooms are often
brightly colored and can produce dyes
of many colors.

579   **Golden Spreading Polypore**
Phellinus chrysoloma (Fr.) Donk
Polyporaceae, Aphyllophorales

Description: *Woody, spreading, with projecting
brownish, stalkless caps with ochre-tawny,
angular or elongated pores.*
Cap: ⅜–3″ (1–7.5 cm) wide; in
densely overlapping clusters or lateral,
partly fused rows; thin and flat to
crustlike, with thin, sharp margin;
tawny to russet-tawny, with brighter
margin, becoming brownish-black;
zoned, with ridges of hair. Flesh 1–3
mm thick; separated from cap surface
hair by black line; tawny or ochre.
Tubes: 2–5 mm long, in annual layers
up to 15 mm deep in old specimens.
Pores (2–5 per mm) round to angular
or wavy; bright ochre-tawny.
Spores: 4.5–5.5 × 4–5 μ; roundish,
smooth, colorless to pale yellowish.
Setae present. Spore print light brown.
Season: Year-round.
Habitat: On living or fallen trunks of conifers.
Range: Canada south to Florida, Midwest,
Colorado, Pacific NW., and California.
Comments: Formerly known as *Fomes pini* var.
abietis and *Phellinus pini* var. *abietis*. The
related Ochre-orange Hoof Polypore (*P.
pini*) produces individual, unattached,
hoof-shaped shelves.

517 Mustard-yellow Polypore
Phellinus gilvus (Schw.) Pat.
Polyporaceae, Aphyllophorales

Description: *Tough, spreading, with projecting, hairy,
ochre to rust-brown, stalkless cap with tiny
brown pores.*
Cap: 1–6″ (2.5–15 cm) wide; often in
overlapping clusters, semicircular to
somewhat convex; ochre to bright rust-
yellow, becoming dark rust-brown or
blackish; velvety to roughly fibrous,
becoming smooth. *Flesh* 0.1–2 cm
thick; corky-tough, *yellowish-brown;*
black with KOH.
Tubes: 1–5 mm long; sometimes in
annual layers. Pores (5–8 per mm)
minute, circular; grayish- to reddish-
brown or darker.
Spores: 4–5 × 2.5–3.5 μ; oblong to
elliptical, smooth, colorless. Setae
present. Spore print whitish.
Season: Year-round.
Habitat: On dead deciduous wood; reported on
dead conifers.
Range: Throughout North America.
Look-alikes: *Inonotus radiatus* is bright ochre, fleshy,
and firm.
Comments: Also known as *Polyporus gilvus.* Under
favorable conditions, this species can
survive for several years.

530 Flecked-flesh Polypore
Phellinus igniarius (Fr.) Quél.
Polyporaceae, Aphyllophorales

Description: *Woody, brownish to dark gray, furrowed,
stalkless cap with brownish pores.*

Cap: 2–10″ (5–25 cm) wide; single to
several, hoof-shaped to convex or
semicircular and flat, with rounded
margin; brown, becoming grayish-
black or black; finely hairy and often
concentrically furrowed, becoming
smooth and usually cracking. Flesh 3–
15 mm thick; rust-brown.
Tubes: 2–5 mm long, in annual layers
up to 10 cm deep in old specimens;

older tube layers become conspicuously
stuffed with white threads. Pores (4–5
per mm) nearly circular, grayish-brown
to brown.

Spores: 5–7 × 4.5–6 μ; almost round,
smooth, thick-walled, colorless. Setae
present. Spore print whitish.

Season: Year-round.

Habitat: On living or dead deciduous trees,
especially birch and aspen.

Range: Throughout North America.

Look-alikes: *P. rimosus** has deeply cracked cap,
grows on black locust. *P. robustus* has
shiny, yellowish-brown flesh. *P.
everhartii,* usually on oak, has brown
spores.

Comments: Also known as *Fomes igniarius.* The
Flecked-flesh Polypore is usually seen as
a complex of closely related species. All
grow on deciduous wood and have
orange-brown flesh, tubes stuffed with
white threads (mycelium), and setae.
These include 2 crustlike fungi: the
Fruit Tree Polypore (*P. pomaceus*) and
the Birch Crust Polypore (*P. laevigatus*).

---

533 Cracked-cap Polypore
Phellinus rimosus (Berk.) Pil.
Polyporaceae, Aphyllophorales

Description: *Woody, brown, concentrically furrowed,
stalkless cap, becoming deeply cracked, with
brownish pores.*
Cap: 2–12″ (5–30 cm) wide;
semicircular, flat to hoof-shaped, with
thick, rounded margin; brown,
becoming blackish; finely hairy, with
concentric furrows, becoming cracked.
Flesh 0.5–2 cm thick; yellow-brown.
Tubes: 2–9 cm long, adding 2–5 mm
per season. Pores (4–6 per mm)
circular, yellow-brown to deep, rich
brown.
Spores: 4–6 μ wide or 4.5–6 × 3.5–5
μ; round to elliptical, smooth,
brownish. Spore print brownish.

Season: Year-round.

Habitat: On living or dead black locust; reported

on other trees in the same family.

Range: New York to Florida; Midwest; Texas; Southwest; California.

Look-alikes: *P. everhartii* has rust-brown flesh, setae in tube surface, and grows on oak. *P. igniarius** has brown flesh, conspicuous, white, threadlike (mycelial) stuffing in older tubes, and does not grow on locust. *P. robustus* has shiny, bright yellow-brown flesh, colorless spores, and grows on oak.

Comments: Also known as *Fomes rimosus.* Although several polypores may develop deep cracks on the cap, few such species occur on black locust.

---

491 Birch Polypore
Piptoporus betulinus (Fr.) Kar.
Polyporaceae, Aphyllophorales

Description: *Fleshy, tough to hard, smooth, whitish cap with inrolled margin and white to brownish pores.*

Cap: 1–10″ (2.5–25 cm) wide; semicircular to kidney-shaped, slightly convex to flat, with rounded, inrolled margin; white to tan, becoming smoky or brown; tough, with soft, smooth, easily dented covering. Flesh 1–3.5 cm thick; white.

Tubes: 2–10 mm long; easily separating from cap flesh. Pores (3–4 per mm) circular to angular, becoming torn with age; white to light brown.

Stalk (when present): rudimentary, lateral, stublike.

Spores: 5–6 × 1.5 μ; sausage-shaped, smooth, colorless. Spore print white.

Season: Annual; overwinters.

Habitat: On living or dead birch trees.

Range: Throughout Canada, south to North Carolina, Midwest, and Washington.

Comments: Also known as *Polyporus betulinus.* Through its varied history this species has been used in the absence of matches to keep fires blazing, as an anesthetic, and as a razor strop.

509 Spring Polypore
Polyporus arcularius Bat. ex Fr.
Polyporaceae, Aphyllophorales

Description: *Fleshy, tough, scaly, brownish, circular
cap with large, angular, whitish pores and
brownish stalk.*
Cap: ⅜–3¼" (1–8 cm) wide; convex to
vase-shaped, circular; golden-brown,
dry, scaly, with fine hairs on margin.
Flesh 1–2 mm thick; white.
Tubes: 1–2 mm long; sometimes
descending stalk. Pores 0.5–1 mm
wide; large, hexagonal or angular,
white to yellowish.
Stalk: ¾–2⅜" (2–6 cm) long, ¹⁄₁₆–⅛"
(1.5–3 mm) thick; central; yellowish-
brown or dark brown, slightly scaly,
smooth or hairy at base.
Spores: 7–11 × 2–3 μ; cylindrical,
smooth, colorless. Spore print white.
Season: May–June; overwinters.
Habitat: On dead deciduous wood; on ground
over buried wood.
Range: Throughout North America.
Look-alikes: *P. brumalis** has smooth, hairy cap and
roundish pores. *Favolus alveolaris** has
kidney-shaped cap and off-center to
lateral, stublike stalk.
Comments: Few mushrooms appear as early in the
year as this species.

543 Black-footed Polypore
Polyporus badius (S.F.G.) Schw.
Polyporaceae, Aphyllophorales

Description: *Leathery, smooth, bay-brown cap with
white to brownish pores and smooth, black
stalk.*
Cap: 2–8" (5–20 cm) wide; convex to
vase-shaped, circular; dark reddish-
brown, becoming blackish with age;
dry, smooth. Flesh 1–7 mm thick;
white.
Tubes: 1–2 mm long; slightly
descending stalk. Pores (5–7 per mm)
minute, circular to angular, whitish to
brownish.

Stalk: ⅜–2⅜″ (1–6 cm) long, ⅛–⅝″ (0.3–1.5 cm) thick; central or off-center; smooth, black or pallid above, black below.

Spores: 6–10 × 3–4 μ; cylindrical to oblong-elliptical, smooth, colorless. Spore print white.

Season: August–October; overwinters.

Habitat: On stumps and logs of deciduous trees; on ground on buried wood; also reported on wood of some conifers.

Range: Throughout Canada, south to Alabama, Ohio, Missouri, and California.

Look-alikes: *P. varius** grows mostly on small, dead branches, is smaller, and cinnamon-buff to tan. *P. melanopus* is smaller, with downy to wrinkled cap and larger pores.

Comments: Also known as *P. picipes*.

375 Winter Polypore
Polyporus brumalis Fr.
Polyporaceae, Aphyllophorales

Description: *Tough, brownish, smooth to hairy cap with round, whitish pores and grayish to light brownish stalk.*

Cap: ⅝–4″ (1.5–10 cm) wide; convex to slightly sunken; circular, with inrolled margin; yellow-brown to dark brown, fading to tan; dry, with dense, short hairs, or roughened when young to smooth. Flesh 1–2 mm thick; white.

Tubes: 1–3 mm long; slightly descending stalk. Pores (2–3 per mm) circular, becoming angular, whitish.

Stalk: 1–2″ (2.5–5 cm) long, 1/16–¼″ (1.5–5 mm) thick; central or off-center; grayish or brownish, minutely hairy or smooth.

Spores: 5–7 × 2–2.5 μ; sausage-shaped to cylindrical, smooth, colorless. Spore print white.

Season: October–June; overwinters.

Habitat: On dead deciduous wood, especially birch.

Range: Nova Scotia to Manitoba, Ohio, and Tennessee.

Look-alikes: *P. arcularius** has scaly, yellowish-brown cap and large, hexagonal pores. *P. badius**, *P. melanopus,* and *P. varius** all have black lower stalk.

Comments: *Polyporus brumalis* means "winter polypore."

---

465 **Bitter Iodine Polypore**
Polyporus hirtus Quél.
Polyporaceae, Aphyllophorales

Description: *Large, fleshy, tough, brownish cap and stalk covered with short, stiff hairs; whitish pores descend stalk.*
Cap: 2–6" (5–15 cm) wide; convex to somewhat irregular, fan-shaped; grayish-brown to dark brown; dry, covered with short, erect, stiff hairs. Flesh 0.3–2 cm thick; tough, white. Odor iodinelike; taste bitter.
Tubes: 2–6 mm long; descending stalk. Pores (1–2 per mm) roundish, becoming angular to irregular; whitish, becoming cream.
Stalk: 1–4" (2.5–10 cm) long, ⅜–1" (1–2.5 cm) thick; usually lateral, solid, hard; grayish-brown to dark brown, with short, erect, stiff hairs.
Spores: 12–17 × 4.5–6 μ; spindle-shaped, smooth, colorless. Spore print white.

Season: September–March.

Habitat: Near trees and stumps, attached to buried wood, especially fir, spruce, Douglas-fir, and hemlock.

Range: Quebec to New York, west to British Columbia and California.

Comments: This polypore is common in the Pacific Northwest and California.

419, 447 **Rooting Polypore**
Polyporus radicatus Schw.
Polyporaceae, Aphyllophorales

Description: *Fleshy, tough, yellow-brown, somewhat scruffy cap with white to yellowish pores descending long, rootlike stalk.*

Cap: 1⅜–10" (3.5–25 cm) wide;
convex to sunken, circular; yellowish-
brown, dry, scaly to velvety, rough or
scruffy. Flesh 3–10 mm thick; white.
Tubes: 1–5 mm long; descending
stalk. Pores (2–3 per mm) angular;
whitish or yellowish.
Stalk: 2⅜–6" (6–15 cm) long, ¼–1"
(0.5–2.5 cm) thick; central; dirty
yellow, scruffy to somewhat scaly, with
black, rootlike underground part.
Spores: 12–15 × 6–8 μ; ovately
elliptical, smooth, colorless. Spore
print white.

Season: August–October.

Habitat: On the ground, probably attached to
buried roots.

Range: E. Canada to Alabama and Midwest.

Look-alikes: *P. melanopus* is tougher, with a shorter,
black, only slightly rootlike stalk.
P. tuberaster has an ochre, scaly cap and
grows from a large, black, tuberlike
structure. Other similar polypores grow
on wood above ground. Boletes have
soft, fleshy texture and detachable
tubes.

Comments: This eastern fall polypore is the only
one with a long, rootlike stalk.

---

507 **Dryad's Saddle**
Polyporus squamosus Fr.
Polyporaceae, Aphyllophorales

Description: *Large, fleshy, tough, scaly, yellowish-brown*
cap with large, white to yellowish pores
descending short stalk.
Cap: 2⅜–12" (6–30 cm) wide; single
or in overlapping clusters; flat to
sunken, almost circular to kidney-
shaped; whitish to dingy yellowish or
brownish; with dense, flat, overlapping
scales. Flesh 0.5–3.5 cm thick; white.
Odor and taste like watermelon rind.
Tubes: 2–8 mm long; large,
descending stalk. Pores angular, white
or yellowish.
Stalk: ⅜–2" (1–5 cm) long, ⅜–1⅛"
(1–4 cm) thick; stublike, lateral or off-

center; black at base.
Spores: 10–16 × 4–6 μ; oblong-
elliptical to cylindrical, smooth,
colorless. Spore print white.

Edibility: Edible.
Season: May–November; most common in May.
Habitat: On living or dead deciduous wood,
such as elm, maple, willow, poplar,
and birch.
Range: E. Canada to North Carolina and
Midwest.
Look-alikes: *P. mcmurphyi,* in California, has smaller
pores and whitish stalk. *P. lentus,* also
called *P. fagicola,* is smaller, with less
scaly cap, smaller pores, and yellowish-
white, central or off-center stalk with
hairy base.
Comments: Also known as "Pheasant's-back
Polypore." The tender edges of the caps
can be pickled, sautéed, or fried.

424, 473 Umbrella Polypore
Polyporus umbellatus Pers. ex Fr.
Polyporaceae, Aphyllophorales

Description: *Fleshy, overlapping, white to brownish,
circular caps with white descending pores,
each with central stalk from a many-
branched, single stalk.*
Cap: ⅜–1⅛" (1–4 cm) wide;
overlapping, circular; white to smoky,
dry, fibrous or smooth.
Tubes: 1–2 mm long; slightly
descending stalk. Pores (2–4 per mm)
angular, whitish, becoming yellowish.
Stalk: 1–3" (2.5–7.5 cm) long, ¾–
1¼" (2–3 cm) thick; short, strongly
branched above ground; rising from
hard, blackish, tuberlike underground
structure with white interior.
Spores: 8–10 × 2.5–3.5 μ;
cylindrical, smooth, colorless. Spore
print white.
Edibility: Choice.
Season: May–October.
Habitat: On ground on buried wood; around
stumps or deciduous trees.
Range: E. Canada to Tennessee; Ohio; Iowa;

reported in Idaho and Washington.

Look-alikes: *Grifola frondosa** has individual, spoon-shaped caps, compound base, and lacks underground tuberous structure.

Comments: The Umbrella Polypore grows in the spring and fall; it is usually found in clusters 3–8″ wide, but can reach 20″ in width. It is a choice edible.

---

453 **Elegant Polypore**
Polyporus varius Fr.
Polyporaceae, Aphyllophorales

Description: *Leathery, cinnamon-buff to tan cap with angular, grayish pores descending stalk; stalk black at base.*
Cap: 2–4″ (5–10 cm) wide; circular to kidney-shaped, flat to concave; cinnamon-buff to tan, weathering whitish; dry, nearly smooth. Flesh 1–5 cm thick; white to pale brown.
Tubes: 1–3 mm long; descending stalk. Pores (4–5 per mm) angular, white to gray or yellowish to light reddish-brown.
Stalk: ¼–3″ (0.5–7.5 cm) long, 1/16–⅝″ (0.15–1.5 cm) thick; nearly central to lateral; tan, with black at base or on lower half; smooth or with fine hairs.
Spores: 7–10 × 2–3.5 μ; cylindrical to sausage-shaped, smooth, colorless. Spore print white.
Season: June–November.
Habitat: On dead deciduous wood; reported on pine.
Range: Throughout North America.
Look-alikes: *P. badius** has thinner flesh and dark, reddish-brown cap. *P. melanopus* has distinct division between pores and stalk. *P. arcularius** and *P. brumalis** lack black on the stalk.
Comments: Also known as *P. elegans*, the Elegant Polypore is sometimes called the "Black-footed Polypore" because of the distinctive black base of its stalk.

581 Orange Poria
Poria spissa (Schw.) Cke.
Polyporaceae, Aphyllophorales

Description: *Crustlike, broadly spreading, stalkless, with yellow-brown margin and brilliant orange pores that darken on bruising.*
Mushroom: 2–10″ (5–25 cm) wide, 6–30″ (15–75 cm) long; spreading, crustlike, thin; with matted, hairy, pale yellowish-brown margin. Flesh 3–5 mm thick; thin, spongy to firm; white to pale brownish.
Tubes: 2–3 mm long; not layered. Pores (6–8 per mm) minute, round to angular; brilliant orange, bruising or drying dark brownish-red.
Spores: 4–5 × 1–1.5 μ; sausage-shaped, smooth, colorless. Spore print white.

Season: July–November.

Habitat: Typically on underside of deciduous wood; rarely on conifers.

Range: Throughout temperate North America.

Look-alikes: *P. mutans* bruises reddish, grows on chestnut. *P. taxicola* has large pores, grows on conifers.

Comments: Unlike most sheeting pore fungi, the Orange Poria is an annual.

489 Pendulous-disc Polypore
Porodisculus pendulus (Schw.) Murr.
Polyporaceae, Aphyllophorales

Description: *Tiny, leathery, hairy, whitish cap with white to brownish pores and knoblike stalk.*
Cap: 1/32–1/4″ (1–5 mm) wide; pendant, semicircular; whitish to brown; leathery to rigid; hairy, finely powdered, or mealy. Flesh 0.5–1 mm thick, white.
Tubes: 0.2–0.5 mm long. Pores (5–6 per mm) very minute, circular, whitish or brownish.
Stalk: 1/8–1/4″ (3–5 mm) long; stublike; attached at top, expanding into cap; whitish, powdery, mealy.
Spores: 3–4 × 1 μ; sausage-shaped, smooth, colorless. Spore print white.

Season: August–October.
Habitat: On freshly dead oak, hickory, and walnut wood; also reported on pine.
Range: Massachusetts to Florida, Ohio, Iowa, and Missouri.
Comments: Also known as *Polyporus pocula,* this is the smallest polypore, but it is often found growing in great numbers. It was originally described as a kind of *Peziza,* or cup fungus.

---

485 Little Nest Polypore
Poronidulus conchifer (Schw.) Murr.
Polyporaceae, Aphyllophorales

Description: *Small, zoned cup followed by small, whitish, shelflike caps with white to yellowish pores.*
Cap: ⅜–2" (1–5 cm) wide; semicircular to kidney-shaped; following, and sometimes attached to, small, white cups ¼–¾" (0.5–2 cm) wide; zoned on upper surface, smooth below; white to grayish-white or yellowish; surface zoned, at least on margin, smooth to slightly hairy, sometimes wrinkled. Flesh 0.5–1 mm thick, white.
Tubes: 1–2 mm long. Pores (2–4 per mm) angular, white to yellowish.
Stalk (when present): lateral, knoblike.
Spores: 5–7 × 1.5–2.5 μ; cylindrical, smooth, colorless. Cup-shaped structures sterile. Spore print white.
Season: June–November; overwinters.
Habitat: On dead elm branches; also reported on other deciduous wood.
Range: E. Canada to Alabama, west to Minnesota and Nebraska.
Comments: Also called *Polyporus conchifer.* The small cups, often found by themselves on dead elm branches, may be misidentified for a cup fungus or a bird's-nest fungus. The small, flat shelves often develop around them, and are not often identified without them.

747 Orange Sponge Polypore

Pycnoporellus alboluteus (Ell. & Ev.) Kotl. &
Pouz.
Polyporaceae, Aphyllophorales

Description: *Soft, spongy, spreading mushroom with*
orange pores breaking into irregular teeth.
Cap: 1–6″ (2.5–15 cm) wide; cushion-
shaped to elongated, spongelike,
spreading; orange, weathering to white.
Flesh 1–3 mm thick; orangish; red
with KOH.
Tubes: 1–3 cm long. *Pores 1–3 mm*
wide, large, angular, *becoming spinelike;*
white to yellowish, becoming orange.
Spores: 6–9 × 2.5–4 μ; cylindrical,
smooth, colorless.

Season: July–October; often overwinters.

Habitat: On lower sides of coniferous logs, such
as fir, pine, spruce, and hemlock.

Range: Pacific NW.; Colorado; Gaspé
Peninsula; rare in East.

Comments: Also known as *Polyporus alboluteus.* On
fallen conifers in the high mountains of
the West, this species looks like a
large, brightly-colored sponge; it can
cover an area of 3′. Only the pore mass
is usually visible; as it ages it
becomes angular, long, and mazelike or
cut-toothed.

512 Cinnabar-red Polypore

Pycnoporus cinnabarinus (Fr.) Kar.
Polyporaceae, Aphyllophorales

Description: *Tough, orange-red, stalkless cap with*
cinnabar- to orange-red pores.
Cap: 1–5″ (2.5–12.5 cm) wide;
semicircular to kidney-shaped or
elongated; flat, with sharp margin;
cinnabar-red to orange, finely hairy and
rough, becoming smooth, aging with
warts and wrinkles. Flesh 0.5–1.5 cm
thick, reddish; black with KOH.
Tubes: 1–6 mm long, occasionally in
2–3 layers. Pores (2–4 per mm)
angular or irregular, cinnabar- to
orange-red.

Spores: 5–6 × 2–2.5 μ; cylindrical,
curved, smooth, colorless. Spore print
whitish.

Season: Year-round.

Habitat: On dead deciduous wood, especially
cherry and oak; reported on coniferous
wood.

Range: E. Canada to Alabama; Midwest;
Colorado; New Mexico; Pacific NW.

Look-alikes: *P. sanguineus,* more southern in range,
is thinner and has smooth surface that
looks as if it has been seared with a hot
iron.

Comments: Also known as *Polyporus cinnabarinus.*
The bright cinnabar-red color of this
polypore makes it conspicuous.

---

585 Split-pore Polypore
Schizopora paradoxa (Fr.) Donk
Polyporaceae, Aphyllophorales

Description: *Hard, broadly spreading, whitish-cream or
pinkish crust, with pores soon breaking up
into teeth.*
Mushroom: 1–6″ (2.5–15 cm) wide,
2–12″ (5–30 cm) long; spreading,
crustlike; up to ¼″ (5 mm) thick and
elongated; with finely hairy, whitish
margin. Flesh 1 mm thick, white to
cream, soft, leathery.
Tubes: very short, sometimes flattened;
occupying entire surface except for
margin; poroid, becoming toothed,
longer in center than along margin.
Pores (1–4 per mm) angular; whitish-
cream to pinkish, drying cream to light
ochre.
Spores: 4–6.5 × 3–4 μ; broadly
elliptical, smooth, colorless.

Season: Annual; overwinters.

Habitat: On fallen twigs and branches of many
deciduous trees, especially oaks; in
woods and parks.

Range: Throughout North America.

Comments: Also known as *Poria versipora,* this very
common, drought-resistant polypore
exhibits a wide variation in its pore
surface; hence the name *Schizopora.*

546 Spongy Toothed Polypore
Spongipellis pachyodon (Pers.) Kotl. & Pouz.
Polyporaceae, Aphyllophorales

Description: *Leathery, tough, spreading, with projecting overlapping, white to ochre, stalkless caps and short, flattened teeth.*
Cap: 1–2″ (2.5–5 cm) wide; single or overlapping, projecting from spreading crust; white, becoming ochre; finely hairy, becoming smooth or sharp-ridged. Flesh 8–10 mm thick; white to pale cream, somewhat 2-layered: densely fibrous below, loose and softer above.
Tubes: 5–10 mm long; descending onto crust; porelike, but soon becoming toothlike; flattened on margin, cylindrical and tapering in center; white to ochre or light brown.
Spores: 5–6.5 μ; round, smooth, colorless.

Season: July–October; overwinters.
Habitat: On dead coniferous wood, such as fir, spruce, pine, and Douglas-fir.
Range: E. Canada to Florida; Michigan; Texas; New Mexico; Colorado; Pacific NW.
Comments: Also known as *Irpex mollis. Spongipellis* means "spongy skin," describing the condition of the very young mushroom.

526 White Spongy Polypore
Spongiporus leucospongia (Cke. & Hark.) Murr.
Polyporaceae, Aphyllophorales

Description: *Soft, spreading, with projecting white, stalkless cap, with thick, cottony covering and white pores becoming torn and toothlike.*
Cap: 1–4″ (2.5–10 cm) wide; semicircular to elongate, convex, thick; downturned margin has cottony roll; pure white; cottony, hairy. Flesh 1–1.5 cm thick; white.
Tubes: 2–6 mm long. Pores (2–3 per mm) angular, even-edged or torn and toothlike, white, discoloring on drying.
Spores: 4–5 × 1–1.5 μ; cylindrically

curved, smooth, colorless.

Season: August—November.

Habitat: On old logs and stumps of conifers, such as fir, larch, pine, spruce, hemlock, and Douglas-fir; at over 5000'.

Range: Rocky Mountains; Pacific NW. to California.

Comments: Also known as *Polyporus leucospongius*, this polypore resembles a cottony, white sponge and is found up to timberline.

---

482 Turkey-tail

Trametes versicolor (Fr.) Pil.
Polyporaceae, Aphyllophorales

Description: *Usually overlapping, small, leathery, thin, stalkless caps with many multicolored zones, alternating hairy and smooth, and with white to yellow pores.*
Cap: 1—4" (2.5—10 cm) wide; usually overlapping or in rosettes; semicircular, kidney- to spoon-shaped, or fused laterally; flat to wavy; multicolored; silky, hairy, or velvety, with alternate smooth zones. Flesh 1—2 mm thick; white.
Tubes: 1—2 mm long. Pores (3—5 per mm) angular, white or yellowish.
Spores: 5—6 × 1.5—2.2 μ; cylindrical to sausage-shaped, smooth, colorless. Spore print white.

Season: May—December; sometimes reviving and persisting several seasons.

Habitat: On dead deciduous wood or in wounds; also on conifers.

Range: Throughout North America.

Look-alikes: *T. hirsuta* has densely hairy, grayish-white cap. *T. velutina* is thicker (1—4 mm), and usually has smoky pore surface. *Stereum ostrea** is parchmentlike and smooth underneath.

Comments: Also known as *Polyporus* and *Coriolus versicolor*. *Trametes* means "flesh" or "fabric" and refers to the connective tissue (hyphae) of the cap projecting into the walls of the tubes.

538   **Violet Toothed Polypore**
Trichaptum biformis (Fr. in Kl.) Ryv.
Polyporaceae, Aphyllophorales

Description: *Tough, spreading, with projecting single to overlapping, white to grayish, hairy, stalkless caps with violet-whitish pores that break into teeth.*
Cap: ⅜–3″ (1–7.5 cm) wide; semicircular, flat; hairy, zoned; white to grayish, brownish or black; becoming smooth with ochre to dark brown zones, violet along margin. Flesh 0.5–1.5 mm thick; white to ochre.
Tubes: 1–10 mm long. *Pores* (2–5 per mm) angular, *becoming toothlike*, white to brownish, usually with violet tint overall or along margin.
Spores: 5–6.5 × 2–2.5 μ; cylindrical, smooth, colorless. Spore print white.

Season: May–December; usually reviving and persisting several seasons.

Habitat: On dead deciduous wood; also reported on conifers.

Range: Throughout North America.

Look-alikes: *T. abietinus* grows on coniferous wood, is smaller, with stiff hairs on cap.

Comments: Also known as *Polyporus* and *Hirschioporus pargamenus,* this very common, widely distributed species is known to occur on at least 65 tree hosts, though rarely on conifers. Hundreds of caps are usually found covering stumps and trunks of dead trees, reducing them to sawdust.

522   **Blue Cheese Polypore**
Tyromyces caesius (Fr.) Murr.
Polyporaceae, Aphyllophorales

Description: *Soft, watery to spongy, finely hairy, grayish-blue, stalkless cap with white to grayish-blue pores.*
Cap: ⅜–3¼″ (1–8 cm) wide; somewhat convex to flat, with sharp margin; grayish-blue overall or along margin or gray to white, sometimes bruising blue;

finely hairy, becoming smooth. Flesh
1−10 mm thick; soft, spongy, and
watery when fresh; white, gray to
yellowish with age. Odor usually
fragrant.
Tubes: 2−8 mm long. Pores (2−4 per
mm) angular to somewhat elongated,
grayish-blue, sometimes white or
bruising blue.
Spores: 4−5 × 0.7−1 μ; cylindrically
curved, smooth, colorless, amyloid.
Spore print pale blue.

Season: August−November.
Habitat: On dead wood.
Range: Throughout North America.
Look-alikes: *T. chioneus** is white and typically
grows on deciduous wood.
Comments: Also called *Polyporus caesius*. This pretty
polypore is not known to be edible.

490 **White Cheese Polypore**
Tyromyces chioneus (Fr.) Kar.
Polyporaceae, Aphyllophorales

Description: *Soft, watery, white, hairy to smooth,
stalkless cap with whitish pores; fragrant.*
Cap: ⅜−4″ (1−10 cm) wide; flat to
slightly convex, semicircular; pure
white or watery white to grayish; finely
hairy to smooth or somewhat warty.
Flesh 2−15 mm thick; soft, spongy,
and watery when fresh; white.
Tubes: 1.5−8 mm long. Pores (3−5 per
mm) angular to round, white to
creamy-white.
Spores: 3.5−5 × 1.5−2 μ; cylindrical
to slightly curved, smooth, colorless.
Season: July−November.
Habitat: On dead deciduous wood; also reported
on conifers.
Range: Throughout North America.
Look-alikes: *T. tephroleucus*, also known as *T. lacteus*,
lacks fragrant odor. *T. fragilis** occurs
on coniferous wood, and bruises brown.
Comments: Also known as *Polyporus albellus*. This is
the most common of a number of
closely related whitish, watery, fleshy,
firm polypores.

589 Staining Cheese Polypore
Tyromyces fragilis (Fr.) Donk
Polyporaceae, Aphyllophorales

Description: *Soft, with projecting, sometimes overlapping, white, stalkless caps covered with soft, white hairs and white pores.*
Cap: ⅜–4" (1–10 cm) wide; sometimes overlapping, often laterally fused; semicircular to elongated, flat to somewhat wavy; white to yellowish, bruising pinkish to brownish; finely hairy to cottony. Flesh 2–12 mm thick; white, reddish, bruising yellow.
Tubes: 2–8 mm long. *Pores (2–4 per mm) angular to irregular; white, bruising yellowish then rust-red.*
Spores: 4.5–5 × 1.5–2 μ; cylindrical to sausage-shaped, smooth, colorless. Spore print whitish.
Season: September–October.
Habitat: On dead coniferous wood.
Range: Canada south to North Carolina, Michigan, Colorado, and California.
Look-alikes: *T. chioneus** does not discolor on bruising. *T. mollis* is much larger, more or less smooth, white, and bruises or ages pinkish. *Amylocystis lapponicus* has hairy cap and bruises reddish when handled.
Comments: Also called *Polyporus fragilis,* this relatively common, widely distributed polypore is often overlooked.

SCHIZOPHYLLUM FAMILY
(Schizophyllaceae)

This small family of 10 genera includes one of the most widely distributed mushrooms in the world, the pretty Common Split Gill. The related Crimped Gill is the only other commonly known species in the family. They, and some related mushrooms, have what appear to be gills, but are actually radiating, branching folds or lobes.

472 Crimped Gill
Plicaturopsis crispa (Pers. ex Fr.) Reid
Schizophyllaceae, Aphyllophorales

Description: *Small, shelflike, white-hairy, stalkless cap*
with whitish, crimped, gill-like folds.
Cap: ⅜–1″ (1–2.5 cm) wide; shelflike, with lobed margin; yellow-brown
to tan, hairy. Flesh membranous,
tough, thin.
Folds: gill-like, crimped, typically
forked; sometimes shallow and veinlike;
well-separated, narrow, whitish.
Spores: 3–4 × 1–1.5 μ; cylindrical to
narrowly elliptical, smooth, colorless.
Spore print white.

Season: Year-round.

Habitat: On branches of deciduous trees,
especially beech, birch, and cherry.

Range: Maine south to Tennessee, west to
North Dakota.

Look-alikes: *Schizophyllum commune** has densely
hairy, white cap and split folds.

Comments: Also known as *Trogia crispa*. The
Crimped Gill is quite common in some
years and absent in others.

487 Common Split Gill
Schizophyllum commune Fr.
Schizophyllaceae, Aphyllophorales

Description: *Stalkless, fan-shaped, white-hairy cap with*
white to pinkish, gill-like folds split on
edges.
Cap: ⅜–1⅝″ (1–4 cm) wide; fan- to
shell-shaped when laterally attached to
wood, vase- to saucer-shaped when
centrally attached; white to gray, dry,
densely hairy. Flesh gray, pliant,
leathery.
Folds: gill-like but split lengthwise,
well-separated; white, gray, or pink-
tinged; hairy.
Spores: 3–4 × 1–1.5 μ; cylindrical,
smooth. Spore print white.

Season: Year-round.

Habitat: On dead branches of deciduous trees;
also planks and boards of hickory,

hornbeam, walnut, and elm.
Range: Maine south to Tennessee, west to
             North Dakota.
Look-alikes: *Plicaturopsis crispa** has crimped folds.
Comments: The Common Split Gill is found
             throughout the world and can usually
             be collected every month of the year. It
             survives loss of moisture by curling
             back the outer sides of its folds and
             rolling the cap margin inward. It
             revives in wet weather.

PARCHMENT FUNGUS FAMILY
(Stereaceae)

The parchment fungi look like thin,
leathery to woody, typically stalkless
polypores, but their undersides (the
fertile surface, or hymenium) are mostly
smooth, instead of being composed of
tubes and pores, and their spores are
produced on the surface, not in tubes,
on clublike structures called basidia.
These mushrooms typically grow on
dead wood and are predominantly late-
summer and fall fungi that decompose
deciduous wood. All polyporelike fungi
with smooth fertile surfaces were once
placed in genus *Stereum,* but a number
of different genera have been split off,
and are so recognized in this guide.
Because these mushrooms are leathery
and tough, none is known to be either
edible or poisonous.

541 **Bleeding Conifer Parchment**
     Haematostereum sanguinolentum (Fr.) Pouz.
     Stereaceae, Aphyllophorales

Description: *Spreading, shell-like, leathery, pinkish-*
             buff, bleeding red.
             Cap: ⅛–⅜″ (0.3–1 cm) wide; often
             overlapping or laterally fused and
             broader, roundish; pinkish- to
             cinnamon-buff; hairs radially flattened;
             indistinctly zoned. Flesh 0.2–0.5 mm
             thick.

Fertile Surface: smooth, brown, "bleeding" red on bruising because of imbedded, color-conducting organs; cracking on drying.

Spores: 8–14 × 3–5 μ (from 2-spored basidia) or 5–7 × 2–3.5 μ (from 4-spored basidia); slightly curved, smooth, colorless. Spore print white.

Season: July–March.

Habitat: On conifers, especially hemlock, pine, and spruce.

Range: N. North America south to Virginia and California.

Comments: Also known as *Stereum sanguinolentum*. This sheetlike, spreading mushroom can cover an area up to 4″ in diameter. The genus *Haematostereum* differs from *Stereum* in possessing blood-red, sap-filled hyphae. *H. gausapatum* is hairy, brownish, and stains red only on bruising.

591 Bristly Parchment
Lopharia cinerascens (Schw.) Cunn.
Stereaceae, Aphyllophorales

Description: *Spreading, shell-like, leathery, ochre-buff, stiff-hairy, zoned, with minutely bristly undersurface.*

Cap: ⅛–⅜″ (0.3–1 cm) wide; often laterally fused; concentrically zoned; cinnamon-buff to brownish, weathering gray. Flesh 5–10 mm thick.

Fertile Surface: powdery to minutely bristly because of large cystidia; often slightly cracked; pinkish-buff to brown.

Spores: 8–15.5 × 5.5–10 μ; oval to cylindrical, smooth, colorless. Cystidia 80–180 × 16–26 μ; conical, coarsely encrusted. Spore print white.

Season: June–February; also overwinters.

Habitat: On dead deciduous wood, especially mulberry and elm, also linden and locust.

Range: Canada to Florida, west to Texas and California; also Mexico.

Comments: Formerly known as *Stereum cinerascens*. It can spread over an area 6″ by 3″.

545 **Crowded Parchment**
Stereum complicatum (Fr.) Fr.
Stereaceae, Aphyllophorales

Description: *Crowded, fan-shaped, leathery, cinnamon-buff, shiny, zoned.*
Cap: ⅛–⅝″ (0.3–1.5 cm) wide; often overlapping or laterally fused, fan-shaped or semicircular; pinkish to cinnamon-buff or grayish; silky-hairy, zoned, radially furrowed to wavy, smooth and shining near margin. Flesh 0.3–0.4 mm thick.
Fertile Surface: smooth or slightly ridged where caps join; *orange,* fading to cream- or cinnamon-buff.
Spores: 5–6.5 × 2–2.5 μ; cylindrical to slightly curved, smooth, colorless. Spore print white.
Season: July–January; overwinters.
Habitat: On dead deciduous twigs and stumps, especially oak.
Range: Throughout North America, except Rocky Mountains.
Comments: Also known as *S. rameale,* this is our most common species of *Stereum.*

537 **Hairy Parchment**
Stereum hirsutum (Willd. ex Fr.) S.F.G.
Stereaceae, Aphyllophorales

Description: *Shell-like, leathery, brownish, densely stiff-hairy, zoned.*
Cap: ¼–¾″ (0.5–2 cm) wide; overlapping and laterally fused, semicircular; coarsely stiff-hairy over thin, golden, horny crust; concentrically zoned and radially lined, cream-buff, becoming grayish. Flesh 0.5–0.7 mm thick.
Fertile Surface: smooth; dull yellowish, buff, or smoky gray, staining yellow.
Spores: 5–8 × 2–3.5 μ; cylindrical, smooth, colorless. Spore print white.
Season: July–November in East; to February on Pacific Coast; persists for several years.
Habitat: On dead deciduous wood, especially birch and beech; sometimes on conifers.

Range: Canada to South Carolina, west to
British Columbia and California;
Mexico.

Comments: Although the cap color varies from
whitish to brown, this mushroom can
be recognized by its smooth, wood-
brown undersurface and hairy cap.

---

536 False Turkey-tail
Stereum ostrea (Blume & Nees ex Fr.) Fr.
Stereaceae, Aphyllophorales

Description: *Petal-like, leathery, with multicolored zones*
and smooth undersurface.
Cap: 3/8–2¾" (1–7 cm) wide; petal- to
fan-shaped or semicircular, often
overlapping, sometimes laterally
attached; densely hairy, multicolored,
with narrow, smooth, shiny, russet to
brownish transverse zones. Flesh 0.4–
0.7 mm thick.
Fertile Surface: smooth; buff to
cinnamon-buff.
Spores: 5.5–7.5 × 2–3 μ; cylindrical
to somewhat flattened on 1 side,
smooth, colorless. Spore print white.

Season: June–January.

Habitat: On logs and stumps of deciduous wood,
especially oak.

Range: Throughout North America.

Comments: Also known as *S. fasciatum, S. lobatum,*
and *S. versicolor.* This, our largest and
most colorful *Stereum,* is frequently
misidentified as the Turkey-tail
(*Trametes versicolor*), a polypore.

---

574 Silky Parchment
Stereum striatum (Fr.) Fr.
Stereaceae, Aphyllophorales

Description: *Shell-like, leathery, silvery, silky-shiny.*
Cap: ¼–3/8" (0.5–1 cm) wide; often
laterally fused, round; attached to wood
by central point; shiny, silvery to pale
gray, with small, radiating, silky
fibers. Flesh 0.25–0.30 mm thick.
Fertile Surface: smooth; light buff to

ochre-buff to brown, whitish with age.
Spores: 6–8.5 × 2–3.5 μ; cylindrical,
smooth, colorless. Spore print white.

Season: Year-round.
Habitat: On twigs and branches of dead
hornbeam, usually on underside; often
in swampy woods.
Range: Northeast to Midwest.
Comments: More commonly known as *S. sericeum.*
Fused caps can spread in areas to 4".

---

572 Ceramic Parchment
Xylobolus frustulatus (Pers. ex Fr.) Boid.
Stereaceae, Aphyllophorales

Description: *Large, irregular patches of small, whitish,
polygonal, tilelike pieces.*
Cap: ⅛–¾" (0.3–2 cm) wide; in
crowded, polygonal, crustlike plates,
with downcurled or free margins;
blackish, smooth. Flesh 0.5–1 mm
thick.
Fertile Surface: smooth, somewhat
convex; pinkish-buff to whitish and
finely powdered; usually appearing as
upper surface of mushroom.
Spores: 3.5–5 × 2.5–3 μ; oval,
smooth, colorless. Cystidia 28–55 ×
3.5–6.5 μ; bottle-brushlike or prickle-
pointed, abundant. Spore print white.
Season: Year-round.
Habitat: On dry, hard oak logs.
Range: Canada to Texas, west to Oregon;
Mexico.
Comments: Previously known as *Stereum frustulosum.*
This clustered, quiltlike species can
spread over areas larger than 2" by 4".

---

436 Stalked Stereum
Cotylidia diaphana (Schw.) Lentz
Stereaceae, Aphyllophorales

Description: *Small, leathery, stalked, vase-shaped,
whitish to pinkish-buff.*
Mushroom: ⅝–1¾" (1.5–4.5 cm)
high, ¼–1¼" (0.5–3 cm) wide at tip;
deeply vase-shaped or split into

spatulalike sections, with even or torn margin; white to pinkish-buff or somewhat darker; translucent, resinous, somewhat radially lined and zoned. Fertile Surface: appearing smooth, but minutely ridged and bristly under hand lens because of protruding cystidia; white to pinkish-buff.
Stalk: ¼–1⅜″ (0.5–3.5 cm) thick; slender, with matted, white down; small ball of white threads (mycelium) at base.
Spores: 4–6 × 2.5–4 μ; elliptical to oval, smooth, colorless. Cystidia 69–100 × 9.5–13 μ; colorless, hairlike, thicker than basidia, projecting beyond them. Spore print whitish.

Season: September–November.
Habitat: On ground, in moist hardwood areas.
Range: New York to Alabama; Mississippi; Iowa; Missouri; West Coast.
Look-alikes: *Craterellus** species are fleshy.
Comments: Formerly known as *Stereum diaphanum.* Other stalked *Stereum* species are best differentiated microscopically.

AGARICS AND BOLETES
(Order Agaricales)

The order Agaricales, as defined here, includes the gilled mushrooms and boletes. Fifteen families are recognized on the basis of similar shape, spore print color, gill attachment, microscopic features, and chemical reactions. Other classifications include certain annual stalked polypores, some agarics that lack gills, and others that have lost the ability to forcibly eject their spores. There are also classifications that exclude some oysterlike mushrooms, the boletes and their allies, and the Russulaceae. The classification used here is somewhat artificial, but a more natural system showing evolutionary relationships would be less useful in a field guide.

AGARICUS AND LEPIOTA FAMILY
(Agaricaceae)

This is the family of our common cultivated mushroom (*Agaricus bisporus*) and many of the common urban and suburban mushrooms. Although many grow in forests, more grow in open woods, grassy areas, lawns, and compost, or along roadsides. They are decomposers of organic debris. Some are among our best edibles, others are poisonous, and a few are deadly. Eight genera are recognized in North America: *Agaricus* (about 200 species), *Chlorophyllum* (1 species), *Cystoderma* (20 species), *Dissoderma* (1 species), *Lepiota* (about 100 species), *Melanophyllum* (1 species), *Phaeolepiota* (1 species), and *Squamanita* (2 species). Typically, these mushrooms have a scaly to granular cap and gills that are free of the stalk. All have membranous veils, most of which become persistent rings on the stalk. In certain genera, such as *Lepiota* and *Melanophyllum*, the ring is often lost. The spore print may be white, dirty green, ochre-brown, chocolate, or reddish- to purplish-brown. These mushrooms are placed in a single family on the basis of a correlation of characteristics that include microscopic and chemical features: the gill flesh is never divergent (as in *Amanita*); the spores are usually smooth, thick-walled, have a pore at the tip, and are nonamyloid (many are dextrinoid).

---

122 Abruptly-bulbous Agaricus
Agaricus abruptibulbus Pk.
Agaricaceae, Agaricales

Description: *Large, white mushroom, bruising yellowish, with bulbous stalk base; odor fragrant, of almonds or anise; in woods.*
Cap: 3–6" (7.5–15 cm) wide; convex to

flat; smooth or with small, flattened fibers; dry, silky, shiny; white, bruising yellowish; yellow with KOH. Flesh thick, bruising yellow. *Odor almond-anise.*

Gills: free, crowded, narrow; *white, becoming pink, then brown.*

Stalk: 3–6" (7.5–15 cm) long, ⅜–⅝" (1–1.5 cm) thick, tapering down to small, abrupt bulb; smooth, white, bruising yellowish.

Veil: partial veil membranous, white, with patches on lower surface; leaving persistent, pendant ring on upper stalk.

Spores: 6–8 × 4–5 μ; elliptical, smooth, purple-brown. Spore print purple-brown.

Edibility: Choice, with caution.

Season: July–October.

Habitat: On the ground, in deciduous and mixed woods.

Range: Ontario and Quebec to North Carolina, west to Michigan.

Look-alikes: *A. silvicola,* as recognized in North America, is not typically abruptly bulbous and has smaller spores. *A. arvensis** grows in pastures and fields. *A. albolutescens,* a northwest species of uncertain edibility, bruises amber. The deadly *Amanita virosa** and related species have saclike cup about stalk base, lack patches on undersurface of partial veil, lack spicy fragrance, and have white gills producing white spore print.

Comments: This is an excellent edible, but a spore print must be made to be sure that one has not collected a poisonous *Amanita.*

156 Horse Mushroom
Agaricus arvensis Schaeff. ex Secr.
Agaricaceae, Agaricales

Description: *Large, white mushroom, bruising yellowish, with spicy odor; in grass.*
Cap: 3–6" (7.5–15 cm) wide; oval, becoming convex; smooth to slightly scaly; margin often with hanging veil

remnants; white, bruising yellowish; yellow with KOH. Flesh thick, bruising yellow. Odor almond-anise.
Gills: free, close, broad; *white to grayish, becoming blackish-brown.*
Stalk: 3–5" (7.5–12.5 cm) long, ⅜–¾" (1–2 cm) thick, sometimes broader at base; smooth, white, bruising yellowish.
Veil: partial veil membranous, white, with patches on lower surface; leaving persistent, flaring to pendant ring on upper stalk.
Spores: 7–9 × 4.5–6 μ; elliptical, smooth, purplish-brown. Spore print blackish-brown.

Edibility: Choice.
Season: June–October; November–April in California.
Habitat: Scattered to numerous, often in fairy rings, in meadows, fields, and grassy waste areas; often near spruce.
Range: Throughout North America.
Look-alikes: *A. abruptibulbus-silvicola* complex are more slender, woodland mushrooms. *A. campestris** has pink gills at first. The poisonous *A. xanthodermus** lacks fragrant odor, has yellow flesh at stalk base.
Comments: Be sure to make a spore print to be certain you do not have an *Amanita*.

150 Prince
Agaricus augustus Fr.
Agaricaceae, Agaricales

Description: *Large, yellowish-brown, fibrous-scaly cap and stalk, bruising yellowish; fragrant.*
Cap: 4–10" (10–25 cm) wide; convex, with flattened center and decurved margin, becoming broadly convex to flat; yellowish-brown, bruising yellowish; yellow with KOH. Odor almond-anise.
Gills: free, close, broad; whitish at first, then pinkish to brown.
Stalk: 3–5" (7.5–12.5 cm) long, ¾–1⅜" (2–3.5 cm) thick, sometimes

enlarging downward; silky above ring, woolly-scaly below; deeply rooted.
Veil: partial veil membranous, white, with patches on lower surface; leaving persistent, pendant ring on upper stalk.
Spores: 8−11 × 5−6 μ; elliptical, smooth, purple-brown. Spore print dark brown.

Edibility: Choice.

Season: July−September; year-round in California in mild, wet weather.

Habitat: Solitary to scattered, in disturbed soil, grassy areas, along paths and roads.

Range: British Columbia to California; Rocky Mountains; reported in East.

Look-alikes: *A. albolutescens* bruises amber.

Comments: The Prince is a robust, meaty, flavorful mushroom.

---

157 **Spring Agaricus**
Agaricus bitorquis (Quél.) Sacc.
Agaricaceae, Agaricales

Description: *Large, brownish, flattened, fibrous cap with double-edged ring on lower stalk; on hard ground.*
Cap: 2−6″ (5−15 cm) wide; broadly convex to flat or with central depression; fibrous, cracking somewhat with age; grayish-brown.
Gills: free, close, narrow; *grayish-pink, becoming dark brown.*
Stalk: 1−2″ (2.5−5 cm) long, ⅜−1¼″ (1−3 cm) thick; short, smooth, whitish.
Veil: partial veil membranous, white; leaving persistent, pendant ring on lower stalk, with flaring upper edge and free lower edge.
Spores: 5−6 × 4−5 μ; round to oval, smooth, purplish-brown. Spore print blackish-brown.

Edibility: Choice.

Season: May−June; also September; November−April in California.

Habitat: Single to several, on packed ground in urban areas.

Range: Throughout North America.

Look-alikes: *A. campestris* * has bright pink gills at first and simple, fragile veil and ring.

Comments: Also known as the "Urban Agaric" and *A. rodmani,* this choice edible is one of the first gilled mushrooms to appear in spring. Because of its double ring, producing what may appear to be both a ring and a cup, a spore print should be made to be sure a poisonous *Amanita* has not been collected.

⊗ 151 **California Agaricus**
Agaricus californicus Pk.
Agaricaceae, Agaricales

Description: *Shiny, white to brownish cap with unpleasant-smelling gills; in urban areas.*
Cap: 2–4″ (5–10 cm) wide; convex to flat; dry, smooth to somewhat scaly, shiny; white or brownish over center; yellow with KOH. *Odor pungent.*
Gills: free, close, narrow; *whitish at first, becoming bright pink and eventually dark brown.*
Stalk: 1–4″ (2.5–10 cm) long, ⅜–⅝″ (1–1.5 cm) thick; whitish, darkening on handling; dry, smooth.
Veil: partial veil membranous, white; leaving persistent, pendant ring on upper stalk.
Spores: 4–7 × 3–6 μ; elliptical, smooth, purple-brown. Spore print dark brown.
Edibility: Poisonous.
Season: September–November; year-round in wet areas.
Habitat: Scattered to several, in urban areas, in lawns and parks.
Range: California.
Look-alikes: *A. campestris* * does not react with KOH or develop a pungent odor.
Comments: A "dead ringer" for the edible Meadow Mushroom (*A. campestris*), this common California mushroom is often mistaken for it; the resultant poisonings are mostly gastric upsets.

153, 154 Meadow Mushroom
Agaricus campestris L. ex Fr.
Agaricaceae, Agaricales

Description: *White to brownish, mostly smooth cap; in grass.*
Cap: 1–4" (2.5–10 cm) wide; convex to nearly flat; dry, smooth or fibrous to woolly; white to grayish or gray-brown.
Gills: free, crowded, broad; *bright pink, becoming chocolate to blackish-brown.*
Stalk: 1–2" (2.5–5 cm) long, ⅜–⅝" (1–1.5 cm) thick, sometimes tapering downward; *short,* smooth to fibrous.
Veil: partial veil membranous, white; leaving thin, evanescent ring on stalk.
Spores: 6–9 × 4–6 μ; elliptical, smooth, purple-brown. Spore print blackish-brown.
Edibility: Choice.
Season: August–September; occasionally in spring; fall–winter in California.
Habitat: Scattered to abundant, in grassy areas.
Range: Throughout North America.
Look-alikes: *A. arvensis** is fragrant and bruises yellow. *A. bitorquis** has double ring. *Lepiota naucina** produces white spore print. The deadly *Amanita virosa** and allies, also white-spored, have saclike cup about stalk base, grow in woods or near trees.
Comments: The Meadow Mushroom may be known better and gathered more than any other wild mushroom in North America. It has a nuttier flavor than the common cultivated mushroom (*A. bisporus,* or *A. brunnescens*). Because deadly white amanitas frequently grow in similar habitats, one should make a spore print to confirm identification.

160 Bleeding Agaricus
Agaricus haemorrhoidarius Schulz. apud Kalchb.
Agaricaceae, Agaricales

Description: *Brownish, fibrous cap with white flesh quickly staining red; in damp woods.*

Cap: 1–4" (2.5–10 cm) wide; convex to flat; dry, fibrous to scaly; grayish- to dark reddish-brown. Flesh thick, white, staining bright red.

Gills: free, close, moderately broad; *whitish, becoming pinkish to purple-brown.*

Stalk: 2–5" (5–12.5 cm) long, ⅜–¾" (1–2 cm) thick, sometimes bulbous; white to pinkish-gray, bruising pinkish. Flesh staining red.

Veil: partial veil membranous, white; leaving persistent, pendant ring on upper stalk.

Spores: 5–7 × 3.5–4 μ; elliptical, smooth, purple-brown. Spore print dark brown.

Edibility: Edible with caution.

Season: July–October; November–February in California.

Habitat: Solitary to several, in damp woods.

Range: Throughout North America.

Look-alikes: *A. silvaticus* is smaller and bruises dull red to reddish-brown.

Comments: This is one of several species of *Agaricus* that stain red, a characteristic best observed by cutting the mushroom in half lengthwise. The Bleeding Agaricus is edible for most people, but, like many species of *Agaricus*, causes gastric upset in some.

⊗ 158 **Felt-ringed Agaricus**
Agaricus hondensis Murr.
Agaricaceae, Agaricales

Description: *Large, white cap with small, pinkish-gray fibers, thick ring on stalk, and foul-smelling, bulbous base; in woods.*

Cap: 3–6" (7.5–15 cm) wide; convex; dry, smooth or with small, flattened fibers; white to pale pinkish-gray, darkening with age; yellow with KOH. Odor mild to creosotelike.

Gills: free, close to crowded, narrow; *grayish-pink, becoming purplish-brown.*

Stalk: 3–6" (7.5–15 cm) long, ⅜–¾" (1–2 cm) thick, with bulbous base; smooth, whitish, darkening with age.

Veil: partial veil membranous, white, with patches on lower surface; leaving persistent, thick, flaring to pendant ring on upper stalk.
Spores: $4.5–5.5 \times 3–3.5 \mu$; elliptical, smooth, purple-brown. Spore print dark brown.

Edibility: Poisonous.

Season: September–October; November–February in California.

Habitat: Scattered to several, in woods.

Range: British Columbia south into California.

Look-alikes: The poisonous *A. meleagris** becomes flat-topped and has darker scales, a sometimes narrowing stalk, thinner ring, and more unpleasant odor.

Comments: This poisonous mushroom's woodland habitat, odor, and staining reaction with KOH allow it to be differentiated from known edible species of *Agaricus*.

⊗ 152 **Western Flat-topped Agaricus**
Agaricus meleagris J. Schaeff.
Agaricaceae, Agaricales

Description: *Large, flat cap with free gills, blackish scales, ring on upper stalk, and strong, unpleasant odor.*
Cap: 2–6" (5–15 cm) wide; rounded at first, becoming convex *with flattish top*, finally becoming flat; dry, with minute grayish or blackish-brown fibers that become grayish scales; yellow with KOH. Flesh whitish, sometimes staining yellowish or pinkish. Odor foul, especially when flesh is crushed.
Gills: free, close, broad; *grayish, becoming pinkish, then chocolate.*
Stalk: 2–7" (5–18 cm) long, 3/8–1" (1–2.5 cm) thick, with slightly bulbous base; silky, buff to pinkish-brown; base flesh yellow when cut.
Veil: partial veil membranous, white, with cottony patches on lower surface; thick; leaving large ring on upper stalk.
Spores: $5–7 \times 3–4 \mu$; oval, smooth, brown. Spore print chocolate.

Edibility: Poisonous.

Season: September–November; September–
March in California; May to mid-
October in Ohio.
Habitat: Single to several, in lawns and grassy
areas, along paths, and in woods.
Range: W. North America; reported in East.
Comments: This species is reportedly poisonous. Do
not eat any *Agaricus* with bad-smelling
gills or with a stalk base that turns
yellow when cut vertically. The Eastern
Flat-topped Agaricus (*A. placomyces*),
possibly poisonous, has a paler cap and
is more slender.

---

161 Wine-colored Agaricus
Agaricus subrutilescens (Kauff.) Hot. & Stuntz
Agaricaceae, Agaricales

Description: *Large, purplish to reddish-brown, fibrous
cap with free gills, sheathlike ring on
cottony, fibrous stalk, and fruity odor.*
Cap: 2–6″ (5–15 cm) wide; convex
with flattened center, becoming flat;
dry; whitish, with purplish-
brown scales; *greenish with KOH.*
Gills: free, close to crowded, narrow;
*white, becoming pinkish, then purplish- to
chocolate-brown.*
Stalk: 2–8″ (5–20 cm) long, ⅜–¾″
(1–2 cm) thick, with enlarged base;
smooth above ring, *covered with woolly,
white patches below.*
Veil: partial veil membranous, white;
leaving large, thin, flaring ring on
upper stalk.
Spores: 5–6 × 3–3.5 μ; elliptical,
smooth, brownish. Spore print
chocolate.
Edibility: Choice, with caution.
Season: September–October; November–
February in California.
Habitat: Single to several, in coniferous woods.
Range: Washington to California.
Look-alikes: Greenish KOH reaction on cap
distinguishes this species from those
that superficially resemble it.
Comments: This delicious edible *Agaricus* can cause
stomach upset in susceptible persons.

⊗ 155 **Yellow-foot Agaricus**
Agaricus xanthodermus Gen.
Agaricaceae, Agaricales

Description: *Large, white to buff, smooth or scaly cap*
with free gills; stalk has whitish ring; base
flesh bright yellow when cut.
Cap: 2⅜−7″ (6−18 cm) wide;
roundish, becoming convex to nearly
flat; margin curved downward; dry,
smooth; whitish to gray, buff at center;
quickly bruising bright yellow; bright
yellow with KOH. Odor strong, foul.
Gills: free, close, moderately broad;
whitish, becoming pink, then chocolate.
Stalk: 2−6″ (5−15 cm) long, ¾−1″ (2−
2.5 cm) thick, sometimes slightly
bulbous; smooth or with small, loose
fibers; whitish, bruising yellow; base
flesh bright yellow when cut.
Veil: partial veil membranous, whitish,
with cottony patches on lower surface;
leaving thick, flaring, pendant,
persistent ring on upper stalk.
Spores: 5−7 × 3−4 μ; elliptical,
smooth, brown. Spore print chocolate.
Edibility: Poisonous.
Season: September−November in Pacific NW.;
November−March in California.
Habitat: Scattered to several, in urban areas:
lawns, pathsides, under hedges.
Range: Pacific NW. to California.
Look-alikes: *A. albolutescens* has almond odor and
bruises amber.
Comments: This is one of a group of poisonous
Agaricus species: all develop an
unpleasant, creosotelike odor and bruise
yellowish. Poisoning, which may last a
day or two, occurs within 2 hours and
typically results in bouts of vomiting,
diarrhea, or both.

⊗ 169, 170 **Green-spored Lepiota**
Chlorophyllum molybdites Mass.
Agaricaceae, Agaricales

Description: *Large, white mushroom with broad, buff*
scales over center, free white gills maturing

sordid gray-green, and ring on stalk.
Cap: 2–12″ (5–30 cm) wide,
doorknob-shaped, becoming flat; dry,
white, with several large, pinkish-buff
patches over center, breaking into many
small scales on expansion.
Gills: free, close, broad; white, slowly
becoming sordid gray-green or darker.
Stalk: 4–10″ (10–25 cm) long, ⅜–1″
(1–2.5 cm) thick, enlarging toward
base; smooth, white, discoloring on
handling.
Veil: partial veil membranous, white,
leaving persistent, double-edged, often
movable, pendant ring on upper stalk.
Spores: 8–13 × 6.5–8 μ; elliptical,
smooth, colorless, thick-walled, with
small pore at tip; *dextrinoid. Spore print
green.*

Edibility: Poisonous.

Season: August–September.

Habitat: Lawns, pastures, meadows; often in
fairy rings.

Range: Florida to California; common in
Denver; reported in New York, New
Jersey, and Michigan.

Look-alikes: *Lepiota** species have white spore print.
*L. procera** has tall, slender, scaly
stalk. *L. rachodes** has flesh bruising
saffron. *L. naucina** has a smooth cap.
*Agaricus** species have brown to purple-
brown spore print.

Comments: Also known as *Lepiota morgani.* This
very common mushroom can be a
drastic sickener, causing one to two or
more days of violent purging.

---

194 **Pungent Cystoderma**
Cystoderma amianthinum (Fr.) Fayod
var. *rugusoreticulatum* (Lor.) A.H.S. & Sing.
Agaricaceae, Agaricales

Description: *Deeply wrinkled, ochre-tawny, granular
cap with attached white gills and evanescent
ring on stalk; pungent odor.*
Cap: 1–2″ (2.5–5 cm) wide; conical to
convex; deeply wrinkled and granular;
ochre-tawny to ochre-buff; dark rust-

brown with KOH. Odor very pungent.
Gills: attached, crowded, narrow, whitish.
Stalk: 1⅝–2¾" (4–7 cm) long, ⅛" (3 mm) thick; smooth and off-white above ring, ochre-buff below.
Veil: partial veil sheaths stalk; ochre-tawny, granular; leaving incomplete ring on upper stalk, with smooth upper surface.
Spores: 5–6 × 3 μ; elliptical, smooth, colorless; *amyloid*. Spore print white.

Season: August–October; November–January in California.

Habitat: Single to several, on moss, humus, or coniferous needles.

Range: Throughout North America.

Comments: The related var. *amianthinum* lacks the powerful, pungent odor of this species, as well as the deeply wrinkled cap surface.

---

191 **Common Conifer Cystoderma**
Cystoderma fallax A.H.S. & Sing.
Agaricaceae, Agaricales

Description: *Rust-brown, granular cap, whitish gills, and sheathing, granular veil leaving flaring ring on stalk.*
Cap: 1–2" (2.5–5 cm) wide; convex to flat; granular, rust-brown to ochre-tawny.
Gills: attached, close, narrow, whitish.
Stalk: 1–3" (2.5–7.5 cm) long, ⅛–¼" (3–5 mm) thick, sometimes slightly bulbous; smooth and off-white above ring.
Veil: partial veil sheaths stalk; reddish-brown, granular; leaving persistent, pendant ring with smooth upper surface.
Spores: 3.5–5.5 × 2.8–3.6 μ; elliptical, smooth, colorless; *amyloid*. Spore print white.

Season: September–October.

Habitat: Scattered to several, on moss, leaf litter, or coniferous needles.

Range: Pacific NW.; Great Lakes area.

Look-alikes: *C. granosum* is larger, bright orange, and occurs on rotten deciduous wood.

Comments: This species belongs to a genus of some 20 similar species that are best differentiated by chemical and microscopic characteristics. All have a granular-warty surface, attached gills, and sheathing veil, which is granular on the outer surface and smooth on the upper surface of the ring.

---

195   **Tuberous Cystoderma**
Dissoderma paradoxum (A.H.S. & Sing.) Sing.
Agaricaceae, Agaricales

Description: *Small, violet-gray mushroom with fibrous cap, attached gills, and scaly stalk mostly sheathed by granular, warty veil.*
Cap: ⅜–1⅜″ (1–3.5 cm) wide; convex to nearly flat, with broad knob; margin incurved at first, somewhat radially grooved, becoming scaly and torn; dry, grayish-violet; with small fibers or scales turning blackish with age.
Gills: attached, distant, broad, thick; violet-gray to pinkish-brown.
Stalk: 2–3″ (5–7.5 cm) long, ¼–⅜″ (0.5–1 cm) thick above, up to 1″ (2.5 cm) thick nearer base; pale violet-gray; *arising from tuberlike structure,* ochre-brown to tawny, granular, warty.
Veil: universal veil brownish-ochre, granular, warty; soon reduced to membranous, granular, warty sheath encasing stalk; warty surface made up of large, spherical cells.
Spores: 8–11 × 4.5–6 μ; elliptical, smooth, colorless, thick-walled; dextrinoid. Spore print whitish.
Season: September–October.
Habitat: On mossy ground in mixed woods.
Range: Pacific NW.
Comments: Also known as *Cystoderma paradoxum* and *Squamanita paradoxum.* This differs from mushrooms in the genus *Cystoderma* by arising from a tuberlike structure, and from those in the genus *Squamanita* by its granular covering.

163 Sharp-scaled Lepiota
Lepiota acutesquamosa (Weinm.) Kum.
Agaricaceae, Agaricales

Description: *Cinnamon cap with brown, pyramidal scales, free gills, and ring on stalk.*
Cap: 2–4" (5–10 cm) wide; rounded to convex, becoming flat or with central knob; dry, covered with cottony-fibrous, pointed to flattened scales; brown to cinnamon-brown.
Gills: free, crowded, narrow, white.
Stalk: 2–4" (5–10 cm) long, ⅜–⅝" (1–1.5 cm) thick, enlarging toward somewhat bulbous base; smooth to fibrous, brownish.
Veil: partial veil cottony-fibrous to weblike, white; leaving often evanescent ring on upper or middle stalk.
Spores: 7–8 × 2.5–3 μ; narrowly elliptical, smooth, colorless; thin-walled, without pore at tip; *dextrinoid.* Spore print white.

Edibility: Edible, but not recommended.
Season: July–October.
Habitat: On the ground in leaf litter.
Range: Throughout North America.
Look-alikes: Other *Lepiotas** with pyramidal cap scales are smaller and whitish.
Comments: This mushroom may actually represent a complex of closely related species, all with pyramidal cap scales and fibrous to weblike veils. Because some might be toxic, none can be recommended.

173, 174 Reddening Lepiota
Lepiota americana Pk.
Agaricaceae, Agaricales

Description: *Large, reddish-brown, scaly cap with free gills and ring on spindle-shaped stalk; bruising and aging dark red.*
Cap: 1–6" (2.5–15 cm) wide; oval, becoming convex to nearly flat or with central knob; smooth or flat, becoming scaly on expansion; pinkish-buff to reddish-brown, white between scales.

Flesh white; bruising yellow-orange
when young; red with handling or age.
Gills: free, close, broad; white, aging
reddish.
Stalk: 3–5″ (7.5–12.5 cm) long,
¼–¾″ (0.5–2 cm) thick at top,
enlarging downward, somewhat
swollen, then narrowed at base; stuffed,
becoming hollow; smooth, white,
bruising and aging reddish. *Cut flesh
yellow when young.*
Veil: partial veil membranous, white;
leaving double-edged, sometimes
movable, sometimes evanescent ring on
upper stalk.
Spores: 8–14 × 5–10 μ; elliptical,
smooth, colorless; thick-walled; pore at
tip; *dextrinoid. Spore print white.*

Edibility: Good.

Season: Late June–October.

Habitat: Clustered or rarely single, in sawdust
piles and in waste areas.

Range: E. North America west to Michigan.

Look-alikes: *L. rachodes** does not redden with age.
L. flammeatincta, a slender-stalked
species under conifers, bruises red
instantly. The poisonous *Chlorophyllum
molybdites** has white gills aging gray-
green and a greenish spore print.

Comments: This is a good edible, often found in
clusters or closely packed dozens.

178 Black-disc Lepiota
Lepiota atrodisca Zel.
Agaricaceae, Agaricales

Description: *Whitish cap with dark fibers and blackish
center, free, white gills, and ring on stalk.*
Cap: ⅜–2″ (1–5 cm) wide; convex,
becoming flat or with broad knob;
margin upturned with age; dry, white,
densely covered with small, dark fibers;
center blackish.
Gills: free, close, narrow, white.
Stalk: 1–3″ (2.5–7.5 cm) long, ¹⁄₁₆–¼″
(1.5–5 mm) thick, enlarging toward
base; dry, smooth, whitish.
Veil: partial veil membranous, white;

leaving thin, fragile ring on upper or
mid-stalk.
Spores: 6–8 × 3–5 μ; elliptical,
smooth, colorless; *dextrinoid*. Spore
print white.

Season: October–November; November–
February in California.
Habitat: On ground or decayed wood.
Range: Washington to California.
Comments: This is representative of a large number
of "regional endemics," species that
may be very common in a given area
but are restricted geographically. There
are more than 50 species of *Lepiota* on
the West Coast; most of these are
not widely distributed. Because some of
the small lepiotas are known to be
dangerously poisonous, none can be
recommended.

⊗ 179 Onion-stalked Lepiota
Lepiota cepaestipes (Sow. ex Fr.) Pat.
Agaricaceae, Agaricales

Description: *Powdery-white cap with radially lined
margin, free gills, and ring on scallion-
shaped stalk; in clusters.*
Cap: 1–3" (2.5–7.5 cm) wide; bell-
shaped, becoming conical to nearly flat,
with central knob; margin with deep
radial grooves; covered with powdery
scales; white, discoloring yellowish or
darker. Flesh thin.
Gills: free, crowded, narrow; white,
with hairy edges.
Stalk: 2–5" (5–12.5 cm) long, ⅛–¼"
(3–5 mm) thick; slender, but swollen
in places and bulbous at base; smooth
to somewhat powdery; white,
discoloring yellowish on handling.
Veil: partial veil membranous, white;
leaving persistent, pendant ring on
upper stalk.
Spores: 6–10 × 6–8 μ; broadly
elliptical, smooth, colorless; thick-
walled, with pore at tip; *dextrinoid*.
Spore print white.
Edibility: Possibly poisonous.

Season: June–September.
Habitat: Clustered, in compost, sawdust, wood chips, and leaf mulch.
Range: Throughout North America.
Comments: Also known as *Leucocoprinus cepaestipes.* Because some people reportedly have been sickened by this species, it cannot be recommended as an edible.

⊗ 176 **Shaggy-stalked Lepiota**
Lepiota clypeolaria (Bull. ex Fr.) Kum.
Agaricaceae, Agaricales

Description: *Small, white cap with brownish scales, free gills, and evanescent ring on shaggy-woolly stalk.*
Cap: 1–3″ (2.5–7.5 cm) wide; bell-shaped, becoming convex to nearly flat or with central knob; margin sometimes upturned with age; fibrous to scaly; brownish over center, ochre-tawny to yellowish toward margin.
Gills: free, close, narrow, white.
Stalk: 2–4″ (5–10 cm) long, ⅛–¼″ (3–5 mm) thick; silky, pale, and hollow above ring; densely woolly-scaly and yellowish-brown below.
Veil: partial veil fibrous, white; leaving evanescent ring on upper stalk.
Spores: 12–20 × 4–5 μ; spindle-shaped, smooth, colorless; thin-walled, without pore at tip; *dextrinoid.* Spore print white.
Edibility: Poisonous.
Season: July–October in Michigan and Ohio; July–November in Florida; November–February in California.
Habitat: On the ground, in coniferous, oak, and mixed woods.
Range: Throughout North America.
Comments: This may represent a species complex in need of more study to be clearly differentiated. The mushroom that is recognized as this species is poisonous.

⊗ 177 **Malodorous Lepiota**
Lepiota cristata (Fr.) Kum.
Agaricaceae, Agaricales

Description: *Small, white cap with concentric rings of reddish-brown scales, free gills, and ring on stalk; pungent odor.*
Cap: ⅝–2″ (1.5–5 cm) wide; broadly convex to nearly flat or with central knob; smooth at first, but soon breaking into concentric rings of small, reddish-brown scales; center brown to reddish-brown. Flesh thin, white, showing between scales toward margin. Odor strong, foul, fishy or spicy.
Gills: free, close, narrow; edges sometimes minutely toothed.
Stalk: 1–3″ (2.5–7.5 cm) long, ¹⁄₁₆–¼″ (1.5–5 mm) thick; slender, becoming hollow; smooth to silky, white to pink.
Veil: partial veil membranous, white; leaving small, evanescent ring on upper stalk.
Spores: 5–7 × 3–4 μ; wedge-shaped, with spurred corner, smooth, colorless; thin-walled, without pore at tip; *dextrinoid*. Spore print white.
Edibility: Possibly poisonous.
Season: June–October; November–February in California.
Habitat: On the ground, in grass and leaf litter along roads and in woods.
Range: Throughout North America.
Comments: The smell of the gills and the spore shape make this species complex distinctive. Slight variations occur throughout North America.

⊗ 162 **Deadly Lepiota**
Lepiota josserandii Bon & Boif.
Agaricaceae, Agaricales

Description: *Small cap with concentric bands of small, flattened, cinnamon-brown scales, free gills, and inconspicuous ring on stalk.*
Cap: 1–2″ (2.5–5 cm) wide; convex, becoming flat with low knob; dry, cinnamon-brown, breaking up into

concentric bands of small, flattened scales, staining reddish-brown with age. Flesh thick, white. Odor musty.
Gills: free or slightly attached, close, broad, whitish to straw-yellow.
Stalk: 1¼–2″ (3–5 cm) long, 2⅜–4⅜″ (6–11 cm) thick; silky, scaly below ring; cinnamon.
Veil: partial veil cobweblike, white, leaving remnants on cap margin and inconspicuous ring on stalk.
Spores: 6–8 × 2.5–4.5 μ elliptical, smooth, colorless; *dextrinoid*. Spore print white.

Edibility: Deadly.

Season: January–March.

Habitat: On the ground, in soil near shrubbery and live oak.

Range: California.

Look-alikes: The equally deadly *L. helveola* has membranous ring and larger spores. *L. cristata** has pungent odor and wedge-shaped spores.

Comments: A number of small lepiotas are known to be poisonous, and some contain the same toxins as the Death Cap and destroying angels. Identification of small lepiotas generally requires microscopic examination, and none should ever be eaten.

⊗ 180 **Lemon-yellow Lepiota**
Lepiota lutea (Bolt. ex Fr.) Quél.
Agaricaceae, Agaricales

Description: *Small, bell-shaped, powdery, yellow mushroom with free gills and ring on stalk.*
Cap: 1–2″ (2.5–5 cm) wide; bell-shaped, becoming conical to nearly flat, with central knob; margin with deep radial grooves; covered by scales or powdery; lemon-yellow, fading to cream-yellow. Flesh very thin.
Gills: free, crowded, narrow; yellowish-white, with hairy edges.
Stalk: 2–3″ (5–7.5 cm) long, ¹⁄₁₆–¼″ (1.5–5 mm) thick, sometimes enlarged toward base; with yellow powder.

Veil: partial veil cottony-fibrous, bright lemon-yellow; leaving movable or evanescent ring on stalk.
Spores: 8–13 × 5–8 μ; elliptical, smooth, colorless; thick-walled, with pore at tip; *dextrinoid*. Spore print white.

Edibility: Poisonous.
Season: July–September; year-round indoors.
Habitat: On ground, in leaf litter and compost; in greenhouses and on house plants.
Range: Throughout S. North America; northward indoors or near buildings.
Comments: Also known as *Leucocoprinus birnbaumii* because it resembles a *Coprinus*. It is closely related to the Onion-stalked Lepiota (*Lepiota cepaestipes*) and is reported to be poisonous.

---

121 Smooth Lepiota
Lepiota naucina (Fr.) Kum.
Agaricaceae, Agaricales

Description: *White lawn mushroom with free gills and bandlike, movable ring on stalk.*

Cap: 2–4″ (5–10 cm) wide; egg-shaped, becoming broadly convex or with slight knob; dry, smooth or rarely somewhat fibrous-scaly; dull white, sometimes grayish. Flesh thick, white.
Gills: free, crowded, broad; white, slowly changing to grayish-pink.
Stalk: 2–6″ (5–15 cm) long, ¼–⅝″ (0.5–1.5 cm) thick, sometimes enlarged at base; smooth to silky or finely hairy, white; stuffed, becoming hollow.
Veil: partial veil membranous, white; leaving persistent, collarlike, double-edged, movable ring on upper stalk.
Spores: 7–9 × 5–6 μ; oval, smooth, colorless; thick-walled, with pore at tip; *dextrinoid*. Spore print white.

Edibility: Edible with caution.
Season: September–October; November–January in California.
Habitat: Scattered to numerous, in lawns and grassy areas.

Range: Throughout North America.
Look-alikes: *Agaricus campestris** has pink gills at
first. The deadly *Amanita virosa** and
related species grow near trees, have
saclike cup about base of stalk, and
nonmovable, flaring, skirtlike ring.
Comments: Also known as *Leucoagaricus naucinus.*
This is one of very few smooth-capped
lepiotas. Although generally considered
choice, some people do become ill after
eating this species—especially its
grayish variant, which often has a
foul, medicinal or industrial odor or
taste.

---

172 Parasol
Lepiota procera (Scop. ex Fr.) S.F.G.
Agaricaceae, Agaricales

Description: *Broad, brownish, densely scaly, knobbed
cap with free gills and movable ring on tall,
scaly stalk.*

Cap: 3–8″ (7.5–20 cm) wide; egg-
shaped, becoming bell-shaped to
convex to nearly flat, with central knob;
at first with brown covering, soon
cracking into many woolly to flattened-
fibrous scales. Flesh thick, white,
showing between scales.

Gills: free, close, broad; white,
darkening somewhat with age, with
woolly edges.
Stalk: 5–16″ (12.5–40 cm) long, ⅜–
½″ (1–1.3 cm) thick; slender, with
bulbous base; covered with pattern of
small, brownish scales, showing white
between.
Veil: partial veil membranous, white;
leaving persistent, double-edged,
movable, bandlike ring on upper stalk.
Spores: 12–18 × 8–12 μ; *very large,*
broadly elliptical, colorless; thick-
walled, with large pore at tip;
dextrinoid. Spore print white.
Edibility: Choice, with caution.
Season: July–October; November–December in
Florida.
Habitat: On the ground, along paths and in

open woods, under oak, aspen, and conifers.

Range: Quebec to Florida, west to Michigan, Minnesota, and Nebraska.

Look-alikes: *L. rachodes** has coarsely woolly cap scales and smooth stalk. *L. americana** grows clustered, has smooth stalk, and bruises and ages reddish. The poisonous *Chlorophyllum molybdites** has smooth stalk, gills maturing grayish-green, and greenish spore print.

Comments: Also known as *Macrolepiota procera*. The Parasol is a favorite edible among experienced mushroom hunters.

171 Shaggy Parasol
Lepiota rachodes (Vitt.) Quél.
Agaricaceae, Agaricales

Description: *Large, coarsely scaly, pinkish- to cinnamon-brown cap with free gills and ring on stalk; cut flesh turns saffron.*
Cap: 3–8″ (7.5–20 cm) wide; convex, becoming nearly flat; smooth at first, soon breaking up into large, coarse, downcurled, fibrous scales; grayish to pinkish- or cinnamon-brown or darker. Flesh thick, white, showing between scales; becoming pinkish to saffron on bruising.
Gills: free, close, broad; white, darkening somewhat with age.
Stalk: 4–8″ (10–20 cm) long, ⅜–1″ (1–2.5 cm) thick; stout, club-shaped, sometimes thickly bulbous with well-defined margin; smooth, white, *bruising brownish*. Flesh white, usually becoming saffron when cut.
Veil: partial veil membranous, white; leaving persistent, double-edged, movable, bandlike ring on upper stalk.
Spores: 6–10 × 5–7 μ; short, elliptical, smooth, colorless; thick-walled, with pore at tip; *dextrinoid*. Spore print white.

Edibility: Choice, with caution.

Season: September–October; November–February in California.

Habitat: Single to several or in fairy rings, on the ground along roads, in gardens.

Range: Throughout North America.

Look-alikes: *L. procera** has tall, very slender, scaly stalk. *L. americana** is usually clustered, has spindle-shaped stalk. The poisonous *Chlorophyllum molybdites** does not bruise saffron on cutting, has white gills maturing grayish-green, and greenish spore print.

Comments: Also known as *Macrolepiota rachodes* or *M. rhacodes*. This choice edible is often abundant. However, the variety with the large, sharply margined bulb that occurs in S. California is reported to cause gastric upset in some. Care must be taken not to gather the toxic Green-spored Lepiota (*Chlorophyllum molybdites*) by mistake.

---

175 Red-tinged Lepiota
Lepiota rubrotincta Pk.
Agaricaceae, Agaricales

Description: *Coral red, slightly scaly cap with free gills and ring on stalk.*
Cap: 1–3″ (2.5–7.5 cm) wide; oval, becoming convex to nearly flat or with knob; smooth and reddish at first, becoming scaly and radially cracked toward margin, torn with age; center remaining smooth and coral-pink or darker, paler toward margin. Flesh thin, white.
Gills: free, close, narrow, white.
Stalk: 2–4″ (5–10 cm) long, ⅛–⅜″ (0.3–1 cm) thick, club-shaped at base; smooth to silky, white; stuffed, becoming hollow.
Veil: partial veil membranous, white; leaving persistent, pendant ring on upper stalk.
Spores: 6–10 × 4–6 μ; elliptical, smooth, colorless; thick-walled, with pore at tip; *dextrinoid*. Spore print white.

Season: July–November.

Habitat: On the ground in deciduous woods;

also in leaf litter and compost.

Range: Throughout North America.

Comments: Also known as *Leucoagaricus rubrotinctus,* this is a strikingly beautiful mushroom when young, unexpanded, and coral red. It cannot be recommended as an edible until more reports establish its harmlessness.

---

159 Red-gilled Agaricus
Melanophyllum echinatum (Roth ex Fr.) Sing.
Agaricaceae, Agaricales

Description: *Small, dark, granular cap with bright red to dark, drab gills and hollow stalk.*
Cap: ⅝–1⅜″ (1.5–4 cm) wide; conical to nearly flat; margin fringed; covered by fragile, spiny scales quickly reduced to powder; dark, dull, grayish-brown.
Gills: free or slightly attached, close, moderately broad; bright red, becoming brownish-drab.
Stalk: 1–3″ (2.5–7.5 cm) long, ¹⁄₃₂–⅛″ (1–3 mm) thick, with bulbous base; tubular and fragile; powdery to smooth, reddish to brown.
Veil: partial veil leaving remnant along cap margin and zone on stalk.
Spores: 5–7 × 2.5–3 μ; narrowly elliptical, smooth, colorless. *Spore print reddish, drying purple-brown.*

Season: July–September.

Habitat: Scattered to many, on leaf litter, in bogs, on compost, in greenhouses.

Range: Throughout North America.

Comments: This species has been described as an *Agaricus,* a *Lepiota,* and a *Cystoderma.*

---

192 Golden False Pholiota
Phaeolepiota aurea (Matt. ex Fr.) Maire ex Kon. & Maub.
Agaricaceae, Agaricales

Description: *Large, golden orange-brown, granular mushroom with attached, brownish gills and ring on stalk.*
Cap: 2–12″ (5–30 cm) wide; rounded

to convex to nearly flat, with central knob; margin often fringed; powdery to granular, golden brown.

Gills: attached, close, broad; pale yellow, becoming orange-brown.

Stalk: 4–10″ (10–25 cm) long, ⅝–2″ (1.5–5 cm) thick, enlarging near base; orange, smooth above ring.

Veil: partial veil sheathing stalk; golden orange-brown, granular; leaving persistent, flaring to pendant ring with smooth upper surface.

Spores: 10–14 × 5–6 μ; elliptical, smooth, yellowish. Spore print yellow-brown to orange-buff.

Edibility: Edible, but not recommended.

Season: September–October.

Habitat: Several to clustered, on compost and leaf litter.

Range: Alaska and Pacific NW.

Comments: Also known as *Pholiota aurea* because of its spore-print color. It is edible for some people but sickens others.

---

193 Knobbed Squamanita
Squamanita umbonata (Sum.) Bas
Agaricaceae, Agaricales

Description: *Ochre-buff, scaly cap, attached gills, and whitish stalk with buff, woolly-scaly zones.*

Cap: 1¼–2⅜″ (3–6 cm) wide; cone-shaped, becoming flattened with conical knob; margin incurved at first; dry, coarsely fibrous to scaly; ochre-buff to ochre-brown.

Gills: attached, close to crowded, moderately broad, white.

Stalk: 1–3″ (2.5–7.5 cm) long, ¼–¾″ (0.5–2 cm) thick, sometimes enlarging at base; cottony-fibrous to scaly, whitish; *arising from cylindrical to club-shaped, grayish-white underground bulb,* sometimes merged with other bulbs. *Area between stalk and bulb has rings of erect, pointed, brownish scales.*

Veil: universal veil leaves rows of erect scales at base of stalk; partial veil leaves zone on upper stalk.

Spores: 6–9 × 3.5–5.5 μ; elliptical, smooth, colorless. Spore print white.

Season: August–September.

Habitat: Single to several or somewhat clustered, on the ground in woods, near trees.

Range: New York, Pennsylvania, and Massachusetts.

Look-alikes: *S. odorata,* a northwestern species, is smaller, with lilac-gray cap and strong, aromatic odor.

Comments: Formerly called *Armillaria umbonata.* This mushroom must be dug up to be identified properly.

AMANITA FAMILY
(Amanitaceae)

This family has 2 genera, *Amanita* and the rare *Limacella.* Most species grow on the ground in forests or woodlands. All develop from an egglike enclosure called a universal veil—a tissue that is quickly outgrown, but which usually leaves patches on the cap and a cup or remnants about the stalk base. A partial veil, which encloses the gills, is also usually present; it typically leaves a skirt or ring on the upper stalk. The gills are usually free of the stalk, although some appear to be slightly attached. All have a white spore print. Most are believed to have a symbiotic, mycorrhizal relationship with trees. Many are stately and colorful, but they can be deadly. The often fatal Death Cap and the Destroying Angel are amanitas, as are the delirium-inducing Fly Agaric and the Panther. Toxic amanitas have caused 90 percent of fatal mushroom poisonings. Because rain and other factors can cause the loss of cap patches, rings, or stalk-base veil remnants, it is vital to know certain microscopic and chemical features to distinguish amanitas from edible look-alikes. The gill cells are divergent and the spores smooth, thin-walled, and lack a pore at the tip. An amyloid reaction may take place on contact with Melzer's solution. Species with a grooved or lined cap margin typically have nonamyloid spores. The amyloid reaction has *nothing* to do with edibility. Although there are a few edible species, none should be eaten or collected by any but the most experienced mushroom hunter.

145 Western Yellow Veil
Amanita aspera (Fr.) Quél.
Amanitaceae, Agaricales

Description: *Brownish cap with yellowish-gray patches, ring on stalk, and loose fragments about bulbous base.*
Cap: 2–4″ (5–10 cm) wide; convex, becoming flat; margin sometimes faintly radially lined; sticky when wet, smooth; dark brown to yellowish-brown, with yellow to gray patches.
Gills: free or slightly attached, close, broad; white to yellowish-white.
Stalk: 3–6″ (7.5–15 cm) long, ⅜–¾″ (1–2 cm) thick, enlarging to bulbous base; somewhat powdery, yellow to brownish-yellow. Flesh sometimes discoloring reddish-brown around insect holes near basal bulb.
Veils: universal veil yellowish-gray or yellow, becoming grayish-yellow; leaving patches on cap and remnants about stalk base; no cup formed. Partial veil membranous, yellow to yellowish-gray; leaving pendant ring on upper stalk.
Spores: 8–10 × 6–7 μ; elliptical, smooth, colorless, *amyloid.* Spore print white.
Season: September–October in Pacific NW.; November–February in California.
Habitat: On the ground under conifers.
Range: Pacific Coast.
Look-alikes: *A. flavorubescens** bruises reddish. *A. rubescens** bruises reddish throughout, is reddish-brown. The poisonous *A. pantherina** has rimmed cup and

nonamyloid spores.

Comments: This striking *Amanita* is not known to be edible.

⊗ 126, 134, **Cleft-foot Amanita**
147 *Amanita brunnescens* Atk.
Amanitaceae, Agaricales

Description: *Brownish cap with whitish patches; stalk with large, vertically split basal bulb; bruising reddish-brown.*
Cap: 1–6″ (2.5–15 cm) wide; convex, becoming flat or with knob; margin sometimes faintly radially lined; sticky, usually with small fibers; dark brown to olive-brown or whitish toward margin, with whitish to pale brown patches. Flesh white, discoloring reddish-brown.
Gills: free or nearly so, close, broad, cream-white.
Stalk: 2–6″ (5–15 cm) long, ⅜–¾″ (1–2 cm) thick, with abrupt, sharp-edged bulb that splits vertically; smooth to scruffy; white, discoloring reddish-brown from base upward.
Veils: universal veil white to pale brownish; leaving patches on cap and rarely about stalk base. Partial veil membranous, whitish; leaving collapsing pendant ring on upper stalk.
Spores: 7–10 μ; round, smooth, colorless, *amyloid*. Spore print white.
Edibility: Possibly poisonous.
Season: July–October.
Habitat: On the ground, in dry deciduous woods, especially among oak.
Range: Quebec to Florida, west to Michigan and E. Texas.
Look-alikes: The deadly *A. phalloides** has saclike cup about stalk base. *A. rubescens** and *A. spissa** have club-shaped stalks without abrupt basal bulb. *A. citrina** has variety with same bulb characteristic, but is greenish-yellow. *A. inaurata** has charcoal patches on its radially furrowed cap.
Comments: A summer variety, which is pure white but also bruises reddish-brown, is

common in eastern oak woods and
called either *A. brunnescens* var. *pallida*
or *A. aestivalis.*

142, 681   American Caesar's Mushroom
Amanita caesarea (Fr.) Schw.
Amanitaceae, Agaricales

Description: *Shiny, reddish-orange cap with radially
lined margin, yellow gills, and ring on
yellowish stalk with white, saclike cup.*
Cap: 2–5¼" (5–13 cm) wide;
hemispherical, becoming convex to
nearly flat, usually with knob; margin
prominently radially lined; sticky,
smooth; deep red to red-orange, aging
yellowish.
Gills: free or nearly so, close, broad,
yellow.
Stalk: 3–8" (7.5–20 cm) long, ⅛–¾"
(0.3–2 cm) thick; smooth to woolly,
scaly; yellowish; stuffed, becoming
hollow.
Veils: universal veil white; leaving
persistent, thick, loose, free-margined,
saclike cup. Partial veil membranous,
yellow; leaving pendant ring on upper
stalk.
Spores: 8–12 × 6–8 μ; oval, smooth,
colorless, nonamyloid. Spore print
white.

Edibility: Good, with caution.

Season: July–October.

Habitat: Single, scattered, or in fairy rings, on
the ground, in dry oak and pine woods.

Range: Quebec to Florida, west to Ohio and
Arizona; also Mexico.

Look-alikes: *A. flavoconia*, A. frostiana,* the
poisonous *A. muscaria*,* and *A.
parcivolvata** all resemble the American
Caesar, but all lack saclike cup and
yellow gills; their caps are not smooth.

Comments: Also known as *A. umbonata,* and
possibly different from the European *A.
caesarea.* It certainly is not as choice.
Extreme care must be exercised to avoid
picking a poisonous look-alike.

138, 671 Coccora
Amanita calyptroderma Atk. & Bal.
Amanitaceae, Agaricales

Description: *Large, orange-brown cap with broad, white patch over center and radially lined margin; stalk with ring and white, saclike cup.*
Cap: 4–12" (10–30 cm) wide; egg-shaped, becoming convex to flat; margin radially lined; sticky when wet, smooth; orange to orange-brown or darker, sometimes yellowish toward margin; large, broad, sheetlike white patch at center.
Gills: free or somewhat attached, close, moderately broad; white to cream-yellow.
Stalk: 4–8" (10–20 cm) long, ¾–1¼" (2–3 cm) thick; smooth, off-white to cream-yellow.
Veils: universal veil white; leaving broad patch over center and large, thick, persistent, saclike cup. Partial veil membranous, cream-yellow; leaving fragile, pendant ring on upper stalk.
Spores: 8–11 × 5–6 μ; elliptical, smooth, colorless, nonamyloid. Spore print white.
Edibility: Choice, with caution.
Season: September–November; sometimes spring.
Habitat: On the ground, under Douglas-fir and oak; reported under madrone.
Range: W. Canada to California, along the Pacific Coast and especially the Oregon-California border.
Look-alikes: *A. velosa** is smaller, thinner. The deadly *A. phalloides** has greenish cap.
Comments: Great care must be taken to avoid gathering a poisonous look-alike by mistake. Two color variants should be avoided: the Green-capped Coccora (*A. calyptrata*), which can be confused with the deadly Death Cap (*A. phalloides*), and the white variant, which can be confused with one of the deadly destroying angels, such as *A. virosa* or *A. ocreata.*

164, 675 Rag-veil Amanita
Amanita cinereopannosa Bas
Amanitaceae, Agaricales

Description: *Large, dull white mushroom with angular patches on cap; ragged ring and basal rows of scales on rooting stalk.*
Cap: 3–6″ (7.5–15 cm) wide; convex, becoming flat to somewhat sunken, with hanging veil remnants on margin; tacky when wet, dull whitish to gray or smoky buff, with many low, angular, wartlike patches.
Gills: free, crowded, narrow to broad, white.
Stalk: 4–6″ (10–15 cm) long, ⅜–⅝″ (1–1.5 cm) thick, with large club- to spindle-shaped rooting bulb up to 2″ (5 cm) wide; solid, white, somewhat torn below, with small fibers.
Veils: universal veil whitish to grayish, leaving angular patches on cap and transverse bands or concentric rows of broad patches about top of bulb. Partial veil membranous, white, leaving ragged, skirtlike, evanescent, grayish ring on upper stalk.
Spores: 9–14 × 5.5–7 μ; elliptical, smooth, colorless, amyloid. Spore print white.

Season: July–August.

Habitat: On the ground near pines, often in sandy soil.

Range: Massachusetts to North Carolina.

Look-alikes: Similar species with bulblike rooting stalks need microscopic examination for identification.

Comments: In wet weather the patches will wash off the cap and the ring readily falls away, but the color and the bulbous rooting stalk should identify this mushroom as one of the number of similar species of *Amanita* related to Coker's Amanita (*A. cokeri*). Because the edibility of this species has not been determined, it should not be eaten.

125 Citron Amanita
Amanita citrina Schaeff. ex S.F.G.
Amanitaceae, Agaricales

Description: *Yellowish cap with buff-white patches and stalk with ring and abruptly bulbous, marshmallowlike base.*
Cap: 2–4" (5–10 cm) wide; convex, becoming flat; margin not radially lined; sticky when wet; lemon- to pale greenish-yellow, with pale patches.
Gills: free or slightly attached, close, moderately broad; cream-white.
Stalk: 3–5" (7.5–12.5 cm) long, ⅜–⅝" (1–1.5 cm) thick, with large, soft, bulbous base; smooth to scruffy, white.
Veils: universal veil whitish to dingy buff; leaving patches on cap and thin covering on stalk bulb. Partial veil membranous, yellowish-white; leaving pendant, collapsing ring on upper stalk.
Spores: 7–10 μ; round, smooth, colorless, amyloid. Spore print white.
Season: August–November.
Habitat: On the ground in oak and pine woods.
Range: E. North America.
Look-alikes: The deadly *A. phalloides** has saclike cup about stalk base. *A. gemmata** is paler, somewhat radially lined, and lacks abruptly bulbous stalk base. *A. brunnescens* var. *pallida* lacks green tint.
Comments: The gills of this species smell like raw potatoes. Because of its similarity in color to some forms of the deadly Death Cap (*A. phalloides*), this mushroom should not be eaten. It has several color and form variants: one has a lavender tint, and one has a vertically split bulb.

---

166 Coker's Amanita
Amanita cokeri (Gil. & Küh.) Gil.
Amanitaceae, Agaricales

Description: *Large, white mushroom with pyramidal patches on cap, and rooting stalk with ring near top and rows of downcurved scales toward base.*

Cap: 3–6" (7.5–15 cm) wide; rounded, becoming convex to nearly flat; margin has hanging veil remnants, lacks radial grooves; tacky when wet, shiny when dry; ivory, with large, high, pyramidal, wartlike patches, more flat and cottony toward margin.

Gills: free or somewhat attached, crowded, broad; white to yellowish or pink-tinged.

Stalk: 5–8" (12.5–20 cm) long, ⅝–¾" (1.5–2 cm) thick, *with large, spindle-shaped, rooting bulb* up to 2" (5 cm) wide; silky, white; often with *concentric, downcurved scales toward base;* solid or stuffed with a cottony web.

Veils: universal veil whitish; leaving conical patches on cap and downcurved scales on lower stalk. Partial veil membranous, white; leaving large, pendant ring on upper stalk; ring grooved above, somewhat torn below.

Spores: 11–13.5 × 7–9 μ; elliptical, smooth, colorless, amyloid. Spore print white.

Season: July–November.

Habitat: On the ground, in oak-pine woods.

Range: New York to North Carolina, west to Indiana and Texas.

Look-alikes: Similar white to grayish *Amanitas** may have distinctive chlorine odor or somewhat differently shaped stalk.

Comments: There are 40 or more amanitas in North America with much the same structure, and all are classified together in section *Lepidella* of the genus *Amanita.* Most occur in the Southeast or in sandy oak-pine areas farther north; other species are found on the Pacific Coast. None should be eaten.

⊗ 127 **Booted Amanita**
Amanita cothurnata Atk.
Amanitaceae, Agaricales

Description: *White cap, often with yellowish center, and with white patches, somewhat radially lined margin, ring on stalk, and rimmed bulb.*

Cap: 1–4" (2.5–10 cm) wide; hemispherical to convex, becoming nearly flat to somewhat sunken; margin faintly radially lined; sticky, smooth; white, often yellowish about center, with thin, white patches.
Gills: free, crowded, moderately broad, white.
Stalk: 2–5" (5–12.5 cm) long, ⅛–⅝" (0.3–1.5 cm) thick, with oval to round basal bulb; woolly to scaly, whitish.
Veils: universal veil white; leaving patches on cap, and with either bandlike, rolled margin or free rim at tip of stalk bulb. Partial veil membranous, white; leaving pendant ring on upper or mid-stalk.
Spores: 8.7–11.8 × 6.3–8.7 μ; round to elliptical, smooth, colorless, nonamyloid. Spore print white.
Edibility: Poisonous.
Season: July–October.
Habitat: On the ground, in oak, oak-pine, and pine woods.
Range: New York to Florida, west to Michigan.
Look-alikes: *A. gemmata** has buff-yellow cap.
Comments: This species is also known as a variety of the poisonous Panther (*A. pantherina*) and believed to be similarly poisonous.

119, 120 **Powder-cap Amanita**
Amanita farinosa Schw.
Amanitaceae, Agaricales

Description: *Powdery gray to brownish-gray mushroom with radially lined to furrowed cap margin and free gills.*
Cap: 1–3" (2.5–7.5 cm) wide; bell-shaped, becoming convex to flat; margin strongly radially lined to pleated; pale gray to brownish-gray, covered with dense gray powder.
Gills: free, close, broad, white.
Stalk: 1–3" (2.5–7.5 cm) long, ⅛–⅜" (0.3–1 cm) thick, with slight basal bulb; grayish to dirty white, powdery.
Veil: universal veil grayish, leaving

mealy, powdery covering on cap and
stalk bulb. No partial veil.

Spores: 6.3–9.4 × 4.5–8 μ; round to
elliptical; smooth, colorless,
nonamyloid. Spore print white.

Season: June–November.

Habitat: On the ground, under coniferous and
deciduous trees; also in grassy wood
borders.

Range: Throughout North America.

Comments: This may not readily be recognized as
an *Amanita* because it has no ring or
partial veil, and the universal veil
disintegrates into the powdery
covering, leaving no cuplike structure.

136, 139 **Yellow Patches**
Amanita flavoconia Atk.
Amanitaceae, Agaricales

Description: *Orange to yellow cap with yellowish*
patches, and ring on stalk with crumbling,
yellow basal patches.
Cap: 1–3″ (2.5–7.5 cm) wide; convex
to nearly flat or with central knob;
margin smooth or faintly radially lined;
sticky, smooth; bright orange to
yellow, with bright yellow, easily lost
patches.
Gills: free or nearly so, close, broad;
whitish to somewhat yellow on edges.
Stalk: 2–4″ (5–10 cm) long, ¼–⅝″
(0.5–1.5 cm) thick, with basal bulb;
smooth to scruffy; white to pale yellow;
stuffed, becoming hollow.
Veils: universal veil yellow; leaving
patches on cap and crumbly remnants
about stalk base. Partial veil
membranous, white to yellow; leaving
pendant ring on upper stalk.
Spores: 7–9.5 × 4.5–5 μ; elliptical,
smooth, colorless, *amyloid*. Spore print
white.

Season: Late June–November.

Habitat: On the ground, in oak and birch woods
and near conifers.

Range: Eastern North America.

Look-alikes: *A. flavorubescens** stains reddish about

stalk base. *A. frostiana*, typically near conifers, has radially lined margin, often collared basal boot, and nonamyloid spores. The poisonous *A. muscaria** is bigger, has concentric bands about stalk base, and nonamyloid spores. *A. caesarea** has saclike, white cup about stalk base. *A. parcivolvata** lacks a ring.

Comments: This is one of the most common eastern amanitas, and has the longest season. Its edibility has not been established and therefore it cannot be recommended.

135, 678 Yellow Blusher
Amanita flavorubescens Atk.
Amanitaceae, Agaricales

Description: *Brownish-yellow cap with yellowish patches; stalk with ring and bulbous base staining reddish, with yellow patches.*
Cap: 2–4″ (5–10 cm) wide; convex; margin not radially lined; sticky; smooth; yellow to brownish-yellow, with yellow patches. Flesh whitish, bruising reddish.
Gills: free or slightly attached, close, moderately broad, cream-white.
Stalk: 2–5″ (5–12.5 cm) long, ⅜–¾″ (1–2 cm) thick, with club-shaped bulb, sometimes narrowing to a point below bulb; minutely hairy, white to yellowish-white.
Veils: universal veil yellow; leaving patches on cap and crumbly remnants about stalk base. Partial veil membranous, whitish to yellowish; leaving pendant ring on upper stalk.
Spores: 7.5–10 × 5.5–6.5 μ; elliptical, smooth, colorless, *amyloid*. Spore print white.
Season: Late June–October.
Habitat: On the ground, under oaks in thin woods or open urban areas.
Range: Quebec to Florida, west to Michigan.
Look-alikes: *A. rubescens** lacks yellow coloration. *A. flavoconia** does not bruise reddish. The poisonous *A. muscaria** does not

discolor and has distinctive base.

Comments: This mushroom is often common in early summer and in fall, but it cannot be recommended because its edibility has not been established.

115, 674   Tawny Grisette
Amanita fulva (Schaeff. ex) Pers.
Amanitaceae, Agaricales

Description: *Tawny cap with radially furrowed margin; stalk has white, saclike cup around base.*
Cap: 2–4″ (5–10 cm) wide; oval, becoming convex to flat or knobbed; margin distinctly furrowed; tacky, smooth; reddish-brown, tawny, or tan, rarely with 1–2 broad, whitish patches.
Gills: free, close, broad; white to cream.
Stalk: 3–6″ (7.5–15 cm) long, ⅛–⅝″ (0.3–1.5 cm) thick; with small, flattened fibers, whitish.
Veil: universal veil white, sometimes with rust stain; leaving persistent, large, loose, saclike cup. No partial veil.
Spores: 8–10 μ; round, smooth, colorless, nonamyloid. Spore print white.

Edibility: Good, with caution.

Season: July–September; January–March in California.

Habitat: On the ground, under deciduous and coniferous trees.

Range: Throughout North America.

Look-alikes: *A. crocea* has flattened scales on stalk, lacks reddish tones in its orange to orange-brown cap.

Comments: Formerly known as *Amanitopsis vaginata* var. *fulva,* the Tawny Grisette is now recognized as a distinct species. It is edible, but the danger of confusing it with the toxic members of the genus should discourage any but the most experienced mushroom hunter from collecting it.

⊗ 128, 129 **Gemmed Amanita**
Amanita gemmata (Fr.) Gill.
Amanitaceae, Agaricales

Description: *Buff-yellow cap with cream patches, ring on stalk, and basal cup attached to stalk or with small, free collar.*
Cap: 1–4″ (2.5–10 cm) wide; rounded, becoming convex to flat; *margin radially lined;* tacky when wet, smooth; pale yellow to buff, with small, irregularly placed, cream patches.
Gills: free or slightly attached, crowded, broad to narrow, white.
Stalk: 2–5″ (5–12.5 cm) long, ¼–¾″ (0.5–2 cm) thick, somewhat enlarged near oval basal bulb; slightly cottony to smooth, white or cream.
Veils: universal veil white to cream; leaving patches on cap and stalk base, or basal cup attached to stalk or with free rim at tip of bulb. Partial veil white; often leaving evanescent, pendant ring on upper stalk.
Spores: 8.7–11 × 5.5–8.5 μ; broadly elliptical, smooth, colorless, nonamyloid. Spore print white.

Edibility: Possibly poisonous.

Season: June–October; November–February in California.

Habitat: On the ground, in oak and pine woods; woods and parks in urban areas.

Range: Throughout North America.

Look-alikes: *A. citrina** has pale greenish-yellow cap, abruptly bulbous base, and amyloid spores.

Comments: The Jonquil Amanita (*A. junquillea*) and Russulalike Amanita (*A. russuloides*) may only be variants of this species. It is thought to have hybridized with the poisonous Panther (*A. pantherina*) in the Northwest, producing mushrooms with characteristics of both species. The Gemmed Amanita often loses its ring and veil remnants, especially after a rain, and can then be mistaken for a *Russula* or *Tricholoma*. This mushroom is sometimes eaten; however, it may well be a species complex, and some of its forms may contain toxins, so it is best avoided.

133 Strangulated Amanita
Amanita inaurata Secr.
Amanitaceae, Agaricales

Description: *Dark brown cap with radially furrowed margin; charcoal patches on cap and base.*
Cap: 2–5" (5–12.5 cm) wide; egg-shaped, becoming convex to flat with low knob; margin radially furrowed; tacky, brownish-black to brownish-gray, with loose, charcoal patches.
Gills: free, close, white.
Stalk: 2–6" (5–15 cm) long, ⅜–⅝" (1–1.5 cm) thick, sometimes enlarged toward base; with grayish hairs, often in chevronlike pattern, and with charcoal veil remnants about base.
Veil: universal veil pale gray to charcoal; leaving patches on cap and evanescent, saclike cup about stalk base, soon disintegrating into patches. No partial veil.
Spores: 11.5–14 μ; round, smooth, colorless, nonamyloid. Spore print white.

Season: July–October; November–March in California.

Habitat: On the ground under conifers, especially eastern hemlock; also in open mixed woods and wooded urban parks.

Range: Throughout North America.

Look-alikes: *A. brunnescens** has whitish patches on cap, no radial furrows, bulbous, chiseled base, amyloid spores. Brown-capped form of *A. vaginata** has few, if any, cap patches, large saclike cup.

Comments: Formerly known as *Amanitopsis strangulata,* its name refers to the tight belt of grayish veil tissue about the stalk base. Like many common fungi, this should not be eaten because it is probably a complex of closely related species, and it is not known if all the forms are edible.

⊗ 143, 144, **Fly Agaric**
680 *Amanita muscaria* var. *muscaria* (L. ex Fr.) Hook.
Amanitaceae, Agaricales

Description: *Blood-red cap with pyramidal, white patches; stalk has ring and bulbous base with rows of cottony patches.*
Cap: 2–10″ (5–25 cm) wide; convex to flat or somewhat sunken; *margin somewhat radially lined,* sometimes with hanging remnants of veil; tacky when wet, smooth; *blood-red* to reddish-orange, with random or concentrically arranged cottony patches.
Gills: free or slightly attached, crowded, broad, whitish.
Stalk: 2–7″ (5–18 cm) long, ⅛–1¼″ (0.3–3 cm) thick, sometimes enlarging toward rounded basal bulb; fibrous to cottony or scaly, white to cream.
Veils: universal veil white; leaving conical to flat patches on cap that are often in concentric rings, and *concentric bands on lower stalk, sometimes as rim at tip of bulb.* Partial veil membranous, white; leaving pendant, fragile, often collapsing ring on upper stalk.
Spores: 9.4–13 × 6.3–8.7 μ; broadly elliptical, smooth, colorless, nonamyloid. Spore print white.
Edibility: Poisonous.
Season: July–October; winter in California.
Habitat: On the ground, under pine, spruce, and birch; also live oak and madrone in California, spruce at higher elevations.
Range: Rocky Mountains and Pacific Coast; rare in East, but reported in Maine, Connecticut, and New York.
Look-alikes: The poisonous *A. muscaria* var. *formosa** has yellow-orange cap. *A. muscaria* var. *flavivolvata* has yellow universal veil.
Comments: This mushroom is called the Fly Agaric because it has been used, mixed in milk, to stupefy houseflies. The Fly Agarics of North America cause delirium, raving, and profuse sweating. Unlike its Siberian relative, this induces no visions.

⊗ 137, 677   **Yellow-orange Fly Agaric**
Amanita muscaria var. *formosa* (Pers. ex Fr.)
Bert. in DeCh.
Amanitaceae, Agaricales

Description: *Orange-red to yellowish cap with pinkish-buff patches; stalk has ring and bulbous base with rows of cottony patches.*
Cap: 2–6" (5–15 cm) wide; convex to flat; *margin slightly radially lined;* tacky when wet, smooth, with loose, cottony patches; *yellow, deep orange, or red-orange.*
Gills: free or nearly attached, crowded, broad to narrow, whitish.
Stalk: 2–6" (5–15 cm) long, ¼–1¼" (0.5–3 cm) thick, sometimes enlarging below to nearly round or slightly rooting basal bulb; fibrous, cottony, or scaly; white, buff, or pale yellow-orange.
Veils: universal veil yellowish-buff to tan; leaving patches on cap and *concentric bands on lower stalk or rim at tip of bulb.* Partial veil membranous, pinkish-buff; leaving pendant, often evanescent or collapsing ring on upper stalk.
Spores: 8.7–13 × 6.3–8 μ; broadly elliptical, smooth, colorless, nonamyloid. Spore print white.

Edibility: Poisonous.

Season: Late June–November; November–February in California.

Habitat: On the ground, often in fairy rings, under spruce, pine, and eastern hemlock; also birch, poplar, and oak.

Range: Widespread; most common in East.

Look-alikes: *A. caesarea** has conspicuously lined cap and large, white, saclike cup. *A. flavoconia** has crumbly veil remnants about stalk base. *A. rubescens** and *A. flavorubescens** bruise red.

Comments: The Yellow-orange Fly Agaric is a very common late summer and fall mushroom in the Northeast. Although it is not deadly, it should not be eaten because it contains toxic compounds that may cause sweating, deep sleep, and disorientation.

⊗ 130, 131, **Panther**
679 *Amanita pantherina* (DC. ex Fr.) Secr.
Amanitaceae, Agaricales

Description: *Brownish cap with white patches; margin has fine radial lines; stalk has ring and basal bulb with free or rolled margin, or rows of cottony scales.*
Cap: 1–6" (2.5–15 cm) wide; round, becoming convex to flat or slightly sunken; margin has fine radial lines; tacky when wet, smooth; usually blackish-, sooty-, or yellowish-brown near margin, sometimes nearly white overall; with whitish, cottony patches.
Gills: free or slightly attached, crowded, broad to narrow, whitish.
Stalk: 2–7" (5–17.5 cm) long, ⅜–1" (1–2.5 cm) thick, sometimes enlarged downward toward roundish basal bulb; fibrous, cottony, or scaly, whitish.
Veils: universal veil white; leaving patches on cap, and *bandlike margin at tip of stalk bulb, typically rolled into stalk, but occasionally free;* sometimes with concentric bands on lower stalk. Partial veil membranous, whitish; leaving *persistent, pendant ring on upper or mid-stalk.*
Spores: 8–14 × 6.3–10 μ; broadly elliptical, smooth, colorless, nonamyloid. Spore print white.
Edibility: Poisonous.
Season: June; September–October; November–February in California.
Habitat: On the ground under conifers, particularly Douglas-fir.
Range: Rocky Mountains to West Coast; rare in East.
Look-alikes: *A. rubescens** lacks rimmed bulb, bruises reddish. *A. spissa** has cap with grayish patches and no radial lines. *A. gemmata** typically lacks well-defined margin on bulb; there are intermediate forms between this and *A. pantherina.* The poisonous *A. cothurnata** is smaller, with a more slender stalk and whitish cap. *A. velatipes* is taller, more robust, with yellowish cap that is

brownish at center.

Comments: There are several varieties of this
species, ranging in color from white to
yellow or brownish; all are spotted like
a panther. Although usually a fall
mushroom, the Panther is sometimes
found in spring, and may fool the
unwary. Its toxins cause delirium,
raving, and a comalike deep sleep.
Because of the color variation of the
cap, the best identifying features are
the ring and structure of the stalk base.

140 False Caesar's Mushroom
Amanita parcivolvata (Pk.) Gil.
Amanitaceae, Agaricales

Description: *Red to orange-yellow cap with yellowish
patches, radially lined or furrowed margin,
and powdery, yellowish-white stalk.*
Cap: 1–5″ (2.5–12.5 cm) wide; round,
becoming convex to flat or slightly
sunken; margin radially lined, often
with hanging veil remnants; tacky
when wet, smooth; *reddish* at center,
orange to yellow-orange near margin,
with cottony, yellowish patches.
Gills: free, crowded, broad, yellowish.
Stalk: 1–5″ (2.5–12.5 cm) long, ⅛–
⅝″ (0.3–1.5 cm) thick, with rounded
basal bulb; *powdery to mealy, yellowish.*
Veil: universal veil yellowish; leaving
numerous cottony-fibrous patches on
cap and stalk base. No partial veil.
Spores: 10–14 × 6.3–8 μ; elliptical,
smooth, colorless, nonamyloid. Spore
print white.
Season: July–September.
Habitat: On the ground in mixed woods; often
on disturbed ground, in lawns, or on
woodland roads under oak.
Range: New Jersey to Florida, west to Ohio.
Look-alikes: *A. caesarea** has well-developed ring
and white, saclike cup.
Comments: Formerly called *Amanitopsis parcivolvata.*
This is a very common summer
mushroom in the Southeast, and one of
the prettiest.

⊗ 113, 673 **Death Cap**
Amanita phalloides (Fr.) Secr.
Amanitaceae, Agaricales

Description: *Smooth, greenish cap with skirtlike ring at top of stalk, and saclike cup about base of stalk.*
Cap: 2½–6″ (6.5–15 cm) wide; convex; margin not radially lined; slightly sticky, smooth, with small, flattened fibers; yellowish-green to greenish-brown, darker at center, often pale on margin; rarely entirely white. Odor becoming foul.
Gills: more or less free, close, broad, white.
egg
Stalk: 3–5″ (7.5–12.5 cm) long, ½– ¾″ (1.5–2 cm) thick; enlarging downward to basal bulb; smooth, whitish to dull greenish-yellow.
Veils: universal veil rarely leaving patch on cap, sheathing basal bulb with large, membranous, saclike cup with free limb. Partial veil leaving membranous, pendant, persistent, white ring at top of stalk. No partial veil.
Spores: 8–11 × 7–9 μ; round, thin-walled, smooth, amyloid. Spore print white.
Edibility: Deadly.
Season: Late September–November in Northeast; November–January in West.
Habitat: Under conifers (spruce, pine, hemlock) and hardwoods (oak in New York, live oak in California); near planted juniper and dogwood in Seattle, oak in Oregon.
Range: Massachusetts to Virginia, west to Ohio; Pacific NW. to California, and spreading.
Look-alikes: The deadly *A. virosa** and its allies are white, taller, and more slender.
Comments: The Death Cap has been mistaken for edible mushrooms to which it bears little resemblance; in almost every case the ring or the cup (volva) was ignored. This deadly species often appears in urban areas under European conifers, as well as under native oaks, and may

produce anywhere from 2 or 3 to several hundred mushrooms. Although the odor soon becomes nauseating, the young mushrooms are described as having a pleasant taste. Symptoms are typically delayed for 10–14 hours or more and begin with severe intermittent vomiting, diarrhea, and cramps. A remission occurs after a day or so, and the victim believes he is better. By the 3rd or 4th day, signs of serious kidney and/or liver dysfunction appear and can, without adequate treatment, cause death in 5–10 days.

⊗ 149   Gray-veil Amanita
Amanita porphyria (A. & S. ex Fr.) Secr.
Amanitaceae, Agaricales

Description: *Brownish cap with grayish patches; stalk with gray ring and abruptly bulbous base.*
Cap: 1–3″ (2.5–7.5 cm) wide; convex to flat; margin incurved, becoming flat, with no radial grooves; tacky when wet, smooth; brown to purple-brown, sometimes with grayish patches.
Gills: free, close, moderately broad, whitish.
Stalk: 2–5″ (5–12.5 cm) long, ⅜–⅝″ (1–1.5 cm) thick, with *large, soft, rounded basal bulb;* whitish, with small, grayish fibers.
Veils: *universal veil grayish;* leaving a few crumbling patches on cap and stalk base. Partial veil membranous, grayish; leaving collapsing ring on upper stalk.
Spores: 7–9 μ; round, smooth, colorless, *amyloid.* Spore print white.

Edibility: Possibly poisonous.
Season: August–October.
Habitat: Under conifers and in mixed woods.
Range: E. United States and Pacific NW.
Look-alikes: *A. brunnescens** has chiseled base. *A. spissa** has clublike base. *A. spreta** has sheathing cup on stalk.
Comments: The Gray-veil Amanita is found every fall and, except for its color, resembles the Citron Amanita (A.

citrina). Though not known to have caused poisonings, it contains toxins and should not be eaten.

132 Blusher
Amanita rubescens (Pers. ex Fr.) S.F.G.
Amanitaceae, Agaricales

Description: *Reddish-brown cap with pinkish patches, and ring on clublike stalk; bruises reddish.*
Cap: 2–6″ (5–15 cm) wide; oval, becoming convex or nearly flat with broad knob at center; margin lacks radial grooves; slightly tacky when wet, smooth; reddish-brown, often with olive tints and cottony patches.
Gills: free or slightly attached, close to crowded, moderately broad; whitish, bruising reddish to reddish-brown.
Stalk: 3–8″ (7.5–20 cm) long, ¼–¾″ (0.5–2 cm) thick, enlarging downward to swollen base; smooth to fibrous, whitish, bruising pink to red.
Veils: universal veil grayish to dirty pink; leaving patches on cap, rarely on stalk base. Partial veil membranous, whitish, staining pinkish; leaving large, collapsing ring on upper stalk.
Spores: 8–10 × 5–6 μ; elliptical, smooth, colorless, amyloid. Spore print white.
Edibility: Good, with caution.
Season: Late June–October; February–April in California.
Habitat: On the ground, in oak woods and wooded urban parks; under white pine.
Range: E. North America and California.
Look-alikes: *A. brunnescens** lacks reddish colors, stains brownish, and has abruptly bulbous, chiseled base. *A. flavorubescens** has yellowish cap and veil. The poisonous *A. pantherina** has cup with well-defined margin, does not bruise reddish although it may discolor.
Comments: When well cooked, this species is tasty, but great care must be taken not to confuse it with any of the similar-looking poisonous or suspected species.

167   Western Woodland Amanita
Amanita silvicola Kauff.
Amanitaceae, Agaricales

Description: *Large, white mushroom with downy, woolly cap and stalk.*
Cap: 2–5″ (5–12.5 cm) wide; convex to flat; margin with no radial grooves; often with hanging veil remnants; tacky when wet, *covered by white, loose tissue.*
Gills: free or slightly attached, crowded, narrow to broad, white.
Stalk: 2–5″ (5–12.5 cm) long, ⅜–1″ (1.5–2.5 cm) thick, with swollen bulb to 2″ (5 cm) thick, rarely rooting; cottony, white.
Veils: universal veil white; leaving soft, flat, cottony patches on cap and stalk base. Partial veil fibrous, white; leaving evanescent ring or zone on upper stalk.
Spores: 8–12 × 4.5–6 μ; elliptical, smooth, colorless, amyloid. Spore print white.

Season: September–October; November–February in California.

Habitat: On the ground, in coniferous woods, particularly under Douglas-fir; in mixed woods in California.

Range: Pacific NW. and California.

Look-alikes: The deadly *A. virosa** and its allies have smooth cap, flaring ring, and saclike cup. *A. smithiana** is usually taller, with wartlike patches on cap.

Comments: Because this species lacks a well-formed ring or cup and often has somewhat attached gills, it may seem to belong to another genus; however, its veils, spore characteristics, and gill tissue structure place it solidly in *Amanita.* Not reliably known to be edible.

165   Smith's Amanita
Amanita smithiana Bas
Amanitaceae, Agaricales

Description: *Large, white mushroom with patches on cap and torn ring on scaly, rooting stalk.*
Cap: 2–5″ (5–12.5 cm) wide; round,

convex, or flat; margin not radially grooved, with hanging veil remnants; tacky when wet; with cottony or feltlike, conical, white patches.
Gills: slightly attached or free, crowded, rather broad; white to cream.
Stalk: 4–8″ (10–20 cm) long, ⅜–1¼″ (1–3 cm) thick, enlarging downward to spindle-shaped bulb up to 2″ (5 cm) wide; ragged, scaly, white.
Veils: universal veil white to pale brown; leaving cottony, wartlike patches on cap and zones on stalk bulb. Partial veil cottony, fibrous, white; leaving ragged ring on upper stalk.
Spores: 10.5–13.5 × 7–8 μ; elliptical, smooth, colorless, amyloid. Spore print white.

Season: September–October.
Habitat: On the ground in coniferous woods.
Range: Pacific NW. and N. California.
Comments: This large, distinctive mushroom is a representative of the group called *Amanita* section *Lepidella,* whose members are most abundant and diverse in the Southeast.

146 Stout-stalked Amanita
Amanita spissa (Fr.) Kum.
Amanitaceae, Agaricales

Description: *Pale brown to dark gray-brown cap with grayish patches, with ring on grayish, clublike stalk.*
Cap: 2–6″ (5–15 cm) wide; convex to flat; margin with no radial grooves; tacky when wet, shiny when dry, smooth; grayish *with brownish center,* with powdery-grayish patches.
Gills: free or sometimes slightly attached, crowded, moderately broad, whitish.
Stalk: 3–5″ (7.5–12.5 cm) long, ⅜–¾″ (1–2 cm) thick, sometimes enlarged downward to clublike or rounded basal bulb; whitish to grayish, with fine grayish, cottony fibers.
Veils: universal veil grayish; leaving

soft, cottony or powdery patches on cap
and stalk base. Partial veil
membranous, whitish; leaving pendant
ring on upper stalk.
Spores: 9–10 × 7–8 μ; oval, smooth,
colorless, amyloid. Spore print white.

Season: Late July–October.

Habitat: On the ground, in oak-pine woods; also
under live oak, hemlock, and yellow
birch.

Range: Quebec to Florida, west to Ohio and
Michigan.

Look-alikes: *A. rubescens** bruises red. The poisonous
A. pantherina* has bulb with well-
formed margin.

Comments: Although this species is reportedly
eaten in Europe, the edibility of
American specimens is uncertain.

---

114 Hated Amanita
Amanita spreta Pk.
Amanitaceae, Agaricales

Description: *Brownish cap with radially lined
margin; whitish stalk with ring and saclike
cup pressed tightly about base.*
Cap: 2–5" (5–12.5 cm) wide; oval,
becoming convex to flat or with central
knob; *margin finely lined;* slightly
tacky when wet, smooth; lead-gray to
grayish-brown, with 1 or more large,
whitish patches.
Gills: free, crowded, rather broad to
narrow, white.
Stalk: 4–6" (10–15 cm) long, ⅜–¾"
(1–2 cm) thick, not bulbous; smooth to
finely fibrous, whitish.
Veils: universal veil white; sometimes
leaving 1–2 large patches on cap and a
membranous, persistent, saclike cup,
often closely pressed to stalk base.
Partial veil membranous, whitish;
leaving thin ring on upper stalk.
Spores: 10–12 × 6–7.5 μ; elliptical,
smooth, colorless, nonamyloid. Spore
print white.

Season: July–September.

Habitat: In sandy woods, under pine and oak.

549

Range: Quebec to Florida, west to Michigan.
Look-alikes: The deadly *A. phalloides** usually has
greenish tones in cap, no radial lines
on margin, bulbous base sheathed by
cup with free limb, and amyloid spores.
Comments: Although this mushroom is called the
Hated Amanita, it is not known to be
poisonous. It doubtless has been
confused with the deadly Death Cap
(*A. phalloides*).

112 Grisette
Amanita vaginata var. *vaginata* (Bull. ex
Fr.) Vitt.
Amanitaceae, Agaricales

Description: *Gray cap with white patches and furrowed
margin; white cup about stalk base.*
Cap: 2−4″ (5−10 cm) wide; oval,
becoming bell-shaped to convex or flat
with central knob; *margin radially
grooved;* tacky when wet, smooth; gray,
sometimes with a few patches.
Gills: free, close to almost distant,
moderately broad, whitish.
Stalk: 4−8″ (10 -20 cm) long, ⅜−⅝″
(1−1.5 cm) thick; cylindrical, smooth
or somewhat mealy, whitish.
Veil: universal veil white; sometimes
leaving patches on cap, and a large,
membranous, persistent, free-
margined, saclike cup. Partial veil
fragile, rarely present.
Spores: 9−12 μ; round, smooth,
colorless, nonamyloid. Spore print
white.
Edibility: Edible with caution.
Season: June−September; November−February
in California.
Habitat: On the ground in open woods; in grass
near trees.
Range: Throughout North America.
Look-alikes: *A. umbrinolutea* has pale yellowish cap
with copper-brown center and margin.
A. pachycolea, on West Coast, is tall
and very broad, with brown cap and
unusually long sheathing cup.
Comments: Formerly known as *Amanitopsis vaginata*.

The Grisette has many different color forms. Some have now been established as varieties based on color, such as var. *alba* (white) and var. *livida* (brownish). Although some people do eat the Grisette, not enough is known about its edibility to recommend it.

---

148 Orange Spring Amanita
Amanita velosa (Pk.) Lloyd
Amanitaceae, Agaricales

Description: *Orange cap with 1 or more white patches and radially furrowed margin; white, saclike cup about stalk base.*
Cap: 2–6″ (5–15 cm) wide; egg-shaped, convex or flat; *margin furrowed;* tacky when wet, smooth; *orange to pinkish-buff,* fading to tan, buff, or whitish; often with loose, large, white, cottony patch over center.
Gills: free or somewhat attached, close, moderately broad, whitish.
Stalk: 2–5″ (5–12.5 cm) long, ⅜–¾″ (1–2 cm) thick, sometimes enlarged toward base; powdery to scaly, whitish; with sheathing cup.
Veil: universal veil white; often leaving large, cottony patch over center of cap, and a *large, membranous, saclike cup.* No partial veil.
Spores: 8.5–12 × 6–11 μ; elliptical, smooth, colorless, nonamyloid. Spore print white.
Edibility: Edible with caution.
Season: February–May.
Habitat: Open areas and lawns, especially near live oak.
Range: California and Oregon.
Look-alikes: *A. fulva** has reddish-brown to tawny cap and rust-staining, saclike cup.
Comments: This species may have a ring zone, created by pressure of the cap against the stalk during the button stage. Once known as *Amanitopsis velosa.* Because amanitas are difficult to distinguish, and because many are toxic, it is best to consider all amanitas inedible.

⊗ 123, 124, **Destroying Angel**
672 *Amanita virosa* Secr.
Amanitaceae, Agaricales

Description: *White mushroom with flaring to ragged ring on stalk; large, saclike cup about base.*
Cap: 2–5" (5–12.5 cm) wide; convex to flat with central swelling, or nearly flat; margin smooth; tacky when wet, smooth, dull to shiny white, may discolor at center of cap with age.
Gills: free or attached, close, narrow to moderately broad, white.
Stalk: 3–8" (7.5–20 cm) long, ¼–¾" (0.5–2 cm) thick, sometimes enlarging downward to basal bulb; cottony to somewhat shaggy, white; with ring and sheathing cup.
Veils: universal veil white; leaving large, membranous, persistent, saclike cup with free limb. Partial veil membranous, white; leaving large, pendant, often torn ring on upper stalk.
Spores: 9–11 × 7–9 μ; nearly round to round, smooth, colorless, amyloid. Spore print white.
Edibility: Deadly.
Season: Late June–early November.
Habitat: On the ground in mixed woods; in grass under or near trees.
Range: Throughout North America.
Look-alikes: Several amanitas are called destroying angels: *A. verna* has smooth stalk and occurs April–June in Pacific NW. (rarely in East). *A. bisporigera,* a smaller, slender-stalked summer mushroom, has 2-spored basidia; it occurs under hardwoods east of the Mississippi from July–October. *A. ocreata,* found in the Southwest under oaks in winter and early spring, has center of cap becoming buff with age. All are most reliably differentiated by their spores, but all are deadly.
Comments: The Destroying Angel, one of the most strikingly beautiful of our mushrooms, is usually found alone or in a small, scattered group, shining white against a

green or brown backdrop. Because the young, unexpanded caps resemble edible *Agaricus* mushrooms, an *Agaricus* must be carefully and completely removed from the ground to make certain no sheathing cup or remnant—indicating an *Amanita*—has been left or overlooked. Symptoms of poisoning by the Destroying Angel are like those of the deadly Death Cap (*A. phalloides*), and include vomiting, diarrhea, and cramps. Kidney and/or liver dysfunction follow and, without treatment, can result in death. Along with the deadly Death Cap, the destroying angels constitute the group of deadly amanitas.

116   Volvate Amanita
Amanita volvata (Pk.) Mar.
Amanitaceae, Agaricales

Description: *White mushroom with cottony or scaly, radially lined cap and thick, sheathing, saclike cup at stalk base.*
Cap: 2–3" (5–7.5 cm) wide; convex to flat; *margin becoming radially lined;* moist to dry, white, covered with cottony, hairlike, scaly veil patches; *bruises reddish-brown to brownish.*
Gills: free, close, moderately broad, white.
Stalk: 2–4" (5–10 cm) long, ¼–⅜" (0.5–1 cm) thick, sometimes slightly enlarged below; white, densely powdery to shaggy; bruises brownish.
Veil: universal veil white, bruises or ages brown; leaving loose, cottony patches, sometimes hairy to powdery, on cap, and thick, large, membranous, saclike cup. No partial veil.
Spores: 9–11 × 6–7 μ; elliptical, smooth, colorless, amyloid. Spore print white.
Season: July–October.
Habitat: On the ground in open mixed woods.
Range: Quebec to Florida, west to Michigan.
Look-alikes: *A. peckiana* has partial veil, thin,

evanescent ring on stalk, smooth cap margin. *A. verna* and *A. virosa**, both deadly, have ringed stalks and smooth, unlined, white caps.

Comments: Because this species lacks a ring, it was formerly known as *Amanitopsis volvata.* Although this mushroom is not known to be poisonous, it should certainly not be eaten.

---

117 **Salmon Amanita**
Amanita wellsii Murr.
Amanitaceae, Agaricales

Description: *Salmon-orange cap with yellowish patches and somewhat radially lined margin; evanescent yellow ring on stalk, with yellowish patches around basal bulb.*
Cap: 1–5″ (2.5–12.5 cm) wide; rounded to convex, becoming flat; margin radially lined with age; tacky when wet; orange to salmon-pinkish or ochre, fading to yellowish; with soft, loose, scalelike patches.
Gills: free or slightly attached, crowded, moderately broad; yellowish to cream-white.
Stalk: 3–6″ (7.5–15 cm) long, ¼–1″ (0.5–2.5 cm) thick, with basal bulb; cottony to nearly smooth, yellowish.
Veils: universal veil yellowish-buff; leaving cottony remnants on cap and stalk base. Partial veil membranous, yellow; leaving evanescent ring on upper stalk.
Spores: 11–14 × 6.3–8.3 μ; elliptical, smooth, colorless, nonamyloid. Spore print white.
Season: August–September.
Habitat: On the ground in mixed woods.
Range: Quebec south to New York; also reported in the mountains of North Carolina.
Comments: This eastern mountain species can be misidentified with age or after a rain, when the ring and veil remnants may disappear, or the color fade.

141 Slimy-veil Limacella
Limacella glioderma (Fr.) Maire
Amanitaceae, Agaricales

Description: *Slimy to tacky, brownish cap with whitish gills; stalk with slimy veil remnants.*
Cap: 1–3″ (2.5–7.5 cm) wide; convex, becoming nearly flat with broad knob; margin incurved at first, becoming upturned; slimy to tacky, dry-granular beneath slime; dark reddish-brown in center, brown elsewhere, fading to pinkish. Odor strongly mealy.
Gills: free or slightly attached, close, broad; whitish, becoming somewhat pinkish-buff.
Stalk: 2–4″ (5–10 cm) long, ¼–⅜″ (0.5–1 cm) thick; smooth to silky and dry, light brown.
Veils: universal veil slimy, leaving slime on cap and in patches on stalk. Partial veil fibrous, evanescent, at times leaving hairy zone on stalk.
Spores: 3–4 μ; round, smooth, colorless, nonamyloid. Spore print white. Gill tissue divergent.

Season: August–October.

Habitat: Single to several, on the ground in woods, especially under hemlock, maple, and birch.

Range: Quebec to Florida, west across N. North America, south to Washington.

Comments: Formerly known as *Lepiota glioderma*, this species was transferred to *Limacella*, which accommodates mushrooms with a slimy universal veil.

---

168 Ringed Limacella
Limacella solidipes (Pk.) H.V. Smith
Amanitaceae, Agaricales

Description: *Whitish mushroom with tacky cap and large, persistent ring.*
Cap: 1–3″ (2.5–7.5 cm) wide; convex to flat; sticky when moist; opaque, dull white, becoming pinkish-buff. Odor strongly mealy.
Gills: slightly attached, crowded,

narrow, whitish.
Stalk: 3–4″ (7.5–10 cm) long, ⅜–⅝
(1–1.5 cm) thick; solid, whitish;
smooth above ring and beaded with
colorless drops, finely hairy below.
Veils: universal veil slimy, leaving film
over cap. Partial veil white,
membranous; leaving large, pendant,
persistent ring on upper stalk.
Spores: 4–5 μ wide; round, smooth,
colorless, nonamyloid. Spore print
white. Gill tissue divergent.
Season: July–September.
Habitat: On the ground in mixed woods and wet
areas.
Range: Widely distributed: New York,
California, Washington; reports spotty.
Look-alikes: *L. lenticularis* develops olive-gray tones
on gills or stalk with age.
Comments: Formerly known as *Lepiota solidipes*.

BOLBITIUS FAMILY
(Bolbitiaceae)

This relatively small family of little
brown mushrooms (LBMs) contains a
few of our very common lawn and wood
mulch fungi. Many of these species can
be found in early spring; some seem to
come up overnight and be gone by
afternoon. Three genera are recognized
here: *Agrocybe, Bolbitius,* and *Conocybe.*
Some *Agrocybe* species are large and
thick-fleshed and resemble pholiotas
more than the other genera in this
family. Some grow on wood or wood
chips, and their mature gills and spore
prints are usually dark brown. *Bolbitius*
species are small, slimy to sticky
mushrooms growing on manure and
compost; they have a cinnamon spore
print. *Conocybe*s grow in lawns, but also
in deep woods, and have conical to
convex caps and cinnamon gills and
spore print. A few *Conocybe* species are
hallucinogenic, and one group, which
is recognized by its stalk ring, has some
species that contain deadly toxins.

These disparate mushrooms are placed together in this family because of microscopic characteristics: the cap cuticle is cellular, the spores are smooth and brownish and have a broad, truncate germ pore, and the gill edges of many species are fringed with bowling-pin-shaped sterile cells (cheilocystidia).

---

223   Maple Agrocybe
Agrocybe acericola (Pk.) Sing.
Bolbitiaceae, Agaricales

Description: *Ochre-yellow cap with brown, attached gills and ring on whitish stalk.*
Cap: 1¼−4″ (3−10 cm) wide; convex, becoming flat, sometimes with low central knob; moist, smooth to wrinkled; ochre-yellow.
Gills: attached, close, narrow; pale off-white, becoming brown.
Stalk: 2−4″ (5−10 cm) long, ¼−⅜″ (0.5−1 cm) thick; whitish, with stringy white threads at base.
Veil: partial veil leaving large, membranous ring on stalk.
Spores: 8.5−10.5 × 5−6.5 μ; elliptical, smooth, with pore at tip.
Spore print cinnamon to rust.
Edibility: Edible with caution.
Season: April−June; August−September.
Habitat: Solitary to scattered, on decayed deciduous logs and wood chips, especially maple.
Range: NE. and NW. North America.
Look-alikes: *A. dura** and *A. praecox** have tan to brown caps. *Agaricus** species have free gills and chocolate spore print. *Hebeloma** species have slimy caps and lighter brown spore print.
Comments: Most agrocybes were formerly placed in the genera *Pholiota* or *Naucoria*. This is one of the first agarics to appear. It is edible, but not particularly tasty; caution must be used because of the danger of misidentification.

207 **Hard Agrocybe**
Agrocybe dura (Fr.) Sing.
Bolbitiaceae, Agaricales

Description: *Tan to buff or brown cap, becoming cracked, with brown gills and evanescent ring on white stalk.*
Cap: 1⅝–4″ (4–10 cm) wide; convex to flat, at times with low central knob; dry, soft like kid leather, smooth, cracking with age; white to tan, becoming buff to brownish.
Gills: attached, close, broad; white, becoming dark brownish.
Stalk: 1⅝–4″ (4–10 cm) long, ¼–⅝″ (0.5–1.5 cm) thick; *solid,* whitish.
Veil: partial veil leaving evanescent ring on stalk, fragments on cap edges.
Spores: 10–14 × 6.5–8 μ; smooth, elliptical, with pore at tip. Spore print dark brown.
Edibility: Edible, but not recommended.
Season: May–June.
Habitat: Single to many, in grass, along shrub borders, in wood chips.
Range: Quebec and N. United States.
Look-alikes: *A. praecox** does not typically develop cracked cap and has smaller spores. *A. acericola** has yellow cap, grows on wood. *Agaricus** species have free gills. *Hebeloma** have sticky, slimy cap.
Comments: This early-spring agrocybe should be avoided: not enough is known about spring mushrooms to be able to recommend any brown-spored species.

47 **Hemispheric Agrocybe**
Agrocybe pediades (Pers. ex Fr.) Fayod
Bolbitiaceae, Agaricales

Description: *Pale yellow-brown, hemispherical cap with brown, attached gills and slender stalk.*
Cap: ⅜–1″ (1–2.5 cm) wide; hemispherical to broadly convex; moist, smooth; typically yellowish to buff.
Gills: attached, close, broad; off-white, becoming rust.
Stalk: ¾–2″ (2–5 cm) long, 1/16–⅛″

(1.5–3 mm) thick; buff to brownish.
Spores: 9–13 × 6.6–7.5 μ; elliptical, smooth, with pore at tip. Spore print dark brown.

Edibility: Edible, but not recommended.

Season: May–June; September.

Habitat: Scattered to numerous, in lawns and grassy areas, waste areas, wood mulch.

Range: Widely distributed in North America.

Look-alikes: *Psathyrella foenisecii** has brown cap fading in concentric zones to tan. *Marasmius oreades** has white gills.

Comments: Formerly called *Naucoria semiorbicularis*. This grass mushroom is especially common after heavy rains. Because it is relatively nondescript, it cannot be recommended as an edible for beginning mushroom hunters.

---

225 **Spring Agrocybe**
Agrocybe praecox (Pers. ex Fr.) Fayod
Bolbitiaceae, Agaricales

Description: *Tan to brown cap with brown, attached gills and ring or ring zone on white stalk.*
Cap: 1¼–3½" (3–9 cm) wide; convex, becoming flat or developing low knob; dry, smooth; brownish.
Gills: attached, close, broad; off-white to brown.
Stalk: 1¼–4" (3–10 cm) long, ⅛–⅝" (0.3–1.5 cm) thick; dry, whitish.
Veil: partial veil leaving persistent, membranous ring or zone on stalk.
Spores: 8–11 × 5–6 μ; elliptical, smooth, with pore at tip. Spore print dark brown.

Edibility: Edible with caution.

Season: April–June.

Habitat: Scattered to numerous, on humus, in lawns, flower beds, and open woods.

Range: Widely distributed in North America.

Look-alikes: *A. acericola** has yellow cap. *A. dura** typically has cracked cap. *Agaricus** species have free gills. *Hebeloma** have sticky, slimy cap.

Comments: This common spring mushroom was formerly known as *Pholiota praecox*.

51 Yellow Bolbitius
Bolbitius vitellinus (Pers. ex Fr.) Fr.
Bolbitiaceae, Agaricales

Description: *Slimy yellow cap with cinnamon-red gills; on manure or rich compost.*
Cap: ¾–2″ (2–5 cm) wide; conical to bell-shaped; margin lined, becoming grooved; sticky, slimy; bright yellow.
Gills: attached, becoming free, narrow; pale yellow, becoming cinnamon-brown, with minutely hairy edges; often liquefying in wet weather.
Stalk: 1¼–3¼″ (3–8 cm) long; ¹⁄₁₆–⅛″ (1.5–3 mm) thick, sometimes slightly enlarged toward base; minutely hairy at top; shiny, pale yellow.
Spores: 10–15 × 6–9 μ; smooth, elliptical to oval, with pore at blunt tip. Spore print rust-ochre.
Edibility: Edible.
Season: May–June; September–early October.
Habitat: Scattered to clustered, on dung, old straw, and rich compost.
Range: Widely distributed in North America.
Look-alikes: *Coprinus** gills liquefy but produce black spores. *Entoloma** species do not liquefy, produce pinkish spore print.
Comments: This mushroom is common in wet weather in open areas, especially pastureland and urban parks.

⊗ 81 Deadly Conocybe
Conocybe filaris Fr.
Bolbitiaceae, Agaricales

Description: *Brown cap with brown gills and large ring midway down long, thin stalk.*
Cap: ¼–1″ (0.5–2.5 cm) wide; conical to convex or flat, often with central knob; moist to dry, smooth; tawny-brown.
Gills: notched, close, broad; off-white, becoming rust.
Stalk: ⅜–1⅝″ (1–4 cm) long; ¹⁄₃₂–¹⁄₁₆″ (1–1.5 mm) thick; yellow- to orange-brown.
Veil: partial veil forms membranous,

movable central ring on stalk.
Spores: 7.5–13 × 3.5–6.5 μ;
elliptical, smooth, with pore at tip.
Spore print cinnamon-brown.

Edibility: Deadly.

Season: August–October.

Habitat: Scattered to numerous, in grass, moss, or on wood chips.

Range: Widely distributed in N. North America.

Look-alikes: Other mushrooms of this species complex can be differentiated only by microscopic characteristics.

Comments: Also known as the "Ringed Conocybe" and *Pholiotina filaris*. This species or group of species has only recently been shown to contain the same lethal toxins as the deadly Destroying Angel (*Amanita virosa*)—additional evidence that LBMs (little brown mushrooms) are not safe to experiment with.

---

5 **White Dunce Cap**
Conocybe lactea (J. Lange) Mét.
Bolbitiaceae, Agaricales

Description: *Dull white, conical cap with cinnamon gills and fragile, white stalk.*
Cap: ⅜–1" (1–2.5 cm) wide; narrowly conical; margin often flared; smooth, with long radial lines; whitish.
Gills: nearly free, close, very narrow; whitish, becoming tawny or reddish-cinnamon.
Stalk: 1⅝–4" (4–10 cm) long, ¹⁄₁₆" (1.5 mm) thick, sometimes with basal bulb; whitish.
Spores: 12–14 × 6–9 μ; elliptical, smooth, with pore at tip. Spore print reddish.

Edibility: Edible.

Season: May–June; September; summer in Southwest.

Habitat: Scattered to numerous, in lawns and grassy areas.

Range: N. North America; Gulf Coast and Mexico; Colorado; California.

Look-alikes: *C. crispa* has crimped gills.

Comments: This early-morning lawn mushroom's stalk is quickly bent and broken by the weight of the cap and the heat of the sun; it often disappears before noon.

⊗ 74 **Bog Conocybe**
Conocybe smithii Wat.
Bolbitiaceae, Agaricales

Description: *Brown cap with cinnamon gills and long, thin stalk; in moss in bogs.*
Cap: ⅛–⅜" (0.3–1 cm) wide; conical, becoming flat with knob; moist, smooth; cinnamon-brown, fading to yellowish.
Gills: attached, close, narrow; off-white, becoming cinnamon-brown.
Stalk: ⅜–2⅜" (1–6 cm) long, ¹⁄₃₂–¹⁄₁₆" (1–1.5 mm) thick; fragile, whitish, turning blue at base with age; covered with minute fibers that soon disappear.
Veil (when present): partial, evanescent.
Spores: 7–9 × 4–5 μ; elliptical, smooth, with pore at tip. Spore print cinnamon-brown.
Edibility: Hallucinogenic.
Season: September–November.
Habitat: Scattered to many, in moss in bogs.
Range: Michigan and Pacific NW.
Comments: Compounds causing altered states of consciousness occur in several agaric families, especially the Strophariaceae. The reactions resulting from ingesting these mushrooms vary considerably; none should be eaten casually.

75 **Brown Dunce Cap**
Conocybe tenera (Schaeff. ex Fr.) Küh.
Bolbitiaceae, Agaricales

Description: *Brown, conical cap with cinnamon gills and long, slender, brownish stalk.*
Cap: ⅜–¾" (1–2 cm) wide; narrowly conical to bell-shaped; dry, with fine radial lines; red- to yellow-brown, fading.
Gills: nearly free, almost distant,

narrow; cinnamon-brown.

Stalk: 1⅝–3⅜″ (4–8.5 cm) long, 1/16″ (1.5 mm) thick; fragile, straight, with fine lines; brownish.

Spores: 11–12 × 6 μ; elliptical, smooth, with pore at tip. Spore print reddish-brown.

Season: May–July; September.

Habitat: Scattered to numerous, in lawns, grassy areas, and in rich soil.

Range: Widely distributed in North America.

Look-alikes: *Panaeolus** species have black, mottled gills and typically grow on manure.

Comments: This cannot be recommended as an edible because too little is known about the complex.

BOLETE FAMILY
(Boletaceae)

King Bolete, Steinpilz, Cep, *Boletus edulis*—all are names given to the best-known bolete, a mushroom that is sold year-round in gourmet shops in strings of dried caps. It is one of more than 200 species of boletes found in North America. Most are edible, though few are as good; some are poisonous unless well cooked. The rule of thumb is to avoid any bolete with orange to red pores, especially any that bruises blue. Boletes are fleshy, stalked mushrooms that usually grow on the ground in woods. Instead of gills, they have a thick, spongelike tube layer, which, except in a very few species, is readily detachable from the cap flesh. Caps are typically brownish, bun-shaped, and dry, slimy, or scaly. The pores, or tube mouths, are most often yellowish to greenish-brown. In most species, the stalk is central. It may have tufts, dots, or a network of ridges or veins; others are smooth, and some become hollow. A partial veil and ring are present in some genera. In many boletes, the flesh, tubes, or pores bruise blue-green, blackish, or reddish on handling or when cut. This is often an important aid to identification, and is noted in the species descriptions when it occurs. The spore print color can be olive-brown, pinkish, yellowish, or black.

407 Shaggy-stalked Bolete
Boletellus betula (Schw.) Gil.
Boletaceae, Agaricales

Description: *Slimy, smooth, apricot-yellow cap with yellowish pores and shaggy stalk.*
Cap: 1¼–3½" (3–9 cm) wide; convex; slimy, smooth; reddish-orange to reddish-brown, becoming apricot-yellow. Flesh green- to orange-yellow.
Tubes: sunken around stalk; dark greenish-yellow. Pores round, pale yellow, becoming greenish-yellow.
Stalk: 4–8" (10–20 cm) long, ¼–¾" (0.5–2 cm) thick; shaggy, red, or yellow above, reddish below; base with white, cottony mat (mycelium).
Spores: 15–18 × 6–9 μ; loosely webbed outer layer and distinctive pore at tip. Spore print olive to olive-brown.
Edibility: Edible.
Season: August–October.
Habitat: On the ground, in pine-oak woods.
Range: North Carolina to Georgia; reported in Michigan, Pennsylvania, Ohio.
Look-alikes: *B. russellii** has scaly to cracked cap.
Comments: This mushroom is most easily recognized by its small, orange-yellow cap and long, ragged, yellow stalk.

403 Russell's Bolete
Boletellus russellii (Frost) Gil.
Boletaceae, Agaricales

Description: *Yellow-brown, woolly cap with large, olive-yellow pores and long, shaggy, red stalk.*
Cap: 1¼–5¼" (3–13 cm) wide, round to convex; margin strongly incurved; hairy to scaly, then cracked; brownish-yellow to olive-gray. Flesh yellowish.

Tubes: becoming sunken around stalk; yellowish-olive to olive-green. Pores large, angular.
Stalk: 4–8″ (10–20 cm) long, ⅜–¾″ (1–2 cm) thick, often curved and sticky near base; solid, deeply ridged, scruffy; dull red.
Spores: 15–20 × 7–11 μ; lined, with deep grooves in outer wall, cleft at tip. Spore print dark olive.

Edibility: Edible.
Season: April; July–September.
Habitat: Single to widely scattered, on the ground under oak; in oak-pine barrens; under hemlock.
Range: E. Canada to N. Florida, west to Minnesota.
Look-alikes: *B. betula** is yellowish, with slimy cap.
Comments: Russell's Bolete is easily identified by its habitat, and jagged, reddish stalk.

---

452 Ash-tree Bolete
Boletinellus merulioides (Schw.) Murr.
Boletaceae, Agaricales

Description: *Brownish, wavy cap with off-centered stalk.*
Cap: 2–4¾″ (5–12 cm) wide; slightly humped to deeply sunken; margin incurved when young, in maturity spreading, wavy-edged, and uplifted; dry, dull, soft; yellow-brown to red-brown. Flesh thin, yellowish, sometimes turning bluish-green.
Tubes: very shallow, sometimes appearing almost gill-like, with crossveins radially arranged, *adhering to cap flesh* and slightly descending stalk; yellow to ochre, bruising dark.
Stalk: ¾–1⅛″ (2–4 cm) long, ¼–1″ (0.5–2.5 cm) thick; off-center, solid; brownish, bruising reddish-brown.
Spores: 7–10 × 6–7.5 μ; oval to almost round, smooth. Spore print olive-brown.

Edibility: Edible.
Season: June–October.
Habitat: On the ground near ash; near maples.

Range: E. Canada to Florida, west to Michigan and Texas.

Comments: Also called *Boletinus porosus* and *Gyrodon merulioides*. Unlike most boletes, it does not have easily detachable tubes.

---

383 Spotted Bolete
Boletus affinis var. *maculosus* Pk.
Boletaceae, Agaricales

Description: *Yellow-spotted, reddish-brown cap with off-white pores and brownish stalk.*
Cap: 1⅜–4″ (3.5–10 cm) wide; convex to nearly flat; margin strongly incurved at first; dry, with white bloom, smooth to pitted, often cracked with age; reddish-brown, becoming yellow-brown with rounded to irregular, pale yellow spots. Flesh white.
Tubes: sunken around stalk with age; white-buff to yellow or tawny. Pores round; off-white, bruising olive-ochre.
Stalk: 1¼–4″ (3–10 cm) long, ⅜–¾″ (1–2 cm) thick, sometimes with enlarged base; dry, smooth to powdery; white to brownish, sometimes reddish.
Spores: 9–16 × 3–5 μ; smooth, elliptical. Spore print yellow-brownish.

Edibility: Edible.

Season: June–September.

Habitat: Under deciduous trees, particularly beech; in mixed conifer stands.

Range: E. Canada to Florida, west to Indiana and N. Michigan.

Comments: The yellow-spotted, reddish-brown cap makes this bolete easy to identify.

---

395 Bay Bolete
Boletus badius Fr.
Boletaceae, Agaricales

Description: *Chestnut-brown cap with small, olive-yellow pores that bruise blue, and a reddish stalk.*
Cap: 1¼–4″ (3–10 cm) wide; convex to nearly flat; sticky in wet weather, but soon dry; yellow-brown to

brownish-red. Flesh whitish, bruising burgundy, rarely blue.

Tubes: becoming sunken or descending stalk. Pores small, round to angular; olive-yellow, bruising blue then brown.

Stalk: 1⅜–3½" (4–9 cm) long, ⅜–¾" (1–2 cm) thick; yellow to dull rose-red, with whitish bloom at first, becoming brownish about base.

Spores: 10–14 × 4–5 μ; smooth, elliptical. Spore print olive-brown.

Edibility: Edible.

Season: Late June–November.

Habitat: Under coniferous and deciduous trees; on rotten wood.

Range: E. Canada to North Carolina, west to Minnesota.

Look-alikes: *B. chrysenteron** has dry, cracking cap. *Gyroporus castaneus** does not bruise blue.

Comments: This mushroom is recognized when young, or after a rain, by its sticky cap.

---

412 Two-colored Bolete
Boletus bicolor Pk.
Boletaceae, Agaricales

Description: *Rose-red cap, yellowish toward margin, with minute, yellow pores and yellowish stalk; all parts slowly bruising blue.*
Cap: 2–6" (5–15 cm) wide; convex, becoming flat; dry, unpolished; rose-red to pinkish, yellowish toward margin. Flesh thick, pale yellow, slowly bruising blue.
Tubes: becoming sunken around stalk; yellow, paler with age, slowly bruising blue. Pores minute, angular; yellow, sometimes reddish in places with age.
Stalk: 2–4" (5–10 cm) long, ⅜–1¼" (1–3 cm) thick; sometimes club-shaped; dry, dull, smooth; yellow, with rosy tint over lower ⅔ of stalk. Flesh yellow, at times slowly bruising blue.
Spores: 8–12 × 3.5–5 μ; smooth, elliptical. Spore print dull olive-brown.

Edibility: Choice.

Season: Late June–October.

Habitat: On the ground under oak; also reported
under aspen and pine.
Range: Nova Scotia to Georgia and Michigan.
Look-alikes: *B. sensibilis,* reportedly poisonous, has
brick-red cap and stains blue instantly.
Comments: This beautiful red and yellow bolete is
common and often abundant in eastern
oak woods. It is a choice edible.

---

413   **Red-cracked Bolete**
Boletus chrysenteron Bull. ex St. Amans
Boletaceae, Agaricales

Description: *Cracked, olive-brownish cap, showing red
between cracks, with yellow pores and
scruffy, yellow to reddish stalk; all parts
bruising bluish.*
Cap: 1¼–3¼" (3–8 cm) wide; convex
to nearly flat; dry and somewhat velvety
when young; olive-brown to reddish-
brown; cracking in quiltlike pattern,
with red flesh in cracks. Flesh white,
then yellow, slowly bruising blue.
Tubes: sunken around stalk; bright
yellow, slowly bruising blue. Pores
irregular, bright yellow, becoming
reddish in some areas.
Stalk: 1⅜–2⅜" (4–6 cm) long, ¼–⅜"
(0.5–1 cm) thick; scruffy, yellowish,
pinkish-red below. Flesh yellow at top,
red at base, blue-green on bruising.
Spores: 9–13 × 3.5–5 μ; smooth,
elliptical. Spore print olive to olive-
brown.
Edibility: Edible.
Season: June–October; overwinters in S.
California.
Habitat: Under deciduous trees, especially oak,
along roads, and on mossy banks.
Range: Widely distributed in North America.
Look-alikes: *Boletellus chrysenteroides* grows on
decaying wood. *Boletus truncatus* has
blunt spores and abruptly tapering stalk
base. *B. subtomentosus** lacks red flesh in
cap cracks.
Comments: This very common urban species, with
no poisonous look-alike, is one of the
first boletes to appear in the East.

405 King Bolete

Boletus edulis Bull. ex Fr.
Boletaceae, Agaricales

Description: *Large, with reddish-brown cap, white to yellowish pores, and whitish to brownish stalk, thickening toward base.*
Cap: 3¼–10" (8–25 cm) wide; convex, becoming nearly flat; moist, sticky in wet weather, smooth to somewhat pitted, sometimes cracking in dry weather; brown, reddish-brown, or cinnamon-buff. Flesh white.
Tubes: sunken around stalk; whitish, becoming greenish-yellow. Pores small and round; white when young, sometimes bruising tawny.
Stalk: 4–10" (10–25 cm) long, ¾–1⅝" (2–4 cm) thick and club-shaped, or 1⅝–4" (4–10 cm) thick and bulbous; whitish to brownish, white-webbed over upper ⅓; webbing below darker, often indistinct. Flesh white.
Spores: 13–19 × 4–6.5 μ; smooth, elliptical. Spore print olive-brown.

Edibility: Choice.

Season: June–October.

Habitat: On the ground, under conifers (pine, hemlock) and deciduous trees (birch, aspen).

Range: Widely distributed throughout North America.

Look-alikes: *B. variipes* has dry, tan cap and grows under oak. *Tylopilus felleus** has dark-webbed stalk, is very bitter.

Comments: Also called the "Cep" or the "Steinpilz." The King is one of the most prized edible mushrooms; it has many varieties of different color, shape, and habitat, but all are good to eat.

408 Frost's Bolete

Boletus frostii Russ. apud Frost
Boletaceae, Agaricales

Description: *Sticky to slimy, blood-red cap with red pores and blood-red, webbed stalk; bruising blue.*
Cap: 2–6" (5–15 cm) wide; round to

convex to flat; margin incurved; at first
with whitish bloom, soon shiny, sticky,
dark blood-red to blackish-red, then
fading with yellowish areas. Flesh
yellowish, instantly bruising blue.
Tubes: becoming sunken around stalk;
yellow to green-yellow, bruising blue.
Pores small, round; deep red to paler,
bruising blue.
Stalk: 1⅝–4¾" (4–12 cm) long, ⅜–1"
(1–2.5 cm) thick; dry, coarsely webbed
to torn-webbed; dark blood-red, often
yellowish at base, slowly bruising blue.
Spores: 11–15 × 4–5 μ; smooth,
elliptical. Spore print olive-brown.

Edibility: Edible, but not recommended.

Season: Late June–October.

Habitat: On the ground in oak woods.

Range: E. Canada to N. Florida, west to
Michigan.

Comments: This spectacular species is edible; but,
while some people can eat any bolete
that has been well cooked, others
experience gastric distress from some
red-pored species that bruise blue.

---

396 Admirable Bolete
Boletus mirabilis Murr.
Boletaceae, Agaricales

Description: *Dark reddish-brown, flattened, scaly cap*
with yellow pores and dark brown stalk.
Cap: 2¾–6" (7–15 cm) wide; convex,
becoming nearly flat; margin inrolled;
moist to sticky, becoming dry, woolly,
and fibrous-scaly, sometimes granular-
scaly, often with hanging fragments;
dark red-brown. Flesh white, rarely
bruising blue.
Tubes: sunken around stalk; yellow to
olive-yellow. Pores round to angular,
yellow to olive-yellow.
Stalk: 3¼–6" (8–15 cm) long, ⅜–1¼"
(1–3 cm) thick, club-shaped, up to 2"
(5 cm) thick at base; moist to dry,
smooth, typically webbed at top; dark
brown, with occasional yellowish
streaks. Flesh dingy pink or yellow.

Spores: 19–24 × 7–9 μ; smooth,
elliptical. Spore print olive-brown.

Edibility: Choice.

Season: September–December.

Habitat: On or near rotting logs of fir, hemlock,
and western redcedar.

Range: Pacific NW. and Michigan.

Look-alikes: *B. projectellus* favors sandy soil under
pine.

Comments: Although a good edible, this
mushroom is sometimes attacked by a
whitish mold, probably a stage in the
life cycle of the Golden Hypomyces
(*Hypomyces chrysospermus*), which renders
it unappetizing.

---

418 Ornate-stalked Bolete
Boletus ornatipes Pk.
Boletaceae, Agaricales

Description: *Dull, velvety, greenish to grayish cap with
a yellowish bloom when young, small, lemon
pores, and shaggy, powdery, yellow stalk.*
Cap: 1⅝–8″ (4–20 cm) wide; convex to
nearly flat; sticky when wet, dull,
velvety; white bloom when young,
yellowish-green to olive-gray. Flesh
chrome-yellow.
Tubes: attached; lemon-yellow,
staining orange-yellow. Pores small;
lemon-yellow, staining orange-brown.
Stalk: 3¼–6″ (8–15 cm) long, ⅜–1⅜″
(1–3.5 cm) thick; yellowish, at first
with powdery covering; network of
ridges bruising dark.
Spores: 9–13 × 3–4 μ; smooth,
elliptical. Spore print olive-brown.

Edibility: Edible.

Season: July–September.

Habitat: On the ground, in open areas near
beech and oak, along roads and wood
borders, on banks.

Range: Quebec to Maryland; Michigan.

Look-alikes: *B. retipes*, in Southeast, has dry, yellow,
powdery cap. *B. griseus* is grayish, with
white tubes staining brown.

Comments: This relatively common summer bolete
has a beautifully patterned yellow stalk.

370 Parasitic Bolete
Boletus parasiticus Bull. ex Fr.
Boletaceae, Agaricales

Description: *Buff-brown cap, pores, and stalk; only found growing on the Pigskin Poison Puffball* (Scleroderma citrinum).
Cap: ¾–3¼" (2–8 cm) wide; convex; dry, dingy yellow to buff-brown. Flesh yellowish.
Tubes: attached, becoming sunken around stalk, yellowish. Pores angular, yellowish, bruising ochre.
Stalk: 1¼–2⅜" (3–6 cm) long, 3¼–5¼" (8–13 cm) thick; yellowish to brownish-yellow.
Spores: 12–18.5 × 3.5–5 μ; smooth, elliptical. Spore print dark olive.
Edibility: Edible with caution.
Season: July–September.
Habitat: On the Pigskin Poison Puffball.
Range: E. Canada to North Carolina, west to Mississippi; common in Southeast.
Comments: This bolete is found only on the puffball that it parasitizes.

398 Peppery Bolete
Boletus piperatus Bull. ex Fr.
Boletaceae, Agaricales

Description: *Tawny cap with yellow, cinnamon, or red pores, and tawny-red stalk with bright yellow base; taste peppery.*
Cap: ⅝–3½" (1.5–9 cm) wide; convex, becoming nearly flat; margin incurved when young; dry to almost sticky, smooth to somewhat scaly; yellow-brown to cinnamon, darker in center. Flesh pale yellow. Taste sharply acrid.
Tubes: attached; ochre, becoming reddish. Pores angular, yellowish, becoming cinnamon, then brick-red.
Stalk: ¾–4¾" (2–12 cm) long, ⅛–⅝" (0.3–1.5 cm) thick; yellow to reddish-cinnamon, *base bright yellow and conspicuous.* Flesh yellowish.
Spores: 8.5–12 × 4–5 μ; smooth, elliptical. Spore print dull cinnamon.

Edibility: Edible with caution.
Season: July—October; September—January on West Coast.
Habitat: On the ground under various conifers; reported in deciduous woods.
Range: Throughout N. North America.
Comments: Like most mushrooms, this peppery species can cause stomach upset if it is not cooked thoroughly.

---

399 Yellow-cracked Bolete
Boletus subtomentosus Fr.
Boletaceae, Agaricales

Description: *Dry, velvety, cracked, brownish cap, showing yellowish between cracks, with yellow pores and variable, yellowish stalk.*
Cap: 2—8″ (5—20 cm) wide; convex; margin incurved at first; dry, somewhat velvety, cracking; olive- to brownish-yellow. Flesh white to pale yellow.
Tubes: becoming sunken or descending stalk; yellow, very slowly bruising greenish-blue. Pores angular, yellow, staining greenish, finally brownish.
Stalk: 1⅝—4″ (4—10 cm) long, ⅜—1¼″ (1—3 cm) thick; occasionally ridged network over upper stalk; yellowish to ochre, sulfur-yellow around base, bruising reddish-brown.
Spores: 10—15 × 3.5—5 μ; smooth, elliptical. Spore print olive-brown.
Edibility: Edible.
Season: July—November.
Habitat: On the ground in mixed woods.
Range: Widely distributed in North America.
Look-alikes: *B. chrysenteron** shows reddish-brown between cracks in cap.
Comments: This species is very common in summer and fall, and in winter in California.

⊗ 409 Red-mouth Bolete
Boletus subvelutipes Pk.
Boletaceae, Agaricales

Description: *Tawny-red to yellowish cap with scarlet to orange pores; yellowish, dotted stalk with*

reddish hairs at base; all parts quickly
bruising blue to blue-black.
Cap: 2⅜–5¼" (6–13 cm) wide; convex
to nearly flat; dry, with minute fibers or
somewhat velvety, patchy with age;
tawny-red to cinnamon-reddish,
yellowish toward margin. Flesh yellow,
instantly bruising blue.
Tubes: attached or descending stalk;
yellowish, quickly bruising blue. Pores
orange to scarlet, bruising blue-black.
Stalk: 1¼–4" (3–10 cm) long, ⅜–¾"
(1–2 cm) thick; dotted in vertical lines,
not webbed, *base usually with reddish
hairs;* tip yellow, reddish below, readily
bruising blue-black.
Spores: 12–18 × 4–6 μ; elliptical,
smooth. Spore print dark olive-brown.

Edibility: Poisonous.

Season: Late June–October.

Habitat: On the ground, in deciduous and
mixed woods, near beech, eastern
hemlock, white spruce, balsam fir.

Range: Quebec to Virginia, west to Michigan.

Look-alikes: *B. luridus* has tubes sunken about stalk,
netted pattern on stalk, and no reddish
hairs at base.

Comments: This common bolete is probably
poisonous. There are several other
similarly colored boletes, but they
either have a netted pattern of veins on
the stalk or lack the dark red hairs at
the base of the stalk. Orange- to red-
pored boletes occur throughout North
America. None should be eaten.

---

414 **Zeller's Bolete**
Boletus zelleri Murr.
Boletaceae, Agaricales

Description: *Black to dark olive-brown, dry, wrinkled
cap with large, irregular, yellowish pores
and reddish to yellowish-tan stalk.*
Cap: 2–4" (5–10 cm) wide; convex to
flat with age; margin incurved when
young; dry, very powdery, wrinkled to
bumpy, becoming smoother; blackish
to dark olive-brown, fading to brown,

often with reddish margin. Flesh white
to yellowish, sometimes bruising blue.
Tubes: often deeply sunken with age;
olive-yellow, bruising blue. Pores
large, yellow, often bruising blue.
Stalk: 2–3¼" (5–8 cm) long, ¼–⅝"
(0.5–1.5 cm) thick; dry, granular to
dotted; tan at first, with red granules;
becoming reddish above, yellowish
below. Flesh yellow, red with age,
sometimes bruising blue.
Spores: 12–15 × 4–5.5 μ; elliptical,
smooth. Spore print olive-brown.

Edibility: Good.
Season: September–October in Pacific NW.;
November–March in California.
Habitat: On the ground, in coastal forests and
under redwood.
Range: Pacific NW. and California.
Look-alikes: *B. chrysenteron** has brownish cap that
typically cracks in a quiltlike pattern.
Comments: This beautiful western bolete has a dark
cap with a powdery bloom.

---

404 Rosy Larch Bolete
Fuscoboletinus ochraceoroseus (Snell) Pom. &
A.H.S.
Boletaceae, Agaricales

Description: *Dry, scaly, rose-pink cap with radial*
yellowish pores; near larch.

Cap: 3–10" (7.5–25 cm) wide; convex,
becoming flat, with incurved margin
typically adorned with veil fragments;
dry, uneven to pitted, with dense,
fibrous scales; rose-red to pinkish,
sometimes yellowish near margin. Flesh
yellowish, pink beneath cuticle; may
bruise bluish. Taste somewhat bitter.
Tubes: attached to or descending stalk,
shallow; yellow, ochre, or brown. Pores
elongated, angular, radially arranged.
Stalk: 1–2" (2.5–5 cm) long, ⅜–1¼"
(1–3 cm) thick, flaring at tip; solid,
with netlike pattern; yellowish,
sometimes red at often swollen base.
Veil: partial veil thin, membranous,
whitish to yellowish, leaving remnants

along cap margin and evanescent ring.
Spores: 7.5–9.5 × 2.5–3.2 μ;
elliptical. Spore print reddish-brown.

Edibility: Edible.

Season: August–October.

Habitat: Scattered to many, under western larch.

Range: Idaho, Washington, and Oregon.

Look-alikes: *F. paluster* is much smaller, with deep red cap, and grows in northern bogs. *Suillus lakei** has olive-brown spore print, is found under western conifers.

Comments: Typical species of *Fuscoboletinus* have a slimy cap; all have a reddish-brown spore print, radially arranged pores, and grow under conifers, often larch.

---

384 Chestnut Bolete
Gyroporus castaneus (Bull. ex Fr.) Quél.
Boletaceae, Agaricales

Description: *Dry, chestnut-brown cap with small, white pores becoming yellowish, and hollow, brownish stalk.*
Cap: 1¼–4″ (3–10 cm) wide; broadly convex to flat; dry, often with white bloom at first; chestnut-brown to tawny-orange. Flesh white.
Tubes: deeply sunken around stalk; white, becoming yellowish. Pores small, round to angular.
Stalk: 1¼–3½″ (3–9 cm) long, ¼–⅜″ (0.5–1 cm) thick, sometimes bulbous, 1″ (2.5 cm) thick near base; dry, smooth; light brown to tawny-orange; loosely stuffed, *becoming hollow.*
Spores: 8–12 × 5–6 μ; elliptical, smooth. Spore print yellow.

Edibility: Choice.

Season: Late June–early October.

Habitat: On the ground, under oak and other hardwoods; reported in beech woods; rarely under live oak on West Coast.

Range: Maine to Florida, west to the Great Lakes and Texas; also on West Coast.

Look-alikes: *G. purpurinus** has dark burgundy cap.

Comments: This is unfortunately a favored host of a deforming fungal parasite, the Golden Hypomyces (*Hypomyces chrysospermus*).

402 Bluing Bolete
Gyroporus cyanescens (Bull. ex Fr.) Quél.
Boletaceae, Agaricales

Description: *Dry, straw-colored cap with white to
yellowish pores and hollow, pale stalk; all
parts instantly bruising deep blue.*
Cap: 1⅝–4¾" (4–12 cm) wide;
convex, sometimes becoming nearly
flat; margin incurved when young; dry,
with minute, flattened fibers; pale straw
to buff. Flesh off-white, but quickly
bruising completely blue.
Tubes: deeply sunken around stalk;
whitish, becoming yellowish, instantly
bruising indigo. Pores small, round.
Stalk: 1⅝–4" (4–10 cm) long, ⅜–1"
(1–2.5 cm) thick, often thicker at base;
dry, woolly, becoming smooth; pale
straw to buff, quickly bruising blue;
loosely stuffed to hollow.
Spores: 8–10 × 5–6 μ; elliptical,
smooth. Spore print yellow.
Edibility: Good.
Season: July–September.
Habitat: On the ground in beech woods; in
sandy soil in mixed woods with
hemlock; under spruce and balsam fir in
Quebec; in apple orchards.
Range: E. Canada to Florida and Minnesota.
Comments: This species is readily identified by its
straw color that instantly bruises deep
blue. It is a good edible.

411 Red Gyroporus
Gyroporus purpurinus (Snell) Sing.
Boletaceae, Agaricales

Description: *Dry, velvety, dark burgundy cap with very
small, white pores, becoming yellow, and
dark reddish, hollow stalk.*
Cap: 1–3½" (2.5–9 cm) wide; convex
to nearly flat, often somewhat sunken;
dry, minutely velvety, dark burgundy.
Flesh white.
Tubes: attached and becoming deeply
sunken around stalk; white, slowly
aging yellow. Pores small, white,

becoming yellow.
Stalk: 1¼–2⅜″ (3–6 cm) long, ⅛–⅜″
(0.3–1 cm) thick; somewhat
roughened, reddish to brownish;
becoming hollow.
Spores: 8–11 × 5–6.5 μ; elliptical,
smooth. Spore print yellow.
Edibility: Edible.
Season: August–September.
Habitat: Under oak and other deciduous trees.
Range: New York to Florida and Minnesota.
Look-alikes: *G. castaneus** is brown to tawny-orange.
Comments: This good edible is distinguished by its
color and hollow stalk.

388 Red-capped Scaber Stalk
Leccinum aurantiacum (Bull. ex St. Amans)
S.F.G.
Boletaceae, Agaricales

Description: *Orange-red cap with off-white pores and
rough stalk with short, rigid projections.*
Cap: 2–8″ (5–20 cm) wide; convex,
becoming flat; dry, somewhat sticky,
generally roughened, but smooth in
places; bright to dull tawny-red. *Flesh
white, slowly bruising burgundy, then
grayish to purplish-black.*
Tubes: sunken around stalk to nearly
free; pale olive-buff, becoming
brownish. Pores minute, whitish,
bruising olive-brown.
Stalk: 4–7″ (10–18 cm) long, ¾–1¼″
(2–3 cm) thick; sometimes club-
shaped; whitish at first, with short,
rigid projections (scabers) aging brown
to black; base may bruise blue-green.
Spores: 13–18 × 3.5–5 μ; elliptical,
smooth. Spore print dark yellow-brown
to dull cinnamon.
Edibility: Good.
Season: August–September.
Habitat: On the ground, under coniferous and
deciduous trees.
Range: Throughout N. North America; also
Colorado and California.
Look-alikes: *L. insigne** grows in aspen or mixed
aspen and birch stands. *L. atrostipitatum*

grows in birch woods; stalk has black
scabers from the first.

Comments: This common and widespread bolete
has a number of look-alikes, all of
which, so far as is known, are edible.

---

389 Aspen Scaber Stalk
Leccinum insigne A.H.S., Thiers, & Wat.
Boletaceae, Agaricales

Description: *Reddish to orange-brown cap with white to
yellow-brown pores and rough stalk with
short, rigid projections; under aspen.*
Cap: 1⅛–6″ (4–15 cm) wide; convex to
broadly convex; dry, minutely granular
to fibrous and scaly, smooth with age;
bright tawny-red, often paler orange-
brown with age. *Flesh white, immediately
bruising violet- to brown-gray.*
Tubes: attached or descending stalk;
off-white, becoming olive-gray,
bruising brownish. Pores whitish,
bruising yellow to olive-brown.
Stalk: 3⅛–4¾″ (8–12 cm) long, ⅜–
¾″ (1–2 cm) thick, enlarged at base to
1″ (2.5 cm); off-white, with short,
brown, rigid projections (scabers),
which darken to blackish.
Spores: 13–16 × 4–4.5 μ; elliptical,
smooth. Spore print yellow-brown.
Edibility: Good.
Season: June–September; late spring–June in
Michigan; August in Rockies.
Habitat: Under aspen or mixed aspen and birch.
Range: Widely distributed in North America.
Look-alikes: *L. aurantiacum** first bruises burgundy.
Comments: This species includes a complex of
varieties that differ in coloration, but
all occur in aspen or aspen-birch woods
and are good edibles.

---

390 Common Scaber Stalk
Leccinum scabrum (Bull. ex Fr.) S.F.G.
Boletaceae, Agaricales

Description: *Brown cap with white to brownish pores
and rough stalk with short, rigid*

projections; sometimes blue-green at base.
Cap: 1⅝–4″ (4–10 cm) wide; convex
to broadly convex or sunken in center;
dry to moist or sticky, smooth; grayish-
brown to yellow-brown. Flesh white,
sometimes bruising slightly brownish.
Tubes: attached and deeply sunken
about stalk; off-white to brown. Pores
small, off-white to brown.
Stalk: 2¾–6″ (7–15 cm) long, ¼–⅝″
(0.5–1.5 cm) thick, enlarged toward
base; off-white, with short, rigid,
blackish projections (scabers),
sometimes red or bluish-green at base.
Spores: 15–19 × 5–7 μ; elliptical,
smooth. Spore print brown.

Edibility: Good.

Season: June–November.

Habitat: On the ground under birch.

Range: Widely distributed in North America.

Look-alikes: Other brownish *Leccinum** species can
be differentiated microscopically.

Comments: One of the most common boletes in
North America, this is frequently
misidentified because of the similarity
of many closely related species, all of
which are believed to be edible.

---

417 **Powdery Sulfur Bolete**
Pulveroboletus ravenelii (Berk. & Curt.) Murr.
Boletaceae, Agaricales

Description: *Powdery, sulfur-yellow cap with yellow
pores and stalk with fragile ring.*
Cap: ⅜–4″ (1–10 cm) wide; convex to
nearly flat; margin incurved; dry,
powdery; bright sulfur-yellow, center
becoming orange to brownish-red.
Flesh white to yellow, slowly bruising
light blue, then yellowish-brown.
Tubes: attached; lemon-yellow to olive-
tinted, bruising greenish-blue. Pores
small, round to angular; bright yellow,
bruising greenish-blue.
Stalk: 1⅝–6″ (4–15 cm) long, ¼–⅝″
(0.5–1.5 cm) thick; dry to moist;
covered with yellow powder at first,
then brilliant yellow with yellow

threads at base.

Veils: universal veil leaving yellow powder on cap and stalk. Partial veil yellow, leaving fragile ring on stalk or fragments about cap margin.

Spores: 8–10.5 × 4–5 μ; elliptical, smooth. Spore print smoky olive.

Edibility: Edible.

Season: July–early September; November–December in California.

Habitat: On the ground, in mixed pine, hemlock, and deciduous stands, spruce woods, and laurel and rhododendron thickets.

Range: E. Canada to N. Florida, west to Texas; reported in California.

Comments: The Powdery Sulfur Bolete is most common in the Southeast.

---

379 Old Man of the Woods
Strobilomyces floccopus (Vahl ex Fr.) Kar.
Boletaceae, Agaricales

Description: *Coarsely scaly, grayish-black cap with*

white to dark gray tubes and woolly-scaly, blackish stalk.

Cap: 1⅝–6″ (4–15 cm) wide; cushion-shaped to flattened; margin with remnants of grayish veil; shaggy-scaly; blackish to grayish. Flesh whitish, bruising reddish, then black.

Tubes: whitish or gray, becoming dark gray; bruising reddish, then black. Pores angular, same color as tubes.

Stalk: 2–4¾″ (5–12 cm) long, ⅝–1″ (1.5–2.5 cm) thick; sometimes enlarged toward base; solid, woolly-scaly, dark grayish. Flesh white, bruising reddish, then black.

Veil: partial veil grayish, leaving 1–2 woolly, sheathlike rings or zones on stalk.

Spores: 9.5–15 × 8.5–12 μ; with full network of ridges. Spore print black.

Edibility: Edible.

Season: July–October.

Habitat: On the ground, among hardwoods (oak) and mixed hardwoods and conifers (pine); also in pine barrens.

Range: Nova Scotia to Florida, west to Michigan and Texas.

Look-alikes: *S. confusus* has erect scales on cap, and spiny spores with incomplete network of ridges.

Comments: This common eastern bolete becomes unappetizing as it ages.

416 Chicken-fat Suillus

Suillus americanus (Pk.) Snell apud Slipp & Snell

Boletaceae, Agaricales

Description: *Reddish-streaked, slimy yellow cap with yellow pores and red- to brownish-dotted upper stalk; under eastern white pine.*
Cap: 1¼–4″ (3–10 cm) wide; convex; margin incurved, with hanging remnants of cottony veil; slimy, bright yellow, with cinnamon to reddish spots or streaks. Flesh yellowish, staining pinkish-brown when cut.
Tubes: attached or descending stalk; yellow, bruising brown. Pores large, angular; yellow, becoming ochre.
Stalk: 1¼–3½″ (3–9 cm) long, ⅛–⅜″ (0.3–1 cm) thick; yellow, covered above with dots that bruise darker.
Veil: partial veil covering young tubes yellowish, extending from cap margin, not attached to stalk, leaving margin remnants but no ring.
Spores: 8–11 × 3–4 μ; elongate-elliptical, smooth. Spore print dull cinnamon.

Edibility: Good.

Season: July–October.

Habitat: Under eastern white pine.

Range: Nova Scotia to North Carolina, west to Michigan.

Look-alikes: *S. granulatus** has buff cap, lacks veil. *S. sibiricus,* common under western white pine, has thicker stalk and sometimes a ring.

Comments: This very common, slimy, yellow bolete, found only under eastern white pine, is a good edible.

387 Short-stalked Suillus
Suillus brevipes (Pk.) Kuntze
Boletaceae, Agaricales

Description: *Slimy, smooth, yellow to brown cap with yellow pores and short, whitish stalk.*
Cap: 2–4″ (5–10 cm) wide; convex to nearly flat; slimy, smooth; brownish, becoming ochre. Flesh white, becoming yellowish.
Tubes: attached or descending stalk, honey-yellow. Pores small, pale yellow.
Stalk: ¾–2″ (2–5 cm) long, ⅜–¾″ (1–2 cm) thick, sometimes longer; solid, white, becoming pale yellow, sometimes becoming dotted.
Spores: 7–10 × 2.8–3.2 μ; elliptical, smooth. Spore print cinnamon.
Edibility: Good.
Season: May–June; September–November.
Habitat: In sandy soil under conifers, especially pine and white spruce.
Range: Quebec to Alabama and Florida, west to Texas; Pacific NW. and California.
Look-alikes: *S. granulatus** has conspicuous dots on stalk. *S. pseudobrevipes* has ring.
Comments: This bolete is often found under 2- and 3-needle pines.

394 Blue-staining Suillus
Suillus caerulescens A.H.S. & Thiers
Boletaceae, Agaricales

Description: *Large, slimy, streaked, reddish-yellow cap with yellow pores and brownish stalk with ring; under Douglas-fir.*
Cap: 2⅜–5½″ (6–14 cm) wide; convex, becoming flat; slimy, with some scales or small fibers, smooth with age; reddish in center, yellowish toward margin. Flesh yellow.
Tubes: attached, yellow, bruising brownish. Pores yellow.
Stalk: 1–3¼″ (2.5–8 cm) long, ¾–1¼″ (2–3 cm) thick; webbed at top, smooth to dotted above ring, smooth to minutely fibrous below; no glandular dots; yellow above ring, reddish-yellow

below, aging dingy brown, staining
brownish on handling. Flesh yellow,
staining blue at base.
Veil: partial veil fibrous, whitish;
leaving ring on stalk.

Edibility: Edible.

Season: October—January.

Habitat: In coastal forests under Douglas-fir.

Range: Pacific NW. and California.

Look-alikes: *S. ponderosus* has slimy, brown to
reddish-brown ring and smooth,
reddish-brown cap.

Comments: This bolete is often abundant.

---

393 Hollow-stalked Larch Suillus
Suillus cavipes (Opat.) A.H.S. & Thiers
Boletaceae, Agaricales

Description: *Dryish, fibrous-scaly, reddish-brown cap*
with yellowish pores and reddish-brown
stalk with evanescent ring; under larch.
Cap: 1¼–4″ (3–10 cm) wide; convex,
expanding to nearly flat; margin
inrolled, with veil remnants; dry, not
sticky, densely fibrous-scaly; yellow to
reddish or reddish-brown. Flesh white
to yellowish, not bruising blue.
Tubes: attached or descending stalk,
yellow to greenish-yellow. Pores
yellowish, angular, and radial.
Stalk: 1¼–3½″ (3–9 cm) long, ¼–¾″
(0.5–2 cm) thick; sometimes club-
shaped, usually reddish-brown;
becoming hollow.
Veil: partial veil fibrous, white,
evanescent; leaving remnants along cap
margin and thin ring on stalk.
Spores: 7–10 × 3.5–4 μ; elliptical,
smooth. Spore print dark olive-brown.

Edibility: Good.

Season: September—October.

Habitat: On the ground under larch.

Range: Nova Scotia west to Washington.

Look-alikes: *S. pictus** has solid stalk and grows
under eastern white pine. *S. tomentosus**
stains blue and occurs under pine.
Fuscoboletinus paluster, another larch bog
species, has purplish-brown spores.

Comments: A number of boletes can be found only under larch; the mushrooms and trees grow in a symbiotic relationship.

376 Dotted-stalk Suillus
Suillus granulatus (L. ex Fr.) Kuntze
Boletaceae, Agaricales

Description: *Slimy, buff to cinnamon cap with small yellow pores and dotted stalk.*
Cap: 2–6" (5–15 cm) wide; broadly convex; slimy when wet, smooth; buff, spotted or streaked with cinnamon, cinnamon with age. Flesh whitish, becoming yellowish.
Tubes: attached, pale yellow. Pores small, cream to pale yellow.
Stalk: 1⅝–3¼" (4–8 cm) long, ⅜–1" (1–2.5 cm) thick; whitish, becoming bright yellow above, paler to dingy cinnamon below, covered with pinkish to brownish dots.
Spores: 7–10 × 2.5–3.5 μ; elliptical, smooth. Spore print dingy cinnamon.
Edibility: Good.
Season: Late June–November; overwinters in California.
Habitat: Under pine, hemlock, spruce.
Range: Nova Scotia to North Carolina; California and Pacific NW.
Look-alikes: *S. albidipes* has cottony roll of tissue about cap margin when young. *S. punctatipes* has radially arranged tubes.
Comments: This very common species or group of species varies considerably, but it always lacks a veil and has conspicuous dots on the stalk.

406 Larch Suillus
Suillus grevillei (Kl.) Sing.
Boletaceae, Agaricales

Description: *Slimy, smooth, bright yellow to red or cinnamon cap with bright yellow pores and mottled, ringed stalk; under larch.*
Cap: 2–6" (5–15 cm) wide; convex to nearly flat; slimy, hairless; bright

yellow to bright red-brown in center.
Flesh yellowish, bruising pinkish.
Tubes: attached and somewhat
descending stalk; bright yellow,
becoming olive-yellow, bruising
brownish. Pores angular, cream to
bright yellow.
Stalk: 1⅝–4″ (4–10 cm) long, ⅜–1¼″
(1–3 cm) thick; pale yellow and
mottled, bruising brownish.
Veil: partial veil yellowish, leaving
cottony ring on stalk.
Spores: 8–10 × 2.8–3.5 μ; elliptical,
smooth. Spore print olive-brown.

Edibility: Edible.

Season: September–November.

Habitat: On the ground under larch.

Range: Nova Scotia to North Carolina;
Midwest; Pacific NW.

Look-alikes: *S. proximus,* if a distinct species, has
stalk base bruising greenish.

Comments: This bolete is a good edible once the
slime layer is removed. Western
specimens are redder, eastern ones more
yellow.

392 Western Painted Suillus
Suillus lakei (Murr.) A.H.S. & Thiers
Boletaceae, Agaricales

Description: *Sticky, yellow cap covered with dry,*
reddish-brown scales, with yellow pores and
yellowish stalk with white ring.
Cap: 2¾–5½″ (7–14 cm) wide;
convex, becoming flat; margin
incurved, often with remnants of veil;
scaly, dry, with slimy underlayer;
yellow, with small, reddish-brown to
pinkish fibers and scales. Flesh yellow,
becoming pinkish when exposed.
Tubes: attached, descending stalk with
age; yellow, bruising reddish. Pores
radially elongated, angular.
Stalk: 1¼–2⅜″ (3–6 cm) long, ⅝–¾″
(1.5–2 cm) thick; dry, fibrous-scaly
toward base; yellowish. Flesh yellowish,
bruising blue.
Veil: partial veil white, leaving zone at

top of stalk or evanescent ring.
Spores: 8–11 × 3–4 μ; elliptical,
smooth. Spore print brownish.

Edibility: Good.

Season: August in high Rocky Mountains;
September–October in Pacific NW.;
November–January in California.

Habitat: On the ground under Douglas-fir.

Range: Pacific NW. and California; Rocky
Mountains.

Look-alikes: *S. pictus** grows only under eastern
white pine.

Comments: Although species in the genus *Suillus*
are characterized by a slimy cap, a few
mushrooms with scaly, dry caps, often
with an underlying slime layer, that
were formerly placed in genus *Boletinus*
are now also included in this group.

---

401   Slippery Jack
Suillus luteus (L. ex Fr.) S.F.G.
Boletaceae, Agaricales

Description: *Slimy, reddish- to yellowish-brown cap with*
white pores, becoming yellow, and purplish,
sheathlike ring on brown-dotted stalk.
Cap: 2–4¾″ (5–12 cm) wide; round,
becoming convex to flat; slimy,
smooth; dark reddish- to yellow-brown.
Flesh white, becoming yellowish.
Tubes: attached; whitish, pale yellow to
olive-yellow. Pores yellow, becoming
brown-dotted.
Stalk: 1¼–3¼″ (3–8 cm) long, ⅜–1″
(1–2.5 cm) thick; brown-dotted.
Veil: partial veil membranous, shiny,
white; leaving a persistent, purplish-
drab, sleevelike ring draping stalk.
Spores: 7–9 × 2.5–3 μ; elliptical,
smooth. Spore print dull cinnamon.

Edibility: Good, with caution.

Season: September–early December.

Habitat: On the ground, under Scots pine, red
pine, and spruce.

Range: E. North America.

Look-alikes: *S. subluteus** has longer, more slender
stalk and less pronounced ring. *S.*
cothurnatus has brown, cigar-bandlike

ring and grows mostly in South under
loblolly and longleaf pines.

Comments: Although this is a favorite edible, it
may cause transient diarrhea if the
slime is not removed.

---

391 Painted Suillus
Suillus pictus (Pk.) A.H.S. & Thiers
Boletaceae, Agaricales

Description: *Dry, fibrous-scaly, red-ochre cap with
yellow pores, whitish cobwebby veil, and
yellowish stalk; under eastern white pine.*
Cap: 1¼–4¾" (3–12 cm) wide;
convex, expanding to nearly flat;
margin incurved, usually with hanging
veil fragments; dry, tacky at times,
coarsely scaly-fibrous; reddish to red-
ochre, fibers often grayish. Flesh
yellow, becoming pink-tinged.
Tubes: attached or descending stalk,
bright yellow. Pores large, yellow,
bruising brownish.
Stalk: 1⅝–4¾" (4–12 cm) long, ⅜–1"
(1–2.5 cm) thick, sometimes wider at
base; typically solid, lower part fibrous.
Veil: partial veil fibrous, whitish;
leaving remnants about cap margin and
grayish ring on stalk.
Spores: 8–12 × 3.5–5 μ; elliptical,
smooth. Spore print olive-brown.
Edibility: Good.
Season: June–November.
Habitat: With eastern white pine; often abundant.
Range: Nova Scotia to North Carolina.
Look-alikes: *S. lakei** grows under Douglas-fir.
*S. cavipes** grows under larch.
Comments: The Painted Suillus is found only under
eastern white pine.

---

372 White Suillus
Suillus placidus (Bon.) Sing.
Boletaceae, Agaricales

Description: *Slimy, ivory cap with yellowish pores and
dotted stalk; under eastern white pine.*
Cap: 1¼–4" (3–10 cm) wide; convex,

becoming flat; slimy, smooth, shiny
when dry; white to ivory, becoming
yellowish with slime, darkening with
age. Flesh white, then yellow, bruising
pale burgundy.
Tubes: attached, rather shallow,
eventually descending stalk; yellowish.
Pores small, yellow, often dotted.
Stalk: 1⅝–4¾" (4–12 cm) long, ¼–1"
(0.5–2.5 cm) thick; white, yellowish
with age, with brownish smears and
dots; sometimes pinkish.
Spores: 7–9 × 2.5–3.2 μ; elliptical,
smooth. Spore print dull cinnamon.

Edibility: Good.

Season: July–October.

Habitat: On the ground under eastern white
pine; often abundant.

Range: Throughout range of eastern white pine.

Look-alikes: *S. acidus* has acidic, slimy coating, veil
leaving ring, and grows under red pine.

Comments: This mushroom is easily recognized and
a good edible.

386 Pungent Suillus
Suillus pungens Thiers & A.H.S.
Boletaceae, Agaricales

Description: *Slimy, white to olive, mottled cap with buff
pores, whitish stalk with red or brown dots;
pungent odor.*
Cap: 1⅝–5½" (4–14 cm) wide; convex
to somewhat flat; margin incurved;
slimy, smooth; white to olive with
olive splotches, mottled, with cottony
roll of white tissue when young. Flesh
white. Odor pungent; taste unpleasant.
Tubes: attached and eventually
descending stalk; white to buff, *with
conspicuous whitish droplets,* darkening to
brownish. Pores angular, buff.
Stalk: 1¼–2¾" (3–7 cm) long, ⅜–¾"
(1–2 cm) thick; dry, white to
yellowish, with large, reddish dots,
becoming brownish.
Spores: 9.5–10 × 2.8–3.5 μ;
elliptical, smooth. Spore print
brownish.

Edibility: Edible.
Season: November—January.
Habitat: On the ground under Monterey pine, póssibly under knobcone pine.
Range: San Francisco Bay area.
Comments: Many species of *Suillus* are restricted to a limited geographical area. None is known to be poisonous.

---

371 **Slippery Jill**
Suillus subluteus (Pk.) Snell apud Slipp & Snell
Boletaceae, Agaricales

Description: *Slimy, yellowish to tawny or dark olive-brown cap with yellowish pores and baggy veil leaving thick ring on dotted stalk.*
Cap: 2–4″ (5–10 cm) wide; convex to flat; slimy, smooth; yellowish, darkening to tawny or dark olive-brown. Flesh whitish to yellowish.
Tubes: attached, yellow to brownish. Pores yellowish with brownish spots, darkening to brown.
Stalk: 1¼–4″ (3–10 cm) long, ¼–⅝″ (0.5–1.5 cm) thick; pinkish-ochre, solid, covered with darkening dots.
Veil: partial veil slimy, membranous, thick, and baggy, with flaring lower margin, leaving a slimy ring on stalk.
Spores: 6–11 × 2–4 μ; elliptical, smooth. Spore print ochre-brown.
Edibility: Good, with caution.
Season: Late August—December.
Habitat: Under 2- and 3-needle pines.
Range: Maine to Georgia, west to Michigan.
Look-alikes: *S. luteus** has a purple-tinted sheathing ring. *S. cothurnatus* has an orange-yellow cap, ochre pores, ochre-yellow flesh, and a brown cigar-bandlike ring; it is more southern in distribution.
Comments: Slippery Jill is a species complex of slimy-capped boletes with a slimy veil and dotted stalk that grows under pine; it includes the common Pitch Pine Bolete (*S. pinorigidus*.) All are edible once the slime layer is removed, although some people may experience transient diarrhea.

415 Tomentose Suillus
Suillus tomentosus (Kauff.) Sing., Snell & Dick
Boletaceae, Agaricales

Description: *Dryish, scaly, yellowish cap with yellow*
pores and brownish-dotted, yellowish stalk.
Cap: 2–4″ (5–10 cm) wide; convex;
margin incurved at first; dry but sticky
beneath yellow to orange-yellow scales.
Flesh yellow, slowly changing to
greenish-blue when cut.
Tubes: attached and descending stalk;
yellowish to olive-yellow, bruising
greenish-blue. Pores dark cinnamon,
becoming yellowish, bruising blue.
Stalk: 1¼–4⅜″ (3–11 cm) long, ⅜–
1¼″ (1–3 cm) thick; sometimes club-
shaped; yellowish, with brownish dots.
Spores: 7–11 × 3.5–4.5 μ; elliptical,
smooth. Spore print dark olive-brown.

Edibility: Edible.

Season: September–October in Pacific NW.;
November–January in California.

Habitat: Under 2- and 3-needle pines in West,
jack pine in Michigan and Nova Scotia.

Range: Pacific NW. and California; Rocky
Mountains; Michigan; Nova Scotia.

Comments: In dry weather, the cap appears dry and
fibrous-scaly; the sticky underlayer is
most conspicuous after rain.

377 Black Velvet Bolete
Tylopilus alboater (Schw.) Murr.
Boletaceae, Agaricales

Description: *Dry, brown to black cap with white bloom,*
white pores becoming pinkish, and gray to
black stalk.
Cap: 1¼–6″ (3–15 cm) wide; convex,
becoming flat; smooth, dark brown to
black, with a hoary bloom at first. Flesh
grayish-white, becoming pinkish.
Tubes: attached and becoming deeply
sunken around stalk; buff to pinkish.
Pores white, becoming pinkish,
bruising black.
Stalk: 1⅝–4″ (4–10 cm) long, ¾–
1⅝″ (2–4 cm) thick; gray to black,

sometimes covered with whitish bloom
at first, discoloring to blackish.
Spores: 7–11 × 3.5–5 μ; elliptical,
smooth. Spore print pinkish.

Edibility: Good.
Season: July–September.
Habitat: Under deciduous trees, especially oak.
Range: New England to Florida, west to
Michigan and Texas.
Look-alikes: *T. eximius** is brownish-purple.
Comments: Easily identified by its velvety cap and
stalk, its pore color, and habitat.

397 Burnt-orange Bolete
Tylopilus ballouii (Pk.) Sing.
Boletaceae, Agaricales

Description: *Bright burnt-orange cap, becoming
brownish, with white pores and yellowish to
brownish stalk.*
Cap: 2–4¾″ (5–12 cm) wide; convex,
becoming nearly flat, usually irregular
in shape; dry, bright orange to
vermilion, darkening to brownish.
Flesh white, bruising pinkish-buff to
brownish-lilac. Taste bitterish.
Tubes: attached and slightly descending
stalk. Pores white, bruising brownish.
Stalk: 1–4¾″ (2.5–12 cm) long, ¼–1″
(0.5–2.5 cm) thick; off-white to
yellowish, bruising or aging brownish.
Spores: 5–11 × 3–5 μ; elliptical,
smooth. Spore print light brown.
Edibility: Edible.
Season: August–September.
Habitat: In sandy soil in oak and beech woods.
Range: Massachusetts to North Carolina, west
along Gulf Coast to E. Texas.
Comments: This common bolete's bitter taste
disappears with thorough cooking.

410 Chrome-footed Bolete
Tylopilus chromapes (Frost) A.H.S. & Thiers
Boletaceae, Agaricales

Description: *Pink to rosy-red, dry cap with white pores
becoming pinkish; stalk cream-white to*

pinkish, chrome-yellow at base.
Cap: 1¼–6" (3–15 cm) wide; convex,
becoming flat; dry, sometimes tacky,
smooth, feltlike; rose-pink, fading.
Flesh white or pink-tinted.
Tubes: attached and sunken around
stalk or nearly free; white, then
yellowish, finally pinkish to brownish.
Pores white when young.
Stalk: 1⅝–6" (4–15 cm) long, ⅜–1"
(1–2.5 cm) thick; dry, cream-white to
pale pink; becoming scaber-dotted,
sometimes pink to red; *yellow at base.*
Spores: 11–17 × 4–5.5 μ; elliptical,
smooth. Spore print rosy-brown.

Edibility: Good.

Season: May–October.

Habitat: Frequently solitary, on the ground
under hardwoods and conifers.

Range: E. Canada to Georgia and Michigan.

Comments: Because this mushroom's stalk becomes
dotted with short, rigid projections
(scabers) with age, this mushroom is
also known as *Leccinum chromapes.*

---

380 Lilac-brown Bolete
Tylopilus eximius (Pk.) Sing.
Boletaceae, Agaricales

Description: *Dry, purplish-tinted, brown cap with
brownish pores and densely granular,
brownish-purple stalk.*
Cap: 2–4¾" (5–12 cm) wide; round,
becoming broadly convex to nearly flat;
dry, tacky to slimy in wet weather;
somewhat pitted; chocolate-brown with
a purplish overtone, with distinct
bloom at first. Flesh whitish, becoming
grayish to pinkish.
Tubes: attached and sunken around
stalk or nearly free; dark brownish-
purple. Pores chocolate-brown with
purplish tint, darkening with age.
Stalk: 1¾–3½" (4.5–9 cm) long, ⅜–
1¼" (1–3 cm) thick; chocolate-brown
with purplish tint, densely marked
with minute, purplish scales, at times
streaked.

Spores: 11–17 × 3.5–5 μ; elliptical,
smooth. Spore print reddish-brown.

Edibility: Good.

Season: June–October, peaking in August.

Habitat: On the ground under eastern hemlock;
also spruce and balsam fir.

Range: Maine to Georgia, west to Michigan.

Look-alikes: *T. alboater** grows in deciduous woods.
T. plumbeoviolaceus* has smoother stalk,
is less brown, and is very bitter.

Comments: The Lilac-brown Bolete is a good edible
species, often found in abundance.

---

382 Bitter Bolete
Tylopilus felleus (Bull. ex Fr.) Kar.
Boletaceae, Agaricales

Description: *Large, tan to brown cap with white pores
becoming pinkish, and pale yellow-brown
stalk with fishnet webbing; bitter tasting.*
Cap: 2–12″ (5–30 cm) wide; round to
convex, becoming flat; dry, but sticky
when wet, smooth to minutely woolly;
dark brown, buff-brown, or tan. Flesh
thick, white. Taste very bitter.
Tubes: attached, deeply sunken around
stalk with age; white, becoming
pinkish, bruising brownish. Pores
white, bruising brownish.
Stalk: 1⅝–4″ (4–10 cm) long, ⅜–1¼″
(1–3 cm) thick; club-shaped to
bulbous; dry, pale brown, typically
brown-webbed over length of stalk,
although sometimes smooth near base.
Spores: 11–15 (17) × 3–5 μ; smooth,
elliptical. Spore print burgundy.

Season: June–November.

Habitat: On the ground, in oak-hemlock woods;
also near spruce, balsam fir, white
spruce, birch, poplar, and on rotting
hemlock logs.

Range: Maine to Florida, Michigan, and Texas.

Look-alikes: *T. indecisus* is not bitter and usually has
stalk webbed only at top. *T.
rubrobrunneus* is very bitter, with
webbing only at stalk top. *Boletus
edulis** is mild, with white webbing.

Comments: When young, its pores are white and

this species is easily mistaken for the King, or Cep (*Boletus edulis*). However, when cooked, it is unpalatable.

---

385 Graceful Bolete
Tylopilus gracilis (Pk.) Henn.
Boletaceae, Agaricales

Description: *Reddish-brown, velvety to fibrous cap with white pores, becoming pink, and slender brown to reddish-orange stalk.*
Cap: 1¼–4″ (3–10 cm) wide; convex to broadly convex; dry, granulose, becoming cracked, chestnut- to reddish- or cinnamon-brown. Flesh white. Odor and taste mild.
Tubes: deeply depressed around stalk, thick, uneven; white, then pinkish-brown. Pores white to pinkish-brown.
Stalk: 2–6″ (5–15 cm) long, ⅜–¾″ (1–2 cm) thick, enlarging toward base; long, slender, granular, longitudinally lined; reddish-brown to cinnamon-tan.
Spores: 10–17 × 5–8 μ; elliptical, pitted (at × 1000). Spore print dark reddish-brown.

Season: Late June–early October.

Habitat: On the ground, in woods under aspen, oak, pine, and hemlock.

Range: Nova Scotia south to North Carolina, west to Michigan.

Comments: Because its spores are ornamented, this has also been placed in *Porphyrellus*.

---

381 Violet-gray Bolete
Tylopilus plumbeoviolaceus (Snell & Dick) Sing.
Boletaceae, Agaricales

Description: *Dry, violet cap with cream pores becoming pinkish, and smooth, violet stalk.*
Cap: 1⅝–6″ (4–15 cm) wide; convex, becoming nearly flat; dry, sometimes with bloom, violet, darkening to gray-purple, then brownish. Flesh white, discoloring slightly if at all. *Taste very bitter.*

Tubes: attached and sunken around stalk; cream, becoming purplish-tan. Pores whitish, becoming pinkish-tan. Stalk: 3¼–4¾" (8–12 cm) long, ⅜–⅝" (1–1.5 cm) thick; smooth, sometimes faintly webbed at top; mottled dark violet, paler toward base. Spores: 10–13 × 3–4 μ; elliptical, smooth. Spore print pinkish-brown.

Season: Late June–September.

Habitat: On the ground, in deciduous woods and borders near oak, hickory, and aspen.

Range: Quebec to Florida, west to Minnesota and Texas.

Look-alikes: *T. alboater** is not purplish. *T. eximius** has scaly-granular stalk.

Comments: This beautiful mushroom is, unfortunately, too bitter to eat.

378 Dark Bolete
Tylopilus pseudoscaber (Secr.) A.H.S. & Thiers
Boletaceae, Agaricales

Description: *Dry, dull, dark brownish cap with dark yellow-brown pores and dark brown stalk; white flesh bruising bright blue.*
Cap: 2–6" (5–15 cm) wide; convex, becoming nearly flat; dry, minutely hairy to fibrous, very dark brown. Flesh white, *bruising bright blue, then reddish-to dull brown.* Odor pungent.
Tubes: attached, deeply sunken around stalk with age; grayish-brown, bruising green-blue. Pores dark yellow-brown. Stalk: 1⅛–4¾" (4–12 cm) long, ⅜–1¼" (1–3 cm) thick; sometimes club-shaped; dry, minutely hairy, brown. Spores: 12–18 × 6–7.5 μ; elliptical, smooth. Spore print reddish-brown.

Season: September–November.

Habitat: On the ground, under coniferous and deciduous trees; along paths in woods.

Range: Pacific NW. and N. California; E. North America and Great Lakes area.

Look-alikes: *T. porphyrosporus* does not bruise blue.

Comments: This is especially common in the Pacific Northwest.

INKY CAP FAMILY
(Coprinaceae)

Members of this family are among our
most common urban and suburban
fungi. Some of these occur so invasively
in gardens and lawns that they are
called "weed fungi." Others are more
common in pastureland, especially on
horse or cow manure; a few grow on
stumps or debris in wooded areas. Some
species, like the Shaggy Mane, are good
edibles. A few cause digestive upsets
or, like the Alcohol Inky, cause an
adverse reaction when eaten before
alcohol is consumed. A few species that
grow on dung are either hallucinogenic
or possibly poisonous. The edibility of
most species in this family is unknown
because they are too small to consider
as edibles or too rarely found.
Coprinus is the genus of the inky caps;
all but a few have gills that liquefy, or
deliquesce, on maturing, dissolving the
cap into a black inky fluid. The spore
print, when obtainable, is black.
Panaeolus, also a black-spored genus, is
generally distinguished by a smooth cap
and a habitat on dung. *Psathyrella*
species have a spore print ranging from
black to purple-black, dark brown, or
brick-red. Sometimes *Panaeolus* and
Psathyrella are split into smaller genera
based on microscopic characteristics.
These 3 genera are placed together in
the family Coprinaceae because of
microscopic characteristics: the cap
cuticle is cellular, the spores are mostly
blackish, smooth, and have a broad
pore at the tip, and in some cases are
distinctly roughened by warts.

---

19 **Alcohol Inky**
Coprinus atramentarius (Bull. ex Fr.) Fr.
Coprinaceae, Agaricales

Description: *Fleshy, gray-brown, radially lined cap
with inky gills.*

Cap: 2–3″ (5–7.5 cm) wide; egg-shaped, becoming bell-shaped to convex; margin pleated; dry, smooth, or silky to somewhat flattened-scaly about center; gray to gray-brown.
Gills: free, crowded, broad; white at first, soon lavender-gray, then blackish, inky, darkening from margin to stalk.
Stalk: 3–6″ (7.5–15 cm) long, ⅜–¾″ (1–2 cm) thick; dry, silky-fibrous; white; hollow.
Veil: partial veil fibrous, white; leaving evanescent ring near stalk base.
Spores: 7–11 × 4–6 μ; elliptical, smooth, with pore at tip.
Pleurocystidia huge. Spore print black.

Edibility: Good, with caution.

Season: May–September; November–April in California.

Habitat: Usually clustered, in grass and wood debris and near buried wood.

Range: Throughout North America.

Look-alikes: *C. insignis* has distinctly warted spores. *C. micaceus** has tawny cap.

Comments: This is a good, meaty edible, but one should take no alcoholic beverages for 1–2 days after eating it. About 30 minutes after drinking alcohol, one may experience a flushing of the face and neck, tingling of fingers and toes, headache, and sometimes nausea. The alcohol causes this transient illness, because the mushroom inactivates an enzyme that detoxifies alcohol in the system. Recovery is spontaneous, usually within a couple of hours.

704 Shaggy Mane
Coprinus comatus (Müll. ex Fr.) S.F.G.
Coprinaceae, Agaricales

Description: *Cylindrical, shaggy-scaly, white cap turning inky from liquefying gills.*
Cap: 1¼–2″ (3–5 cm) wide, 1⅝–6″ (4–15 cm) high; cylindrical, gradually expanding as gills liquefy, leaving only stalk; dry, covered with flat scales becoming down-curled; white

with light reddish-brown scales.
Gills: free or nearly so, very crowded;
white, becoming black and inky from
margin to stalk top.
Stalk: 2⅜–8″ (6–20 cm) long, ⅜–
¾″ (1–2 cm) thick; bulbous, white;
hollow, with central strand of minute
fibers.
Veil: partial veil leaving ring on lower
part of stalk.
Spores: 11–15 × 6.3–8.5 μ; smooth,
elliptical, blunt, with pore at tip. Spore
print black.

Edibility: Choice.
Season: May–early June; September–October;
November–January in Southeast.
Habitat: Scattered to clustered and common, in
grass, wood chips, and hardpacked soil.
Range: Throughout North America.
Comments: This mushroom can be gathered by the
bushel in urban and suburban areas.
When unexpanded and white, it is
an excellent substitute for asparagus; it
can also be pickled. Fragile as this
mushroom is, it has the remarkable
ability to push up through asphalt.

4 Non-inky Coprinus
Coprinus disseminatus (Pers. ex Fr.) S.F.G.
Coprinaceae, Agaricales

Description: *Small, brownish-gray, bell-shaped, grooved
cap on thin stalk.*
Cap: ¼–⅝″ (0.5–1.5 cm) wide; bell-
shaped to convex; deeply pleated,
minutely scruffy; buff to honey-brown
over center, becoming grayish toward
margin.
Gills: attached, nearly distant, broad;
white, becoming ashy to black, not
inky.
Stalk: ¾–1¼″ (2–3 cm) long, ¹⁄₆₄–¹⁄₃₂″
(0.5–1 mm) thick; smooth to minutely
hairy, fragile; white; hollow.
Spores: 7–10 × 4–5 μ; elliptical,
smooth, with pore at tip. Spore print
blackish.
Edibility: Edible.

Season: May–October; November–March in S. California.

Habitat: In great numbers, on deciduous wood debris, often in lawns and grassy areas.

Range: Widespread in E. North America.

Look-alikes: Gills of most other *Coprinus* liquefy. *Psathyrella** caps are radially lined, not deeply grooved. *Panaeolus** are sturdier, without pleated caps.

Comments: Also known as "Little Helmets." This delicate little mushroom always appears in masses on buried wood, on lawns, or in grassy areas. It has also been called a *Psathyrella* and a *Pseudocoprinus*.

705 Woolly-stalked Coprinus
Coprinus lagopus (Fr.) Fr.
Coprinaceae, Agaricales

Description: *Small, curling, grayish cap covered with loose fibers; with inky gills and long, thin, fragile, soft-hairy, whitish stalk.*
Cap: ⅜–2″ (1–5 cm) wide, ¾–3″ (2–7.5 cm) high; conical; margin curling up and back with age; dry, densely covered with evanescent, white fibers; becoming radially furrowed; grayish.
Gills: free, close, narrow; grayish, becoming black and inky, but drying, not dissolving.
Stalk: 2–8″ (5–20 cm) long, ¹⁄₁₆–¼″ (1.5–5 mm) thick; slender, fragile; white, covered with whitish, hairy scales; hollow.
Veils: universal veil leaving loose fibers on cap. Partial veil (when present) evanescent.
Spores: 10–12 × 6–7 μ; elliptical, smooth, with pore at tip. Spore print black.

Season: July–September; October–February in California.

Habitat: Single to scattered, on wood debris.

Range: Widely distributed in North America.

Look-alikes: *C. lagopides* has smaller spores.

Comments: The Woolly-stalked Coprinus, a woodland species, is distinguished by its habitat, size, and curled-back, dry cap.

42 Mica Cap
Coprinus micaceus (Bull. ex Fr.) Fr.
Coprinaceae, Agaricales

Description: *Bell-shaped, tawny brown, radially furrowed cap with somewhat inky gills and fragile, white stalk.*
Cap: 1–2″ (2.5–5 cm) wide; egg-shaped, becoming bell-shaped to convex; covered at first by glistening granules that are soon lost and usually not seen; margin radially lined; furrowed nearly to center; tawny, becoming grayish.
Gills: attached, close, moderately broad; off-white to gray, then black, somewhat inky, not entirely dissolving.
Stalk: 1–3″ (2.5–7.5 cm) long, ⅛–¼″ (3–5 mm) thick; smooth to silky-fibrous, fragile; white; hollow.
Veil: universal veil leaving evanescent, micalike granules on cap.
Spores: 7–10 × 4–5 μ; elliptical, smooth, with pore at tip. Spore print blackish.

Edibility: Edible.

Season: April–October; year-round in S. California.

Habitat: Densely clustered, about stumps or wood debris.

Range: Throughout North America.

Look-alikes: *C. atramentarius** is fleshier, grayish, with gills turning lavender-gray, then black. *C. disseminatus** is smaller, not inky.

Comments: Also known as the "Common Inky Cap." What is called *C. micaceus* in North America is actually a complex of species.

3 Japanese Umbrella Inky
Coprinus plicatilis (W. Curtis ex Fr.) Fr.
Coprinaceae, Agaricales

Description: *Grayish-brown, grooved cap with smooth, circular, sunken center, blackish gills, and fragile, white stalk.*
Cap: ⅜–1″ (1–2.5 cm) wide; conical,

becoming flat to somewhat upturned;
radially grooved with brown, disc-
shaped, sunken center; gray to gray-
brown toward margin.
Gills: *free and attached to collar about
stalk,* distant, narrow; grayish,
becoming black but not inky.
Stalk: 2–3″ (5–7.5 cm) long, ⅟₃₂–⅛″
(1–3 mm) thick, sometimes with basal
bulb; dry, fragile; white; hollow.
Spores: 10–13 × 6.5–10 μ; broadly
oval, smooth. Spore print black.

Edibility: Edible.
Season: May–September.
Habitat: Single to numerous, in grass and lawns.
Range: Widely distributed in North America.
Comments: This is one of the more common species
in urban and suburban areas.

703 Scaly Inky Cap
Coprinus quadrifidus Pk.
Coprinaceae, Agaricales

Description: *Grayish cap covered with flaking, buff
patches; with inky gills and brownish
strands at base of stalk.*
Cap: 1–3″ (2.5–7.5 cm) wide, 2–3″
(5–7.5 cm) high; egg-shaped,
becoming bell-shaped; gray to gray-
brown, covered with loose, flaky, white
to buff patches.
Gills: free, crowded, broad; white, then
purplish-brown to black, inky.
Stalk: 2–5″ (5–12.5 cm) long, ¼–⅜″
(0.5–1 cm) thick; woolly, scaly; white,
with brownish runners (rhizomorphs) at
base; hollow.
Veils: universal veil leaving flaky
patches on cap. Partial veil white,
leaving evanescent ring on lower stalk.
Spores: 7.5–10 × 4–5 μ; elliptical,
smooth, with pore at tip. Spore print
blackish.

Edibility: Edible with caution.
Season: June–July or August.
Habitat: In large clusters, on hardwood debris,
especially ash and elm.
Range: Atlantic Coast to Great Lakes.

Look-alikes: *C. atramentarius** has small, flattened scales. *C. micaceus** lacks cap patches.

Comments: Sometimes this large, clustered *Coprinus* has an unpleasant odor and taste; it is then reported to cause gastric upset.

---

41 Orange-mat Coprinus
Coprinus radians (Desm.) Fr.
Coprinaceae, Agaricales

Description: *Brownish, radially lined cap covered with mealy particles; with inky gills and mat of yellow-orange threads about stalk base.*
Cap: ¾–1¼" (2–3 cm) wide; egg-shaped, becoming conical or bell-shaped; covered by evanescent, brown, mealy granules, finely wrinkled-lined to center; tawny brown, fading to grayish.
Gills: attached or nearly free, crowded, broad; white, becoming brown, then black and inky.
Stalk: 1–3" (2.5–7.5 cm) long, ¹⁄₁₆–¼" (1.5–5 mm) thick; smooth to mealy, fragile; white, with yellow-orange mass of threads (mycelium); hollow.
Veil: universal veil leaving mealy particles on cap.
Spores: 8–10 × 4–6 μ; oval to elliptical, smooth, with pore at top. Spore print black.
Season: May–October.
Habitat: Scattered on wet wood; in basements.
Range: Widely distributed in North America.
Look-alikes: *C. laniger* is smaller, with large cap scales.
Comments: This species is one of a handful of household mushrooms that crop up where wood has become wet.

---

87 Bell-cap Panaeolus
Panaeolus campanulatus (Bull. ex Fr.) Quél.
Coprinaceae, Agaricales

Description: *Brownish, bell-shaped cap with white, toothlike veil remnants dotting margin, black gills, and long, thin stalk.*
Cap: ¾–2" (2–5 cm) wide; bell-shaped;

margin toothed with fragments of veil tissue; smooth, moist to dry, dark reddish-brown fading to gray-buff.
Gills: notched, nearly distant, broad, pale grayish, then gray-black, mottled.
Stalk: 2⅜–5½" (6–14 cm) long, ¹⁄₁₆–¼" (1.5–5 mm) thick; brownish, with dense gray fuzz.
Veil: partial veil leaving toothlike remnants about cap margin.
Spores: 13–16 × 8–11 μ; elliptical, smooth, with pore at tip. Spore print blackish.

Season: June–September; overwinters in S. California.

Habitat: Single to several on horse or cow dung.

Range: Widely distributed in North America.

Look-alikes: *P. retirugis* has meshlike cap surface.

Comments: This pretty, common mushroom appears throughout the growing season.

20 Semi-ovate Panaeolus
Panaeolus semiovatus (Sow. ex Fr.) Lund. & Nannf.
Coprinaceae, Agaricales

Description: *Whitish, egg-shaped cap with blackish-brown gills and white stalk with ring.*
Cap: 1¼–3½" (3–9 cm) wide; egg-shaped, becoming broadly conical; smooth, tacky, shiny, tan to whitish.
Gills: notched, nearly distant, broad, off-white to brown or black, mottled.
Stalk: 3¼–6" (8–15 cm) high, ¼–⅝" (0.5–1.5 cm) thick, sometimes enlarged at base; whitish to buff.
Veil: partial veil leaving membranous, evanescent ring around middle of stalk.
Spores: 15–20 × 8–11 μ; elliptical, smooth, with pore at tip. Spore print blackish.

Season: June–September.

Habitat: Single to several, on horse dung.

Range: Widely distributed in North America.

Look-alikes: *P. phalaenarum* lacks ring and has solid, twisted-lined stalk.

Comments: This common species is also known as *P. separatus* and *Anellaria separata*.

⊗ 91 **Girdled Panaeolus**
Panaeolus subbalteatus (Berk. & Br.) Sacc.
Coprinaceae, Agaricales

Description: *Tawny, broadly conical to flat cap with dark belt around margin, brown gills, and hairy reddish stalk.*
Cap: 1¼–2″ (3–5 cm) wide; conical to expanded, with central knob; smooth to roughened, moist, dark, and tawny fading to tan, with broad, dark-belted zone around margin.
Gills: attached, close, broad; brownish, becoming blackish, mottled.
Stalk: 1⅝–4″ (4–10 cm) long, ⅛–¼″ (3–5 mm) thick; reddish-brown, covered with whitish hairs, sometimes bruising blue at base.
Spores: 11–14 × 7–9 μ; elliptical, smooth, with pore at tip. Spore print blackish.
Edibility: Hallucinogenic.
Season: June–July.
Habitat: Scattered to numerous, on dung or manured soil; often in gardens.
Range: Widely distributed in North America.
Look-alikes: *Psathyrella foenisecii** grows in grass. *Panaeolus acuminatus* is reddish-brown and does not bruise blue.
Comments: This hallucinogen is common enough to be considered a weed.

25 **Common Psathyrella**
Psathyrella candolleana (Fr.) Maire
Coprinaceae, Agaricales

Description: *Buff-brown to white cap with marginal veil remnants, purplish- to gray-brown gills, and fragile, white stalk.*
Cap: 1–4″ (2.5–10 cm) wide; conical, becoming convex to nearly flat and knobbed; *margin with hanging veil remnants at first;* smooth, brownish, fading to white.
Gills: attached, crowded, narrow; white, becoming grayish or lavender-tinted, then dark brown.
Stalk: 2–4″ (5–10 cm) long, ⅛–⅜″

(0.3–1 cm) thick; silky, fibrous, fragile; white; hollow.

Veil: partial veil white, leaving evanescent ring on upper stalk; sometimes membranous and persistent.

Spores: 7–10 × 4–5 μ; elliptical, smooth, with pore at tip. Spore print purplish-brown.

Edibility: Edible.

Season: May–September.

Habitat: Single to scattered, about stumps or in grassy areas, on buried wood.

Range: Widely distributed in North America.

Comments: This mushroom is actually a complex of species. The complex is recognized by its color, fragility, marginal veil remnants, form of growth, and habitat.

107 Red-gilled Psathyrella
Psathyrella conissans (Pk.) A.H.S.
Coprinaceae, Agaricales

Description: *Brownish cap with reddish gills and white stalk.*

Cap: 1–2″ (2.5–5 cm) wide; convex, with inrolled margin, expanding to nearly flat; margin radially lined; moist, smooth, sometimes wrinkled; dark brown, fading to buff-brown or reddish-tinged.

Gills: attached, crowded, narrow; pale brown, becoming reddish.

Stalk: 1–2″ (2.5–5 cm) long, ⅛–¼″ (3–5 mm) thick; smooth to finely hairy at tip; rigid, white; hollow.

Spores: 6.5–8 × 3.5–5 μ; oblong to elliptical, smooth, with indistinct pore at tip. *Spore print pinkish-red.*

Season: August–October.

Habitat: Clustered at base of trees and stumps.

Range: E. Canada to Tennessee, west to Michigan.

Look-alikes: *P. sublateritia,* possibly just a variety, has brick-red spore print and is somewhat larger. *P. spadicea* has brown to purplish-brown gills and spore print.

Comments: In all respects but spore print color, this is typical of its genus.

21 Parasitic Psathyrella
Psathyrella epimyces (Pk.) A.H.S.
Coprinaceae, Agaricales

Description: *Dingy white, dark-gilled mushroom; on Shaggy Mane* (Coprinus comatus).
Cap: 1–2″ (2.5–5 cm) wide; convex to nearly flat; margin with hanging veil remnants at first; dry, silky, fibrous; white to dingy.
Gills: attached, close, narrow; off-white, becoming blackish-brown.
Stalk: 1–3″ (2.5–7.5 cm) long, ¼–⅝″ (0.5–1.5 cm) thick; woolly, mealy, white; stuffed, becoming hollow.
Veil: partial veil white; leaving evanescent ring near base of stalk.
Spores: 7–9 × 4–5 μ; oval to elliptical, smooth, *pore at tip not distinct.* Spore print blackish.
Season: August–September.
Habitat: Parasitic on Shaggy Mane.
Range: N. North America south to New York, west to Minnesota and Wisconsin.
Comments: This is one of very few gilled mushrooms known to parasitize another mushroom.

26 Lawn Mower's Mushroom
Psathyrella foenisecii (Fr.) A.H.S.
Coprinaceae, Agaricales

Description: *Brownish, conical cap, fading in bands, cracking with age, with brown gills.*
Cap: ⅜–1¼″ (1–3 cm) wide; conical to convex, or nearly flat, often with central knob; smooth, moist, dark brown, fading to tan, with zones of color from center to margin, broadly cracking on drying.
Gills: attached, nearly distant, broad, dark brown to violet-brown.
Stalk: 1⅝–3¼″ (4–8 cm) long, ¹⁄₁₆–⅛″ (1.5–3 mm) thick; pallid, smooth.
Spores: 12–15 × 6.5–9 μ; elliptical, ornamented with warts, with pore at tip. Spore print dark purple-brown.
Edibility: Edible.

Season: May–July; September; overwinters in
S. California.

Habitat: Scattered to numerous in grass.

Range: Widely distributed in North America.

Look-alikes: *Agrocybe pediades** has yellowish-buff
cap, lighter gills and spore print. *P.
candolleana** is larger and more fragile,
with large partial veil. *P. castaneifolia*
has darker gills, thicker stalk, and
strong odor.

Comments: This spring agaric comes up in great
numbers in late May and early June; it
is a mediocre edible and may be a mild
hallucinogen in large quantities.

93 Clustered Psathyrella
Psathyrella hydrophila (Fr.) Maire
Coprinaceae, Agaricales

Description: *Fading, dark brown cap with brownish
gills and fragile white stalk.*
Cap: 1–2″ (2.5–5 cm) wide; conical to
convex, becoming flat; margin with
hanging veil remnants at first; moist,
smooth; dark rust-brown, fading to
honey-brown or tan.
Gills: attached, crowded, narrow to
moderately broad; buff, becoming
brownish to dark reddish-brown.
Stalk: 1–6″ (2.5–15 cm) long, $\frac{1}{16}$–$\frac{3}{8}$″
(0.15–1 cm) thick; fibrous, fragile,
whitish; hollow.
Veil: partial veil whitish; sometimes
leaving a thin zone of small fibers on
upper stalk.
Spores: 4.5–6 × 3–3.5 μ; elliptical,
smooth, with inconspicuous pore at tip.
Spore print dark brown.

Season: July–September; November–March in
S. California.

Habitat: In dense clusters, on decaying
deciduous stumps and logs, at the base
of trees or stumps, or on buried wood.

Range: Widely distributed in North America.

Look-alikes: Closely related species of *Psathyrella* can
be differentiated microscopically.

Comments: This very common form is often seen in
large clusters.

96 Ringed Psathyrella
Psathyrella longistriata (Murr.) A.H.S.
Coprinaceae, Agaricales

Description: *Fading, dark brown cap with dark gills and persistent ring on fragile white stalk.*
Cap: 1–4" (2.5–10 cm) wide; conical to convex, with incurved margin, becoming broadly convex to flat or with broad knob; moist, smooth to wrinkled about center; covered at first with evanescent, fibrous patches; dark rust-brown, fading to pale cinnamon-buff.
Gills: attached, close, moderately broad; pale buff, then purplish-brown.
Stalk: 2–4" (5–10 cm) long, 1/4–3/8" (0.5–1 cm) thick; fragile, whitish; sheathed up to ring by thin, fibrous, white layer, often breaking into woolly scales; hollow.
Veils: universal veil leaving fibrous patches on cap. Partial veil white; leaving persistent, membranous ring on upper stalk, with upper surface lined in wet weather, lower surface woolly.
Spores: 7–9 × 4–5 μ; elliptical, smooth, with pore at tip. Spore print dark purplish-brown.

Season: September–November.

Habitat: Scattered to several, on ground under both coniferous and deciduous trees.

Range: Pacific NW. to California.

Comments: This coastal species is unusual because it has a persistent, membranous ring. Typically, *Psathyrella* species have a partial veil that leaves, at most, hanging veil remnants on the cap margin and a fibrous zone on the stalk.

---

92 Corrugated-cap Psathyrella
Psathyrella rugocephala (Atk.) A.H.S.
Coprinaceae, Agaricales

Description: *Deeply wrinkled, brownish cap with dark gills and fragile, off-white stalk.*
Cap: 2–4" (5–10 cm) wide; convex to nearly flat or with broad knob; margin upturned with age; coarsely

radially wrinkled, hairless; watery-brownish to tawny, fading to tannish.
Gills: attached, close, moderately broad; whitish, becoming purplish-brown to black.
Stalk: 3−5″ (7.5−12.5 cm) long, ¼−⅜″ (0.5−1 cm) thick; smooth; off-white above, darkening to brownish below; hollow.
Veil: partial veil leaving ring zone on upper stalk.
Spores: 9−11 × 6−7.5 μ; oval to elliptical; warted, with snoutlike, protruding pore at tip. Spore print dark purplish-brown.

Edibility: Edible.

Season: July−September.

Habitat: Several to numerous, on and around deciduous stumps.

Range: Atlantic Coast to Great Lakes.

Look-alikes: *P. delineata,* nearly indistinguishable, has rust- to reddish-brown cap turning dirty tan, and smaller, smooth spores.

Comments: Both this species and the Wrinkled-cap Psathyrella (*P. delineata*) are unusually large and fleshy for the genus.

99 Velvety Psathyrella
Psathyrella velutina (Pers. ex Fr.) Sing.
Coprinaceae, Agaricales

Description: *Brownish, scaly-hairy cap with brownish gills and hairy ring zone on upper part of scaly-hairy stalk.*
Cap: 2−4″ (5−10 cm) wide; rounded to convex, becoming flat or with knob; margin with hanging remnants of veil; densely scaly-hairy; ochre to tawny.
Gills: attached, crowded, moderately broad; yellow to rust-brown, mottled; edges often with droplets of moisture.
Stalk: 2−4″ (5−10 cm) long, ⅛−⅝″ (0.3−1.5 cm) thick; fibrous to woolly-scaly; whitish above ring zone, tawny below; hollow.
Veil: partial veil whitish to buff; leaving soft, fibrous-cottony zone of small fibers on upper stalk, soon

darkened by spores.

Spores: 9–12 × 6–7 μ; elliptical, warted, with snoutlike, protruding pore at tip. Spore print blackish-brown.

Edibility: Edible.

Season: June–September.

Habitat: Single to numerous, in lawns and grassy areas; also along paths and in gravelly ground.

Range: Widely distributed in North America.

Look-alikes: *Inocybe** species produce a light brown to rust-brown spore print.

Comments: This is edible, provided the area in which one collects it has not been sprayed or otherwise polluted.

CORTINARIUS FAMILY
(Cortinariaceae)

This is the largest family of brown-spored gilled mushrooms. Most grow in forests or woodlands in a mycorrhizal relationship with trees. This family includes the Gypsy, one of the better edible mushrooms, but also hundreds of unknown edibility, many poisonous ones, and some deadly species.

Some of these are quite colorful, but most are brownish. Many have a veil covering the immature gills. This veil is usually cobwebby (a cortina—hence the family and genus names), and rarely leaves more than a zone of small fibers on the upper stalk. Some species, however, have a membranous veil that leaves a ring on the stalk, and some have no veil at all. A universal veil, covering the entire mushroom when first formed, is often present, but rarely leaves any remnants as the mushroom expands. The spore print may be bright rust, orange-brown, or gray-brown; none is purple-brown. These mushrooms are grouped together on the basis of their spores, which are typically roughened and lack a germ pore, and of the cap surface, which is not cellular.

30 Brown Alder Mushroom
Alnicola melinoides (Fr.) Küh.
Cortinariaceae, Agaricales

Description: *Honey-brown cap with radially lined*
margin, brown gills and stalk.
Cap: ⅜–1″ (1–2.5 cm) wide; convex,
becoming flat and somewhat knobbed;
margin becoming radially lined and
wavy; moist at first, minutely scruffy-
hairy; tan, becoming brownish at
center, honey- to ochre-brown toward
margin. Odor mild to strongly acidic.
Gills: attached, almost distant, broad,
yellow to rust-brown.
Stalk: 1–1⅜″ (2.5–4 cm) long, ⅟₃₂–
⅟₁₆″ (1–1.5 mm) thick; brittle, pale
yellowish to brownish, sometimes with
scattered, minute yellow fibers at first.
Spores: 10–15 × 5.3–7.3 μ; elliptical,
warty, brownish. Spore print brownish.
Season: July–November.
Habitat: Under alder in boggy areas.
Range: Widely distributed; most common in
Pacific NW.
Look-alikes: *A. scolecina* has dark reddish-brown cap.
Comments: Formerly placed in the genus *Naucoria.*
The genus *Alnicola* is reserved for those
species with roundish cap cuticle cells
and warty spores, growing near alder.

342 Silvery-violet Cort
Cortinarius alboviolaceus (Fr.) Fr.
Cortinariaceae, Agaricales

Description: *Grayish-violet, dry cap with purplish gills*
and club-shaped, grayish-violet stalk.
Cap: 1¼–2⅜″ (3–6 cm) wide; bell-
shaped, becoming convex to broadly
knobbed; margin curved under; dry,
silky-shiny, silvery-violet.
Gills: attached, close, moderately broad;
pale violet, then cinnamon-brown.
Stalk: 1⅝–3¼″ (4–8 cm) long, ¼–⅜″
(0.5–1 cm) thick, enlarging down to
bulbous base to 1″ (2.5 cm); violet.
Veils: universal veil white, leaving
flattened, sheathlike covering on stalk,

easily overlooked. Partial veil
cobwebby, white, leaving no ring.
Spores: 7.5–10 × 5–6 μ; elliptical,
minutely roughened, brownish. Spore
print rust.

Edibility: Edible.

Season: August–October.

Habitat: Single to several, in mixed woods,
especially under beech and oak.

Range: Widely distributed in North America.

Look-alikes: *C. iodes** has slimy cap, nonbulbous
base. *C. obliquus* has abruptly bulbous
base, lacks outer veil. *Clitocybe nuda**
lacks veil, produces pinkish-buff spore
print.

Comments: When still young, but after all trace of
the veil has been lost, this common fall
mushroom is difficult to distinguish
from the Blewit (*Clitocybe nuda*) without
a spore print.

---

331 Bracelet Cort
Cortinarius armillatus (Fr.) Fr.
Cortinariaceae, Agaricales

Description: *Large, reddish-brown mushroom with club-
shaped stalk with irregular, reddish bands.*

Cap: 2–5″ (5–12.5 cm) wide; bell-
shaped, becoming convex to nearly flat;
margin curved under, sometimes with
hanging remnants of veil; moist, tawny
red to reddish-orange, with minute
fibers or nearly smooth.
Gills: attached, distant, broad; pale
cinnamon, becoming dark rust.
Stalk: 3–6″ (7.5–15 cm) long, ⅜–¾″
(1–2 cm) thick, club-shaped to
bulbous, up to 1⅜″ (3.5 cm) thick at
base; solid, fibrous, brownish, with 1–
4 reddish bands around stalk.
Veils: universal veil reddish, leaving
remnants on margin and bands on
stalk. Partial veil cobwebby, whitish,
leaving zone on upper stalk.
Spores: 9–12 × 5.5–7.5 μ; elliptical,
warty, brownish. Spore print rust.

Edibility: Edible.

Season: August–October.

Habitat: Single to several, in mixed deciduous and coniferous woods, especially near birch.

Range: Quebec to New Jersey, west to Idaho.

Comments: In eastern woods this is one of the most commonly encountered corts and is often the dominant mushroom in late summer. The similar Ringed Cort (*C. haematochelis*) of the Pacific Northwest differs in spore shape.

332 Cinnabar Cort
Cortinarius cinnabarinus Fr.
Cortinariaceae, Agaricales

Description: *Bright cinnabar to cherry- or rust-red cap, gills, and stalk.*
Cap: 1¼–2⅜" (3–6 cm) wide; bell-shaped to knobbed or flat; dry, silky, shiny, sometimes split on margin; bright cinnabar, aging to rust-red.
Gills: attached, nearly distant, moderately broad; cinnabar, becoming dark rust-red.
Stalk: 1¼–2¾" (3–7 cm) long, ⅛–⅜" (0.3–1 cm) thick; fibrous, shiny, cinnabar-red.
Veil: partial veil cobwebby, cinnabar-red, leaving no ring on stalk.
Spores: 8–10 × 4.5–5.5 μ; elliptical, slightly roughened, brownish. Spore print rust.

Season: August–October.

Habitat: Several to numerous, under coniferous and deciduous trees, especially oak and beech.

Range: Widely distributed in E. North America.

Look-alikes: *C. sanguineus** is dark blood-red.

Comments: Also known as *Dermocybe cinnabarina*. Like all dermocybes, this differs from other corts by the presence of distinctive pigments: these species are generally recognized by their bright red, orange, or yellow colors and by their lack of purplish tones, bulbous stalk, and slime layer. Some are reportedly poisonous.

276 Slimy-banded Cort
Cortinarius collinitus (Fr.) S.F.G.
Cortinariaceae, Agaricales

Description: *Slimy, tawny-orange cap with rust gills at maturity; stalk with off-white to bluish, slimy bands.*

Cap: 1¼–4″ (3–10 cm) wide; bell-shaped to convex, flat, or somewhat knobbed; margin incurved; slimy, smooth, ochre-tawny.
Gills: attached, close, broad; off-white to grayish, becoming rust-cinnamon.
Stalk: 2–6″ (5–15 cm) long, ¼–⅝″ (0.5–1.5 cm) thick; cylindrical, rigid, whitish, covered by slime layer.
Veils: universal veil slimy, thick, breaking across stalk, leaving off-white to rusty or bluish bands. Partial veil cobwebby, leaving collapsed band on upper stalk.
Spores: 12–15 × 7–8 μ; elliptical, warted, brownish. Spore print rust.

Season: Late August–October; overwinters in California.

Habitat: Scattered to many, under conifers and deciduous trees, especially aspen and beech; under tanoak and madrone in California.

Range: Widely distributed in North America.

Look-alikes: *C. vanduzerensis* has purple slime veil, grows under conifers in Oregon.

Comments: Corts with a slimy cap and stalk form a small group in this huge genus. Before the veil breaks, these mushrooms appear to be encased in a bubble.

---

299 Saffron-colored Cort
Cortinarius croceofolius Pk.
Cortinariaceae, Agaricales

Description: *Small, dry, brownish-cinnamon cap with saffron to orange gills and saffron stalk.*
Cap: 1–2″ (2.5–5 cm) wide; convex to nearly flat or somewhat knobbed; dry, slightly fibrous; brownish-cinnamon, saffron on margin.
Gills: attached, close, narrow; saffron to

orange, becoming brownish-cinnamon.
Stalk: 1–2″ (2.5–5 cm) long, ¼–⅝″
(0.5–1.5 cm) thick, slightly enlarged
below; fibrous, hollow, saffron.
Veil: partial veil cobwebby, saffron.
Spores: 6–7.5 × 4–5 μ; elliptical,
slightly roughened, brownish. Spore
print rust.

Season: September–December.
Habitat: On the ground in coniferous woods.
Range: New York to Pacific NW.
Look-alikes: *C. cinnamomeus* has longer stalk and
greenish-yellow to rust-tinted gills.
Comments: Also known as *Dermocybe croceofolia*.
This cort is often found in late fall
when few other mushrooms are out.

⊗ 298 **Deadly Cort**
Cortinarius gentilis (Fr.) Fr.
Cortinariaceae, Agaricales

Description: *Deep orange-brown cap and stalk with*
remnants of yellow veil; under conifers.
Cap: 1–2″ (2.5–5 cm) wide; conical,
becoming bell-shaped to broadly
knobbed; moist, smooth, orange- to
yellow-brown, fading.
Gills: attached, almost distant, broad,
yellowish-brown, becoming rust.
Stalk: 1–3″ (2.5–7.5 cm) long, ⅛–¼″
(3–5 mm) thick; yellow- to orange-
brown.
Veils: universal veil yellow, evanescent.
Partial veil cobwebby, bright yellow,
leaving faint zone on stalk.
Spores: 7–9 × 5.5–7 μ; oval,
minutely roughened, brownish. Spore
print rust.
Edibility: Deadly.
Season: July–August in Rocky Mountains;
September–October in North.
Habitat: Abundant, under conifers.
Range: Widespread in N. North America.
Comments: Also ironically known as the "Gentle
Cort." In Europe this mushroom is
reported to contain deadly toxins; it is
related to species such as the Poznan
Cort (*C. orellanus*) that are known to

have caused fatalities. Consequently, the mushroom hunter is warned not to eat any "LBMs," or little brown mushrooms. Eating this mushroom and closely related species typically involves life-threatening kidney failure; symptoms usually do not appear until 3 days to 2 weeks after ingestion.

347 Bulbous Cort
Cortinarius glaucopus (Schaeff. ex Fr.) S.F.G.
Cortinariaceae, Agaricales

Description: *Tacky, reddish-brown cap with wavy-bent margin, purplish gills becoming brownish, and purplish, bulbous stalk.*
Cap: 2–5" (5–12.5 cm) wide; convex, becoming flat, margin curved under, wavy; slimy-tacky, grayish, reddish-brown in center and in streaks.
Gills: attached, close to crowded, moderately broad; purplish at first, becoming cinnamon-brown.
Stalk: 2–4" (5–10 cm) long, ⅝–1" (1.5–2.5 cm) thick; pale purplish, solid, bulbous.
Veil: partial veil cobwebby, leaving faint zone on stalk.
Spores: 6–10 × 4.5–5 μ; elliptical, minutely roughened, brownish. Spore print rust.

Season: July–August in Rocky Mountains; September–October in Pacific NW.; November–March in S. California.

Habitat: Single to many, under coniferous and deciduous trees, especially spruce, fir and oak.

Range: Widespread in W. North America.

Look-alikes: *C. alboviolaceus** has dry, silvery-lavender cap. *C. pseudoarquatus* has abruptly bulbous base.

Comments: This species is representative of a large number of corts that have a slimy-tacky cap and an abruptly bulbous, dry stalk. Although none is known to be poisonous, exact identification is difficult and none can be recommended as edible.

344 Viscid Violet Cort
Cortinarius iodes Berk. & Curt.
Cortinariaceae, Agaricales

Description: *Slimy-tacky, purple cap becoming yellow-spotted around or in center, with purplish gills becoming rust.*
Cap: 1–2" (2.5–5 cm) wide; bell-shaped to convex; slimy to tacky, smooth, dark purplish, becoming paler, yellow-spotted or yellowish at center.
Gills: attached, close, moderately broad; violet, becoming grayish-cinnamon.
Stalk: 2–3" (5–7.5 cm) long, ¼–⅜" (0.5–1 cm) thick, sometimes enlarged below; slimy-tacky, solid, purplish.
Veil: partial veil cobwebby, pale violet, evanescent.
Spores: 8–12 × 5–6.5 μ; elliptical, minutely roughened, brownish. Spore print rust.
Edibility: Edible.
Season: August–September.
Habitat: Several to many, under deciduous trees in low woods.
Range: Widespread in E. North America; reported in Olympic Mountains.
Look-alikes: *C. iodeoides* has bitter-tasting cap surface and smaller spores. *C. heliotropicus,* now recognized as a variant, has discrete yellow to cream spots on cap.
Comments: Although this is edible, like most corts it looks better than it tastes.

305 Variable Cort
Cortinarius multiformis (Fr. ex Secr.) Fr.
Cortinariaceae, Agaricales

Description: *Tacky, ochre cap with whitish gills becoming rust-brown, and off-white to tannish, bulbous stalk.*
Cap: 2–4" (5–10 cm) wide; convex to flat or with knob; margin incurved; hoary white, becoming sticky, ochre, buff, or reddish-orange.
Gills: attached, close, narrow; whitish, becoming tannish-cinnamon.

Stalk: 1⅜–3½" (4–9 cm) long, ⅜–¾"
(1–2 cm) thick, with marginate bulb;
white, becoming tannish.
Veil: partial veil cobwebby, white.
Spores: 7–10 × 4–5.5 μ; elliptical,
minutely roughened, brownish. Spore
print brownish.

Season: July–August in Rocky Mountains;
September–December on Pacific Coast.
Habitat: Single to scattered under conifers or oak.
Range: Widespread in N. North America.
Comments: This common, variable cort is best
identified when very young, when it
has whitish gills and an evanescent
white bloom on the cap.

---

333 Blood-red Cort
Cortinarius sanguineus (Wulf. ex Fr.) S.F.G.
Cortinariaceae, Agaricales

Description: *Dark blood-red cap, gills, and stalk.*
Cap: 1–2" (2.5–5 cm) wide; bell-
shaped, becoming broadly knobbed;
dry, silky, dark blood-red.
Gills: attached, close, broad, dark
blood-red.
Stalk: 2–4" (5–10 cm) long, ⅛–¼"
(3–5 mm) thick; slender, blood-red.
Veil: partial veil cobwebby, reddish.
Spores: 6–9 × 4–5 μ; elliptical,
roughened, reddish-brown. Spore print
rusty reddish-brown.

Season: Late August–November.
Habitat: Scattered on moss, in coniferous woods.
Range: Widespread in N. North America.
Look-alikes: *C. cinnabarinus** is bright red and often
found in deciduous woods.
Comments: Also known as *Dermocybe sanguinea.*

---

300 Red-gilled Cort
Cortinarius semisanguineus (Fr.) Gill.
Cortinariaceae, Agaricales

Description: *Yellowish cap and stalk with red gills.*
Cap: 1–2⅜" (2.5–6 cm) wide; bell-
shaped to convex or nearly flat; moist,
silky to fibrous, tawny yellow,

sometimes with somewhat shiny zones.
Gills: attached, crowded, narrow;
cinnabar- to blood-red.
Stalk: 1¼–2⅜" (3–6 cm) long, ⅛–¼"
(3–5 mm) thick; solid-fibrous, yellow.
Veil: partial veil cobwebby, yellowish.
Spores: 6–8 × 4–5 μ; elliptical,
roughened, brownish. Spore print rust.

Season: Late July–November.
Habitat: Several to many, on moss and ground
near coniferous and deciduous trees.
Range: Widely distributed in N. North
America.
Comments: Also known as *Dermocybe semisanguinea*.

---

341 **Pungent Cort**
Cortinarius traganus (Weinm. ex Fr.) Fr.
Cortinariaceae, Agaricales

Description: *Dry, purplish-lilac cap and stalk with*
cinnamon gills and a strong odor.
Cap: 2–5" (5–12.5 cm) wide; convex to
nearly flat or with knob; dry, with
minute fibers, lilac, sometimes with
pale wedges on margin. Odor pungent.
Gills: attached, almost distant, broad,
cinnamon to reddish cinnamon-brown.
Stalk: 2–5" (5–12.5 cm) long, ⅜–2"
(1–5 cm) thick, enlarging toward base;
with minute fibers, solid, purplish.
Flesh rusty cinnamon-brown.
Veil: partial veil cobwebby, grayish-
lilac, leaving minute fibers on stalk.
Spores: 7–10 × 4.5–6 μ; elliptical,
warty, brownish. Spore print rust.
Season: September–November.
Habitat: Several to many under conifers; often in
moss.
Range: Widely distributed in North America;
common in Pacific NW.
Look-alikes: Similar species have bluish gills when
young. *C. camphoratus* has fetid odor.
C. pyriodorus and *C. fragrans* smell like
overripe pears; the latter has whitish
stalk flesh.
Comments: The odor of this distinctive species is
sometimes faint, described as either
pleasant or unpleasant.

340 Violet Cort
Cortinarius violaceus (Fr.) S.F.G.
Cortinariaceae, Agaricales

Description: *Dark violet, dry, scaly cap, gills, and*
stalk; near conifers.
Cap: 2–4¾" (5–12 cm) wide; convex
to flat with slight knob; margin fibrous
to fringed; dry, covered with tufts of
minute, erect fibers; dark violet,
sometimes becoming shiny. Flesh
thick, dark violet.
Gills: attached, broad, nearly distant;
dark violet.
Stalk: 2¾–7" (7–18 cm) long, ⅜–1"
(1–2.5 cm) thick, often enlarged at
base; conspicuously fibrous, dark violet.
Veil: partial veil cobwebby, leaving a
high, minutely fibered ring on stalk.
Spores: 13–17 × 8–10 μ; elliptical to
oblong, minutely warted. Spore print
rust-cinnamon.
Edibility: Edible.
Season: September–October.
Habitat: In groups on the ground in coniferous
forests, often near decomposed logs.
Range: Widely distributed in North America.
Look-alikes: *C. iodes** has sticky cap. *C.*
*alboviolaceus** has dry, silvery-lilac cap.
Comments: This stately species is not as common as
some other purplish corts.

⊗ 228 Deadly Galerina
Galerina autumnalis (Pk.) A.H.S. & Sing.
Cortinariaceae, Agaricales

Description: *Fading, brownish, tacky cap with yellowish*
gills becoming rust; ring on brownish stalk;
on decaying wood.
Cap: 1–2½" (2.5–6.5 cm) wide;
convex, becoming flat or with slight
knob, margin translucent and radially
lined when moist; sticky, smooth, dark
brown to ochre-tawny, unevenly fading
to yellowish or buff.
Gills: attached, close, broad; yellowish,
becoming rust.
Stalk: 1–4" (2.5–10 cm) long, ⅛–⅜"

(0.3–1 cm) thick, sometimes enlarging toward base; hollow, somewhat longitudinally lined; off-white above brownish to blackish base; lower stalk and base with dense, white threads (mycelium).

Veil: partial veil membranous, white, leaving evanescent ring on upper stalk.

Spores: 8.5–10.5 × 5–6.5 μ; elliptical, roughened, with smooth depression. Spore print rust.

Edibility: Deadly.

Season: October–November; May–June.

Habitat: Scattered to abundant, on well-decayed coniferous and deciduous logs.

Range: Throughout North America.

Look-alikes: The deadly *G. marginata* (if a distinct species) has moist, not tacky, cap. *Pholiota mutabilis** has down-curled scales on stalk, grows in massive, dense clusters.

Comments: Also known as "Autumn Galerina" and formerly known as *Pholiota autumnalis*. This galerina is one of a few very common, harmless-looking little brown mushrooms that are deadly. Symptoms typically occur 10 or more hours after ingestion and follow the same sequence as those for the destroying angels: vomiting and diarrhea, cramps, then a short remission followed by kidney and/or liver dysfunction or failure, coma, and death.

---

43 Sphagnum-bog Galerina
Galerina tibiicystis (Atk.) Küh.
Cortinariaceae, Agaricales

Description: *Tall, slender-stalked, tawny-brownish mushroom; in sphagnum bogs.*
Cap: ⅜–1⅜″ (1–3.5 cm) wide; conical, becoming somewhat knobbed; margin incurved at first, radially lined; moist, smooth, tawny, fading to ochre, then ochre-buff.
Gills: attached, close to nearly distant, narrow to moderately broad; tawny to ochre.

Stalk: 2–8″ (5–20 cm) long, ¹⁄₁₆–⅛″
(1.5–3 mm) thick; very long and thin,
fragile, hollow; tawny to pale buff.
Spores: 8.5–14 × 5–7 μ; elliptical,
roughened with smooth, pool-like
depression. Spore print rust.

Season: May–November.

Habitat: Abundant in sphagnum bogs.

Range: N. North America; south in bogs.

Look-alikes: This is one of many similar, moss-
inhabiting *Galerinas*.

Comments: Little is known about the edibility of
most galerinas. These mushrooms are
usually found growing in bogs and are
thought to be most abundant in dry
years, when other mushrooms are
scarce. They are difficult to identify to
species.

⊗ 39 Deadly Lawn Galerina
Galerina venenata A.H.S.
Cortinariaceae, Agaricales

Description: *Small mushroom with moist, reddish-brown*
cap, fading to buff; in lawns.
Cap: ⅜–1⅜″ (1–3.5 cm) wide; convex,
becoming flat or with depressed center
and somewhat torn, arched margin;
moist, smooth; reddish- to cinnamon-
brown, fading to dingy yellowish-white
or pinkish-buff. Flesh moderately
thick. Odor mealy; taste bitterish.
Gills: attached, nearly distant, broad;
yellowish-brown, becoming cinnamon-
brown.
Stalk: 1¼–1⅝″ (3–4 cm) long, ⅛–¼″
(3–5 mm) thick, enlarging somewhat
toward base; brownish, smooth, with
cottony, white mycelium about base.
Veil: partial veil leaving a small, thin
ring pressed against upper stalk.
Spores: 8–11 × 5–6.5 μ; oval,
roughened with smooth, pool-like
depression at base. Spore print rust.

Edibility: Deadly.

Season: November–January.

Habitat: In lawns.

Range: Washington and Oregon.

Look-alikes: *G. autumnalis** and *G. marginata*, both deadly, grow on decaying wood.

Comments: This species usually grows in lawns, but it may also occur on buried decomposed wood. It has reportedly caused life-threatening poisoning similar to that caused by the Deadly Galerina (*G. autumnalis*) and the destroying angels.

---

301 Little Gym
Gymnopilus penetrans (Fr. ex Fr.) Murr.
Cortinariaceae, Agaricales

Description: *Small, yellowish, smooth cap and yellowish stalk with whitish, evanescent veil.*
Cap: 1–2" (2.5–5 cm) wide; bell-shaped to convex or nearly flat; moist, smooth, chrome- to golden-yellow, fading with age. Taste bitter.
Gills: attached, close, moderately broad; yellowish, then rust-spotted.
Stalk: 1⅝–2⅜" (4–6 cm) long, ⅛–¼" (3–5 mm) thick; sometimes enlarging downward; yellowish, base white-hairy.
Veil: partial veil fibrous, white, leaving no ring on stalk.
Spores: 6.5–9.5 × 4–5.5 μ; elliptical, warty, brownish. Spore print orange-brown.
Season: August–October.
Habitat: On wood of coniferous and deciduous trees.
Range: Widely distributed in North America.
Look-alikes: *G. sapineus* has scaly cap and yellowish veil.
Comments: Formerly known as *Flammula penetrans*. This is one of more than 30 *Gymnopilus* species that are similar microscopically and lack a ring.

⊗ 214 Big Laughing Gym
Gymnopilus spectabilis (Fr.) A.H.S.
Cortinariaceae, Agaricales

Description: *Large, clustered, yellowish-orange mushroom with ringed stalk; on wood.*

Cap: 3¼–7" (8–18 cm) wide; convex and knobbed to nearly flat; smooth, slightly silky-fibrous or with minute scales; light orange-yellow to ochre-orange. Flesh thick, yellowish. Odor aniselike or mild; taste very bitter. Gills: attached, crowded, narrow to moderately broad, pale yellow to rust. Stalk: 2–8" (5–20 cm) long, ⅜–1¼" (1–3 cm) thick; solid, pale yellow to ochre, streaked with minute fibers. Veil: partial veil thin and membranous to fibrous, pale yellow; leaving indistinct, evanescent ring or zone on stalk; ring sometimes persistent, high. Spores: 7.5–10.5 × 4.5–6 μ; oval to elliptical to almond-shaped, wrinkled. Spore print orange to rust-orange.

Edibility: Hallucinogenic.

Season: August–October.

Habitat: Clustered on wood, stumps, or on ground over buried wood.

Range: Widely distributed in North America.

Look-alikes: *Armillariella mellea** is honey-brown and has white to cream spore print. The poisonous *Omphalotus olearius** lacks veil, has white to cream spore print.

Comments: In Japan this species was named the Big Laughing Gym because its ingestion has caused unmotivated laughter and foolish behavior.

⊗ 259 **Poison Pie**

Hebeloma crustuliniforme (Bull. ex St. Amans) Quél.

Cortinariaceae, Agaricales

Description: *Cream, slimy cap with cinnamon-buff center, pale brownish gills, and strong, radishlike odor.*

Cap: 1¼–3½" (3–9 cm) wide; convex, with inrolled margin, becoming flat or uplifted; smooth and sticky, varnished when dry. Flesh at center thick, white. Odor of gills or crushed flesh radishy. Gills: attached, crowded, narrow, becoming moderately broad; off-white, becoming tannish to gray-brown, edges

white, toothed, beaded at first with
moisture, brown-dotted when dry.
Stalk: 1⅛–2¾″ (4–7 cm) long, ¼–⅜″
(0.5–1 cm) thick; white, solid; base
surrounded by abundant white threads,
top covered with fine, flaky particles.
Spores: 9–13 × 5.5–7 μ; almond-
shaped, minutely dotted or appearing
smooth. Spore print brown to pale rust.

Edibility: Poisonous.

Season: September–November; through May in
California.

Habitat: Scattered, in groups, or in fairy rings,
frequently near residential areas;
mycorrhizal with conifers and
hardwoods.

Range: Widely distributed; most common in
West.

Look-alikes: *Tricholoma** species have white spores.
*Entoloma** have salmon-pink spores.
*Pluteus** have free gills and dark spores.

Comments: Many hebelomas are hard to identify to
species, but they can be recognized to
genus by the sticky cap, pale gill color,
flaky stalk, and radishy odor. Some
species have a veil; a few have a stalk
ring. None should be eaten.

278 Corpse Finder
Hebeloma syriense Kar.
Cortinariaceae, Agaricales

Description: *Slimy-tacky, reddish-brown cap with
whitish gills becoming cinnamon-brown;
growing near corpses.*
Cap: 1–2″ (2.5–5 cm) wide; convex to
nearly flat; margin incurved at first;
slimy, smooth, brick-red, fading to
ochre-brown.
Gills: attached or notched, close,
moderately broad; whitish to reddish-
or cinnamon-brown; edges fringed.
Stalk: 1⅝–2⅜″ (4–6 cm) long, ⅛–¼″
(3–5 mm) thick; cottony-scaly above,
smooth below, elastic-tough; whitish,
becoming brownish above.
Spores: 8–10.5 × 5–6 μ; elliptical,
minutely roughened, brownish. Spore

print pale cinnamon.

Season: September—October.

Habitat: Single to several or clustered, on the ground in deciduous woods.

Range: E. North America west to Minnesota, south to Texas.

Comments: This species once brought a crime to light when human remains were found beneath it; in another instance, the mushroom was discovered growing out of a buried box containing a baby's bones. A related Japanese species is known to grow from the remains of animal carcasses.

⊗ 15 **White-disc Fiber Head**
Inocybe albodisca Pk.
Cortinariaceae, Agaricales

Description: *Grayish-brown, silky cap with creamy white, knobbed center, whitish gills becoming grayish-brown, and grayish stalk.*
Cap: ⅝—1⅜" (1.5—3.5 cm) wide; convex, becoming flat with broad knob; moist, smooth at center, elsewhere becoming minutely fibrous to radially cracked; grayish- to pinkish-brown, white to cream at center. *Odor spermatic.*
Gills: attached, close, narrow; whitish, becoming grayish-brown; edges fringed.
Stalk: 1—2" (2.5—5 cm) long, ⅛—¼" (3—5 mm) thick, with bulb; solid, grayish, tinged pinkish at times, minutely white-haired.
Spores: 6—8 × 4.5—6 μ; angular-warty, brownish. Spore print brown.

Edibility: Poisonous.

Season: August—November.

Habitat: On the ground, near coniferous and deciduous trees, especially hemlock, aspen, beech, and birch.

Range: Widely distributed in N. North America.

Look-alikes: *I. umbratica* is whitish to grayish.

Comments: This mushroom is most common in the Pacific Northwest. It is one of the many inocybes with a spermatic odor. Nearly all are poisonous.

⊗ 314 Caesar's Fiber Head
Inocybe caesariata (Fr.) Kar.
Cortinariaceae, Agaricales

Description: *Dry, densely fibrous-scaly, yellow to tawny cap and stalk; gills rust-ochre.*
Cap: 1–2" (2.5–5 cm) wide; broadly convex to nearly flat, with margin incurved at first; dry, densely woolly-fibrous, becoming partly scaly; yellow-ochre or yellow-buff to tawny.
Gills: attached, close, broad; dull ochre-yellow, becoming rust-ochre, edges white-fringed.
Stalk: ⅝–1⅝" (1.5–4 cm) long, 1/16–¼" (1.5–5 mm) thick; densely cottony-fibrous to scaly; yellow-ochre to tawny.
Veil: partial veil thin, evanescent.
Spores: 9–12.5 × 5–7 μ; elliptical, smooth, brownish. Spore print brown.
Edibility: Poisonous.
Season: July–October.
Habitat: On the ground in grass, near deciduous trees along roads.
Range: N. North America south to Virginia.
Look-alikes: There are many scaly *Inocybes** that can only be differentiated microscopically.
Comments: Many inocybes are quite scaly and resemble pholiotas.

⊗ 269 Green-foot Fiber Head
Inocybe calamistrata (Fr.) Gill.
Cortinariaceae, Agaricales

Description: *Dry, densely scaly, dark brown cap with brownish gills; stalk scaly, dark; smoky greenish-blue near base.*
Cap: ⅜–1⅝" (1–4 cm) wide; bell-shaped to convex; dry, densely scaly, dark brown. *Odor spermatic.*
Gills: attached, close, broad; brown to rust, edges white-fringed.
Stalk: 1⅝–3¼" (4–8 cm) long, 1/16–¼" (1.5–5 mm) thick; firm, solid, covered with down-curled, fibrous scales; dark brown; smoky greenish-blue below.
Spores: 9–12 × 4.5–6.5 μ; elliptical, smooth, brownish. Spore print brown.

Edibility: Poisonous.

Season: August–November.

Habitat: Single to scattered, in deciduous and coniferous woods.

Range: Widely distributed in North America.

Look-alikes: *I. hirsuta* var. *maxima* has reddish-brown tints in cap, flesh bruising somewhat red, and strong, fishy odor.

Comments: This may not be recognized as an *Inocybe* at first, but it has an odor typical of the genus and its greenish base distinguishes it from others. Care should be taken not to confuse this poisonous species with a *Psilocybe,* which has purple-brown spores and, in some species, bruises blue.

⊗ 318 **Straw-colored Fiber Head**
Inocybe fastigiata (Schaeff. ex Fr.) Quél.
Cortinariaceae, Agaricales

Description: *Dry, bell-shaped, yellowish cap with radial fibers, whitish gills becoming grayish-brown, and pale, fibrous stalk.*
Cap: 1–3" (2.5–7.5 cm) wide; conical to bell-shaped to nearly flat, with broad central knob with age; margin upturning or split on edge; dry, with *minute radial fibers,* yellowish to slightly darker in center. *Odor spermatic.*
Gills: attached, becoming nearly free, close to crowded, narrow; whitish, becoming olive-grayish to pale brown.
Stalk: 1⅝–3¼" (4–8 cm) long, ⅛–⅜" (0.3–1 cm) thick; fibrous, whitish to somewhat yellowish.
Spores: 8–18 × 5–7.5 μ; elliptical, smooth, brownish. Spore print brown.

Edibility: Poisonous.

Season: August–November.

Habitat: Single to numerous, in coniferous and deciduous woods.

Range: Widely distributed in North America.

Look-alikes: *I. fastigiella* has darker, brownish cap. The poisonous *I. sororia** has paler cap and strong odor of green corn.

Comments: This is a typical *Inocybe* in its cap shape and surface features.

⊗ 316 **Black-nipple Fiber Head**
Inocybe fuscodisca (Pk.) Mass.
Cortinariaceae, Agaricales

Description: *Moist to tacky, conical to flat, brownish cap with blackish, nipplelike knob.*
Cap: ⅜–1″ (1–2.5 cm) wide; conical to bell-shaped, becoming nearly flat, with prominent nipple; moist, smooth to minutely fibrous, brownish, nipple blackish. *Odor spermatic.*
Gills: attached, close, broad; whitish, becoming grayish-brown.
Stalk: 2–3″ (5–7.5 cm) long, ¹⁄₁₆–⅛″ (1.5–3 mm) thick; with minute fibers, tip minutely hairy, off-white to gray.
Spores: 7–10 × 4.5–6 μ; elliptical, smooth, brownish. Spore print brown.
Edibility: Poisonous.
Season: August–November.
Habitat: Scattered to numerous, with conifers.
Range: Pacific NW.; Southeast.
Comments: This is one of the few easily recognized inocybes among the more than 500 species in North America.

⊗ 16 **White Fiber Head**
Inocybe geophylla (Sow. ex Fr.) Kum.
Cortinariaceae, Agaricales

Description: *Small, dry, silky-glossy, white cap with white to gray-brown gills and white stalk.*
Cap: ⅝–1¼″ (1.5–3 cm) wide; conical to nearly flat with a central knob; dry, silky-glossy, whitish. *Odor spermatic.*
Gills: attached, close, broad, whitish to grayish-brown.
Stalk: 1–2″ (2.5–5 cm) long, ¹⁄₁₆–⅛″ (1.5–3 mm) thick; firm, silky, minutely hairy at tip, white.
Veil: partial veil cobwebby, evanescent.
Spores: 7.5–11 × 5–7 μ; elliptical, smooth, brownish. Spore print brown.
Edibility: Poisonous.
Season: July–November.
Habitat: Several to many, with coniferous and deciduous trees.
Range: Widely distributed in North America.

Comments: This is one of the most common and
widely distributed inocybes. When
young, the cap is smooth to slightly
silky and does not show characteristic
radial fibers.

⊗ 315 **Torn Fiber Head**
Inocybe lacera (Fr.) Kum.
Cortinariaceae, Agaricales

Description: *Dry, brown, fibrous cap, becoming scaly,*
with whitish gills, becoming brownish, and
fibrous whitish stalk; along roads.
Cap: ⅜–1⅝" (1–4 cm) wide; convex,
becoming flat or with slight knob; dry,
fibrous, becoming somewhat scaly and
torn, ragged near margin, dark brown.
Gills: attached, close, broad; whitish,
becoming grayish-brown.
Stalk: 1¼–1⅝" (3–4 cm) long, ⅛–¼"
(3–5 mm) thick; solid, fibrous to nearly
smooth, whitish above, brownish
toward base.
Veil: partial veil cobwebby, evanescent.
Spores: 10–16 × 4.5–6 μ; elliptical,
smooth, brownish. Spore print brown.
Edibility: Poisonous.
Season: May–November.
Habitat: On ground, in sandy, gravelly, or
burned soil; along paths under aspen
and conifers.
Range: N. North America.
Look-alikes: Other similar brown *Inocybes** have
smaller spores.
Comments: This species is often found in May and
June, before most gilled ground
mushrooms start appearing.

⊗ 271 **Woolly Fiber Head**
Inocybe lanuginosa (Bull. ex Fr.) Kum.
Cortinariaceae, Agaricales

Description: *Dry, conical, scaly, brown cap.*
Cap: ¾–1¼" (2–3 cm) wide; bell-
shaped to nearly flat with knob; dry,
densely woolly-scaly, brown to tawny.
Gills: attached, close, broad; grayish-

white, becoming bright cinnamon;
edges white-fringed.
Stalk: 1–2" (2.5–5 cm) long, ⅛"
(3 mm) thick; fibrous, brownish.
Veil: partial veil cobwebby, grayish-
white, evanescent.
Spores: 8–10.5 × 5–7 μ; elliptical,
with warts or nodules, brownish. Spore
print yellowish-brown.

Edibility: Poisonous.

Season: August–October.

Habitat: On ground or wood debris, sometimes
on rotten wood; in coniferous and
deciduous woods.

Range: Widely distributed in North America.

Look-alikes: Most similar *Inocybes** grow only on soil.

Comments: This debris-dwelling inocybe is
unusual, and its habit of growth may
help to identify it. It is more
important, however, to recognize it as
an *Inocybe* than identify it to species, as
many in the genus are poisonous.

⊗ 13 Lilac Fiber Head

Inocybe lilacina (Bond.) Kauff.
Cortinariaceae, Agaricales

Description: *Dry, conical to bell-shaped, silky, pale
lilac-brown cap and stalk, with whitish
gills becoming grayish-brown.*
Cap: ⅜–1¼" (1–3 cm) wide; conical or
bell-shaped to convex or nearly flat,
knobbed; dry, silky-glossy; pale
grayish-lilac with pinkish-brown tones.
Odor spermatic.
Gills: attached, close, broad; whitish to
grayish-brown.
Stalk: 1–2" (2.5–5 cm) long, ¹⁄₁₆–⅛"
(1.5–3 mm) thick; slightly bulbous at
base; silky, gray-lilac to pink-brown.
Veil: partial veil cobwebby, evanescent.
Spores: 7–9 × 4–5.5 μ; elliptical,
smooth, brownish. Spore print brown.

Edibility: Poisonous.

Season: August–November.

Habitat: On the ground, under deciduous and
coniferous trees.

Range: Widely distributed in North America.

Look-alikes: *I. violacea* is deep violet-purple. Similar *Cortinarius** have rust spore print.

Comments: This mushroom is often found growing with the poisonous White Fiber Head (*I. geophylla*), from which it differs only in color. Some mycologists call it *I. geophylla* var. *lilacina*.

⊗ 14 **Blushing Fiber Head**
Inocybe pudica Küh.
Cortinariaceae, Agaricales

Description: *Small, white mushroom, bruising reddish or orange, with gills becoming brownish.*
Cap: 1–3" (2.5–7.5 cm) wide; conical to bell-shaped, becoming nearly flat with small knob; dry, minutely fibered, smoothish at first; white, bruising pinkish-red to orange. *Odor spermatic.*
Gills: attached, close, broad; white, becoming grayish-brown.
Stalk: 2–3" (5–7.5 cm) long, ¼–⅜" (0.5–1 cm) thick; minutely fibered, solid, minutely hairy at tip; white, bruising pinkish-red to orange.
Veil: partial veil cobwebby, whitish, evanescent.
Spores: 8–10 × 4.5–6 μ; elliptical, smooth, brown. Spore print gray-brown.
Edibility: Poisonous.
Season: Late August–December.
Habitat: Scattered to many, under pine and Douglas-fir.
Range: Pacific NW. to C. California.
Comments: This mushroom can be mistaken for an agaricus when very young, but reddish-staining species of *Agaricus* lack a spermatic odor and have free gills and dark brown spores.

⊗ 317 **Pungent Fiber Head**
Inocybe sororia Kauff.
Cortinariaceae, Agaricales

Description: *Dry, pale yellowish, conical to bell-shaped cap with radial fibers.*

Cap: 1–3″ (2.5–7.5 cm) wide; conical to bell-shaped, becoming nearly flat, but knobbed; dry, with radial fibers; creamy to straw-yellow or slightly darker; margin becoming upturned and split on edge. *Odor pungent, of green corn.*

Gills: attached, close to crowded, narrow; whitish, becoming yellowish, edges white-fringed.

Stalk: 1–4″ (2.5–10 cm) long, ¹⁄₁₆–¼″ (1.5–5 mm) thick, with slightly bulbous base; fibrous, silky, minutely hairy at tip, solid, whitish, becoming dingy with age.

Spores: 9–16 × 5–8 μ; elliptical, smooth, brownish. Spore print brown.

Edibility: Poisonous.

Season: Late August–December.

Habitat: Single to scattered, with hardwoods in mixed woods.

Range: Widely distributed in North America.

Comments: This mushroom is very common in the Pacific Northwest in the fall. On the Olympic Peninsula and in areas of heavy rainfall, this species, like many others, can grow to sizes far exceeding the normal range.

⊗ 183 **Scaly Fiber Head**
Inocybe terrigena (Fr.) Küh.
Cortinariaceae, Agaricales

Description: *Dry, scaly, brownish cap with gray to brown gills; stalk with dense, down-curled scales and fibrous-scaly ring at top.*

Cap: 1¼–2¾″ (3–7 cm) wide; convex to nearly flat; dry, densely scaly, golden- to cinnamon-brown. *Odor strongly pungent.*

Gills: attached, close, moderately broad; golden-yellow to olive-brown.

Stalk: 1⅝–2⅜″ (4–6 cm) long, ⅛–⅜″ (0.3–1 cm) thick; firm, with dense, down-curled scales; golden-brown, smooth above ring.

Veil: partial veil golden, leaving fibrous-scaly, rimlike ring on upper stalk.

Spores: 9–12 × 4.5–6.5 μ; elliptical, smooth, brownish. Spore print brown.

Edibility: Poisonous.

Season: July–September.

Habitat: On the ground with conifers and aspen.

Range: Rocky Mountains; probably widely distributed in West.

Comments: This was once called a *Pholiota* because of its more or less persistent ring.

---

88 Pretty Phaeocollybia
Phaeocollybia fallax A.H.S.
Cortinariaceae, Agaricales

Description: *Slimy, greenish, conical cap with purplish gills becoming rust, and rootlike stalk.*
Cap: ⅜–2″ (1–5 cm) wide; conical, becoming flat, knobbed; margin inrolled at first; slimy to tacky, smooth, olive fading to light olive or buff. Odor radishy.
Gills: free or slightly attached, close, narrow; violet to violet-gray, becoming rust-brown.
Stalk: 3–5″ (7.5–12.5 cm) long, ⅛–⅜″ (0.3–1 cm) thick, tapering to long, rootlike base; hollow, brittle; bluish-gray to tawny or brownish.
Spores: 7–9 × 4.5–5.5 μ; oval, with distinctive porelike beak at tip, warty, brownish. Spore print dark rust.

Season: September–October.

Habitat: Scattered to many, under Douglas-fir.

Range: Pacific NW.

Comments: This genus of mushrooms grows mostly with conifers. All but a few of the 20 or so species in North America are found along the northwest Pacific Coast, often under Sitka spruce. None is reliably known to be edible.

---

89 Kit's Phaeocollybia
Phaeocollybia scatesiae A.H.S. & Trappe
Cortinariaceae, Agaricales

Description: *Slimy, cinnamon-brown, conical cap with long, tapering, rootlike stalk; in clusters.*

Cap: ¾–2⅜" (2–6 cm) wide; conical, becoming flat, knobbed; margin inrolled; slimy, smooth, cinnamon-brown, darker in center, fading to grayish-brown.

Gills: attached, crowded, narrow, pale to dark cinnamon.

Stalk: 3–7" (7.5–18 cm) long, ⅛–¼" (3–5 mm) thick, tapering down to rootlike base 2–5" (5–12.5 cm) deep in ground; smooth, shiny, grayish to red.

Spores: 8–9.5 × 5–6 μ; oval, with distinctive porelike beak at tip, warty, brownish. Spore print dark rust.

Season: October–November.

Habitat: In massive clusters, with conifers.

Range: Pacific NW.

Look-alikes: *P. californica* grows in arcs and rings, and lacks grayish cap tones and yellowish gills.

Comments: Two or 3 members of this genus can be found in the Northeast; these probably occur under conifers across all of northern North America.

---

201 **Gypsy**
Rozites caperata (Fr.) Kar.
Cortinariaceae, Agaricales

Description: *Ochre cap with white bloom, attached brownish gills, and ring around midstalk.*
Cap: 2–6" (5–15 cm) wide; oval, expanding to convex, broadly knobbed; wrinkled, silky or silky-scaly; moist, with superficial hoary coating at first; ochre-orange. Flesh thick, white.

Gills: attached, close, broad; off-white, becoming dull tawny.

Stalk: 2–4¾" (5–12 cm) long, ⅜–¾" (1–2 cm) thick; firm, solid, whitish, with persistent ring around middle.

Veils: universal veil leaves frosting on cap and evanescent fragments at base of stalk. Partial veil leaves persistent ring around middle stalk.

Spores: 11–14 × 7–9 μ; almond-shaped, roughened. Spore print rust.

Edibility: Choice.

Season: September—November.
Habitat: Single to many, scattered or in groups, on humus in coniferous and hardwood forests.
Range: E., N., and NW. North America.
Look-alikes: *Agaricus** species have free gills and purple-brown spores. *Cortinarius** lack ring. *Pholiota** have smooth spores.
Comments: This good edible is often found in large quantity.

CREPIDOTUS FAMILY
(Crepidotaceae)

This is a family of primarily tropical mushrooms with at least 3 genera represented in North America: *Crepidotus* (more than 100 species), *Simocybe* (1 species), and *Tubaria* (6 species). These brown-spored mushrooms do not fit well into any other family, but do share certain features on which their placement together is based. The species included here typically grow on wood or wood debris, and have smooth spores that lack a pore at the tip.
Crepidotus is a large genus of mostly small, stalkless, shelflike mushrooms that grow on wood. *Tubaria* is a small genus that is not well understood in North America; it includes little brown mushrooms (LBMs) with scruffy caps that grow on sticks and debris or in grass. *Simocybe,* a tropical genus with just a single North American representative, often has an off-center stalk and grows on wood. No family member is reported to be poisonous, but none is a well-known edible.

494   **Flat Crep**
Crepidotus applanatus (Pers. ex Pers.) Kum.
Crepidotaceae, Agaricales

Description: *Small, shell-shaped, stalkless, white to cinnamon mushroom with white gills*

becoming brownish; on deciduous wood.
Cap: ⅜–1⅝" (1–4 cm) wide; shell-shaped to petal-like; margin radially lined when wet; moist, smooth or minutely hairy, white, becoming brownish to cinnamon; attached to wood by hairlike strands.
Gills: radiating from point of attachment, close to crowded, narrow; *white, becoming brownish.*
Spores: 4–5.5 μ; round, minutely spiny, brownish; no pore at tip. Spore print brownish to cinnamon-brown.

Season: July–September.

Habitat: In overlapping clusters, on dead deciduous wood.

Range: Widely distributed in North America.

Look-alikes: *C. malachius* has broad gills. *Pleurotus** species have white spore print.

Comments: This is one of a group of common, whitish, stalkless summer mushrooms that cover deciduous logs. Although none is known to be poisonous, their edibility is undetermined. Care should be taken not to confuse the larger, thicker-fleshed, white-spored oyster mushrooms (*Pleurotus*) with species of this genus.

---

504 Jelly Crep
Crepidotus mollis (Fr.) Stde.
Crepidotaceae, Agaricales

Description: *Stalkless, hairy-scaly, brownish cap with gelatinous flesh and whitish gills becoming brownish; on wood.*
Cap: ⅜–3¼" (1–8 cm) wide; kidney-shaped to semicircular; brown, fibrous to scaly or mostly smooth, moist; olive-brown drying to pale ochre. Flesh gelatinous.
Gills: radiating from point of attachment, crowded to close, narrow to moderately broad; *white, becoming cinnamon-brown.*
Spores: 7–11 × 4.5–6.5 μ; elliptical, smooth, brownish, no pore at tip. Spore print yellowish-brown.

Season: June–October.
Habitat: In overlapping clusters, on dead deciduous wood, especially oak; rarely on conifers.
Range: Widely distributed in North America.
Look-alikes: *Panellus** and *Pleurotus** species have white spores, lack gelatinous flesh.
Comments: Once known as *C. fulvotomentosus.*

---

29 American Simocybe
Simocybe centunculus (Fr.) Kar.
Crepidotaceae, Agaricales

Description: *Small, brownish-olive, velvety cap with yellowish-green to olive-brown gills and curved, sometimes off-center stalk; on wood.*
Cap: ⅜–1⅜" (1–3.5 cm) wide; convex, becoming flat, margin somewhat wavy; velvety, silky to powdery or appearing smooth, translucent and radially lined when moist; dingy to brownish-olive.
Gills: attached, crowded, broad, thick; yellow-gray to olive-brown; edges somewhat toothed.
Stalk: ⅜–1¾" (1–4.5 cm) long, ¹⁄₁₆–¼" (1.5–5 mm) thick; often off-center; curved, hollow, whitish, minutely hairy at top, olive below, with white threads (mycelium) about base.
Spores: 6–8 × 4–5 μ; elliptical, smooth, yellowish-brown; no pore at tip. Spore print ochre to olive-brown.
Season: July–November.
Habitat: On decayed wood, usually deciduous.
Range: Widely distributed in North America.
Look-alikes: *Pluteus californicus* has pink spore print.
Comments: Formerly known as *Naucoria centuncula.*

---

224 Ringed Tubaria
Tubaria confragosa (Fr.) Küh.
Crepidotaceae, Agaricales

Description: *Little brown mushroom with hoary cap, gills slightly descending ringed stalk, and mat of cottony white hairs about base.*
Cap: ⅜–2" (1–5 cm) wide; convex to nearly flat; moist, covered at first with

small, whitish fibers (hoary); dark reddish-cinnamon, fading to pale cinnamon-buff.

Gills: broadly attached or slightly descending stalk, close, moderately broad; off-white, becoming cinnamon to reddish-brown.

Stalk: 1–4″ (2.5–10 cm) long, $\frac{1}{16}$–$\frac{1}{4}$″ (1.5–5 mm) thick, sometimes enlarged downward; off-white and silky above ring, brownish below; dense, white cottony threads (mycelium) at base.

Veil: partial veil whitish, leaving persistent, flaring ring on upper stalk.

Spores: 6.5–9 × 4–6 μ; elliptical, smooth, brownish, with very thin wall, and no pore at tip. Spore print dark reddish-cinnamon.

Season: August–October.

Habitat: On rotting deciduous and coniferous logs.

Range: Widespread in N. North America.

Comments: Once known as *Pholiota confragosa* and *Phaeomarasmius confragosus*. Unlike other tubarias, this has a persistent ring and darker spore print.

---

78 Fringed Tubaria

Tubaria furfuracea (Pers. ex Fr.) Gill.
Crepidotaceae, Agaricales

Description: *Little brown mushroom with scattered whitish fibers on cap, gills slightly descending stalk, and mat of cottony white hairs about base.*

Cap: $\frac{3}{8}$–1$\frac{1}{4}$″ (1–3 cm) wide; convex to flat; moist, brown to cinnamon-brown, fading to buff or pinkish-buff, usually with minute, whitish fibers and small, white patches on cap margin.

Gills: broadly attached or slightly descending stalk; close, moderately broad; light dull yellow, then tawny.

Stalk: 1–2″ (2.5–5 cm) long, $\frac{1}{32}$–$\frac{1}{16}$″ (1–1.5 mm) thick, sometimes slightly enlarged downward; thin, fragile, with minute fibers, silky at top; brownish or dirty white, base with dense, cottony white threads (mycelium).

Veil: partial veil fibrous, whitish, evanescent.

Spores: $6-8.5 \times 4-6 \mu$; elliptical, smooth, very pale brown, with very thin wall collapsing at maturity; no pore at tip. Spore print pale ochre to ochre-brown.

Season: April–November; November–March in California; April–May in mountains.

Habitat: Scattered to numerous, along paths, on wood debris, sticks, or humus.

Range: Widely distributed in North America.

Look-alikes: *T. pellucida* has smaller cap and spores.

Comments: This mushroom is often very common in spring and fall. Identification should be confirmed microscopically.

ENTOLOMA FAMILY
(Entolomataceae)

Most of the mushrooms in this large family are difficult to identify to species. Few are known to be edible and several are known to cause stomach upset. All grow on the ground or on very rotten wood and occur from early spring to fall, being most common and varied after summer rains. The species differ considerably in size, shape, and color, but all have attached gills, a pink to salmon spore print, and spores that are angular to warted. More than half a dozen genera have been described, but only four are recognized here: *Entoloma, Clitopilus, Pouzarella,* and *Rhodocybe.* *Clitopilus* and *Rhodocybe* both have gills that descend the stalk, but their spores differ; in *Clitopilus* they are longitudinally ribbed and angular in end view, while in *Rhodocybe* they are somewhat bumpy to warted. *Pouzarella* species are dark and scaly-hairy, with coarsely stiff hairs about the stalk base. *Entoloma* (called *Rhodophyllus* in Europe) is a complex of several distinct groups of species, some of which have been recognized as distinct genera. The

large, fleshy *Tricholoma*like species
represent the core of the genus
Entoloma; the slender, conical, thin-
fleshed species are sometimes referred to
as *Nolanea;* many of the small, often
bluish, flat caps are placed by some
mycologists in *Leptonia.*

242 Sweetbread Mushroom
Clitopilus prunulus (Scop. ex Fr.) Kum.
Entolomataceae, Agaricales

Description: *Dry, white to grayish cap with white to*
pinkish gills descending stalk.
Cap: 2–4" (5–10 cm) wide; convex to
flat; dry, felty, dull white to ash-gray;
margin wavy with age. Flesh firm,
white. Odor and taste of bread dough.
Gills: nearly descending stalk; white at
first, becoming pale pink.
Stalk: 1⅝–3¼" (4–8 cm) long, ⅛–⅝"
(0.3–1.5 cm) thick; often off-center,
hairless, dull white.
Spores: 10–12 × 5–7 μ; elliptical,
smooth, with longitudinal ridges,
angular in end view. Spore print
salmon-pink.
Edibility: Good.
Season: July–September; November in
Southwest.
Habitat: On the ground in open woods.
Range: Widely distributed in North America.
Look-alikes: *C. orcellus,* if a distinct species, has
sticky, whitish cap. *Entoloma abortivum**
has cucumberlike odor, typically
appears with aborted material. The
poisonous *Clitocybe dealbata** has white
or pale yellow-pink, smooth spores.
Comments: As its common name implies, this is an
excellent edible mushroom.

253, 670 Aborted Entoloma
Entoloma abortivum (Berk. & Curt.) Donk
Entolomataceae, Agaricales

Description: *Gray-brown cap with pinkish gills and*
whitish stalk; found with whitish, bumpy,

roundish to irregular masses.

Cap: 2–4" (5–10 cm) wide; convex, with inrolled margin, becoming flat or knobbed; fibrous to scaly, somewhat zoned; grayish to gray-brown. Odor somewhat cucumberlike.

Gills: attached or descending stalk, close, narrow to moderately broad; pale gray, becoming buff- to salmon-pink.

Stalk: 1–4" (2.5–10 cm) long, ¼–⅝" (0.5–1.5 cm) thick, enlarged at base; solid, somewhat scruffy; white to grayish, with white threads at base.

Aborted Form: 1–4" (2.5–10 cm) wide, 1–2" (2.5–5 cm) high; roundish to elongated, fused clusters of tissue; firm to spongy; whitish and bumpy outside, whitish inside, with marbled, reddish-pink veins.

Spores: 8–10 × 4.5–6 μ; elliptical, angular, 6-sided. Spore print cinnamon to salmon-pink.

Edibility: Good, with caution.

Season: August–December.

Habitat: Several to abundant, on the ground or in leaf litter, or on or about stumps and rotten wood.

Range: E. North America west to Texas.

Look-alikes: None, if gilled mushroom is found with adjacent aborted forms; otherwise it can be mistaken for other species of *Entoloma**.

Comments: Formerly known as *Clitopilus abortivus*. Both the gilled and aborted forms can be found alone. When firm, both are edible and choice, but it is ill-advised to eat the gilled form because of the risk of collecting a poisonous species of *Entoloma*.

---

369 **Midnight-blue Entoloma**
Entoloma madidum (Fr.) Gill.
Entolomataceae, Agaricales

Description: *Tacky, bluish-gray cap with white to bluish-gray gills becoming pinkish.*
Cap: 2–5" (5–12.5 cm) wide; conical to convex or knobbed; sticky, streaked

by small, flattened fibers; bluish-gray or darker, knob brownish. Odor mealy.
Gills: attached, close, broad; white, then bluish-gray or violet to pinkish.
Stalk: 2–4" (5–10 cm) long, ¼–⅝" (0.5–1.5 cm) thick, tapering somewhat downward; solid, hard, fibrous-lined; bluish above, white to yellowish below.
Spores: 7–9 × 6–7.5 μ; round to oval, angular, 5- to 6-sided. Spore print salmon-pink to cinnamon.

Season: September–October.

Habitat: On the ground near conifers; near madrone and tanbark oak in California.

Range: North Carolina; California and Pacific NW.

Comments: This species might be mistaken for a *Russula,* but russulas have white, yellow, or ochre spores and brittle flesh. Although this is reportedly edible, no entoloma except the aborted form of the Aborted Entoloma (*E. abortivum*) can be recommended.

50 Yellow Unicorn Entoloma
Entoloma murraii (Berk. & Curt.) Sacc.
Entolomataceae, Agaricales

Description: *Small, yellow mushroom with long, sharp point on conical cap.*
Cap: ⅜–1¼" (1–3 cm) wide; bell-shaped, becoming conical, with spikelike point on center; silky, shiny; yellow or yellow-orange, fading.
Gills: attached, almost distant, broad; pale yellow, becoming pinkish.
Stalk: 2–4" (5–10 cm) long, 1/16–¼" (1.5–5 mm) thick; dry, yellow.
Spores: 9–12 × 8–10 μ; clearly 4-sided, smooth. Spore print salmon-pink.

Season: June–October.

Habitat: On the ground in damp woods and swamps.

Range: Maine to Alabama, west to Great Lakes.

Comments: Also known as *Nolanea muraii* and *Entoloma cuspidatum.*

49 Salmon Unicorn Entoloma
Entoloma salmoneum (Pk.) Sacc.
Entolomataceae, Agaricales

Description: *Salmon mushroom with bell-shaped to sharply conical cap and slender stalk.*
Cap: ⅜–1⅝″ (1–4 cm) wide; bell-shaped to sharply conical; moist, smooth; salmon, fading with age.
Gills: attached, almost distant, broad, salmon.
Stalk: 2–4″ (5–10 cm) long, ⅟₁₆–⅛″ (1.5–3 mm) thick; fragile, salmon; base with white threads (mycelium).
Spores: 10–12 × 10–12 μ; nearly round when young, then clearly 4-sided. Spore print salmon-pink.
Season: August–October.
Habitat: On the ground in leaf litter, in moss under conifers, on rotten logs, among rhododendron, in mixed woods.
Range: Quebec to North Carolina and Ohio.
Look-alikes: *E. murraii** is yellowish. *Hygrophorus** species have waxy gills and white spores. *Conocybe** have cinnamon-brown spores.
Comments: Also known as *Nolanea salmonea*.

101 Blue-toothed Entoloma
Entoloma serrulatum (Fr.) Hes.
Entolomataceae, Agaricales

Description: *Small, slender, blue mushroom with blue-black, toothed gill edges.*
Cap: ⅜–1⅝″ (1–4 cm) wide; convex, expanding, center becoming sunken and scaled; bluish-black, fading to pale violet-gray.
Gills: attached or slightly descending stalk, close, moderately broad, off-white, then bluish-gray to pinkish; edges dark blue-black, toothed.
Stalk: 1⅝–3¼″ (4–8 cm) long, ⅟₁₆–⅛″ (1.5–3 mm) thick; scruffy near top, smooth below; bluish-black, base with white threads (mycelium).
Spores: 9–12 × 6–8 μ; oval, angular, 5- to 7-sided. Spore print salmon-pink.

Season: June–September.
Habitat: Single to several, on the ground in rich humus or on moss-covered logs.
Range: Quebec to North Carolina; Midwest; Pacific NW.; Mexico.
Look-alikes: Other small blue to blue-black mushrooms with pinkish spores lack blue-black, toothed gill edges. Some species of *Psilocybe** bruise blue.
Comments: Also known as *Leptonia serrulata.* This little blue mushroom is frequently encountered in wet summer woods.

⊗ 275, 367 **Lead Poisoner**
Entoloma sinuatum (Bull. ex Fr.) Kum.
Entolomataceae, Agaricales

Description: *Lead-gray or dirty brownish, slippery cap; flesh with mealy odor and taste.*
Cap: 2¾–6″ (7–15 cm) wide; convex to flat, knobbed at times; margin inrolled, spreading; smooth, slightly hoary, slippery; dull dirty brown to grayish. Flesh thickish near stalk, white to blackish-brown. Odor and taste strongly cucumberlike.
Gills: attached, close to almost distant, broad; pale grayish to yellowish when young, pinkish at maturity.
Stalk: 1⅝–4¾″ (4–12 cm) long, ⅜–1″ (1–2.5 cm) thick; sparsely fibrous, pale grayish; hollow.
Spores: 7–10 × 7–9 μ; angular, roundish, with many oil drops. Spore print salmon-pinkish.
Edibility: Poisonous.
Season: August–September.
Habitat: Scattered or in groups, under hardwoods and conifers.
Range: Widely distributed in North America.
Look-alikes: *Tricholoma** species have white spores. *Clitocybe nuda** lacks cucumberlike odor and taste. *Clitopilus** have gills descending stalk.
Comments: Also called *E. lividum.* This mushroom, like several other species of *Entoloma,* can cause severe gastric upset, vomiting, and diarrhea for 1–2 days.

⊗ 274   **Straight-stalked Entoloma**
Entoloma strictius (Pk.) Sacc.
Entolomataceae, Agaricales

Description: *Gray-brown, knobbed cap with pinkish*
gills and lined, straight stalk.
Cap: 1–2″ (2.5–5 cm) wide; conical,
becoming bell-shaped to convex or flat
with central knob; margin lined when
moist; smooth, drying grayish-brown.
Gills: attached or nearly free,
moderately close and broad; white at
first, becoming pinkish.
Stalk: 2–4″ (5–10 cm) long, ¹⁄₁₆–¼″
(1.5–5 mm) thick; silky, long,
straight, with longitudinal, twisted
lines; off-white to grayish; base covered
with white threads (mycelium).
Spores: 10–13 × 7.5–9 μ; elliptical,
angular, 5- to 6-sided. Spore print
salmon-pink.
Edibility: Poisonous.
Season: April–September.
Habitat: On the ground, in deep humus or on
rotting logs; in wet areas of woods.
Range: E. Canada to Florida, west to Great
Lakes.
Look-alikes: Similar species of *Entoloma** can only be
differentiated microscopically.
*Melanoleuca** species have white spores.
Comments: Also known as *Nolanea strictior.* This
eastern *Entoloma* is reportedly
poisonous.

⊗ 282   **Early Spring Entoloma**
Entoloma vernum Lund.
Entolomataceae, Agaricales

Description: *Small, brown to tan cap with nipple-shaped*
knob, brown gills and stalk.
Cap: 1–2″ (2.5–5 cm) wide; conical to
bell-shaped, with distinct nipple-
shaped knob, becoming broadly convex
at times; margin incurved at first,
becoming wavy; moist, shiny, smooth,
center sometimes wrinkled; dark
brownish to tan, opaque.
Gills: attached, nearly distant, broad;

grayish-buff to pinkish-brown.
Stalk: 1–4" (2.5–10 cm) long, ⅛–⅜"
(0.3–1 cm) thick; scruffy at tip, fibrous
below; brownish; base with white
threads (mycelium).
Spores: 8–11 × 7–8 μ; elliptical,
angular, 6-sided. Spore print salmon.
Edibility: Poisonous.
Season: April–early June.
Habitat: On the ground under conifers, in oak-
hickory woods; along paths.
Range: Canada to New York and Wisconsin.
Look-alikes: Other spring Entolomas* can only be
differentiated microscopically.
Comments: Also known as Nolanea verna. This little
mushroom is one of the first gilled
mushrooms in the Northeast. It can
cause severe gastric upset.

368 Violet Entoloma
Entoloma violaceum Murr.
Entolomataceae, Agaricales

Description: *Dry, scaly, violet- to gray-brown cap with
attached white gills becoming pinkish-
brown.*
Cap: 1–2" (2.5–5 cm) wide; convex to
nearly flat; margin incurved; dry,
densely fibrous to scaly; gray-brown,
with violet tint.
Gills: attached, close, broad; white,
then pinkish-buff to pinkish-brown.
Stalk: 2–4" (5–10 cm) long, ¼–⅜"
(0.5–1 cm) thick, enlarged downward;
somewhat powdery at top, fibrous-scaly
below; violet-gray, base with cottony
white mass (mycelium).
Spores: 8–10 × 5.5–7.5 μ; oval,
angular, 6-sided. Spore print orange-
cinnamon to pinkish- to buff-brown.
Season: May–October.
Habitat: On the ground in leaf litter, in
deciduous or mixed woods.
Range: Massachusetts to North Carolina.
Look-alikes: *Pluteus** species have distinctly free
gills, grow mostly on wood, and have
smooth spores. *Cortinarius violaceus** has
veil over young gills, rust spore print.

Comments: This species appears in northeastern
woods in May before the trees leaf out.
It may be the same as *E. porphyrophaeum.*

---

100 **Hairy-stalked Entoloma**
Pouzarella nodospora (Atk.) Mazz.
Entolomataceae, Agaricales

Description: *Dark brown mushroom with scaly cap and*
hairy stalk with coarse, stiff hairs at base.
Cap: ⅜–2″ (1–5 cm) wide; bell-shaped
to conical, expanding to nearly flat or
with knob; margin incurved, becoming
flat or wavy; dry, fibrous- to densely
scaly; dark gray- to cinnamon-brown.
Gills: deeply attached or nearly free,
almost distant to distant, broad;
brownish.
Stalk: 2–4″ (5–10 cm) long, 1/16–¼″
(1.5–5 mm) thick; dry, hairy; pale
grayish to dark grayish-brown, with
dark brown, coarse, stiff hairs at base.
Spores: 13–16 × 7–9 μ; elliptical,
angular, 6- to 8-sided. Caulocystidia
long, thick-walled, bristlelike. Spore
print pink- to cinnamon-brown.
Season: July–early October.
Habitat: Single to several, on leaf litter or rotten
wood in deciduous woods.
Range: Massachusetts to North Carolina, west
to Ohio.
Look-alikes: *P. dysthales* is much smaller and has
scaly cap and stalk. *P. babingtonii* and
P. strigosissima are smaller and have
bristlelike hairs on cap as well as stalk.
Scaly *Inocybe** species produce brown
spore prints.
Comments: Also known as *Nolanea nodospora.* Like
true entolomas, this species has
pinkish, angular spores.

---

283 **Cracked-cap Rhodocybe**
Rhodocybe mundula (Lasch) Sing.
Entolomataceae, Agaricales

Description: *Dingy white cap, concentrically cracked,*
with pinkish gills descending stalk.

Cap: 1–2" (2.5–5 cm) wide; convex to
flat or sunken, becoming concentrically
cracked or lined; margin inrolled at
first, then wavy; dry, smooth,
sometimes obscurely zoned; dingy
white, becoming ash. *Odor cucumberlike;
taste bitter.*
Gills: deeply descending stalk,
crowded, narrow; off-white to
brownish-gray, then salmon-tinged.
Stalk: 1–2" (2.5–5 cm) long, ¹⁄₁₆–¼"
(1.5–5 mm) thick; flexible, finely
hairy; dingy white, base with white
threads (mycelium).
Spores: 4–6 × 4–4.5 μ; round to oval,
nearly smooth to somewhat bumpy,
angular at ends. Spore print salmon.
Season: July–October.
Habitat: Several to clustered, in woods and open
areas.
Range: Maine to North Carolina, west to
Minnesota; possibly Gulf States.
Comments: Formerly called *Clitopilus noveboracensis.*

GOMPHIDIUS FAMILY
(Gomphidiaceae)

Throughout the fall, a number of dark-
spored mushrooms with a slimy cap and
thick, waxy gills grow under conifers,
especially across the North. These are
the Gomphidiaceae. Some of these
mushrooms are restricted in their
growth to an area near specific conifers.
Many are edible, and none is known to
be poisonous.
Two genera are recognized here:
Gomphidius and *Chroogomphus.* These
genera differ in a number of ways, most
noticeably in the color of the flesh,
which is white in *Gomphidius* and
pinkish to orange in *Chroogomphus.*
Microscopically, both genera are similar
to boletes; some mycologists group this
family together with the Paxillaceae
and the Boletaceae in a single order,
the Boletales.

220 Brownish Chroogomphus
Chroogomphus rutilus (Schaeff. ex Fr.) O.K.M.
Gomphidiaceae, Agaricales

Description: *Slimy, brownish cap with brownish gills descending ochre stalk; under conifers.*
Cap: 1–5″ (2.5–12.5 cm) wide; peglike, expanding to convex, with a small, sharp knob; margin incurved, becoming pendant; slimy to tacky; reddish- or ochre-brown. Flesh pinkish.
Gills: descending stalk, distant, broad, thick; ochre-buff, darkening to cinnamon-brown.
Stalk: 2–7″ (5–18 cm) long, ¼–¾″ (0.5–2 cm) thick, narrowing toward base; dryish, ochre-buff to pink-tinged.
Veil: partial veil hairy, dry; leaving evanescent, hairy zone on upper stalk.
Spores: 14–22 × 6–7.5 μ; long, elliptical, smooth. Spore print smoky-gray to blackish.
Edibility: Edible.
Season: August–October; December–January in Southwest.
Habitat: Single to abundant, under pines.
Range: Widely distributed in N. North America; Southwest; Mexico.
Look-alikes: *C. vinicolor** is smaller, more conical, and has thick-walled cystidia.
Comments: Formerly known as *Gomphidius rutilus*. All *Chroogomphus* mushrooms are believed to be edible, and this is one of the better-tasting species.

---

221 Woolly Chroogomphus
Chroogomphus tomentosus (Murr.) O.K.M.
Gomphidiaceae, Agaricales

Description: *Dry, ochre-orange cap, gills, and stalk; under conifers.*
Cap: 1–3″ (2.5–7.5 cm) wide; peglike, expanding to convex or flat; dry, woolly; ochre-orange. Flesh orange.
Gills: descending stalk, distant, broad, thick; yellow-orange, becoming gray.
Stalk: 2–7″ (5–18 cm) long, ⅜–⅝″ (1–1.5 cm) thick, narrowing toward

base; dry, scruffy; ochre-orange.
Veil: partial veil thin, leaving no ring.
Spores: 15–25 × 6–9 μ; long,
elliptical, smooth. Spore print smoky-
gray to blackish.

Edibility: Edible.

Season: Late August–December.

Habitat: Single to scattered under conifers,
especially hemlock, Douglas-fir, and
grand fir.

Range: Rocky Mountains and West Coast.

Look-alikes: *C. leptocystis* has grayer cap.

Comments: Formerly called *Gomphidius tomentosus*.
Its dry cap is unusual in this family.

---

219 **Wine-cap Chroogomphus**
Chroogomphus vinicolor (Pk.) O.K.M.
Gomphidiaceae, Agaricales

Description: *Slimy, ochre to burgundy cap with ochre
gills descending stalk; under conifers.*
Cap: 1–3″ (2.5–7.5 cm) wide; peglike,
expanding to convex or turban-shaped,
often with a small, sharp knob; margin
incurved, becoming long-hanging;
slimy to tacky, soon dry; ochre,
becoming burgundy-red. Flesh orange
to flushed red.
Gills: descending stalk, distant, broad,
thick; ochre, becoming smoky-ochre.
Stalk: 2–4″ (5–10 cm) long, ¼–¾″
(0.5–2 cm) thick, narrowing toward
base; dry, ochre-buff, then reddish.
Veil: partial veil hairy, dry; leaving
evanescent, hairy zone on upper stalk.
Spores: 17–23 × 4.5–7.5 μ; long,
elliptical, smooth. Spore print smoky-
gray to blackish.

Edibility: Edible.

Season: Late August–late October; November–
March in Southwest.

Habitat: Scattered to numerous, under pine.

Range: Widely distributed in North America.

Look-alikes: *C. rutilus** is brownish to reddish-
brown. *C. pseudovinicolor* has much
thicker stalk and dry cap.

Comments: Formerly known as *Gomphidius vinicolor*.

218 Slimy Gomphidius
Gomphidius glutinosus (Schaeff. ex Fr.) Fr.
Gomphidiaceae, Agaricales

Description: *Slimy, brownish cap with grayish,*
descending gills and yellowish stalk base;
under conifers.
Cap: 1–4″ (2.5–10 cm) wide; peglike,
expanding to convex or flat; margin
upturned; slimy, smooth; gray-brown
to purple-gray or reddish-brown, often
spotted or stained blackish.
Gills: descending stalk, close, broad,
thick; grayish-white, becoming smoky.
Stalk: 2–4″ (5–10 cm) long, ¼–¾″
(0.5–2 cm) thick, narrowing toward
base; white, base yellow.
Veil: partial veil slimy, colorless;
leaving slimy ring, blackening from
spores, on upper stalk.
Spores: 15–21 × 4–7.5 μ; elliptical,
smooth. Spore print smoky-gray to
blackish.
Edibility: Edible.
Season: June–November.
Habitat: Under conifers, especially spruce.
Range: Throughout N. North America; south
in mountainous coniferous woods.
Look-alikes: *G. oregonensis* is pinkish-orange to red-
brown, grows in clusters, usually set
deep in soil. *G. largus* is much larger
and has pinkish to reddish cap.
Comments: This is the most common and widely
distributed species of *Gomphidius*. There
are several varieties, which differ in cap
color but are similar microscopically.

216 Clustered Gomphidius
Gomphidius oregonensis Pk.
Gomphidiaceae, Agaricales

Description: *Slimy, pinkish-orange cap with yellowish*
lower stalk; clustered under conifers.
Cap: 1–6″ (2.5–15 cm) wide; peglike,
expanding to convex or flat; slimy,
smooth, pinkish-orange to reddish-
brown, sometimes becoming splotchy
with age.

Gills: slightly descending stalk, almost distant to distant, broad, thick, some forked; whitish, becoming grayish.
Stalk: 2–5″ (5–12.5 cm) long, 3/8–2″ (1–5 cm) thick, narrowing toward base; whitish above, yellowish below; set deep in soil.
Veil: partial veil slimy, colorless; leaving slimy ring, blackening from spores, on upper stalk.
Spores: 10–13 × 4.5–8 μ; long, elliptical, smooth. Spore print smoky-gray to black.

Edibility: Edible.
Season: August–November.
Habitat: Clustered, rarely single, under conifers.
Range: Rocky Mountains to West Coast.
Look-alikes: *G. glutinosus** has purple to brownish cap, not typically clustered.
Comments: Like other species of *Gomphidius,* this resembles a *Hygrophorus.*

217 Rosy Gomphidius
Gomphidius subroseus Kauff.
Gomphidiaceae, Agaricales

Description: *Slimy, reddish cap with whitish gills descending whitish stalk with slimy veil remnants; under conifers.*
Cap: 1⅝–2⅜″ (4–6 cm) wide; peglike, expanding to convex, flat, or somewhat sunken; slimy, smooth; pinkish to red.
Gills: descending stalk, close to almost distant, broad; whitish, then smoky.
Stalk: 1⅜–3″ (3.5–7.5 cm) long, ¼– ⅝″ (0.5–1.5 cm) thick, narrowing toward base; white to amber at base.
Veil: partial veil slimy, colorless; leaving thin, slimy ring, blackening from spores, on upper stalk.
Spores: 15–20 × 4.5–7 μ; elliptical, long, smooth. Spore print blackish.

Edibility: Edible.
Season: June–October in Rocky Mountains; September–December on West Coast.
Habitat: Under conifers, especially Douglas-fir.
Range: Widely distributed in N. North

America; south in mountains to North
Carolina and California.

Comments: This is primarily a western species, but
it can be found in eastern mountainous
coniferous woods.

HYGROPHORUS FAMILY
(Hygrophoraceae)

These are mostly small, waxy, brightly
colored, white-spored mushrooms that
grow abundantly on moss and on the
ground in wet, open places as well as in
deciduous and coniferous forests. Many
are very common and widely
distributed. Some appear in late spring,
while some others occur in late fall and
are often the last ground mushrooms of
the year. The caps are typically conical
to convex, and mostly thin-fleshed. The
gills, which are attached to and
somewhat descend the stalk, are
triangular, thick at the base and
narrowing to the edge; most are waxy.
In general, these mushrooms fall into 2
broad groups: brightly colored species
with gill tissue that, microscopically,
appears parallel; and white, gray, or
brown species, some of which have
veils, and all of which have divergent
gill tissue. None is known to be
poisonous, although several are too
bitter to eat or are otherwise
unpalatable. Only one genus,
Hygrophorus (200 species), is recognized
here. Although it is sometimes divided
into six or more distinct genera, all
species can be identified in the field as
hygrophori by their color, texture, or
habitat.

---

360 **Gray Almond Waxy Cap**
Hygrophorus agathosmus Fr.
Hygrophoraceae, Agaricales

Description: *Slimy, ash-gray cap with white to grayish
gills somewhat descending stalk; fragrant.*

Cap: 1–4" (2.5–10 cm) wide; convex, with inrolled margin, becoming flat, slightly sunken, or knobbed; slimy to sticky, smooth, ash-gray. Flesh white. Odor fragrant, of bitter almonds.
Gills: attached or slightly descending stalk, close to distant, narrow, thin; waxy, white, becoming dirty gray.
Stalk: 2–4" (5–10 cm) long, ¼–¾" (0.5–2 cm) thick; dry or moist, smooth, white, ash-gray with age.
Spores: 7–10.5 × 4.5–5.5 μ; elliptical, smooth. Spore print white.

Edibility: Edible.

Season: August–January.

Habitat: Scattered under spruce, pine, and Douglas-fir; also in mixed woods.

Range: N. North America south to Maryland and California.

Look-alikes: *H. bakerensis** has cinnamon-buff cap.

Comments: This species appears in the late fall in coniferous forests.

273 Tawny Almond Waxy Cap
Hygrophorus bakerensis A.H.S. & Hes.
Hygrophoraceae, Agaricales

Description: *Slimy, brownish cap with white gills descending whitish stalk; fragrant.*
Cap: 2–6" (5–15 cm) wide; blunt, with incurved, cottony margin, becoming flat; slimy when wet, with small, flattened fibers beneath slime; brown to tawny-olive or cinnamon-buff, fading toward whitish margin. Odor fragrant, of bitter almonds.
Gills: descending stalk, close to almost distant, narrow to broad, moderately thick but thin-edged; waxy, cream.
Stalk: 2–6" (5–15 cm) long, ⅜–1" (1–2.5 cm) thick; dry, unpolished, cottony; white to pinkish-buff.
Spores: 7–10 × 4.5–6 μ; elliptical, smooth. Spore print white.

Edibility: Edible.

Season: September–December.

Habitat: Scattered to numerous, under conifers.

Range: Widespread in W. North America.
Look-alikes: *H. agathosmus** is ash-gray.
Comments: Though this species' fragrance is pleasing, its flavor is very poor.

---

365 Dusky Waxy Cap
Hygrophorus camarophyllus (Fr.) Dumée
Hygrophoraceae, Agaricales

Description: *Slimy to dry, brownish-gray cap with white to grayish gills somewhat descending dry, brownish-gray stalk.*
Cap: 2–5″ (5–12.5 cm) wide; blunt to convex or slightly knobbed to flat; margin downy; sticky but soon dry, smooth but appearing streaked with small fibers; brownish-gray. Faint odor of coal tar.
Gills: slightly descending stalk, close to almost distant, moderately broad, thin; waxy, white or slightly grayish.
Stalk: 1–6″ (2.5–15 cm) long, ⅜–¾″ (1–2 cm) thick; dry, silky above, smooth below; brownish-gray.
Spores: 7–9 × 4–5 μ; elliptical, smooth. Spore print white.
Edibility: Edible.
Season: July–November.
Habitat: Scattered to numerous, under conifers.
Range: N. North America.
Look-alikes: *H. calophyllus* is always slimy, has pale pinkish gills and slightly fragrant odor.
Comments: This *Hygrophorus* can be found with pine and spruce near snowbanks in the fall.

---

56 Chanterelle Waxy Cap
Hygrophorus cantharellus (Schw.) Fr.
Hygrophoraceae, Agaricales

Description: *Small, dry, red-orange cap with orange-yellow gills descending dry stalk.*
Cap: ⅜–1⅜″ (1–3.5 cm) wide; convex to flat, becoming sunken in center; dry, silky to finely scruffy; scarlet, ochre, or orange, paler with age.
Gills: descending stalk, almost distant

to distant; broad, thick but thin-edged; waxy, orange-yellow, paler than cap. Stalk: 1–4" (2.5–10 cm) long, ⅟₁₆–¼" (1.5–5 mm) thick; dry, smooth; color of cap or paler, with whitish base. Spores: 7–12 × 4–8 μ; elliptical, smooth. Spore print white.

Edibility: Edible.

Season: July–October.

Habitat: Several to somewhat clustered, on soil and decaying or moss-covered logs.

Range: E. North America.

Look-alikes: *H. miniatus** has scarlet gills not descending stalk. Chanterelles have forked, blunt-edged, gill-like ridges.

Comments: Also known as *Hygrocybe cantharellus*. This mushroom often grows in moss and is one of the few hygrophori that grow on decaying wood.

---

260 **Golden-spotted Waxy Cap**
Hygrophorus chrysodon (Fr.) Fr.
Hygrophoraceae, Agaricales

Description: *Slimy, white cap with golden granules, and white gills descending white stalk.*
Cap: 1–3" (2.5–7.5 cm) wide; convex to slightly knobbed; sticky when wet, shiny when dry; margin inrolled, tufted-hairy, white; center white, with golden granules.
Gills: descending stalk, distant; broad, thick but thin-edged; waxy, white or with edges powdered yellow.
Stalk: 1–3" (2.5–7.5 cm) long, ¼–⅝" (0.5–1.5 cm) thick; white, with golden granules at top.
Spores: 7–10 × 3.5–5 μ; elliptical, smooth. Spore print white.

Edibility: Edible.

Season: July–September in western mountains; overwinters in California.

Habitat: Single to scattered, on soil in coniferous woods; with live oak, madrone, and manzanita in California.

Range: Coniferous areas of North America.

Comments: The scattered golden granules make this an easy species to recognize.

65   **Scarlet Waxy Cap**
      Hygrophorus coccineus (Fr.) Fr. sensu Rick.
      Hygrophoraceae, Agaricales

Description: *Moist, blood-red cap and stalk and red to*
               yellow-orange gills.
               Cap: 1–2″ (2.5–5 cm) wide; conical,
               knobbed; margin incurved to
               spreading; moist, smooth, blood-red.
               Gills: attached, close to almost distant;
               broad, thick but thin-edged; waxy, red
               to yellow-orange.
               Stalk: 1–3″ (2.5–7.5 cm) long, ⅛–⅜″
               (0.3–1 cm) thick; moist, blood-red to
               yellowish near base; hollow.
               Spores: 7–10.5 × 4–5 μ; elliptical,
               smooth. Spore print white.
Edibility: Edible.
Season: July–November.
Habitat: Deciduous and coniferous woods.
Range: E. North America; Texas and Gulf
              Coast; West Coast.
Look-alikes: *H. puniceus,* perhaps a variant, has
              sticky cap.
Comments: Also known as *Hygrocybe coccinea.* This
              edible species reportedly favors
              magnolia in the South and redwood in
              California, but it also occurs in
              clearings and meadows in the
              Northeast.

⊗ 71, 72   **Witch's Hat**
            Hygrophorus conicus (Fr.) Fr.
            Hygrophoraceae, Agaricales

Description: *Slightly sticky, reddish-orange, conical cap*
               and paler stalk, bruising black.
               Cap: ¾–3½″ (2–9 cm) wide; peaked or
               conical with knob; sticky when moist,
               smooth or slightly streaked with small
               fibers; red to orange, bruising black.
               Gills: nearly free, close, broad; white or
               pale yellow, becoming yellow, pinkish-
               yellow, or olive-orange, bruising black.
               Stalk: ¾–8″ (2–20 cm) long, ¼–⅝″
               (0.5–1.5 cm) thick; not sticky,
               somewhat longitudinally lined and
               twisted; orange-yellow to whitish at

base, bruising black; hollow.
Spores: 8–14 × 5–7 μ; elliptical, smooth. Spore print white.

Edibility: Possibly poisonous.

Season: July–September in East; November–April in California.

Habitat: Single to scattered, in coniferous woods; reported under redwood.

Range: Widely distributed in North America.

Look-alikes: *H. acutoconicus* does not bruise black. *Entoloma salmoneum** does not bruise black, and has pinkish, angular spores.

Comments: Also known as *Hygrocybe conica.* Although eaten by many people, this species is reported to have caused 4 deaths in China; some believe it is also hallucinogenic.

236 White Waxy Cap
Hygrophorus eburneus (Fr.) Fr.
Hygrophoraceae, Agaricales

Description: *Slimy, white cap with waxy, white gills descending slimy, white stalk.*
Cap: 1–4″ (2.5–10 cm) wide; blunt to convex with inrolled margin, becoming flat to knobbed or with sunken center and with elevated margin; slimy to sticky, smooth to silky; white to pale yellowish-pink.
Gills: descending stalk, nearly distant to distant; moderately broad, thick but thin-edged; waxy, pure white.
Stalk: 2–6″ (5–15 cm) long, ⅟₁₆–⅝″ (0.15–1.5 cm) thick, sometimes gradually tapering; slimy, pure white.
Spores: 6–9 × 3.5–5 μ; elliptical, smooth. Spore print white.

Edibility: Edible.

Season: August–January.

Habitat: Single to scattered, in coniferous woods, beech woods, and grassy areas; also in oak-pine woods.

Range: N. North America south to North Carolina, Colorado, and California.

Comments: This is a widely distributed, commonly collected edible. Any look-alikes will lack the slimy cap and stalk.

48 Golden Waxy Cap
Hygrophorus flavescens (Kauff.) A.H.S. & Hes.
Hygrophoraceae, Agaricales

Description: *Slimy, orange-yellow cap with yellow gills and orange-yellow stalk with white base.*
Cap: 1–2¾" (2.5–7 cm) wide; broadly convex, with incurved margin, becoming flat or sunken in center, or knobbed; margin lined; sticky, then dry and shiny; orange to yellow.
Gills: attached, close to nearly distant; broad, thick but thin-edged; waxy, yellowish.
Stalk: 2–3" (5–7.5 cm) long, ¼–⅝" (0.5–1.5 cm) thick, often compressed; smooth, moist; yellowish to orange, with whitish base.
Spores: 7–9 × 4–5 μ; elliptical, smooth. Spore print white.

Edibility: Edible.

Season: June–November; January in California.

Habitat: Scattered to numerous, on soil and humus in deciduous and mixed woods; often in wet, mossy areas.

Range: E. North America to Michigan, Texas, and West Coast.

Look-alikes: *H. chlorophanus* is slimy and lacks whitish stalk base.

Comments: Also known as *Hygrocybe flavescens*. This species seems to be associated with leaf litter rather than trees.

261 Yellow-centered Waxy Cap
Hygrophorus flavodiscus Frost apud Pk.
Hygrophoraceae, Agaricales

Description: *Slimy, ochre-yellowish cap with pinkish-white gills descending slimy white stalk.*
Cap: 1–3" (2.5–7.5 cm) wide; convex to flat; margin white-fibrous; sticky, drying in radial streaks, smooth; white, with reddish or pale yellow center, drying ochre-orange or orange-buff.
Gills: descending stalk, nearly distant; medium broad, thick but thin-edged; waxy, pinkish, fading to whitish.
Stalk: 1–3" (2.5–7.5 cm) long, ¼–⅝"

(0.5—1.5 cm) thick; enclosed in slimy
sheath; fibrous to scaled at top; white or
sometimes yellow-stained.
Veil: universal veil leaving slimy white
sheath on stalk.
Spores: 6—8 × 3.5—5 μ; elliptical,
smooth. Spore print white.

Edibility: Edible.

Season: October—November.

Habitat: Scattered to clustered, in pine woods.

Range: Maine to Pennsylvania; Pacific NW.

Look-alikes: *H. gliocyclus* has cream-buff cap and
yellowish gills.

Comments: This is a late fall *Hygrophorus* that often
does not appear until after a frost. It
should be cleaned before it is cooked.

---

304 Late Fall Waxy Cap
Hygrophorus hypothejus (Fr.) Fr.
Hygrophoraceae, Agaricales

Description: *Slimy, brownish to olive-yellow cap; whitish
gills descend slimy, yellowish stalk.*
Cap: 1—3″ (2.5—7.5 cm) wide;
somewhat knobbed to convex or flat, or
with sunken center; slimy, smooth, or
with small, firmly attached fibers near
margin; brown to dark olive-brown in
center; greenish-yellow to ochre,
then bright ochre-orange to scarlet.
Gills: descending stalk, nearly distant
to distant; narrow, thick but thin-
edged; waxy, white to pale yellow.
Stalk: 1—6″ (2.5—15 cm) long, ⅛—⅝″
(0.3—1.5 cm) thick, tapering
downward; slimy and dark brown to
olive-yellow below; yellow above, with
fibrous, evanescent zone.
Veil: partial veil cottony, evanescent.
Spores: 7—9 × 4—5 μ; elliptical,
smooth. Spore print white.

Edibility: Edible.

Season: June—December; common in late fall.

Habitat: Scattered to numerous, in bogs and
under conifers, often 2-needle pines.

Range: Widely distributed in North America.

Look-alikes: *H. fuligineus* has blackish-brown to
brownish-olive cap.

Comments: Once the slime is removed, this
mushroom is edible and digestible.

---

366 Inocybelike Waxy Cap
Hygrophorus inocybiformis A.H.S.
Hygrophoraceae, Agaricales

Description: *Dry, fibrous, dark gray cap with whitish*
gills somewhat descending dry, whitish
stalk, covered with grayish-brown fibers.
Cap: 1–3″ (2.5–7.5 cm) wide; conical,
with incurved margin, expanding to
knobbed, flat, or with sunken center;
margin fringed; dry, with small, dark
fibers or somewhat scaly; pale to dark
gray or gray-brown.
Gills: somewhat descending stalk,
nearly distant; broad, thick but thin-
edged; waxy, off-white to grayish-buff.
Stalk: 1–3″ (2.5–7.5 cm) long, ¼–⅝″
(0.5–1.5 cm) thick; smooth to fibrous;
streaked with small, dark fibers from
base to zone left by broken veil; white
and smooth to silky above.
Veil: partial veil leaving fibrous zone on
cap margin and upper stalk.
Spores: 9–14 × 5–8 μ; elliptical,
smooth. Spore print white.
Edibility: Edible with caution.
Season: July–November.
Habitat: Scattered to numerous, under spruce
and balsam fir.
Range: NW. North America.
Look-alikes: Grayish *Inocybe** species have brownish
spore print. Grayish *Tricholoma** have
notched, thin, nonwaxy gills.
Comments: Because of its resemblance to poisonous
inocybes and tricholomas, beginners
should avoid this species.

---

53 Orange-gilled Waxy Cap
Hygrophorus marginatus Pk.
Hygrophoraceae, Agaricales

Description: *Moist, orange-yellow cap, fading to yellow,*
with bright orange gills.
Cap: ⅜–2″ (1–5 cm) wide; conical to

convex or bell-shaped, sometimes flat or
with low knob; moist, smooth; deep
chrome to orange, fading to yellow.
Gills: attached, nearly distant; broad,
thick but thin-edged; waxy, brilliant
orange.
Stalk: 2−4″ (5−10 cm) long, ⅛−¼″
(3−5 mm) thick; smooth, buff to pale
orange-yellow.
Spores: 7−10 × 4−6 μ; elliptical,
smooth. Spore print white.

Edibility: Edible.

Season: June−October.

Habitat: Single to scattered, on soil and humus
in mixed woods.

Range: E. and NW. North America.

Look-alikes: *H. miniatus** has fading scarlet gills. *H.
flavescens** has yellow gills.

Comments: Also known as *Hygrocybe marginata* and
Humidicutis marginata. This species
retains its bright orange gill color.

---

69 Fading Scarlet Waxy Cap
Hygrophorus miniatus Fr.
Hygrophoraceae, Agaricales

Description: *Small, dry, red to orange-red mushroom.*
Cap: ¾−1⅜″ (2−4 cm) wide; broadly
convex, with incurved margin,
becoming flat or with sunken center;
dry, smooth to minutely fibrous; *scarlet,*
fading to orange or pale yellow.
Gills: attached or slightly descending
stalk, close to nearly distant; broad,
thick but thin-edged; waxy, scarlet to
orange, fading to yellow.
Stalk: 1−2″ (2.5−5 cm) long, ⅛″ (3
mm) thick; smooth, scarlet, fading.
Spores: 6−10 × 4−6 μ; elliptical,
smooth. Spore print white.

Edibility: Edible.

Season: July−November; winter in California.

Habitat: Scattered on soil, among mosses, or on
rotting logs, in deciduous and mixed
woods.

Range: Widely distributed in North America.

Look-alikes: *H. ruber* has somewhat slimy, conical
cap. *H. cantharellus** has gills strongly

descending stalk. *Cantharellus** species
have forked, thick-edged, gill-like
ridges descending stalk.

Comments: Also known as *Hygrocybe miniata*. This
variable species is edible but flavorless.

---

199 Slimy-sheathed Waxy Cap
Hygrophorus olivaceoalbus Fr.
Hygrophoraceae, Agaricales

Description: *Slimy, smoky-gray cap with thick, whitish
gills slightly descending slime-sheathed
stalk.*
Cap: 1–5" (2.5–12.5 cm) wide; convex
or knobbed, becoming flat; slimy, with
minute, smoky-gray to black-streaked
fibers beneath slime; amber to black in
center, ash-gray toward margin.
Gills: attached or slightly descending
stalk, close to nearly distant;
moderately broad, thick but thin-
edged; waxy, pure white to ash.
Stalk: 3–6" (7.5–15 cm) long, ⅜–1¼"
(1–3 cm) thick; white above ring;
slimy, with blackish markings below.
Veils: universal veil leaving cap and all
but top of stalk slimy. Partial veil
fibrous, blackish; leaves sheath from
base to near top.
Spores: 9–12 × 5–6 μ; elliptical,
smooth. Spore print white.
Edibility: Edible with caution.
Season: July–December.
Habitat: Scattered to numerous or clustered,
under Sitka spruce on Pacific Coast;
near Engleman spruce in Rockies.
Range: N. North America south to New York,
Michigan, Rockies, N. California.
Comments: This mushroom is often abundant in
late summer.

---

309 Salmon Waxy Cap
Hygrophorus pratensis Fr.
Hygrophoraceae, Agaricales

Description: *Dry, orange to buff cap with pale orange to
buff gills descending whitish stalk.*

Cap: 1–3″ (2.5–7.5 cm) wide; blunt, convex, flat or with central knob; dry, smooth to minutely fibrous; salmon-orange to tawny, fading to ochre-buff.
Gills: descending stalk, nearly distant to distant; moderately broad, thick but thin-edged; waxy, pinkish-buff to off-white.
Stalk: 1–3″ (2.5–7.5 cm) long, ¼–¾″ (0.5–2 cm) thick, sometimes tapering downward; dry, smooth; whitish to orange.
Spores: 5.5–8 × 3.5–5 μ; elliptical to almost round, smooth. Spore print white.

Edibility: Choice.

Season: May–December; November–March in California.

Habitat: Single to scattered or clustered, in open places or grassy areas; also in woods.

Range: Widely distributed in North America.

Look-alikes: Chanterelles have forked, blunt-edged, gill-like ridges.

Comments: Also known as *Camarophyllus pratensis*. This mushroom is a choice edible.

103 Parrot Mushroom
Hygrophorus psittacinus Fr.
Hygrophoraceae, Agaricales

Description: *Small, slimy, green cap and stalk fading to ochre, yellow, or pink.*
Cap: ⅜–1¼″ (1–3 cm) wide; conical to bell-shaped or flat, margin translucent and radially lined; slimy to sticky, shiny when dry; dark green, fading to ochre-buff, tawny, or pinkish.
Gills: attached or slightly descending stalk, almost distant; narrow to broad, thick but thin-edged; waxy, greenish to reddish or yellow.
Stalk: 1–3″ (2.5–7.5 cm) long, 1/16–¼″ (1.5–5 mm) thick; slimy, sticky; green, becoming yellow or orange, then pink.
Spores: 6.5–10 × 4–6 μ; elliptical, smooth. Spore print white.

Edibility: Edible.

Season: June—September; December—February
in California.

Habitat: Scattered to numerous, in coniferous
and mixed woods, in pastures, along
roads, and in mossy banks.

Range: Widely distributed in North America.

Comments: Also known as *Hygrocybe psittacina*. This
highly distinctive mushroom changes
color with age; older specimens do not
show any green pigment and are likely
to be misidentified.

---

246   Turpentine Waxy Cap
Hygrophorus pudorinus Fr.
Hygrophoraceae, Agaricales

Description: *Slimy, pinkish-buff cap with white to
pinkish-tinged gills descending dry, white to
pinkish-tinged stalk; odor resinous.*
Cap: 2—4″ (5—10 cm) wide; convex to
nearly flat; margin inrolled, downy,
white; sticky, smooth; pale tan to
pinkish-buff. Taste turpentinelike.
Gills: slightly descending stalk, nearly
distant; narrow, thick but thin-edged;
waxy, white to pinkish-tinged.
Stalk: 2—4″ (5—10 cm) long, ⅜—¾″
(1—2 cm) thick, sometimes tapering
downward; stout, dry; white to buff or
pink, with white tufts above, becoming
reddish with age or on drying;
yellowish to orange with KOH.
Spores: 6.5—9.5 × 4—5.5 μ; elliptical,
smooth. Spore print white.

Edibility: Edible.

Season: August—October; to January in
California.

Habitat: Scattered to numerous, with spruce or
hemlock; sometimes in bogs.

Range: N. North America south to New York,
Arizona, and N. California.

Look-alikes: Similar *Hygrophorus** species lack odor
and taste of turpentine and KOH
reaction on stalk.

Comments: This common, widely distributed
species has several varieties, which are
distinguished by color and odor. The
cooked mushroom is palatable.

345 Russulalike Waxy Cap
Hygrophorus russula (Fr.) Quél.
Hygrophoraceae, Agaricales

Description: *Large, slimy, pink to purplish-red cap with white to pinkish gills somewhat descending pink-streaked, whitish stalk.*
Cap: 2–5″ (5–12.5 cm) wide; convex, with inrolled margin, expanding to nearly flat; sticky, becoming dry, smooth to flattened-fibrous; pink to purplish- or pinkish-red, often streaked with small, purple-red fibers.
Gills: bluntly attached or descending stalk, close; narrow to broad, thick but thin-edged; waxy, white, becoming pink-tinged or spotted purplish-red.
Stalk: 1–3″ (2.5–7.5 cm) long, ⅝–1⅜″ (1.5–3.5 cm) thick, sometimes tapering downward; dry, smooth; white, staining or streaking pinkish.
Spores: 6–8 × 3–5 μ; elliptical, smooth. Spore print white.
Edibility: Good.
Season: September–October in East; November–February in California.
Habitat: Scattered to numerous, with oak.
Range: Widely distributed in North America.
Look-alikes: *H. erubescens* has nearly distant gills, bruises yellow. *H. purpurescens* has partial veil leaving evanescent ring. *H. amarus,* pink- and yellow-tinged, is very bitter. All grow under conifers.
Comments: Formerly known as *Tricholoma russula.* The gills of this meaty edible do not look waxy, but when rubbed between finger and thumb, they usually have a waxy feel.

325 Larch Waxy Cap
Hygrophorus speciosus Pk.
Hygrophoraceae, Agaricales

Description: *Slimy, orange-red cap with whitish gills somewhat descending slimy, sheathed, whitish to orange stalk.*
Cap: 1–2″ (2.5–5 cm) wide; convex but expanding, sometimes broadly

knobbed; slimy-sticky, *bright orange to orange-red or golden-yellow*.

Gills: attached or descending stalk, distant to almost distant; narrow, thick but thin-edged; waxy, white to yellow.

Stalk: 2–4″ (5–10 cm) long, ⅛–⅜″ (0.3–1 cm) thick; covered by slimy veil; white to yellow-orange or orange.

Veil: universal veil leaving cap and most of stalk slimy.

Spores: 8–10 × 4.5–6 μ; elliptical, smooth. Spore print white.

Edibility: Edible.

Season: August–October.

Habitat: Single, scattered, or clustered, under conifers, especially larch, in bogs.

Range: E. North America, Rocky Mountains, and Pacific NW.

Comments: A variety, *kauffmanii*, is somewhat larger and has a fibrous partial veil that leaves an evanescent ring zone.

PAXILLUS FAMILY
(Paxillaceae)

The Paxillaceae are traditionally defined as a family of brown-spored mushrooms with a gill layer that separates readily from the flesh of the cap. This feature, atypical of gilled mushrooms, allies the Paxillus family to the boletes. *Phylloporus,* once called *Paxillus,* is transferred by some to the family Boletaceae because its spores are like those of the boletes and it produces pores as well as gills. Some white-spored genera have been transferred to the Paxillaceae because of microscopic and chemical features. Three genera are recognized here. Although some people eat these mushrooms, none can be recommended. Some innocuous species have poisonous look-alikes in other families. One species, *Paxillus involutus,* is well liked in some places but is known to have a cumulative toxic effect that causes a life-threatening illness.

311 False Chanterelle

Hygrophoropsis aurantiaca (Wulf. ex Fr.)
Maire
Paxillaceae, Agaricales

Description: *Dark orange cap with forked, orange gills descending orange to orange-brown stalk.*

Cap: 1–2⅜" (2.5–6 cm) wide; convex, expanding to flat or sunken, with inrolled margin at first; dry, somewhat feltlike, especially in center; dark orange to orange-brown.
Gills: descending stalk, crowded, narrow, forked, typically blunt-edged; bright orange.
Stalk: 1–4" (2.5–10 cm) long, ¼–⅝" (0.5–1.5 cm) thick, thickened toward base; sometimes off-center; spongy, dry, somewhat hairy, orange-brown to yellowish.
Spores: 5–8 × 3–4.5 μ; elliptical, smooth; many dextrinoid. Spore print white to cream.

Edibility: Edible with caution.

Season: August–November; overwinters in California.

Habitat: Single to scattered, on ground or decayed coniferous wood, often pine.

Range: Widely distributed in North America.

Look-alikes: The poisonous *Omphalotus olearius** is smooth, orange, with sharp-edged, unforked gills and grows in united clusters on wood. *Cantharellus cibarius** is yellow, with thick-edged gills and fruity aroma. *Hygrophorus pratensis** is smooth, with waxy gills, no brown pigment; usually in grassy areas.

Comments: This edible mushroom was once listed as poisonous, probably because it was mistaken for the poisonous Jack O'Lantern (*Omphalotus olearius*).

434 Fragrant Hygrophoropsis

Hygrophoropsis olida (Quél.) Mét.
Paxillaceae, Agaricales

Description: *Strongly fragrant, pinkish mushroom; under conifers.*

Cap: ⅜–1⅜" (1–3.5 cm) wide; convex
to flat and sunken in center, becoming
funnel-shaped; margin turned in and
down, becoming wavy; moist to dry,
smooth, pinkish, fading to buff. Odor
strongly fragrant.
Gills: descending stalk, close, narrow,
thick, sharp-edged, some forked, with
crossveins; pinkish to whitish.
Stalk: ⅜–1¼" (1–3 cm) long, 1⁄16–¼"
(1.5–5 mm) thick, narrowing toward
base; central or off-center, pinkish.
Spores: 3–5 × 2.5–3 μ; elliptical,
smooth; dextrinoid. Spore print white.

Season: July–October.

Habitat: Several to many, on the ground under
conifers, including spruce, fir, pine,
and hemlock.

Range: N. North America south to Colorado
and California in mountains.

Comments: Formerly called *Cantharellus morgani*
and *Clitocybe morgani.*

---

322 **Velvet-footed Pax**
Paxillus atrotomentosus (Bat. ex Fr.) Fr.
Paxillaceae, Agaricales

Description: *Dry, brown cap with yellowish gills
descending thick, dark, tough, velvety-hairy
stalk.*
Cap: 1⅝–4¾" (4–12 cm) wide; convex
to flat or sunken in center; margin
incurved; dry, with matted blackish-
brown wool. Flesh tough.
Gills: attached or descending stalk,
close, narrow, forked or porelike near
stalk, yellowish.
Stalk: 1–4" (2.5–10 cm) long, ⅜–1¼"
(1–3 cm) thick; off-center to nearly
lateral; stout, solid, tough, covered
with blackish-brown, velvety, matted
hair.
Spores: 5–6 × 3–4 μ; oval, smooth,
some dextrinoid. Spore print yellowish.

Season: Late July–October.

Habitat: Single to scattered or clustered, on
conifer stumps or decayed wood,
especially pine

Range: Coniferous areas of N. North America and south along coastal mountains.

Look-alikes: The poisonous *P. involutus** grows on ground, lacks velvety-hairy stalk.

Comments: This frequently collected mushroom is technically edible but is, in fact, unpalatable.

⊗ 287 **Poison Paxillus**
Paxillus involutus (Bat. ex Fr.) Fr.
Paxillaceae, Agaricales

Description: *Dry to slimy, brownish cap with inrolled margin and yellowish gills descending smooth, brownish stalk.*
Cap: 1⅝–4¾" (4–12 cm) wide; convex to flat, becoming sunken in center; margin remaining inrolled; dry, slimy in wet weather, covered with matted fibers; spotted, brownish.
Gills: descending stalk, crowded, broad, forked, sometimes with pores near base; yellowish-olive, bruising brown; separable from cap.
Stalk: 1⅝–4" (4–10 cm) long, ⅜–¾" (1.5–2 cm) thick; dry, hairless, yellowish-brown, streaked or stained darker brown.
Spores: 7–9 × 4–6 μ; elliptical, smooth. Spore print clay-brown.

Edibility: Poisonous.

Season: July–November.

Habitat: Single to several, near or on wood, in mixed woods.

Range: Widely distributed in North America.

Look-alikes: *P. atrotomentosus** has dark, velvety stalk. *P. vernalis* has thicker stalk and darker spore print and occurs under aspen. *Lactarius** species contain milky latex and produce a white spore print.

Comments: Although this species is eaten in some places, in other parts of its range it can have a decidedly acid-sour taste. There are reports that it can produce a gradually acquired hypersensitivity that causes kidney failure.

499 Stalkless Paxillus
Paxillus panuoides (Fr. ex Fr.) Fr.
Paxillaceae, Agaricales

Description: *Stalkless, shelflike mushroom with smooth, ochre cap and forked, yellowish gills.*
Cap: 1–4" (2.5–10 cm) wide; petal-shaped, with thin, wavy, incurved margin; dry, downy to smooth, olive-yellow to yellow-brown.
Gills: radiating from base, close, forked, with crossveins, often crimped near base; yellow; separable from cap.
Spores: 4–6 × 3–4 μ; elliptical, smooth, many dextrinoid. Spore print yellowish.

Season: May–November.

Habitat: In overlapping clusters, on dead stumps, logs, and timbers.

Range: Widely distributed in North America.

Look-alikes: *P. corrugatus* has orange-yellow, corrugated gills. *Phyllotopsis nidulans** has densely hairy cap, unpleasant odor, and pink spore print. *Pleurotus** species have white or lilac spore print. *Crepidotus** species have brown spore print, unforked, unseparable gills.

Comments: This mushroom is common on fallen trees and structural timbers.

324, 326 Gilled Bolete
Phylloporus rhodoxanthus (Schw.) Bres.
Paxillaceae, Agaricales

Description: *Dry, brownish cap with bright yellow gills descending stalk.*
Cap: 1–3" (2.5–7.5 cm) wide; convex to expanded, somewhat turban-shaped or sunken; dry, somewhat matted-woolly, often cracked; reddish-yellow to reddish-brown, rarely olive-brown. Flesh yellow.
Gills: descending stalk, distant, broad, sometimes forked, with crossveins; porelike near stalk; bright yellow, bruising blue; separable from cap.
Stalk: 1⅝–3¼" (4–8 cm) long, ¼–⅜" (0.5–1 cm) thick, narrowing toward

base; reddish to pinkish-yellow.
Spores: 9—12 × 3–5 μ; narrowly
elliptical, smooth. Spore print
yellowish to olive-green.

Edibility: Edible.
Season: June—October.
Habitat: Under conifers or in mixed woods.
Range: Widely distributed in N. North
America; also Florida and C. California,
and in North Carolina above 4000 feet.
Look-alikes: *Boletus** species and allies have tubular
undersurface. *Hygrophorus** species have
waxy gills that are not easily separable.
Comments: This is a species complex with great
variation in cap color, gill and pore
development, and spore print color.

PLUTEUS FAMILY
(Pluteaceae)

Members of this family have salmon- to
brownish-pink spores, and usually have
distinctly free gills. Two common
genera are included here: *Pluteus* (100+
species) and *Volvariella* (20+ species).
Pluteus species grow on wood or buried
wood or sawdust; some can occur from
early spring to late fall. *Volvariella*
species grow on wood or on the ground;
they develop from an egglike universal
veil, which leaves a persistent, saclike
cup about the stalk base. In addition to
the free gills and pinkish spore print,
these 2 genera have smooth spores and a
unique microscopic character: the gill
tissue is convergent, rather than
divergent (as in *Amanita*) or parallel to
interwoven, as in most other
mushrooms. Most are known to be
edible; none is known to be poisonous.

229 Yellow Pluteus
Pluteus admirabilis (Pk.) Pk.
Pluteaceae, Agaricales

Description: *Yellow cap with pink, free gills and*
slender, yellow stalk; on wood.

Cap: ⅜–1¼" (1–3 cm) wide; convex, becoming flat, knobbed; moist, smooth to wrinkled at center, yellow to ochre.
Gills: free, close, broad, off-white to pale yellow, becoming pinkish.
Stalk: 1¼–2⅜" (3–6 cm) long, ⅟₃₂–⅛" (1–3 mm) thick; fragile, smooth, yellow.
Spores: 5.5–7 × 4.5–6 μ; broadly elliptical, smooth. Spore print salmon.

Edibility: Edible.

Season: June–September.

Habitat: Single to several, on decayed deciduous wood.

Range: E. North America.

Look-alikes: *P. leoninus* has smooth cap and white stalk. *P. flavofuligineus* has smoky-yellow, velvety cap and pinkish-tinged stalk. *P. aurantiorugosus** has brilliant orange-red cap.

Comments: This mushroom's pinkish gills make it easy to identify.

---

233 Black-edged Pluteus
Pluteus atromarginatus (Kon.) Küh.
Pluteaceae, Agaricales

Description: *Dark brownish cap with pink, free gills and white, dark-tinged stalk; on wood.*
Cap: 1¼–4" (3–10 cm) wide; convex to flat or with low central knob; blackish-brown, streaked with small, flattened black fibers.
Gills: free, close, broad; white, then salmon *with dark brown to black edges.*
Stalk: 2–4" (5–10 cm) long, ¼–⅝" (0.5–1.5 cm) thick; whitish, coated with small, dark fibers.
Spores: 6.5–8 × 4.5 μ; elliptical, smooth. Pleurocystidia horned. Spore print salmon-pink.

Edibility: Edible.

Season: August–October.

Habitat: Single to scattered, on coniferous logs and debris.

Range: Widespread in N. North America.

Look-alikes: *P. umbrosus* lacks horned pleurocystidia. *P. cervinus** is larger, paler, and lacks

dark gill edges. *Entoloma** species have
attached gills and angular spores.

Comments: This thick-fleshed edible may easily be
mistaken for the Fawn Mushroom (*P. cervinus*).

---

230 Golden Granular Pluteus
Pluteus aurantiorugosus (Trog) Sacc.
Pluteaceae, Agaricales

Description: *Bright red to orange-red cap with pink, free gills and whitish stalk; on wood.*
Cap: 1–2¼" (2.5–5.5 cm) wide;
convex to nearly flat with small central
knob; granular, bright red to orange,
fading to bright yellow.
Gills: free, close, broad; whitish,
becoming pinkish.
Stalk: 1¼–2⅜" (3–6 cm) long, ⅛–⅜"
(0.3–1 cm) thick; with thin, fibrous
lines; whitish to yellowish, becoming
reddish near base.
Spores: 6–7 × 4.5–5 μ; elliptical,
smooth. Spore print salmon-pink.
Edibility: Edible.
Season: August–October.
Habitat: Single to several or clustered, on
decayed deciduous wood.
Range: NE. North America west to Rockies.
Comments: Also called *P. caloceps* and *P. coccineus*.
This is perhaps the most beautiful
pluteus. It may be more wide-ranging
than is thought, but it is not common.

231, 232 Fawn Mushroom
Pluteus cervinus (Schaeff. ex Fr.) Kum.
Pluteaceae, Agaricales

Description: *Brownish cap with white to pink, free gills and whitish stalk.*

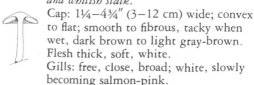

Cap: 1¼–4¾" (3–12 cm) wide; convex
to flat; smooth to fibrous, tacky when
wet, dark brown to light gray-brown.
Flesh thick, soft, white.
Gills: free, close, broad; white, slowly
becoming salmon-pink.
Stalk: 2–4" (5–10 cm) long, ¼–⅜"

(0.5–1 cm) thick; firm, solid, smooth or with some small fibers; white, sometimes tinged brownish or black. Spores: 5.5–7 × 4–5 μ; elliptical, smooth. Pleurocystidia abundant, conspicuous, thick-walled, spindle-shaped, with hornlike projections. Spore print salmon- to brownish-pink.

Edibility: Edible.

Season: May–October; November–May in Southwest.

Habitat: On wood, sawdust piles, or on ground over buried wood.

Range: Widely distributed in North America.

Look-alikes: *P. magnus* is stouter, with darker, coarsely wrinkled cap. *P. atromarginatus** has dark-edged gills. *Entoloma** species have attached gills and angular spores. *Tricholomopsis platyphylla** has attached gills that remain white, and white spore print.

Comments: Also known as *P. atricapillus*. This is the most frequently collected pluteus and one of the most common mushrooms in North America. It is edible, but good only when very young. *P. cervinus* var. *alba* is white and usually clustered on sawdust or buried wood.

234 Pleated Pluteus
Pluteus longistriatus Pk.
Pluteaceae, Agaricales

Description: *Brownish-gray cap, deeply pleated with age, with free, pink gills and pale, lined stalk; on wood.*
Cap: 1–2″ (2.5–5 cm) wide; convex to nearly flat or sunken in center; radially lined, becoming furrowed, granular; brownish-gray.
Gills: free, close, broad; off-white, becoming pinkish.
Stalk: ¾–3¼″ (2–8 cm) long, ¹⁄₁₆–⅛″ (1.5–3 mm) thick; slender, with lines of small fibers.
Spores: 6–7.5 × 5.5 μ; almost round, smooth. Spore print salmon-pink.

Edibility: Edible.
Season: August–September.
Habitat: Single to several on deciduous wood debris.
Range: E. North America; California.
Look-alikes: *P. seticeps* is smaller, with longer, darker, radially lined cap and short stalk with short, dark hairs. *Entolomas** have attached gills and angular spores.
Comments: A fragile, slender pluteus with a deeply pleated cap, it is easily overlooked.

109 Tree Volvariella
Volvariella bombycina (Schaeff. ex Fr.) Sing.
Pluteaceae, Agaricales

Description: *Large, white to yellow-tinged cap with free pink gills, smooth stalk, and large, saclike cup around stalk base; sunk in wood.*
Cap: 2–8″ (5–20 cm) wide; egg-shaped at first, becoming bell-shaped to convex; dry, becoming silky-fibered; yellowish. Flesh thin, soft, white.
Gills: free, crowded, broad; white, becoming pinkish.
Stalk: 2⅜–8″ (6–20 cm) long, ⅜–¾″ (1–2 cm) thick, enlarged at base; dry, hairless.
Veil: universal veil leaving a thick, deep cup.
Spores: 6.5–10.5 × 4.5–6.5 μ; elliptical, smooth. Spore print salmon to pink.
Edibility: Edible.
Season: July–September, into October.
Habitat: Solitary, in wounds on hardwood trees and on stumps and logs, especially maple, beech, oak, and elm.
Range: Canada to Florida, west to Michigan.
Look-alikes: *Pluteus** species lack cup. *Amanitas** have white spore print, grow on ground.
Comments: This species projects out of wounds in trees; it is not common, but is easily identified when found. The related Straw Mushroom (*V. volvacea*), which is cultivated commercially in the Orient, is also a good edible.

110 Tiny Volvariella
Volvariella pusilla (Pers. ex Fr.) Sing.
Pluteaceae, Agaricales

Description: *Tiny, white cap with free, pink gills and*
saclike cup about stalk base; on ground.
Cap: ¼–1¼" (0.5–3 cm) wide; convex;
margin radially lined; dry, with
flattened fibers, white.
Gills: free, close to almost distant,
broad; white, becoming pinkish.
Stalk: ⅜–2" (1–5 cm) long, ¹⁄₃₂–⅛"
(1–3 mm) thick; smooth, white to
grayish.
Veil: universal veil leaving a thick,
deep, often lobed cup.
Spores: 5.5–8.5 × 4–6 μ; elliptical,
smooth. Spore print salmon-pink.

Season: July–August.

Habitat: Single to several, in lawns, gardens,
and greenhouses.

Range: Quebec and E. United States.

Comments: This tiny species is often concealed by
grass and may be overlooked.

---

118 Smooth Volvariella
Volvariella speciosa (Fr.) Sing.
Pluteaceae, Agaricales

Description: *Large, sticky, white cap with free pink*
gills and whitish, saclike cup around stalk
base; on ground.
Cap: 2–6" (5–15 cm) wide; convex to
flat; sticky, smooth; dull white to light
grayish, shiny when dry.
Gills: free, crowded, broad; white,
slowly becoming pinkish.
Stalk: 3½–8" (9–20 cm) long, ⅜–¾"
(1–2 cm) thick; smooth, dry, white.
Veil: universal veil leaving thick, deep
cup.
Spores: 11.5–21 × 7–12.5 μ; broadly
elliptical, smooth. Spore print salmon-
to brownish-pink.

Edibility: Edible with caution.

Season: June–July in N. North America;
October–April in South and California.

Habitat: Single to scattered, in open areas, along

roads, on dung, in lawns, and in woods.

Range: Widely distributed in North America.

Look-alikes: The deadly *Amanita virosa** and related species have white gills, white spore print, and persistent ring. *Pluteus** species lack cup.

Comments: This must be identified with great care because of its resemblance to the young stages of the deadly amanitas.

111 **Parasitic Volvariella**
Volvariella surrecta (Knapp) Sing.
Pluteaceae, Agaricales

Description: *Small, whitish cap with free, pink gills; thick, saclike cup about stalk base; on cap of Cloudy Clitocybe* (Clitocybe nebularis).
Cap: 1–3¼" (2.5–8 cm) wide; convex, with silky fibers, whitish.
Gills: free, close, broad; white, becoming pink.
Stalk: 1⅝–3½" (4–9 cm) high, ⅛–⅝" (0.3–1.5 cm) thick; with flattened fibers, whitish.
Veil: universal veil leaving deep cup.
Spores: 5.5–7.5 × 3.5 μ; elliptical, smooth. Spore print salmon-pink.

Season: September–October.

Habitat: Single to several on Cloudy Clitocybe.

Range: W. North America.

Comments: This is one of very few gilled mushrooms to grow parasitically on other mushrooms.

RUSSULA FAMILY
(Russulaceae)

Two of the most abundant, varied, and widespread genera of gilled mushrooms are *Lactarius* and *Russula*. They differ from other gilled mushrooms by having distinctively brittle flesh. In *Russula* species, the cap margin is often broken and the gills flake apart on handling. *Lactarius* species are more compact; they contain a substance known as

latex, which exudes from any bruised or
cut part of the fruiting body, especially
the gills. This latex can be clear,
milky, or colored; in areas of moderate
or plentiful rainfall it is often
abundant. Both genera are ground-
inhabiting forest mushrooms, and all
are believed to be mycorrhizal with
specific trees. Many species are edible,
although a few are reportedly
poisonous. The rule of thumb for eating
russulas is not to eat any with a sharp,
acrid taste. For lactarii, one should
know the species and not experiment.
Some species of both genera are
parasitized by molds in the genus
Hypomyces; some russulas are attacked
by the gilled fungus *Asterophora.*
The brittle cap, gill, and stalk flesh is
composed throughout of two kinds of
tissue: long, slender, interconnecting
filamentous cells (typical of all
mushrooms) and unique nests of large,
roundish cells called sphaerocysts. In
addition, all species of both genera have
spores that are colorless (white to
yellow in mass) and ornamented with a
pattern of amyloid ridges and warts.

256 Burnt-sugar Milky
Lactarius aquifluus Pk.
Russulaceae, Agaricales

Description: *Cinnamon-brown cap with watery spots and
burnt-sugar odor.*
Cap: 1¼–6″ (3–15 cm) wide; convex to
flat or sunken, sometimes slightly
knobbed, with margin inrolled at first;
moist, obscurely zoned or with dark,
watery spots; fawn to cinnamon. Flesh
fawn to cinnamon; *latex watery.* Odor
fragrant, like burnt sugar or curry.
Gills: attached or slightly descending
stalk, close; white, becoming light
pinkish-cinnamon.
Stalk: 1⅝–3¼″ (4–8 cm) long, ⅜–¾″
(1–2 cm) thick; becoming hollow; paler
than cap, with white bloom.

Spores: 6–9 × 5.5–7.5 μ; elliptical, ornamented, amyloid. Spore print buff.

Season: July–October.

Habitat: On the ground in woods.

Range: Nova Scotia to New York, west to Minnesota, south to West Virginia.

Comments: This northern species is common in late summer.

---

295 Aromatic Milky
Lactarius camphoratus (Fr.) Fr.
Russulaceae, Agaricales

Description: *Small, reddish-brown cap with sharp knob, reddish-brown gills, and fragrant odor.*
Cap: ⅝–2″ (1.5–5 cm) wide; conical to broadly convex-sunken, usually with knob, becoming flat to sunken; smooth, moist, then dry; red-brown. *Latex white, unchanging.* Odor fragrant.
Gills: attached or slightly descending stalk, close to crowded, narrow; pale pinkish-cinnamon or wine-red.
Stalk: ⅝–2⅜″ (1.5–6 cm) long, ⅛–⅝″ (0.3–1.5 cm) thick, sometimes irregular; hoary, moist to dry, reddish-brown; base with coarse, stiff hairs.
Spores: 7–8.5 × 6–7.5 μ; round to broadly elliptical, ornamented, amyloid. Spore print yellowish.

Season: August–October.

Habitat: On the ground in coniferous and mixed forests.

Range: NE. North America.

Look-alikes: *L. rufulus* grows in Southwest, lacks fragrant odor and knob on cap. *L. fragilis,* in Southeast, has yellow gills.

Comments: This species' fragrant odor is best detected when a large collection is placed in a bag or dried.

---

247 Willow Milky
Lactarius controversus (Fr.) Fr.
Russulaceae, Agaricales

Description: *Large, dingy off-white cap with pinkish-cream gills.*

Cap: 1⅝–9" (4–23 cm) wide; convex to sunken to vase-shaped, margin inrolled, zoned; flattened-fibrous, whitish, with lavender to pinkish stains, sticky, then dry. Flesh white; *latex white, unchanging.* Taste slowly but strongly acrid.

Gills: attached, close to crowded; pale pink to purplish-fawn.

Stalk: 1–3¼" (2.5–8 cm) long, ⅝–2" (1.5–5 cm) thick; white, somewhat tacky, sometimes spotted.

Spores: 6–7.5 × 4.5–5 μ; elliptical, ornamented, amyloid. Spore print cream to pale pinkish-cinnamon.

Season: July–October.

Habitat: Under poplar and willow.

Range: Maine south to North Carolina, west to Alaska, and south to New Mexico.

Comments: Like many fungi, this species only grows close to specific trees.

---

293   Corrugated-cap Milky
Lactarius corrugis Pk.
Russulaceae, Agaricales

Description: *Reddish-brown, corrugated cap; latex white, abundant; odor becoming fishlike.*
Cap: 2–8" (5–20 cm) wide; convex to somewhat sunken, with margin sometimes conspicuously wrinkled; velvety or with sheen, dry, corrugated; orange- to reddish-brown. Flesh white, staining brown; *latex white, abundant.*
Gills: attached, close; ochre to buff.
Stalk: 2–4⅜" (5–11 cm) long, ⅝–1¼" (1.5–3 cm) thick; brown to reddish-brown, dry, velvety.
Spores: 9–12 × 8.5–12 μ; roundish, ornamented, amyloid. Spore print white.

Edibility: Choice.

Season: June–September.

Habitat: On soil in deciduous and mixed woods.

Range: E. North America west to Michigan and Texas.

Look-alikes: *L. volemus** has more shallowly wrinkled, orange-brown cap, and whitish to creamy gills bruising brown.

L. hygrophoroides * has orange-brown cap
and whitish to creamy distant gills.
Comments: This milky often grows in the same
areas as the Voluminous-latex Milky (*L. volemus*); both species are choice.

---

248 Deceptive Milky
Lactarius deceptivus Pk.
Russulaceae, Agaricales

Description: *Common large white mushroom with cottony inrolled margin and acrid taste.*
Cap: 2–10″ (5–25 cm) wide; convex, becoming sunken to vase-shaped; with conspicuously inrolled margin of soft, thick, cottony tissue when young; dry, smooth, dull, scaly with age; whitish, staining brownish, often with yellowish areas. Flesh white; *latex white, unchanging, staining flesh brownish.* Odor pungent; taste strongly acrid.
Gills: attached, close to nearly distant; cream to ochre, bruising darker.
Stalk: 1⅝–3½″ (4–9 cm) long, ⅜–1⅜″ (1–3.5 cm) thick, sometimes tapering; white, staining brownish, velvety, sometimes scaly, dry.
Spores: 9–13 × 7.5–9 μ; broadly elliptical, ornamented, amyloid. Spore print white to pale ochre-buff.
Season: June–September.
Habitat: On soil, in deciduous woods near oak; in mixed woods near hemlock.
Range: E. North America west to Michigan and Ohio.
Look-alikes: *L. piperatus* * has extremely close gills. *L. vellereus* lacks cottony roll of tissue.
Comments: This common mushroom is too acrid to be palatable, but it is nonetheless edible if thoroughly cooked.

---

312 Orange-latex Milky
Lactarius deliciosus (Fr.) S.F.G.
Russulaceae, Agaricales

Description: *Orange, zoned cap staining green with age.*
Cap: 2–5½″ (5–14 cm) wide; broadly

convex, with sunken center and inrolled
margin, becoming vase-shaped with
age; white bloom at first, becoming
smooth, slimy in wet weather; orange,
with several zones, staining greenish
with age. Flesh cream-yellowish,
staining greenish; *latex orange, fading to
orange-yellow, then greenish.* Odor fruity,
peachlike; taste mild.
Gills: attached, crowded; orange to pale
orange-yellow, dull green on bruising
and aging.
Stalk: 1¼–2¾″ (3–7 cm) long, ⅜–1″
(1–2.5 cm) thick; orange with white
bloom, sometimes pitted; staining
greenish with age.
Spores: 7–10 × 6–7 μ; roundish,
ornamented, amyloid. Spore print
bright cream.

Edibility: Good.
Season: August–October.
Habitat: Scattered to many, often abundant;
under conifers, especially pine.
Range: Widely distributed in North America.
Look-alikes: *L. thyinos* has sticky stalk and grows in
northern arborvitae bogs.
Comments: Several varieties of the Orange-latex
Milky occur, differing somewhat in
coloration; all are basically orange, with
orange latex that stains or spots green
with age. All are edible, although not
as good as the name *deliciosus* suggests.

362 Gerard's Milky
Lactarius gerardii Pk.
Russulaceae, Agaricales

Description: *Velvety-brown cap with very distant gills.*
Cap: 1⅜–4¾″ (3.5–12 cm) wide;
convex, becoming flat to sunken; dry,
velvety, unzoned, dark brown to
yellow-brown. Flesh white,
unchanging; *latex white, unchanging.*
Gills: attached, almost distant to very
distant, thickish; white to pale cream.
Stalk: 1⅜–3¼″ (3.5–8 cm) long, ⅜–
¾″ (1–2 cm) thick; pale to dark brown,
dry, velvety.

Spores: 7–10 × 7.5–9 μ; round to broadly elliptical, ornamented, amyloid. Spore print white.

Edibility: Edible.

Season: July–September; sometimes into November.

Habitat: On soil in deciduous and coniferous woods.

Range: Quebec to North Carolina and Michigan.

Look-alikes: *L. lignyotus** has close gills, flesh bruising reddish to violet when cut. *L. fallax* grows in Northwest and has crowded gills and white latex that stains flesh reddish-brown.

Comments: Gerard's Milky, a complex of forms, is relatively common, especially in the Southeast. A variety, *fagicola,* has flesh that bruises brown, and another variety, *subrubescens,* bruises pinkish to red; both are found in the East.

---

290 Hygrophorus Milky
Lactarius hygrophoroides Berk. & Curt.
Russulaceae, Agaricales

Description: *Orange-brown cap with very distant gills.*
Cap: 1¼–4" (3–10 cm) wide; convex, becoming flat to sunken or somewhat vase-shaped; dry, unzoned, minutely velvety, wrinkled at times; orange-brown. Flesh white; *latex abundant, white.*
Gills: attached, distant, broad; white, becoming cream-colored.
Stalk: 1¼–2" (3–5 cm) long, ¼–⅝" (0.5–1.5 cm) thick; dry, orange-brown or paler, with white bloom.
Spores: 7.5–10.5 × 6–7.5 μ; elliptical, ornamented, amyloid. Spore print white.

Edibility: Choice.

Season: June–September.

Habitat: On soil in deciduous woods.

Range: E. North America to Michigan and Texas.

Look-alikes: *L. volemus** has close gills, brown-staining latex. *L. corrugis** has dark reddish-brown cap, close, ochre gills.

Comments: One of the most beautiful and best edible species of *Lactarius,* this milky has a number of varieties.

---

337 **Indigo Milky**
Lactarius indigo (Schw.) Fr.
Russulaceae, Agaricales

Description: *Indigo, fading with age, bruising green.*
Cap: 2–6″ (5–15 cm) wide; convex to sunken, with margin inrolled at first; sticky, hairless, usually zoned, dark blue when young, fading to pale gray-blue with greenish stains. Flesh whitish, turning blue when cut, then greenish; *latex scanty, dark blue, slowly becoming dark green on exposure to air.*
Gills: attached, close; dark blue, becoming paler with age.
Stalk: ¾–3¼″ (2–8 cm) long, ⅜–1″ (1–2.5 cm) thick, sometimes tapered; dark blue to silver-blue; becoming hollow, dry.
Spores: 7–9 × 5.5–7.5 μ; broadly elliptical to round, ornamented, amyloid. Spore print cream.

Edibility: Edible.

Season: July–late October.

Habitat: Few to many, on soil in oak and pine woods.

Range: E. North America to Michigan and Texas.

Look-alikes: *L. paradoxus** has wine-brown latex.

Comments: Indigo is an unusual color for mushrooms, and so this milky is quite easy to identify.

---

363 **Chocolate Milky**
Lactarius lignyotus Fr.
Russulaceae, Agaricales

Description: *Often wrinkled, velvety, chocolate-brown, dry cap with close gills.*
Cap: ¾–4″ (2–10 cm) wide; convex, knobbed, with margin incurved at first, becoming flat to sunken; dull, dry,

velvety, often wrinkled, blackish to blackish-brown but fading. Flesh white, staining rose on cut surfaces; *latex white, abundant, discoloring only cut surfaces.*
Gills: attached, close to almost distant; white, pale ochre when spores mature.
Stalk: 2–4¾″ (5–12 cm) long, ⅛–¾″ (0.3–2 cm) thick; often grooved, velvety, blackish, with whitish base.
Spores: 9–10.5 × 9–10 μ; roundish, ornamented, amyloid. *Spore print bright ochre.*

Season: August–October.

Habitat: On soil and moss in coniferous woods and sphagnum bogs.

Range: E. North America west to Wisconsin.

Look-alikes: *L. fallax* has crowded gills. *L. gerardii** has very distant gills and white spores. *L. fumosus* has yellow-brown cap and white latex that stains cut surfaces red.

Comments: Despite its name, *lignyotus,* this usually grows on moss, not wood. There are several variants of this species.

250 **Buff Fishy Milky**
Lactarius luteolus Pk.
Russulaceae, Agaricales

Description: *Velvety, dry, whitish-buff cap; odor fishy.*
Cap: 1–3¼″ (2.5–8 cm) wide; convex to flat or sunken; velvety, dry, unzoned, whitish to buff, with white bloom. Flesh whitish, staining brown from latex; *latex abundant, white, unchanging, but staining flesh brown; sticky.* Odor mild, becoming fishlike.
Gills: attached, close; white, becoming yellowish to brown where bruised.
Stalk: 1–2⅜″ (2.5–6 cm) long, ¼–⅝″ (0.5–1.5 cm) thick; dry, fuzzy-velvety, white to buff, staining brown.
Spores: 7–9 × 5.5–7 μ; elliptical, ornamented, amyloid. Spore print white to cream.

Edibility: Edible.

Season: June–November.

Habitat: On soil in deciduous and mixed woods.

Range: Quebec to North Carolina, west to
Michigan and Texas.
Look-alikes: *L. volemus** is orange-brown.
Comments: Despite its odor, this is edible. In the
Northeast this species is one of the first
Lactarius mushrooms to appear.

364 Slimy Lactarius
Lactarius mucidus Burl.
Russulaceae, Agaricales

Description: *Slimy, brownish to pinkish-buff cap.*
Cap: 1¼–3½" (3–9 cm) wide; convex
to sunken, with central knob; smooth,
persistently slimy to sticky, unzoned,
dark, dingy, chocolate-brown to
pinkish-buff. Flesh white; *latex white.*
Gills: attached, close; white to cream,
staining blue-green to gray with latex.
Stalk: 1⅝–3¼" (4–8 cm) long, ¼–⅜"
(0.5–1 cm) thick; sticky, smooth, pale.
Spores: 7.5–10 × 6–8 μ; elliptical,
ornamented, amyloid. Spore print
white.
Season: June–November.
Habitat: On the ground under conifers.
Range: Quebec to North Carolina west to
Michigan and Idaho.
Look-alikes: *L. pseudomucidus* has slimier cap and
stalk. *L. kauffmanii* has blackish to
blackish-brown cap and is twice as
large; both occur in the Northwest.
Comments: This species is actually a complex
encompassing several varieties and look-
alikes. All share the dark, slimy
cap, pale stalk, and acrid white latex.

349 Silver-blue Milky
Lactarius paradoxus Beard. & Burl.
Russulaceae, Agaricales

Description: *Silver-blue, staining greenish with age.*
Cap: 2–3¼" (5–8 cm) wide; convex,
with inrolled margin, becoming sunken
and vase-shaped; smooth, sticky when
wet, zoned, gray-blue with greenish
tones and a silvery sheen, fading with

age. Flesh pale, green-tinged; *latex dark wine-brown, scanty, staining green.*
Gills: attached, close; dark wine-brown, green on bruising.
Stalk: 1–1¼" (2.5–3 cm) long, ⅜–⅝" (1–1.5 cm) thick, narrowing to rootlike base; dry, somewhat hoary, gray-blue, staining green with age.
Spores: 7–9 × 5.5–6.5 μ; broadly elliptical, ornamented, amyloid. Spore print buff to yellow.
Edibility: Edible.
Season: August–September in Northeast; October–February in South.
Habitat: On soil under palmetto palm, live oak and red oak, and in lawns in Florida; under pine in Tennessee and Michigan.
Range: E. North America to Michigan, Texas.
Look-alikes: *L. subpurpureus** has pinkish cap. *L. indigo** has indigo latex.
Comments: This belongs to a group of mushrooms that bruise or age greenish and have latex that is colored from the first.

292 **Peck's Milky**
Lactarius peckii Burl.
Russulaceae, Agaricales

Description: *Usually zoned, dark brick-red cap with sheen, with reddish-brown gills.*
Cap: 2–6" (5–15 cm) wide; convex, with inrolled margin, becoming sunken, sometimes vase-shaped with age; dry, becoming shiny, dark bay-red, typically zoned with darker lines. Flesh pale brownish; *latex abundant, white, unchanging.* Taste acrid.
Gills: attached and somewhat descending stalk, close; pale buff, becoming reddish- to dark brown.
Stalk: ¾–2⅜" (2–6 cm) long, ⅜–1" (1–2.5 cm) thick, sometimes abruptly narrowed below; reddish-brown or paler, with white bloom, spotting brownish on handling.
Spores: 6–7.5 μ in diameter; round, ornamented, amyloid. Spore print white.

Season: July–September.
Habitat: On the ground in deciduous woods, especially near oak.
Range: Massachusetts to Alabama.
Look-alikes: *L. volemus** lacks sheen and well-marked zones, and has fishlike odor.
Comments: Peck's Milky is too acrid to be considered palatable.

---

240 Peppery Milky
Lactarius piperatus (Fr.) S.F.G.
Russulaceae, Agaricales

Description: *Common, white mushroom with densely crowded, narrow gills.*
Cap: 2–6" (5–15 cm) wide; convex, becoming sunken to vase-shaped; dry, unzoned, smooth to somewhat wrinkled; white, staining dingy tan with age. Flesh white, unchanging; *latex white, abundant. Taste very acrid.*
Gills: attached, very crowded; white to pale cream.
Stalk: ¾–3¼" (2–8 cm) long, ⅜–1" (1–2.5 cm) thick; dry, white, with white bloom.
Spores: 4.5–7 × 5–5.5 μ; elliptical, ornamented, amyloid. Spore print white.
Edibility: Edible with caution.
Season: July–September.
Habitat: On soil in deciduous woods.
Range: E. North America west to Michigan.
Look-alikes: *L. deceptivus** has almost distant gills and cottony marginal roll. *L. vellereus* has distant gills.
Comments: This edible is too acrid to be tasty without first being parboiled. A variety, *glaucescens,* has latex that dries greenish; it is reportedly poisonous.

⊗ 284 Northern Bearded Milky
Lactarius representaneus Britz. sensu Neu.
Russulaceae, Agaricales

Description: *Yellow to orange-yellow cap, bruising purplish, with a bearded margin.*

Cap: 2⅜–8" (6–20 cm) wide; convex-sunken to flat, becoming vase-shaped; sticky in wet weather, unzoned; yellow to orange-yellow, developing rusty tones over center, staining purple where bruised. Flesh white, staining dull lilac; *latex abundant, white to cream, becoming violet.* Taste bitter.

Gills: attached, close to crowded, cream to ochre, developing orange cast and spots; lavender where bruised.

Stalk: 2–4¾" (5–12 cm) long, ⅜–1¾" (1–4.5 cm) thick, sometimes club-shaped; surface sticky, pitted, white to yellow-orange, staining lilac.

Spores: 8–12 × 6.5–9 μ; broadly elliptical, ornamented, amyloid. Spore print yellowish.

Edibility: Possibly poisonous.

Season: August–September.

Habitat: On the ground under spruce.

Range: Quebec to Vermont; Pacific NW.; also Colorado and Arizona in mountains.

Look-alikes: *L. speciosus* is duller, strongly zoned, and grows in deciduous woods.

Comments: This mushroom may be poisonous and should not be eaten.

⊗ 289 **Red-hot Milky**
Lactarius rufus (Fr.) Fr.
Russulaceae, Agaricales

Description: *Bay-red cap with acrid taste.*
Cap: 1⅝–4¾" (4–12 cm) wide; convex, with incurved margin, becoming flat to sunken to broadly vase-shaped; hoary, becoming dull, dry; bay-red. Flesh dingy purplish; *latex abundant, white, unchanging, acrid.*

Gills: attached or slightly descending stalk, crowded; whitish when young, becoming buff to wine-tan with age.

Stalk: 2–4" (5–10 cm) long, ⅜–⅝" (1–1.5 cm) thick; hoary, becoming dingy purplish; base whitish, dry.

Spores: 7.5–10.5 × 5–7.5 μ; broadly elliptical, ornamented, amyloid. Spore print whitish to pale yellow.

Edibility: Poisonous.
Season: August–November.
Habitat: On soil under pine; in sphagnum bogs.
Range: Widely distributed in North America.
Look-alikes: *L. rufulus* has white flesh and pinkish-yellow gills; it grows under live oak in late winter in California.
Comments: This is a fairly common and widespread mushroom for which other species are often mistaken. There are several varieties; in some, the taste is mild at first but soon becomes very acrid. In North America, this is reportedly poisonous, but the Scandinavian variety is canned and sold commercially.

⊗ 285 **Spotted-stalked Milky**
Lactarius scrobiculatus (Fr.) Fr.
Russulaceae, Agaricales

Description: *Large, yellow-orange, scaly cap with inrolled, hairy margin; stalk pitted, ochre.*
Cap: 3–10″ (7.5–25 cm) wide; domed, becoming funnel-shaped, margin edge long, inrolled, hairy, becoming smooth; very sticky, shiny when dry, becoming scaly; yellow-orange. Flesh white; *latex abundant, white, quickly turning sulfur-yellow.* Odor fruity; taste burning-acrid.
Gills: attached or descending stalk, crowded; white to yellow, with pink-tinted bruised area.
Stalk: 1¼–2⅜″ (3–6 cm) long, ⅝–1⅜″ (1.5–3.5 cm) thick, sometimes narrowing to rootlike base; pitted with glazed ochre spots, otherwise tawny, with white bloom.
Spores: 6.5–9 × 5–7.5 μ; broadly elliptical, ornamented, amyloid. Spore print white to creamy yellow.
Edibility: Poisonous.
Season: August–October.
Habitat: Under conifers, especially pine.
Range: N. North America; Colorado; California.
Comments: This milky is a complex of species that have similar cap color, margin, pitted

stalk, and sulfur latex. Reportedly
poisonous. Do not eat any *Lactarius*
with white latex that turns yellow.

361 Dirty Milky
Lactarius sordidus Pk.
Russulaceae, Agaricales

Description: *Brownish-yellow, nearly smooth cap with
greenish tints; stalk brownish, spotted.*
Cap: 2–4″ (5–10 cm) wide; convex to
sunken or vase-shaped; nearly smooth,
pale brownish-yellow with greenish
hues, somewhat tacky. Flesh white,
pink-tinged; *latex white.* Taste acrid.
Gills: attached, close; white to
yellowish.
Stalk: 1⅝–2¾″ (4–7 cm) long, ⅜–1″
(1–2.5 cm) thick; brownish, with
pitted, dingy, greenish-brown spots.
Spores: 5.5–7.5 × 5–6 μ; elliptical,
ornamented, amyloid. Spore print dull
white.

Season: August–October.
Habitat: On the ground under conifers.
Range: Nova Scotia to North Carolina, west to
Michigan, Idaho, and Alaska.
Look-alikes: *L. atroviridis,* more often found in
Southeast, is dark green.
Comments: This very common eastern species is not
known to be edible.

244 Variegated Milky
Lactarius subpurpureus Pk.
Russulaceae, Agaricales

Description: *Pinkish, becoming mottled and greenish.*
Cap: 1¼–4″ (3–10 cm) wide; convex,
becoming flat and somewhat sunken;
smooth, wine-red, somewhat zoned,
spotted with green. Flesh white to
pink, staining red when first cut, then
greenish; *latex wine-red, scanty.* Taste
vaguely peppery.
Gills: attached, nearly distant; wine-
red.
Stalk: 1¼–3¼″ (3–8 cm) long, ¼–⅝″

(0.5—1.5 cm) thick; smooth, somewhat
slimy when wet; dull wine-red, spotted
dark reddish.
Spores: 8—11 × 6.5—8 μ; elliptical,
ornamented, amyloid. Spore print
cream.

Edibility: Good.
Season: July—October.
Habitat: On soil under hemlock and pine.
Range: Maine to North Carolina.
Look-alikes: *L. paradoxus** has bluish-green cap.
Comments: This species is widely distributed in
eastern North America, but is most
common in the Southeast.

⊗ 251 **Pink-fringed Milky**
Lactarius torminosus (Fr.) S.F.G.
Russulaceae, Agaricales

Description: *Pinkish cap with white, bearded margin.*
Cap: 2—4¾″ (5—12 cm) wide; convex
to sunken or somewhat shallowly vase-
shaped, with cottony, bearded,
incurved margin; smooth, sometimes
zoned, pale pink to whitish at margin.
Flesh white to pink-tinged; *latex white,
unchanging. Taste instantly acrid.*
Gills: attached, close to crowded;
whitish, becoming tan.
Stalk: 1¼—2¾″ (3—7 cm) long, ¼—¾″
(0.5—2 cm) thick; smooth to fuzzy,
sometimes spotted, dry; pink or paler.
Spores: 7.5—10 × 6—7.5 μ; elliptical,
ornamented, amyloid. Spore print
cream.
Edibility: Poisonous.
Season: August—November.
Habitat: On the ground, usually under birch.
Range: E. North America to Idaho; Colorado.
Look-alikes: *L. pubescens* has smaller spores.
Comments: Although generally considered
poisonous in North America, this
species is sold commercially in Finland.
It is sometimes parasitized by a
brownish mold, *Hypomyces torminosus.* A
variety or subspecies of this milky,
called *nordmanensis,* has white latex that
slowly changes to yellow.

⊗ 338 **Common Violet-latex Milky**
Lactarius uvidus (Fr.) Fr.
Russulaceae, Agaricales

Description: *Sticky, lilac-drab cap.*
Cap: 1¼–4" (3–10 cm) wide; convex
to flat, finally somewhat sunken;
margin incurved at first; smooth,
sticky, becoming dry, typically
unzoned; pale lilac-drab, darkening
somewhat; green with KOH. Flesh
whitish; *latex white, becoming dingy cream
and staining flesh dull lilac.* Taste slowly
bitter.
Gills: attached, close; cream, staining
dull lilac on bruising, then dingy tan.
Stalk: 1¼–2¾" (3–7 cm) long, ⅜–⅝"
(1–1.5 cm) thick; off-white, sticky at
first, shiny when dry, rarely spotted;
often stained ochre around base.
Spores: 7.5–9.5 × 6.5–7.5 μ; round
to broadly elliptical, ornamented,
amyloid. Spore print pale yellow.
Edibility: Poisonous.
Season: July–October.
Habitat: On the ground under aspen, birch, and
pine.
Range: Maine to Tennessee; Michigan;
Colorado.
Look-alikes: *L. maculatus* is strongly zoned.
Comments: Like all others in its genus with
purplish-staining latex or flesh, this is
believed to be poisonous.

⊗ 288 **Yellow-latex Milky**
Lactarius vinaceorufescens A.H.S.
Russulaceae, Agaricales

Description: *Buff to cinnamon-pink, zoned cap.*
Cap: 1⅝–4¾" (4–12 cm) wide;
convex, with margin inrolled at first;
smooth, sticky, zoned with watery
spots or bands; buff, becoming pinkish-
cinnamon or darker. Flesh white,
staining yellow; *latex white, becoming
bright yellow.*
Gills: attached, close; wine-buff,
becoming spotted and stained pinkish,

reddish-brown with age.
Stalk: 1⅝–2¾" (4–7 cm) long, ⅜–1"
(1–2.5 cm) thick; smooth above, with
coarse, stiff hairs at base; pale pinkish-
buff, becoming darker.
Spores: 6.5–9 × 6–7 μ; round to
broadly elliptical, ornamented,
amyloid. Spore print white to
yellowish.

Edibility: Possibly poisonous.
Season: August–October.
Habitat: On soil under pine.
Range: Nova Scotia to North Carolina, west to
Wisconsin.
Look-alikes: *L. chrysorheus* is a paler, southeastern
species. *L. croceus* has bright orange cap.
Comments: In the Northeast, this is the most
common milky with white latex that
turns yellow. Possibly poisonous.

---

291   **Voluminous-latex Milky**
Lactarius volemus (Fr.) Fr.
Russulaceae, Agaricales

Description: *Orange-brown cap with fishlike odor.*
Cap: 2–5¼" (5–13 cm) wide; convex,
with margin incurved at first,
becoming flat to sunken or vase-shaped;
surface unzoned, dry, minutely velvety,
orange-brown. Flesh whitish, staining
brownish when cut; *latex white,
abundant, slowly turning brown, staining
flesh brownish.* Odor becoming fishlike
with handling or age.
Gills: attached, close; white to cream,
staining brown to dark brown.
Stalk: 2–4" (5–10 cm) long, ⅜–¾"
(1–2 cm) thick, sometimes tapering
downward; orange-brown, smooth to
minutely velvety, longitudinally lined.
Spores: 7.5–10 × 7.5–9 μ; round,
ornamented, amyloid. Spore print
white.

Edibility: Choice.
Season: June–September.
Habitat: On soil in deciduous woods.
Range: E. North America west to Michigan,
Wisconsin, Minnesota, and Texas.

Look-alikes: *L. corrugis** is more red-brown, has corrugated cap and ochre gills. *L. hygrophoroides** has broad, distant gills and is odorless. *L. luteolus** is similar but smaller, bleached-looking.

Comments: Despite its odor, this species is one of the best edibles in its genus. The latex stains fingers as well as gills, so gloves should be used when cleaning it.

---

357 Tacky Green Russula
Russula aeruginea Lindbl. ex Fr.
Russulaceae, Agaricales

Description: *Tacky, greenish cap with yellowish-white gills and stalk.*
Cap: 2–3⅜" (5–8.5 cm) wide; cushion-shaped to convex or flat with a sunken center; margin radially lined to roughened, fragile; slightly sticky or dry, velvety or smooth, sometimes slightly cracked; grayish-olive to yellow-green, sometimes spotted yellowish. Flesh brittle, white.
Gills: attached or slightly descending stalk, close, broad; yellowish-white.
Stalk: 1⅝–2⅜" (4–6 cm) long, ⅜–¾" (1–2 cm) thick, sometimes tapering; dry, dull, mostly smooth; yellowish-white, darker below.
Spores: 6.2–8.3 × 5.1–6.8 μ; broadly elliptical to oval or round, ornamented, amyloid. Spore print orange-yellow.

Edibility: Good.

Season: July–September; November–March in California.

Habitat: On the ground in deciduous and coniferous woods, especially under oak, live oak, aspen, and lodgepole pine.

Range: Widely distributed in North America.

Look-alikes: *R. crustosa** has dry, quilted cap pattern. *R. variata** usually has purplish tones in cap and conspicuously forked gills. *R. olivacea* grows under western conifers, has red tones in cap.

Comments: This widespread russula is relatively easy to identify and is often abundant.

252 Short-stalked White Russula
Russula brevipes Pk.
Russulaceae, Agaricales

Description: *Large, dry, dull white cap, gills, and typically short stalk, all staining brownish.*
Cap: 4–8" (10–20 cm) wide; broadly convex, with sunken center; unlined margin inrolled at first; dry, minutely woolly, white, staining dull yellow to brown. Flesh white, brittle.
Gills: attached or somewhat descending stalk, close to crowded; white, staining cinnamon to brownish.
Stalk: 1–3" (2.5–7.5 cm) long, 1–1⅛" (2.5–4 cm) thick; dry, smooth, dull white with brown stains.
Spores: $8–11 \times 6.5–10 \ \mu$; broadly elliptical, ornamented, amyloid. Spore print white to cream.
Season: July–October.
Habitat: On the ground in mixed woods.
Range: Widely distributed in North America.
Look-alikes: *R. cascadensis* grows under northwestern conifers, is smaller, has more acrid taste. *Lactarius vellereus* species complex have more membranous gills; fresh specimens exude latex when cut.
Comments: Previously known as *R. delica.* Although disagreeable-tasting, when this species is parasitized by the Lobster Mushroom (*Hypomyces lactifluorum*), it is transformed into a choice edible.

321 Graying Yellow Russula
Russula claroflava Groves
Russulaceae, Agaricales

Description: *Yellow cap with flesh, gills, and stalk becoming ash-gray with age.*
Cap: 2–4" (5–10 cm) wide; convex, becoming flat to slightly sunken; slightly sticky when moist, smooth, yellow, becoming ash-gray with age. Flesh white, then gray. Taste mild.
Gills: attached, close, moderately broad; white, becoming yellowish, then ash-gray.

Stalk: 2–3″ (5–7.5 cm) long, ⅝–¾″ (1.5–2 cm) thick; smooth, white or yellow, becoming ash-gray.
Spores: 8–11 × 7.5–9 μ; nearly round, ornamented, amyloid. Spore print pale yellow.

Edibility: Edible.

Season: July–September.

Habitat: On the ground in coniferous and mixed woods.

Range: Widely distributed in North America.

Look-alikes: *R. lutea* has a sticky yellow cap, ochre gills, and unchanging flesh. *R. ochroleuca* has sticky, shiny, yellow-orange cap and yellowish gills.

Comments: Also known as *R. flava.* Although *Russula* is typically a genus of red-capped mushrooms, there are species with white, yellow, or greenish caps.

286 Firm Russula
Russula compacta Frost
Russulaceae, Agaricales

Description: *Dry, dull reddish-orange to ochre cap, off-white gills and stalk, all bruising rusty to yellowish-ochre.*
Cap: 1⅜–7″ (3.5–18 cm) wide; cushion-shaped with incurved margin, becoming convex to nearly flat with sunken center or vase-shaped; margin unlined; sticky and shiny when wet; smooth to somewhat cracked; pale orange-yellow. Flesh firm, hard, becoming brittle; white. Odor slight, becoming rank, fishy; taste undistinctive to somewhat disagreeable.
Gills: attached, crowded to almost distant, narrow to broad; white to pale yellow.
Stalk: 1–5″ (2.5–12.5 cm) long, ⅜–1¾″ (1–4.5 cm) thick, sometimes enlarging or tapering toward base; dry, dull, mostly smooth to wrinkled; white. Solid, becoming hollow.
Spores: 7.6–10 × 6.3–8.6 μ; broadly elliptical to oval, ornamented, amyloid. Spore print white.

Edibility: Edible.
Season: August—September.
Habitat: On the ground in beech-maple woods
             and under northern conifers.
Range: E. North America west to Michigan.
Look-alikes: *R. decolorans* has flesh bruising ash-gray,
             yellowish gills and spore print, and
             most often grows under conifers.
Comments: This large, hard-fleshed russula does
             not shatter easily.

356 Green Quilt Russula
Russula crustosa Pk.
Russulaceae, Agaricales

Description: *Greenish, cracked cap with a white bloom;*
             gills and stalk yellowish.
             Cap: 2–6″ (5–15 cm) wide; cushion-
             shaped to convex or flat with sunken
             center, becoming vase-shaped with age;
             margin fragile, rough-lined; sticky
             when wet, becoming dry, dull, with a
             white bloom, cracked, quiltlike;
             grayish- to yellowish-green, sometimes
             with orange-yellow tones. Flesh
             yellowish-white, brittle.
             Gills: attached, close to almost distant,
             moderately broad; white, becoming
             pale yellow.
             Stalk: 1⅜–3⅜″ (3.5–8.5 cm) long,
             ⅝–1¾″ (1.5–4.5 cm) thick; dry, dull,
             smooth to feltlike, longitudinally
             wrinkled, yellowish-white; stuffed,
             becoming hollow.
             Spores: 5.8–10.2 × 4.5–7.7 μ;
             elliptical to oval, ornamented, amyloid.
             Spore print pale orange-yellow.
Edibility: Good.
Season: July—September.
Habitat: Single to many, on the ground in
             woods, usually with oak.
Range: E. North America west to Michigan
             and Texas.
Look-alikes: *R. virescens* lacks orange tones in cap
             and has yellowish-white spore print. *R.*
             aeruginea* lacks cracked cap surface.
Comments: An excellent edible, found in large
             quantities after August rains.

249 Red-and-black Russula
Russula dissimulans Shaffer
Russulaceae, Agaricales

Description: *Whitish to grayish-brown cap and stalk,*
bruising bright red, slowly turning black.
Cap: 2½–7″ (6.5–18 cm) wide; convex
to nearly flat and slightly sunken,
becoming somewhat funnel-shaped
with age; margin incurved, arching
with age; tacky to dry, matted and
feltlike; chalk-white, becoming
brownish, aging and decaying black.
Flesh turns bright reddish when cut or
bruised, then slowly brown to black.
Gills: attached or slightly descending
stalk; close to almost distant, narrow to
moderately broad, brittle; yellowish-
buff to dingy, bruising reddish, then
brown to black.
Stalk: 1⅝–3¼″ (4–8 cm) long, ¾–
1¾″ (2–4.5 cm) thick; smooth, rigid,
dull whitish, becoming brown to black;
bruising red, then black.
Spores: 7.7–10.8 × 6.5–9 μ;
elliptical, ornamented, amyloid. Spore
print white.

Season: July–September in East; October–
December in Pacific NW.

Habitat: On the ground in woods.

Range: Widespread in N. North America.

Look-alikes: *R. nigricans* has distant gills and smaller
spores. *R. densifolia* has close to
crowded gills and somewhat slimy cap
cell layer. Other similar *Russulas** turn
black directly upon bruising.

Comments: This mushroom and allied red-staining
species, as well as some that stain only
black, are often found covered with
numerous small, white Powder Caps
(*Asterophora lycoperdoides*).

⊗ 328 Emetic Russula
Russula emetica (Schaeff. ex Fr.) S.F.G.
Russulaceae, Agaricales

Description: *Slimy-tacky, reddish cap with off-white*
gills and stalk; acrid taste.

Cap: 1–3" (2.5–7.5 cm) wide; cushion-shaped, becoming convex, flat with sunken center or vase-shaped; margin incurved, fragile, becoming rough-lined; sticky, smooth, sometimes minutely cracked; red to reddish-orange or deep pink. Flesh white, brittle. Taste strongly acrid.

Gills: attached, close, moderately broad; yellowish-white.

Stalk: 2–4" (5–10 cm) long, ¼–1" (0.5–2.5 cm) thick, enlarging toward base; dry, dull, longitudinally wrinkled; white to yellow-white; stuffed, becoming partly hollow.

Spores: 8–11.3 × 6.7–9 μ; elliptical to oval, ornamented, amyloid. Spore print white to yellowish-white.

Edibility: Poisonous.

Season: August–September.

Habitat: Singly or in groups, usually in sphagnum moss, rarely on very rotten wood, in boggy areas in coniferous or mixed woods.

Range: Widely distributed in North America.

Look-alikes: Most other red *Russulas** are best distinguished microscopically. *R. silvicola* grows in dry woodlands and is more common.

Comments: Because this species can be emetic, it should not be eaten.

343 **Fragile Russula**
Russula fragilis (Pers. ex Fr.) Fr.
Russulaceae, Agaricales

Description: *Tacky, reddish cap with mixed multicolored tones, grayish-olive about center; with whitish gills and stalk; taste acrid.*
Cap: 1–3" (2.5–7.5 cm) wide; cushion-shaped, becoming convex to flat to concave; margin incurved, becoming rough-lined, fragile; sticky and slimy when moist, smooth to velvety, roughened, or cracked; typically multicolored, red or grayish- to purplish-red and often grayish-green over center, with lavender and

yellowish-pink tints toward margin.
Flesh brittle. Odor fruity or unpleasant.
Gills: attached or almost free, close to
almost distant, broad; white to
yellowish-white.
Stalk: 1–3″ (2.5–7.5 cm) long, ¼–⅝″
(0.5–1.5 cm) thick; dry to moist, dull,
smooth or somewhat finely hairy,
wrinkled lengthwise; white to yellow-
white; solid, then stuffed or hollow.
Spores: 5.7–9 × 4.5–7.7 μ; broadly
elliptical, ornamented, amyloid. Spore
print yellowish-white.

Season: July–September.

Habitat: Single to several, on the ground, rotten
wood, or moss, in coniferous and mixed
woods.

Range: Widely distributed in North America.

Look-alikes: Because any of the cap color tints can
predominate, it can be mistaken for
several *Russulas**.

Comments: This species is best recognized when
most of the cap colors are present in a
single mushroom.

329 Blackish-red Russula
Russula krombholzii Shaffer
Russulaceae, Agaricales

Description: *Tacky to dry, reddish cap, reddish-black
over center, with yellowish-white gills and
whitish stalk.*
Cap: 2⅜–4¾″ (6–12 cm) wide; convex
to nearly flat or with sunken center,
becoming somewhat vase-shaped;
margin wavy at times; sticky and shiny
when wet, becoming dry, smooth or
velvety, sometimes wrinkled toward
center; dark grayish-purple; purplish-
red to blackish-red over center, paler
toward margin. Flesh brittle. Taste
slightly acrid or not distinctive.
Gills: attached, close to almost distant,
moderately broad, yellowish-white to
pale orange-yellow.
Stalk: 2–4″ (5–10 cm) long, ⅝–1¾″
(1.5–4.5 cm) thick, sometimes
enlarged toward base; dry, dull, smooth

to feltlike, hairy at base, longitudinally
wrinkled; whitish, bruising ash;
yellowish-brown near base.
Spores: 7–10.4 × 5.2–7.7 μ; elliptical
to oval, ornamented, amyloid. Spore
print yellowish-white.

Season: May–October.
Habitat: Single to several, on the ground in
          deciduous and mixed woods.
Range: NE. North America west to Michigan.
Look-alikes: *R. fragilis** is smaller, more fragile.
Comments: Also known as *R. atropurpurea.* This is
          one of the very first ground-inhabiting
          gilled mushrooms to appear.

306 Almond-scented Russula
Russula laurocerasi Melz.
Russulaceae, Agaricales

Description: *Slimy, yellowish cap, gills, and stalk, with*
             fragrant, almondlike to fetid odor.
             Cap: 1–5″ (2.5–12.5 cm) wide;
             cushion-shaped, expanding to convex or
             flat with sunken center to somewhat
             vase-shaped with age; margin incurved,
             conspicuously rough-lined; slimy and
             shiny when wet, becoming dry and
             dull, then smooth to slightly wrinkled,
             sometimes streaked; yellow to yellow-
             brown or yellow-orange. Flesh brittle.
             Odor almondlike; taste unpleasant.
             Gills: attached, close to almost distant,
             broad, yellowish-white to pale orange.
             Stalk: 1–4″ (2.5–10 cm) long, ⅜–1¼″
             (1–3 cm) thick, sometimes tapering
             toward base; dry, dull, smooth to finely
             hairy at top, yellowish-white to pale
             orange-yellow.
             Spores: 6.8–10.7 × 6.8–9 μ; broadly
             elliptical to oval, ornamented, amyloid.
             Spore print pale orange-yellow.
Season: July–September.
Habitat: Single to several, on the ground,
          especially in woods with beech, maple,
          and hemlock.
Range: E. Canada south to North Carolina,
          west to Michigan.
Look-alikes: *R. fragrantissima,* despite its name, is

705

usually strongly fetid, also darker and larger. *R. subfoetens* is smaller, with dark brown to brownish-orange cap and odor more fetid than fragrant. *R. ventricosipes* grows in sand under pines and has stalk swollen in middle, narrowing abruptly above and below.

Comments: This species and its look-alikes become fetid and have a disagreeable flavor.

---

339 **Purple-bloom Russula**
Russula mariae Pk.
Russulaceae, Agaricales

Description: *Dry, purplish to reddish-purple cap with white bloom, whitish gills, and white stalk flushed with red at center.*
Cap: 1–2″ (2.5–5 cm) wide; cushion-shaped to convex or flat, slightly sunken; margin fragile; dry, velvety, with a bloom; purplish to dark reddish-purple or maroon. Flesh white, brittle.
Gills: attached, close to almost distant, somewhat narrow; white, then yellow.
Stalk: 1 3″ (2.5–7.5 cm) long, ⅜–1″ (1–2.5 cm) thick, sometimes tapering downward; white with rosy blush; spongy-stuffed.
Spores: 7–10 × 5.5–8.5 μ; round, ornamented, amyloid. Spore print whitish to cream-yellow.

Edibility: Good.
Season: June–October.
Habitat: Single to several, on the ground in deciduous woods, especially near oak; also in grassy areas near oak.
Range: Canada to Florida, west to Michigan.
Comments: Some divide this russula into 2 species, one with a red cap, the other lilac-purple; the latter is called *R. alachuana*.

---

334 **Rosy Russula**
Russula rosacea (Pers. ex Secr.) Fr.
Russulaceae, Agaricales

Description: *Tacky, red cap with yellowish-white gills and rosy-red stalk; taste acrid.*

Cap: 1–4" (2.5–10 cm) wide; convex to flat or slightly sunken; margin fragile; sticky when wet, smooth; dark red to bright, vivid red, fading to pink. Flesh white, brittle. Taste acrid.
Gills: attached or slightly descending stalk, close, narrow to broad; cream-white to pale yellow.
Stalk: 2–4" (5–10 cm) long, ⅜–⅝" (1–1.5 cm) thick, sometimes tapering toward base; smooth, dry; white, flushed rosy-red; becoming hollow.
Spores: 7–9 × 6–8 μ; round to oval, ornamented, amyloid. Spore print pale yellow.

Season: September–October.
Habitat: Under conifers, especially pine.
Range: Pacific Coast; Rocky Mountains.
Look-alikes: The poisonous *R. emetica** usually grows in sphagnum moss.
Comments: This mushroom should not be eaten.

---

348 Variable Russula
Russula variata Ban. apud Pk.
Russulaceae, Agaricales

Description: *Greenish cap with purplish tones and conspicuously forked gills.*
Cap: 2–6" (5–15 cm) wide; cushion-shaped, with incurved margin, becoming convex to flat with sunken center, then somewhat vase-shaped; margin sometimes wavy or split, fragile; slightly sticky or dry, dull, smooth or radially wrinkled, streaked or cracked; *extremely variable in color,* usually greenish with yellow and purple tones, sometimes purple or pinkish. Flesh white, brittle.
Gills: attached or slightly descending stalk, close, narrow, forked; white, becoming yellowish-white.
Stalk: 1–4" (2.5–10 cm) long, ⅜–1¼" (1–3 cm) thick; dry, dull, feltlike to smooth, wrinkled; white to yellowish-white, sometimes bruising slightly darker; solid, becoming hollow.
Spores: 7.3–11.4 × 5.7–9.4 μ;

broadly elliptical, ornamented, amyloid. Spore print white.

Edibility: Edible.

Season: July—early October.

Habitat: Single to several, on the ground in deciduous woods, especially under oak and aspen; also in mixed woods.

Range: Canada to Florida, west to Michigan.

Comments: Also known as *R. cyanoxantha* var. *variata* and the "Forked-gill Russula."

330 Shellfish-scented Russula
Russula xerampelina (Schaeff. ex Secr.) Fr.
Russulaceae, Agaricales

Description: *Reddish cap and yellow gills with distinctive odor of shellfish.*
Cap: 1–6" (2.5–15 cm) wide; broadly convex to flat with sunken center; margin radially lined and fragile; sticky to dry; purplish-red or lighter, or with greenish tones. Flesh white, brittle. Odor fishlike; taste mild.
Gills: attached, close to almost distant, broad; yellowish to orange-yellow.
Stalk: 2–3" (5–7.5 cm) long, ⅝–1¼" (1.5–3 cm) thick, enlarging toward base; wrinkled to grooved; white to pink-flushed, bruising olive-yellowish; hollow with age.
Spores: 7.5–11.5 × 6–8.5 μ; elliptical to round, ornamented, amyloid. Spore print yellowish.

Edibility: Good.

Season: July—September.

Habitat: Under conifers, especially hemlock, Douglas-fir, and pine; in mixed woods.

Range: Widely distributed in North America.

Comments: This may be a complex of closely related species; the odor separates them from others in this genus.

STROPHARIA FAMILY
(Strophariaceae)

This is the principal family of the so-called hallucinogenic mushrooms—

those fungi that, when eaten, cause transient changes in vision and sense of space and time. Actually, very few are hallucinogenic; some are edible, others are poisonous, and most are of unknown edibility. These mushrooms are decomposers rather than symbionts. They grow on the ground in lawns, humus, and dung, or on wood.
The family may be divided into 2 groups of dark-spored mushrooms: those with purple-brown to purple-black spore prints, and those with brown spore prints. Both of these groups have attached gills and share certain microscopic characteristics: the cap cuticle is typically filamentous and the spores are smooth and usually have a pore at the tip. Many species have similarly shaped sterile cells (chrysocystidia) on the gill surfaces; the amorphous matter in these cells stains yellow with KOH.

62 Smoky-gilled Naematoloma
Naematoloma capnoides (Fr.) Kar.
Strophariaceae, Agaricales

Description: *Orange, convex cap with grayish-brown gills; in clusters on coniferous wood.*
Cap: 1–3″ (2.5–7.5 cm) wide; convex to broadly convex; margin incurved at first, often with marginal veil remnants; smooth, tawny-orange.
Gills: attached, close, broad; grayish, becoming purple-brown.
Stalk: 2–3″ (5–7.5 cm) long, ⅛–⅜″ (0.3–1 cm) thick, pale to brownish.
Veil: partial veil cobwebby, leaving zone of fibrils on upper stalk.
Spores: 6–7 × 4–4.5 μ; elliptical, smooth, with pore at tip.
Chrysocystidia present. Spore print purple-brown.
Edibility: Good.
Season: September–December.
Habitat: Scattered to clustered, on coniferous wood, especially Douglas-fir.

Range: Widely distributed in North America.
Look-alikes: The poisonous *N. fasciculare** has yellowish-green gills at first. *N. sublateritium** has brick-red cap.
Comments: This species is a good edible, but not choice. Great care should be taken not to confuse it with the poisonous Sulfur Tuft (*N. fasciculare*).

77 Dispersed Naematoloma
Naematoloma dispersum (Fr.) Kar.
Strophariaceae, Agaricales

Description: *Tawny cap with brownish gills and long, thin stalk.*
Cap: ⅜–1⅝" (1–4 cm) wide; bell-shaped to expanded with central knob; moist, smooth; tawny, but fading.
Gills: attached, close, broad; off-white, then olive-gray, then purple-brown.
Stalk: 2–4" (5–10 cm) long, 1/16–¼" (1.5–5 mm) thick; tough, silky; brownish below, yellowish above.
Veil: partial veil cobwebby, leaving zone of fibrils on upper stalk.
Spores: 7–10 × 4–5 μ; elliptical, smooth, with pore at tip. Chrysocystidia present. Spore print purple-brown.
Season: August–November.
Habitat: Single to scattered or in groups, on conifer debris.
Range: Northwest and Rocky Mountains.
Look-alikes: *N. udum* grows in bogs, has larger spores. The mildly hallucinogenic *Psilocybe pelliculosa** bruises blue.
Comments: This species frequently can be seen in logging areas in the western mountains.

⊗ 61 Sulfur Tuft
Naematoloma fasciculare (Huds. ex Fr.) Kar.
Strophariaceae, Agaricales

Description: *Yellowish, convex to flat cap with greenish-yellow gills; in clusters on wood.*
Cap: 1–3¼" (2.5–8 cm) wide; convex to flat, with slight central knob; moist,

smooth; orange- to sulfur- or olive-
yellow. Taste bitter.
Gills: attached, crowded, narrow;
sulfur-yellow, becoming greenish, then
tinted purple-brown.
Stalk: 2–4¾" (5–12 cm) high, ¹⁄₁₆–⅜"
(0.15–1 cm) thick, sometimes narrowed
at base; yellowish, becoming tawny.
Veil: partial veil leaves fibrous zone.
Spores: 6.5–8 × 3.5–4 μ; elliptical,
smooth, with pore at tip. Spore print
purple-brown.

Edibility: Poisonous.
Season: May–June; late August–November;
overwinters in California.
Habitat: Scattered to clustered, on logs and
stumps; also in grass or on the ground
above buried wood.
Range: Widely distributed in North America.
Look-alikes: N. *capnoides** has grayish gills. N.
*sublateritium** has brick-red cap.
*Armillariella mellea** has off-white gills,
whitish spore print, and well-formed
ring. The deadly *Galerina autumnalis**
has rust-brown spore print.
Comments: This bitter mushroom is reported to
have caused deaths in Europe and the
Orient. In North America, some people
have experienced gastric distress from
eating it, but nothing more severe.

36 Brick Tops
Naematoloma sublateritium (Fr.) Kar.
Strophariaceae, Agaricales

Description: *Brick-red, convex cap with grayish-purple
gills; in clusters on deciduous wood.*
Cap: 1⅛–4" (4–10 cm) wide; convex
to broadly convex or flat; margin
inrolled at first; moist, brick-red, paler
at margin.
Gills: attached, close, narrow; whitish
to purplish-gray.
Stalk: 2–4" (5–10 cm) high, ¼–⅝"
(0.5–1.5 cm) thick; whitish, staining
darker.
Veil: partial veil cobwebby, leaving
zone of fibrils on upper stalk.

Spores: 6–7 × 4–4.5 μ; elliptical, smooth, with pore at tip. Chrysocystidia present. Spore print purple-brown.

Edibility: Good.

Season: Late August–November.

Habitat: Clustered on deciduous stumps or logs.

Range: E. North America; also Mexico.

Look-alikes: *N. capnoides** has orange cap and grows on conifers. The poisonous *N. fasciculare** has yellowish cap and yellow-green gills at first.

Comments: One of the better edibles of late fall in the Northeast, this species is only tasty when very young.

---

190 Powder-scale Pholiota

Phaeomarasmius erinaceellus (Pk.) Sing.
Strophariaceae, Agaricales

Description: *Scaly-powdery, rust-brown cap and stalk with yellowish gills; on wood.*
Cap: ⅜–1⅝" (1–4 cm) wide; conical to convex or nearly flat; dry, granular-scaly, breaking into powder; dark rust-brown, becoming yellowish-brown. Taste metallic, bitter.
Gills: notched, crowded, broad; off-white to yellowish or cinnamon-buff, becoming ochre-tawny.
Stalk: 1¼–2⅜" (3–6 cm) long, ⅛–¼" (3–5 mm) thick; covered with granular, rust-brown scales below veil; buff and finely powdered above.
Veil: partial veil fibrous; leaving temporary zone of minute fibers or slight ring on upper stalk.
Spores: 6–8 × 4–4.5 μ; elliptical, smooth, no pore at tip. Spore print dingy cinnamon.

Season: August–October.

Habitat: Single to several, on rotting deciduous logs.

Range: N. North America south to Maryland.

Comments: Formerly known as *Pholiota erinaceella*, this was transferred to *Phaeomarasmius* because of its powdery cap and lack of microscopic *Pholiota* characteristics.

66 **Bitter Pholiota**
Pholiota astragalina (Fr.) Sing.
Strophariaceae, Agaricales

Description: *Slimy, bright red-orange cap with yellowish gills and stalk; bitter.*
Cap: 1–2" (2.5–5 cm) wide; bell-shaped, becoming broader, with knob; margin with veil remnants; sticky to slimy when wet, becoming dry; red-orange, fading to pinkish-orange, *discoloring blackish* in places. Taste bitter.
Gills: almost free, close, moderately broad; orange-yellow.
Stalk: 2–4" (5–10 cm) long, ¼" (5 mm) thick; fibrous; pale yellow, drab orange at base, bruising brownish; hollow.
Veil: partial veil evanescent.
Spores: 5–7 × 3.8–4.5 μ; oval to elliptical, smooth, with minute pore. Chrysocystidia present. Spore print brown.
Season: Late August–October.
Habitat: Scattered or in small clusters, on dead coniferous wood.
Range: Canada to North Carolina, west to Michigan; Pacific NW. and California.
Comments: When young and fresh, the Bitter Pholiota is one of the prettiest mushrooms to appear in coniferous and mountainous areas.

186 **Golden Pholiota**
Pholiota aurivella (Fr.) Kum.
Strophariaceae, Agaricales

Description: *Slimy-glutinous, ochre cap with spotlike scales, brownish gills, and evanescent ring on scaly, ochre stalk; clustered on wood.*
Cap: 2–6" (5–15 cm) wide; bell-shaped to convex or with broad knob; sticky to slimy, covered with large, flattened, spotlike scales, sometimes lost; ochre-orange to tawny.
Gills: attached, close, moderately broad; pale yellow to tawny-brown.

Stalk: 2–4" (5–10 cm) long, ¼–⅝" (0.5–1.5 cm) thick; dry, cottony above ring, fibrous or with downcurled scales below; yellowish to yellow-brown.
Veil: partial veil off-white; leaving evanescent ring or zone on upper stalk.
Spores: 7–10 × 4.5–6 μ; elliptical, smooth, with pore at tip.
Chrysocystidia absent; pleurocystidia present. Spore print brownish.

Edibility: Good, with caution.

Season: September–November.

Habitat: Usually clustered on living trunks and on logs of both deciduous and coniferous trees, especially maple, elm, beech, and birch.

Range: Canada to Tennessee, west to Washington and New Mexico.

Look-alikes: *P. squarrosoides** has erect, pointed scales. *P. limonella* has narrower spores. *P. hiemalis* is a reportedly poisonous northwestern species that appears in late fall on coniferous wood; it has gelatinous scales on stalk.

Comments: The mushroom known as the Golden Pholiota may really be a group of species that cannot be differentiated in the field. Once the slimy cap has been cleaned and the mushroom cooked, this becomes a fine edible species.

189 Burnt-ground Pholiota
Pholiota carbonaria A.H.S.
Strophariaceae, Agaricales

Description: *Slimy, yellowish cap with reddish-brown scales and stalk with reddish veil; on burned ground.*
Cap: 1–2" (2.5–5 cm) wide; convex, becoming flat, slightly sunken, or with small knob; margin incurved at first, with hanging remnants of veil; slimy to tacky; yellowish to buff, becoming brownish at center, with concentric rows of small, reddish-brown scales.
Gills: attached, crowded, narrow; white, becoming brownish.
Stalk: 1–2" (2.5–5 cm) long, ¼" (5

mm) thick; fibrous, yellowish, covered
with small, reddish-brown scales from
veil; tip finely powdery.
Veil: partial veil fibrous, cinnabar-red,
fading to ochre; leaving remnants along
cap margin and on stalk, but no ring.
Spores: 5–8 × 3.5–4.5 μ; elliptical,
smooth, with pore at tip. Spore print
tawny-brown.

Season: August–December.

Habitat: On burned ground, roots, or wood.

Range: Pacific NW. and N. California.

Look-alikes: *P. fulvozonata* has darker tawny cap and
russet veil. *P. highlandensis* has brown
cap, broad gills, and yellowish veil.

Comments: A number of pholiotas grow only on
burned sites.

---

208 Destructive Pholiota
Pholiota destruens (Brond.) Gill.
Strophariaceae, Agaricales

Description: *Large, tacky, scaly, whitish cap with white
to brownish gills and ring on scaly stalk.*
Cap: 3–8″ (7.5–20 cm) wide; convex to
broadly convex; margin shaggy with
veil remnants; tacky, white to cream,
sometimes darker; covered with whitish
to buff, woolly-scaly patches.
Gills: attached, close, broad; white,
becoming brownish.
Stalk: 2–6″ (5–15 cm) long, ⅜–1¼″
(1–3 cm) thick, enlarging downward to
base; solid, hard; silky above ring zone,
woolly-scaly or with woolly patches
below; white, darkening with age.
Veil: partial veil white, copious;
leaving evanescent ring on upper stalk.
Spores: 7–9.5 × 4–5.5 μ; elliptical to
oval, smooth, with pore at tip. Spore
print cinnamon-brown.

Edibility: Edible.

Season: September–October, or to November.

Habitat: Single to clustered, usually at ends of
cottonwood, aspen, and poplar logs.

Range: Canada to New York; Great Lakes
states; Colorado; Pacific NW.; New
Mexico.

Look-alikes: *Pleurotus dryinus** has white gills
descending stalk that do not discolor.
Comments: The Destructive Pholiota is conspicuous
on poplars in late fall, even after frost
has killed most terrestrial mushrooms.

---

185 **Yellow Pholiota**
Pholiota flammans (Fr.) Kum.
Strophariaceae, Agaricales

Description: *Slimy, scaly, bright yellow cap with*
yellowish gills and ring on scaly stalk.
Cap: 1⅝–4" (4–10 cm) wide; conical,
becoming broadly convex with knob;
sticky, with top layer of downcurled,
fibrous scales; bright yellow.
Gills: attached, close, moderately
broad; bright yellow.
Stalk: 2–5" (5–12.5 cm) long, ¼–⅜"
(0.5–1 cm) thick, sometimes enlarged
at base; densely scaly, silky above ring
zone; bright yellow.
Veil: partial veil yellow; leaves
evanescent zone of fibers on upper stalk.
Spores: 4–5 × 2.5–3 μ; oblong to
elliptical, smooth. Chrysocystidia
present. Spore print ochre-brown.
Season: Late August–October.
Habitat: Single or in small clusters, on
coniferous logs and stumps.
Range: Quebec and N. United States south to
Tennessee; also Pacific NW.
Look-alikes: The poisonous *P. squarrosa** has dry,
dull yellowish cap. *P. squarrosoides**, on
deciduous wood, is ochre-tawny.
Comments: This pholiota may be readily identified
by its color and downcurled scales.

---

227 **Changing Pholiota**
Pholiota mutabilis (Schw. ex Fr.) Kum.
Strophariaceae, Agaricales

Description: *Fading, smooth, brown cap with cinnamon*
gills and ringed, conspicuously scaly stalk;
in dense clusters on dead wood.
Cap: ⅝–2⅜" (1.5–6 cm) wide; blunt,
becoming conical with incurved

margin, to convex, flat, or knobbed;
moist to sticky, smooth; reddish-brown
to tawny, fading as it dries from center
to margin to ochre or yellowish-buff.
Gills: attached or slightly descending
stalk, close, broad to narrow; off-white
to dull cinnamon.
Stalk: 2–4″ (5–10 cm) long, 1/16–5/8″
(0.15–1.5 cm) thick; silky above ring,
with downcurled scales below; off-white
to brownish.
Veil: partial veil off-white; leaving
membranous ring, or sometimes only
fibrous zone, on upper stalk.
Spores: 5.5–7.5 × 3.7–5 μ; oval,
smooth, with blunt tip from broad
pore. Chrysocystidia present. Spore
print cinnamon-brown.

Edibility: Good, with caution.

Season: September–November.

Habitat: In dense clusters on coniferous and
deciduous logs and stumps.

Range: Widely distributed in North America.

Look-alikes: *P. vernalis* grows near snowbanks in
spring and lacks scales on stalk. *P.
veris,* another spring mushroom, has
thicker stalk lacking downcurled scales.
*Galerina autumnalis** and *G. marginata*,
both deadly, lack stalk scales, rarely
grow in massed clusters, and have
roughened spores.

Comments: Also known as *Kuehneromyces mutabilis.*
This good edible is easy to recognize: it
grows clustered on stumps and has
conspicuously scaly stalks and fading
caps. To be safe, examine the spores
microscopically.

⊗ 187 **Scaly Pholiota**
Pholiota squarrosa (Müll. ex Fr.) Kum.
Strophariaceae, Agaricales

Description: *Scaly, dry, yellow-brown cap with
brownish gills and ring on scaly stalk; in
clusters at bases of trees or on wood.*
Cap: 1–4″ (2.5–10 cm) wide; blunt to
convex, becoming broader, knobbed or
nearly flat; margin sometimes with veil

remnants; dry, covered with downcurled scales; tawny scales on yellow surface, sometimes greenish-yellow along margin with age.

Gills: attached, close, narrow; pale yellow, becoming greenish-tinged, then brown to rust-brown.

Stalk: 2–4" (5–10 cm) long, ⅛–⅝" (0.3–1.5 cm) thick, sometimes tapering to base; dry, covered with downcurled scales; yellowish-tawny.

Veil: partial veil yellowish; leaving ring zone or persistent, membranous ring on upper stalk.

Spores: 5–8 × 3.8–4.5 μ; elliptical to oval, smooth, with pore at tip. Chrysocystidia present. Spore print brown.

Edibility: Poisonous.

Season: July–October.

Habitat: In clusters at base of deciduous and coniferous trees, especially aspen and birch; also on logs and stumps.

Range: Widely distributed in North America.

Look-alikes: *P. squarrosoides** has slime layer.

Comments: Although the Scaly Pholiota was long regarded as edible, it is now known to cause gastric discomfort in some people within an hour after having been eaten.

184, 188 Sharp-scaly Pholiota
Pholiota squarrosoides (Pk.) Sacc.
Strophariaceae, Agaricales

Description: *Sticky, ochre cap with pointed, dry scales, brownish gills, and ring on scaly, dry stalk.*

Cap: 1–4" (2.5–10 cm) wide; blunt, becoming convex with knob or nearly flat; margin with veil remnants; sticky, with dry, downcurled or pointed, tawny scales; white to cinnamon.

Gills: attached, close, moderately broad; whitish, becoming rust-brown.

Stalk: 2–6" (5–15 cm) long, ¼–⅝" (0.5–1.5 cm) thick; dry, silky above ring, downcurled, ochre-tawny scales below ring; buff.

Veil: partial veil off-white; leaving
fibrous or membranous, often
evanescent ring on upper stalk.
Spores: $4-5.5 \times 2.5-3.5 \ \mu$; oval to
broadly elliptical, smooth.
Chrysocystidia absent; pleurocystidia
abundant. Spore print brown.

Edibility: Edible.

Season: September–October.

Habitat: Clustered on deciduous wood, such as
birch, alder, beech, and maple.

Range: Widely distributed in North America.

Look-alikes: *P. squarrosa** is dry. *P. aurivella** has
slimy cap and flattened scales.

Comments: This edible is often abundant.

---

181   Ground Pholiota
Pholiota terrestris Overholts
Strophariaceae, Agaricales

Description: *Slimy, brown cap with dry scales, gray-*
brown gills, and ring on scaly stalk.
Cap: ¾–4" (2–10 cm) wide; conical to
convex, becoming knobbed or nearly
flat; margin with veil remnants;
fibrous-scaly, with slimy layer beneath;
dark gray-brown.
Gills: attached, crowded, narrow; pale,
becoming grayish-brown to brown.
Stalk: 1–4" (2.5–10 cm) long, ⅛–⅜"
(0.3–1 cm) thick; finely hairy above
ring, sheathlike below, with
downcurled scales; dark brownish.
Veil: partial veil off-white; leaving ring
zone or ring on upper stalk.
Spores: $4.5-6.5 \times 3.5-4.5 \ \mu$;
elliptical, smooth, with pore at tip.
Chrysocystidia present. Spore print
brown.

Edibility: Edible.

Season: June–January.

Habitat: Clustered on ground, in lawns, grassy
areas, and along roads.

Range: Widespread in W. North America.

Look-alikes: *P. squarrosoides**, on wood, is ochre.

Comments: This mushroom's mycelium may be
attached to buried wood, making it
appear to be growing from ground.

⊗ 86 **Potent Psilocybe**
Psilocybe baeocystis Sing. & A.H.S.
Strophariaceae, Agaricales

Description: *Sticky, conical, brown cap with brownish*
gills and off-white stalk; bruising blue.
Cap: ⅝–2¼″ (1.5–5.5 cm) wide;
conical with incurved margin,
expanding to convex or flat; sticky,
olive- to buff-brown, bruising and
aging greenish about margin.
Gills: attached, close, broad; grayish,
becoming dark purplish-gray.
Stalk: 2–2¾″ (5–7 cm) long, ¹⁄₁₆–⅛″
(1.5–3 mm) thick; whitish, covered
with small, whitish fibers.
Veil: partial veil evanescent.
Spores: 10–13 × 6.3–7 μ; elliptical,
smooth, with pore at tip. Spore print
dark purplish.
Edibility: Hallucinogenic.
Season: September–November.
Habitat: Scattered to numerous, in wood chips,
on decayed wood, and decaying moss.
Range: Pacific NW.
Look-alikes: *P. strictipes* has long, brittle, straight
stalk. The hallucinogenic *P. cyanescens**
has broad, wavy, knobbed cap.
Comments: This species is a potent hallucinogen
that contains several active compounds.
Its side effects are not well known.

⊗ 33 **Blue-foot Psilocybe**
Psilocybe caerulipes (Pk.) Sacc.
Strophariaceae, Agaricales

Description: *Brownish, knobbed cap, fading to yellow,*
with brown gills; whitish stalk bluish at
base; on decayed wood.
Cap: ⅜–1⅜″ (1–3.5 cm) wide; conical
to convex with incurved margin,
becoming flat or broadly knobbed;
sticky, becoming dry, smooth; watery-
cinnamon to yellowish; bruising
greenish or bluish, sometimes slowly.
Gills: attached, close to crowded,
narrow; brownish to rust-cinnamon.
Stalk: 1¼–2⅜″ (3–6 cm) long, ¹⁄₁₆–⅛″

(1.5–3 mm) thick, enlarging to base;
whitish, staining greenish-blue.
Veil: partial veil evanescent.
Spores: 7–10 × 4–5.5 μ (but 10–12
× 5.7 μ from 2-spored basidia);
elliptical, smooth, with pore at tip.
Spore print dark purple-brown.

Edibility: Hallucinogenic.
Season: August–October.
Habitat: Single or in small clusters, on
deciduous wood and wood mulch,
especially birch and maple.
Range: Maine to North Carolina, west to
Michigan.
Comments: Often overlooked or ignored as just
another little brown mushroom, this
hallucinogenic species turns blue on
handling, usually after several minutes.

⊗ 82 **Dung-loving Psilocybe**
Psilocybe coprophila (Bull. ex Fr.) Kum.
Strophariaceae, Agaricales

Description: *Sticky, brownish cap with brown gills and
yellowish-brown stalk.*
Cap: ⅜–1¼″ (1–3 cm) wide; convex to
broadly convex or flat; margin with
whitish patches at first; smooth, dark
reddish-brown, fading to grayish-
brown.
Gills: attached, nearly distant, broad;
whitish to brown or purplish-brown.
Stalk: ¾–1⅝″ (2–4 cm) long, ¹⁄₁₆–¼″
(1.5–5 mm) thick; whitish, darkening
to brown, but not bruising blue.
Veil (when present): partial veil
evanescent.
Spores: 11–14 × 7–8.5 μ; elliptical,
smooth, with pore at tip. Spore print
purplish-brown.
Edibility: Hallucinogenic.
Season: June–October.
Habitat: Single to numerous, on horse or cow
dung.
Range: Widely distributed in North America.
Look-alikes: *P. merdaria* has central ring zone on
stalk. *Stropharia semiglobata** is ringed,
yellowish. *Panaeolus** species have black

spores. *Coprinus** species liquefy.

Comments: This weak hallucinogen is the most widespread psilocybe in North America.

⊗ 226 **Common Large Psilocybe**
Psilocybe cubensis (Ear.) Sing.
Strophariaceae, Agaricales

Description: *Large, fleshy, yellowish cap with brown gills and a persistent ring on stalk; bruising blue; on cow manure.*
Cap: ⅝–3½" (1.5–9 cm) wide; conical or bell-shaped, becoming convex to flat with central knob; sticky, hairless; white with brownish-yellow center, becoming entirely brownish-yellow, bruising and aging bluish.
Gills: attached, close, narrow; gray, becoming deep violet-gray, then black; edges whitish.
Stalk: 1⅜–6" (3.5–15 cm) long, ⅛–⅝" (0.3–1.5 cm) thick, becoming enlarged below; smooth, grooved at top; white, bruising blue.
Veil: partial veil membranous; leaving persistent white ring (soon blackish from falling spores) on upper stalk.
Spores: 11.5–17 × 8–11.5 μ; oval to elliptical, smooth, thick-walled, blunt, with distinct pore at tip. Spore print purple-brown.
Edibility: Hallucinogenic.
Season: Nearly year-round.
Habitat: On cow and horse dung in pastures.
Range: Gulf Coast.
Comments: This is an abundant member of the Gulf Coast pastureland flora.

⊗ 31 **Bluing Psilocybe**
Psilocybe cyanescens Wkfld.
Strophariaceae, Agaricales

Description: *Tacky, wavy, brown cap, fading to yellowish, with brownish gills and whitish stalk; bruising blue.*
Cap: ¾–1⅝" (2–4 cm) wide; convex, becoming nearly flat with undulating or

wavy margin; sticky to moist, smooth;
dark chestnut-brown, fading to
yellowish, bruising blue.
Gills: attached, nearly distant, broad;
cinnamon-brown, becoming darker.
Stalk: 2⅜–3¼" (6–8 cm) long, ⅛–¼"
(3–5 mm) thick, sometimes enlarged at
base; curved, whitish, bruising blue.
Veil: partial veil white, evanescent.
Spores: 9–12 × 5.5–8.3 μ; elliptical,
smooth, with pore at tip. Spore print
purple-brown.

Edibility: Hallucinogenic.
Season: September–November.
Habitat: Several to many, in coniferous mulch.
Range: British Columbia to San Francisco.
Look-alikes: The hallucinogenic *P. baeocystis** and *P. strictipes* lack wavy margin.
Comments: When ingested in large quantity, this
can be strongly hallucinogenic.

83 Mountain Moss Psilocybe
Psilocybe montana (Fr.) Quél.
Strophariaceae, Agaricales

Description: *Small, dark brown mushroom; in moss.*
Cap: ¼–1" (0.5–2.5 cm) wide; conical,
becoming broadly convex to knobbed;
margin lined; moist, smooth; dark
reddish-brown, drying yellowish-brown.
Gills: attached, almost distant,
moderately broad; pale brown,
becoming dark reddish-brown.
Stalk: 1–2" (2.5–5 cm) long, ¹⁄₁₆" (1.5
mm) thick, sometimes enlarged at base;
dry, smooth, dark reddish-brown.
Veil: partial veil evanescent.
Spores: 5.5–8 × 4–5 μ; elliptical,
smooth, with pore at tip. Spore print
purple-brown.

Season: July–September.
Habitat: Single to several, in moss.
Range: W. North America; eastward in
mountains.
Comments: *Psilocybe* is a large genus of mostly
small, brownish mushrooms with
purple-brown spore prints. Relatively
few are known to be hallucinogenic.

⊗ 44 **Conifer Psilocybe**
Psilocybe pelliculosa (A.H.S.) Sing. & A.H.S.
Strophariaceae, Agaricales

Description: *Sticky, dark brown, conical cap with*
brown gills and off-white, hairy stalk.
Cap: ¼–¾" (0.5–2 cm) wide; conical
to bell-shaped; sticky, smooth; dark
brown, fading to tan, bruising blue.
Gills: attached, close, narrow;
cinnamon-brown, then darkening.
Stalk: 2⅜–3¼" (6–8 cm) long, ¹⁄₁₆"
(1.5 mm) thick; whitish, darkening;
covered with small, grayish fibers.
Veil: partial veil evanescent.
Spores: 9.3–11 × 5.5 μ; elliptical,
smooth, with pore at tip. Spore print
purple-brown.
Edibility: Hallucinogenic.
Season: September–November.
Habitat: Several to many, separately or in
clusters, on conifer mulch in woods.
Range: British Columbia to N. California.
Look-alikes: The hallucinogenic *P. semilanceata**,
found in manured grass, has smooth
stalk. *P. silvatica* has smaller spores.
Comments: This species, often confused with the
Liberty Cap (*P. semilanceata*), lacks its
narrowly conical cap and is only weakly
hallucinogenic.

⊗ 85 **Liberty Cap**
Psilocybe semilanceata (Fr. ex Secr.) Kum.
Strophariaceae, Agaricales

Description: *Slimy, narrowly conical, brown to tan cap*
with brownish gills and smooth, off-white
stalk; in pastures and manured areas.
Cap: ⅜–1" (1–2.5 cm) wide; sharply
conical, often peaked, and not
expanding; sticky, smooth; brownish,
fading to tan, bruising blue on margin.
Gills: attached, close, broad; grayish,
becoming dark brown.
Stalk: 2–4" (5–10 cm) high, ¹⁄₁₆" (1.5
mm) thick; very thin, whitish.
Veil: partial veil evanescent.
Spores: 11–14 × 7–8 μ; elliptical,

smooth, with pore at tip. Spore print
purple-brown.

Edibility: Hallucinogenic.

Season: Late August–November.

Habitat: Scattered to numerous, in tall grass and
grassy hummocks in cow pastures.

Range: Widely distributed; common in Pacific
NW.; also reported in Quebec.

Look-alikes: The hallucinogenic *P. pelliculosa** and
P. silvatica grow in wood chips or
mulch, and have conical caps.

Comments: This species is one of the most
familiar hallucinogens of the Oregon
coast.

---

55 Scaly-stalked Psilocybe
Psilocybe squamosa (Pers. ex Fr.) Orton
Strophariaceae, Agaricales

Description: *Sticky, brownish cap with grayish- to
purple-brown gills and densely scaly stalk.*
Cap: 1–3″ (2.5–7.5 cm) wide; conical,
expanding to broadly knobbed; sticky,
with evanescent marginal scales,
becoming smooth; brown to tawny or
olive-brown with age.
Gills: attached, close to almost distant,
broad; whitish to gray or purple-brown.
Stalk: 2⅜–4¾″ (6–12 cm) high, ⅛–
⅜″ (0.3–1 cm) thick; brownish,
densely scaly below ring (scales and
ring readily lost on handling or
weathering).
Veil: partial veil leaving evanescent
ring on stalk.
Spores: 12–14 × 6–7.5 μ; elliptical,
smooth, with pore at tip. Spore print
purple-brown.

Season: Late August–October.

Habitat: Single to several, on decayed wood in
mixed forests.

Range: Widespread in N. United States.

Look-alikes: *P. thrausta* has red cap.

Comments: This species has also been placed in the
genera *Stropharia* and *Naematoloma*
because it shares characteristics of all 3
groups. In fact, some specialists hold
that these genera should be combined.

⊗ 84, 95 **Stuntz's Blue Legs**
Psilocybe stuntzii Guzman and Ott
Strophariaceae, Agaricales

Description: *Sticky, brownish cap with brownish gills and brownish, ringed stalk; bruising blue.*
Cap: ⅝–1⅛" (1.5–4 cm) wide; conical, expanding to broadly convex with central knob, or nearly flat; becoming somewhat wavy and uplifted; sticky to moist, smooth; dark to yellow-brown, often green-tinged on margin.
Gills: attached, close to almost distant, broad; off-white, becoming brownish.
Stalk: 1¼–2⅜" (3–6 cm) long, ⅛" (3 mm) thick, sometimes enlarged at base; yellowish, smooth to fibrous.
Veil: partial veil leaves fragile ring that becomes bluish zone on upper stalk.
Spores: 8–12.5 × 6–8 μ; elliptical, smooth, with pore at tip. Spore print purple-brown.

Edibility: Hallucinogenic.
Season: September–December.
Habitat: Several to clustered, in coniferous wood-chip mulch; reported in lawns
Range: Pacific NW.
Look-alikes: The deadly *Galerina autumnalis** has tawny cap fading to yellow, brown gills, and rust-brown spore print. *Stropharia** species do not bruise blue.
Comments: Also known as the "Washington Blue Veil." Like some other blue legs, this does not blue conspicuously. To avoid confusing it with the Deadly Galerina (*Galerina autumnalis*), be sure to take a spore print.

198 **Green Stropharia**
Stropharia aeruginosa (Curt. ex Fr.) Quél.
Strophariaceae, Agaricales

Description: *Slimy, green to blue-green cap with lilac-brown gills; stalk with evanescent ring.*
Cap: 1–3" (2.5–7.5 cm) wide; bell-shaped, becoming broadly convex with knob or nearly flat; margin with veil remnants; slimy to tacky, slightly scaly

to smooth; green to distinctly blue-green, fading to yellowish-green.
Gills: attached, close, broad; whitish, then dingy to purple or gray-brown.
Stalk: 1–3″ (2.5–7.5 cm) long, ⅛–⅝″ (0.3–1.5 cm) thick; somewhat sticky, cottony-scaly, becoming smooth; white above, greenish-blue below.
Veil: partial veil off-white; leaving evanescent ring on upper stalk.
Spores: 6–9 × 4–5 μ; elliptical, smooth, with pore at tip.
Chrysocystidia present. Spore print purple-brown.

Season: Late August–October.

Habitat: Single to several, in grassy areas and on wood debris.

Range: Great Lakes east and south; Pacific NW. and S. California.

Look-alikes: *Pholiota subcaerulea* is more blue than green, with no purple tint to gills or spores. The hallucinogenic *Psilocybe cyanescens** has brown cap when fresh; on bruising or aging, it strongly resembles older specimens.

Comments: This beautiful mushroom has long been suspected of being poisonous, but there is no evidence to support this belief.

197 Questionable Stropharia
Stropharia ambigua (Pk.) Zel.
Strophariaceae, Agaricales

Description: *Slimy yellow cap with evanescent white scales, lilac-brown gills, and evanescent ring on off-white stalk; in coniferous woods.*
Cap: 2–6″ (5–15 cm) wide; blunt to convex, becoming nearly flat; margin with veil remnants; sticky, smooth, yellow.
Gills: attached, close, broad; whitish, becoming grayish to purple-brown.
Stalk: 3–6″ (7.5–15 cm) long, ⅜–¾″ (1–2 cm) thick; silky above ring, fibrous to cottony-scaly below, becoming smooth; whitish, with white runners (rhizomorphs) at base.
Veil: partial veil white; leaving

membranous, evanescent ring or ring zone on upper stalk.

Spores: $11-14 \times 6-7.5 \mu$; elliptical, smooth, with pore at tip.

Chrysocystidia present. Spore print dark purple-brown.

Season: Late August–November.

Habitat: Scattered to numerous, in coniferous or mixed woods.

Range: Pacific NW.; California coast.

Comments: A tall, elegant mushroom when fresh, but of uncertain edibility.

⊗ 215 **Garland Stropharia**
Stropharia coronilla (Bull. ex Fr.) Quél.
Strophariaceae, Agaricales

Description: *Slimy-tacky, yellow cap with purple-brown gills and ring on short, white stalk; in lawns.*

Cap: $1-2''$ (2.5–5 cm) wide; convex to nearly flat; margin with veil remnants; sticky in wet weather, becoming tacky to dry, smooth; pale yellow.

Gills: attached, close, broad; whitish, becoming purple-brown.

Stalk: $1-2''$ (2.5–5 cm) long, $\frac{1}{8}-\frac{3}{8}''$ (0.3–1 cm) thick; dry, woolly-fibrous, becoming smooth; whitish, with white runners (rhizomorphs) at base.

Veil: partial veil white; leaving persistent, membranous ring, lined on upper surface, on upper to mid-stalk.

Spores: $7-10 \times 4.5-5.5 \mu$; elliptical, smooth, with pore at tip.

Chrysocystidia present. Spore print dark purple-brown.

Edibility: Poisonous.

Season: July–November; year-round in California.

Habitat: Single to numerous, in grassy areas.

Range: New England to North Carolina; Colorado; Idaho; West Coast.

Look-alikes: *Agaricus** species have free gills. *S. ambigua** is much taller, larger, and grows in woodlands. *S. riparia* has buff cap and grows in aspen woods or along streams and seepage areas.

Comments: This can be mistaken for an agaricus. There is at least 1 documented case of poisoning as a result.

202   Hard's Stropharia
Stropharia hardii Atk.
Strophariaceae, Agaricales

Description: *Slimy-tacky, dark-spotted, ochre cap with purple-brown gills and with ring on scaly, yellowish stalk.*
Cap: 1–4″ (2.5–10 cm) wide; convex, becoming flat; sticky, becoming dry; bright ochre to brownish-yellow with darkened areas.
Gills: attached, close, moderately broad; gray-brown, then purple-brown.
Stalk: 2–3″ (5–7.5 cm) long, ¼–⅝″ (0.5–1.5 cm) thick; slimy but drying; slightly scaly to woolly-scaly; whitish to pale yellow, with white runners (rhizomorphs) at base.
Veil: partial veil white; leaving persistent, membranous ring on stalk.
Spores: 5–9 × 3–5 μ; elliptical, smooth, with pore at tip.
Chrysocystidia present. Spore print purple-brown.

Season: July–August in Southeast; September–October in Northeast and Ohio.

Habitat: Single to several, on the ground in deciduous woods, especially near beech.

Range: Massachusetts to North Carolina, west to Ohio; most common in Southeast.

Look-alikes: *Agaricus** species have free gills. *S. coronilla**, on lawns, has yellow cap.

Comments: This is one of the larger but less frequently seen species of its genus.

200   Lacerated Stropharia
Stropharia hornemannii (Fr. ex Fr.) Lund. & Nannf.
Strophariaceae, Agaricales

Description: *White-scaly, slimy, brownish cap with lilac-brown gills and ring on scaly stalk.*
Cap: 2–6″ (5–15 cm) wide; bell-shaped

to broadly convex to flat; margin with veil remnants; slimy when wet, smooth; grayish- or purple-brown to smoky- or reddish-brown, then paler.
Gills: attached, close, broad; pale gray, becoming dull purple-brown.
Stalk: 2–5″ (5–12.5 cm) long, ¼–¾″ (0.5–2 cm) thick; silky above ring, densely cottony-scaly below; white; white runners (rhizomorphs) at base.
Veil: partial veil white; leaving persistent, flaring to skirtlike, membranous ring on upper stalk.
Spores: 10–14 × 5.5–7 μ; elliptical, smooth, with pore at tip. Chrysocystidia present. Spore print purple-brown.

Season: Late August–November.
Habitat: Single to numerous, under conifers or on well-decayed coniferous wood.
Range: Widespread in N. North America.
Look-alikes: *Pholiota** species lack purple tints in their spores and typically grow on wood. *Agaricus** species have free gills.
Comments: This is a common fall stropharia.

203, 204 Wine-cap Stropharia
Stropharia rugosoannulata Farlow ex Murr.
Strophariaceae, Agaricales

Description: *Red, red-brown to tan cap with lilac- to gray-black gills and lined ring on stalk.*

Cap: 2–6″ (5–15 cm) wide, bell-shaped, becoming convex to flat; dry, smooth but cracking with age; wine-red to reddish-brown, fading to tan or grayish-white.
Gills: attached, crowded, broad; *white, becoming grayish-lilac to purplish-black.*
Stalk: 4–6″ (10–15 cm) long, ⅜–¾″ (1–2 cm) thick, with widened or bulbous base; smooth to fibrous; whitish, discoloring, with white runners (rhizomorphs) at base.
Veil: partial veil white, deeply wrinkled to segmented below; leaving persistent, membranous ring, with lined upper surface, on upper stalk.

Spores: $10-13 \times 7.5-9$ μ; elliptical, smooth, with pore at tip. Chrysocystidia present. Spore print deep purple-brown, almost black.

Edibility: Choice.

Season: May—October.

Habitat: Scattered to numerous, on wood chips and mulch.

Range: Widespread in N. North America.

Look-alikes: *Agaricus** species have free gills. *Agrocybe** species have brown gills.

Comments: This flavorful edible can be gathered week after week on compost or wood chips, where it produces hundreds of large, firm, fleshy mushrooms.

---

46 Round Stropharia
Stropharia semiglobata (Batsch ex Fr.) Quél.
Strophariaceae, Agaricales

Description: *Slimy, yellowish cap with lilac-brown gills, with ring on off-white stalk; on dung.*
Cap: ⅜–2″ (1–5 cm) wide; round to convex; slimy, smooth, pale yellow.
Gills: attached, close to nearly distant, broad; pale grayish, becoming grayish- to purple-brown.
Stalk: 2–3″ (5–7.5 cm) long, 1⁄16–¼″ (1.5–5 mm) thick; slimy, whitish to pale yellow.
Veil: partial veil slimy; leaving ring zone, darkened by spores, on upper stalk.
Spores: $15-19 \times 7.5-10$ μ; large, long, elliptical, smooth, with pore at tip; purple-brown. Chrysocystidia present. Spore print purple-brown.

Edibility: Edible.

Season: June—September; overwinters in California.

Habitat: Single to several, on horse dung.

Range: Widely distributed in North America.

Look-alikes: The hallucinogenic *Psilocybe coprophila** has dark cap and lacks ring. *Agrocybe pediades**, in grass, has brown gills. Species of *Agaricus** have free gills.

Comments: This common mushroom is edible, but not particularly tasty.

TRICHOLOMA FAMILY
(Tricholomataceae)

This very large and diverse group contains all the gilled mushrooms that cannot be included in any other family. More than 75 genera, with more than 1000 species, have been placed here: many genera are narrowly defined and contain only a very few species, while others include more than 100. There is no concise, overall description of these mushrooms except that most have white or pale spore prints and attached gills. Rather, groups of related genera are differentiated by a correlation of characteristics, including microscopic and chemical features.

Most of the species in this family fall into 1 of 3 broad divisions: mushrooms that grow on wood and are stalkless or have an off-center or lateral stalk; mushrooms that are fleshy, grow usually on the ground, and have gills descending the stalk; and mushrooms that are thin-fleshed, usually grow on the ground, and have attached gills that do not descend the stalk.

196 Shaggy-stalked Armillaria
Armillaria albolanaripes Atk.
Tricholomataceae, Agaricales

Description: *Bright yellow cap with brownish center, white to yellow gills, and whitish stalk sheathed by yellow-tipped scales.*
Cap: 2–5" (5–12.5 cm) wide; convex to broadly convex with knob, becoming flat or slightly upturned; margin with veil remnants; tacky to dry, with small, flattened fibers or scales; bright yellow, *becoming brownish over center,* yellow or whitish toward margin.
Gills: attached or nearly free, close, moderately broad; white to yellow.
Stalk: 1–3" (2.5–7.5 cm) long, ⅜–1" (1–2.5 cm) thick; sheathed by fibrous scales often in concentric zones; below

ring, scales white with yellow to brown tips; smooth, white above ring; stuffed. Veil: partial veil leaving ragged, cottony ring on upper stalk. Spores: 6–8 × 4–5 μ; elliptical, smooth, colorless, weakly amyloid. Spore print white.

Season: Summer in Rocky Mountains; fall in Pacific NW.; November–February in California.

Habitat: Scattered to numerous, under conifers.

Range: Pacific NW., N. California, Colorado, and New Mexico.

Look-alikes: *A. straminea* is brighter yellow, with scalier cap.

Comments: This is a fairly common western armillaria of mountain woods.

---

209 Fragrant Armillaria
Armillaria caligata (Viv.) Gil.
Tricholomataceae, Agaricales

Description: *Brownish fibers and scales over cap with white gills; stalk smooth and white above ring, brownish-scaly and sheathlike below.* Cap: 2–5″ (5–12.5 cm) wide; convex; margin at times uplifted with age, often with veil remnants; dry, cinnamon-brown top layer breaking into patches or small scales, pale flesh showing through cracks. Odor sometimes spicy-fruity; taste nutty. Gills: attached, close, narrow to moderately broad; white, staining brownish with age. Stalk: 2–4″ (5–10 cm) long, ¾–1¼″ (2–3 cm) thick; solid white above ring; *below ring, sheathed by cinnamon-brown veil breaking into zones and patches.* Veil: partial veil leaving membranous, persistent, flaring ring on upper stalk; ring white above, cinnamon below. Spores: 6–7.5 × 4.5–5.5 μ; broadly elliptical, smooth, colorless, nonamyloid. Spore print white.

Edibility: Choice.

Season: July–October, or into November.

Habitat: On the ground under deciduous trees,

particularly oak, in East; under conifers in West; at times in sandy soil.

Range: Widely distributed in North America.

Look-alikes: *A. ponderosa** has white cap when young.

Comments: Also known as *Tricholoma caligatum.* This species has several varieties. Two of these, also found under oak, may have a pungent odor and disagreeable taste, or a slight odor and bitter taste and may stain bluish-green. Another variety occurs under conifers, especially spruce and fir in the West, and has a fragrant, cinnamonlike odor.

---

213 White Matsutake

Armillaria ponderosa (Pk.) Sacc.

Tricholomataceae, Agaricales

Description: *Large, thick-fleshed, white mushroom with reddish-brown scales on cap and stalk, white ring, and discoloring white gills.* Cap: 2–8″ (5–20 cm) wide, convex, becoming flat; cottony margin inrolled at first, uplifted with age; tacky to dry, smooth and *white at first, becoming fibrous-scaly; scales and spots dark reddish-brown over center,* paler toward white margin, often streaked with small, brownish fibers. Odor spicy-sweet.
Gills: attached, close to crowded, narrow to broad; whitish, staining pinkish-brown.
Stalk: 2–6″ (5–15 cm) long, ¾–1⅝″ (2–4 cm) thick, sometimes tapering below; solid, sheathed from base to ring by soft, white veil that breaks into patches, becoming pinkish-brown; white and cottony above ring.
Veil: partial veil leaving membranous, thick, persistent, soft ring on upper stalk; ring white above, often pinkish-brown on underside.
Spores: 5–7 × 4.5–5.5 μ; broadly elliptical to nearly round, smooth, colorless, nonamyloid. Spore print white.

Edibility: Choice.

Season: August–November; December–

February in California.

Habitat: Scattered to numerous, in coastal areas, in sandy soil, and under conifers.

Range: N. North America; Rockies.

Look-alikes: *Catathelasma ventricosa** has double ring, lacks fragrant-spicy odor.

Comments: Also known as *Tricholoma ponderosum*. This is sold fresh, in season, in some Japanese grocery stores. In N. New England, the Pungent Armillaria (*A. viscidipes*) has a pungent, potatolike odor and taste and is common on mossy hummocks under hemlock.

---

205 Scaly Yellow Armillaria
Armillaria straminea (Kromb.) Kum. var. *americana* Mit. & A.H.S.
Tricholomataceae, Agaricales

Description: *Scaly, yellow cap with yellowish gills and white stalk with shaggy, yellow scales in concentric zones below ring.*
Cap: 2–7" (5–18 cm) wide; conical to convex, becoming broadly knobbed to flat; margin incurved at first, becoming straight, with remnants of cottony yellow veil; dry, *with flattened, fibrous, deep yellow or darker scales, becoming pyramidal and downcurled and arranged in concentric circles;* becoming smooth and straw-yellow, sometimes fading to whitish.
Gills: attached, moderately close, broad; whitish, then lemon-yellow.
Stalk: 2–5" (5–12.5 cm) long, ⅝–1" (1.5–2.5 cm) thick, sometimes with thick bulb; smooth and white above veil; shaggy below, with yellowish scales arranged in concentric zones.
Veil: partial veil leaving thick, yellowish, cottony ring on upper stalk.
Spores: 6–8 × 4–5 μ; elliptical, smooth, colorless, weakly amyloid. Spore print white.

Season: July–August in Rocky Mountains; September–October in Pacific NW.

Habitat: On the ground under aspen and in mixed woods.

Range: Rocky Mountains and Pacific NW.
Look-alikes: *A. albolanaripes** is less scaly and has cap with brownish center.
Comments: Also known as *A. luteovirens*. There is a white form of this species that also has large, downcurled scales.

---

210 Fetid Armillaria
Armillaria zelleri Stuntz & A.H.S.
Tricholomataceae, Agaricales

Description: *Thick-fleshed, orange-brown, scaly cap with off-white gills and whitish stalk with orange-brown scales below ring; odor unpleasant.*
Cap: 2–6″ (5–15 cm) wide; rounded, becoming flat or broadly convex with knob; margin incurved at first, with veil remnants; slimy to tacky, fibrous to minutely scaly; orange to orange-brown, sometimes tinted greenish. Odor and taste mealy, disagreeable.
Gills: attached, close to crowded, narrow to moderately broad; whitish, staining rust-brown.
Stalk: 2–5″ (5–12.5 cm) long, ⅜–1″ (1–2.5 cm) thick, tapering to pointed base; stalk below ring with sheath breaking into orange scales or fibrous zones; white and cottony above ring.
Veil: partial veil leaving membranous, ragged ring on upper stalk; ring white above, orange on underside.
Spores: 4–5 × 3–4 μ; elliptical, smooth, colorless, nonamyloid. Spore print white.
Edibility: Edible.
Season: July–August in Rocky Mountains; December–February in California.
Habitat: Scattered under pine and aspen.
Range: N. United States and W. Canada, especially Pacific NW.; also reported in Maine and Tennessee.
Look-alikes: *Tricholoma aurantium** has evanescent veil and flat ring zone.
Comments: Also known as *Tricholoma zelleri*. This mushroom is abundant at times in the Northwest.

182, 206   Honey Mushroom
Armillariella mellea (Vahl ex Fr.) Kar.
Tricholomataceae, Agaricales

Description: *Yellow-brown, sticky cap with erect black hairs over center, discoloring whitish gills, and stalk with white to cream ring; on wood.*
Cap: 1–4" (2.5–10 cm) wide; oval, becoming convex to nearly flat; sticky to dry; yellow- to rust-brown, often with blackish, erect, hairlike scales.
Gills: attached or slightly descending stalk, nearly distant, narrow; whitish, staining yellow to reddish.
Stalk: 2–6" (5–15 cm) long, 1/4–3/4" (0.5–2 cm) thick; fibrous, whitish, discoloring yellowish to rust-brown; stuffed to hollow.
Veil: partial veil leaving membranous, usually persistent ring on upper stalk.
Spores: 7–10 × 5–6 μ; elliptical, smooth, colorless, nonamyloid. Spore print white.

Edibility: Choice, with caution.

Season: August or September–November; December–February in California.

Habitat: Clustered at bases and near stumps of trees; also in open areas.

Range: Widely distributed in North America.

Look-alikes: *A. tabescens** has dry cap and no ring. The poisonous *Omphalotus olearius** is orange, lacks ring. The hallucinogenic *Gymnopilus spectabilis** is orange, very bitter, has orange-brown spores. The deadly *Galerina autumnalis** is smaller, has smooth cap, evanescent ring, and brown spores. The poisonous *Naematoloma fasciculare** is smaller, has smooth cap, greenish-yellow gills when young, and purple-brown spores.

Comments: Also known as *Armillaria mellea.* While this is an excellent edible, it must be well cooked—even so, some people may experience stomach upset. This is a complex of nearly identical species, many of which have toxic look-alikes. The most dangerous of these are listed above. The Honey produces stringlike

runners, or rhizomorphs, which can
extend as far as 100 m, and from
which additional fruiting bodies grow.
One specimen was estimated to be
more than 450 years old.

272, 313 Ringless Honey Mushroom
Armillariella tabescens (Scop. ex Fr.) Sing.
Tricholomataceae, Agaricales

Description: *Yellow-brown, dry, scaly cap with white to
discolored gills and stalk; clustered on wood
or over buried wood.*
Cap: 1–4″ (2.5–10 cm) wide; convex
to flat or sunken; margin uplifted with
age; dry, yellow-brown, with reddish-
tawny to brown, flat to erect scales.
Gills: somewhat descending stalk,
nearly distant, narrow to broad;
whitish, staining pinkish to brownish.
Stalk: 3–8″ (7.5–20 cm) long, ¼–⅝″
(0.5–1.5 cm) thick, tapering toward
base; fibrous-scruffy; off-white to
brownish; stuffed to hollow.
Spores: 6–10 × 5–7 μ; elliptical,
smooth, colorless, nonamyloid. Spore
print white.
Edibility: Good, with caution.
Season: September–November.
Habitat: In large clusters, at bases of trees or
over buried wood, especially oak.
Range: NE. North America to Florida, west to
Kansas and Texas.
Look-alikes: *A. mellea** has tacky cap and ring. The
poisonous *Omphalotus olearius** is
orange, with smooth cap and close gills
descending stalk. The hallucinogenic
*Gymnopilus spectabilis** is orange, with
smooth cap and ring, orange-brown
spores, and bitter taste.
Comments: Also known as *Clitocybe monadelpha*.
This species is related to the Honey
Mushroom (*A. mellea*) because it has
runners (rhizomorphs), which can climb
the bark of trees. It is very tasty, but
has been known to cause stomach
upset, and should be thoroughly
cooked.

12 Powder Cap
Asterophora lycoperdoides (Bull. ex. Mér.) Dit. ex Fr.
Tricholomataceae, Agaricales

Description: *Small, dry, white cap becoming powdery-brown, with white gills and stalk; on decayed mushrooms.*

Cap: ⅜–¾″ (1–2 cm) wide; nearly round; dry, cottony; *white, becoming covered with dense brown powder.* Odor and taste mealy.

Gills: attached, distant, narrow, thick; sometimes forked, often malformed; whitish.

Stalk: ¾–1¼″ (2–3 cm) long, ⅛–⅜″ (0.3–1 cm) thick; stout, silky to minutely hairy; white, becoming brownish; stuffed to hollow.

Spores: 5–6 × 3.5–4 μ; oval, smooth, colorless, nonamyloid. Spore print white. Chlamydospores (12–15 μ) round, bumpy and winged, thick-walled, brownish.

Season: July–November.

Habitat: On rotting mushrooms, especially *Lactarius* and *Russula* species.

Range: Widely distributed in North America.

Look-alikes: *A. parasitica* has nonpowdery, dirty white, silky cap and smooth chlamydospores. *Collybia tuberosa** and its relatives have well-formed gills and usually some kind of tuber.

Comments: Also known as *Nyctalis asterophora*. This mushroom occurs on species of *Russula* and *Lactarius*, especially the *Russula nigricans-dissimulans* group. It is most frequently seen after the host mushroom has decayed and been reduced to a misshapen blackish mass.

9 Conifer-cone Baeospora
Baeospora myosura (Fr.) Sing.
Tricholomataceae, Agaricales

Description: *Small, tan to whitish cap with crowded, white gills and white to brownish stalk; on fallen conifer cones.*

Cap: ¼–¾" (0.5–2 cm) wide; convex, becoming broadly convex; moist, *smooth;* tan, drying whitish-tan.
Gills: attached, crowded, narrow; white.
Stalk: ⅝–2¼" (1.5–5.5 cm) long, 1/32–1/16" (1–1.5 mm) thick; dry, minutely hairy, with long, coarse hairs at base; whitish, becoming brownish; hollow.
Spores: 3–4 × 1–2 μ; small, elliptical, smooth, colorless, amyloid. Spore print white.

Season: September–October.

Habitat: On cones of eastern white pine, Norway and Sitka spruce, and probably other conifers.

Range: New York to Alaska.

Look-alikes: *Strobilurus** species are most reliably differentiated microscopically. *Collybia** species do not grow on cones.

Comments: Formerly known as *Collybia conigena.* Just as some mushrooms grow only on oak or under pine, some occur only on conifer cones.

105 Lavender Baeospora
Baeospora myriadophylla (Pk.) Sing.
Tricholomataceae, Agaricales

Description: *Small, lavender mushroom, aging ochre-brown, with lavender gills; on decayed wood.*
Cap: ⅜–1⅝" (1–4 cm) wide; convex to flat, sunken in center at times; margin incurved at first, wavy and uplifted with age; moist, smooth; lavender to brownish-lavender, becoming ochre-brown to pale buff.
Gills: attached or nearly free, close to crowded, narrow; lavender, becoming tinged ochre-brown or paler.
Stalk: ⅜–2" (1–5 cm) long, 1/32–⅛" (1–3 mm) thick; pliant-tough, dry, often flattened to grooved, becoming smooth, with long, coarse hairs at base; lavender, becoming somewhat brownish; stuffed to hollow.
Spores: 3.5–4.5 × 2–3 μ; small,

broadly elliptical, smooth, colorless, amyloid. Spore print white.

Season: June–October.

Habitat: Several to clustered, on decayed wood of both deciduous and coniferous trees, especially poplar and hemlock.

Range: N. North America.

Look-alikes: *Collybia iocephala** is violet, with distant gills, has an unpleasant odor, and grows on the ground. *Laccaria amethystina*, also ground-dwelling, has thick, distant gills. *Mycena lilacifolia*, also on wood, has slimy, lilac cap fading to light yellow, gills descending stalk, and nonamyloid spores.

Comments: Formerly known as *Collybia myriadophylla*.

---

280 Pink Calocybe
Calocybe carnea (Bull. ex Fr.) Küh.
Tricholomataceae, Agaricales

Description: *Small, fleshy, pinkish cap with white gills; in lawns.*
Cap: ⅝–1" (1.5–2.5 cm) wide; convex to nearly flat or with knob; dry, smooth to minutely hairy; *brick-red to pinkish*, becoming whitish-tan.
Gills: attached or slightly descending stalk, *crowded*, narrow; *white*.
Stalk: ⅝–1" (1.5–2.5 cm) long, ¹⁄₁₆–⅛" (1.5–3 mm) thick; fibrous, somewhat finely hairy; pinkish; hollow.
Spores: 4–6 × 2–3 μ; oval, smooth, colorless; basidia with carminophilous granules. Spore print white.

Season: August–October.

Habitat: On the ground in lawns and open places; also in open woods.

Range: Widespread in N. North America.

Look-alikes: *C. persicolor* has peach to pinkish-buff cap and fused stalks. *C. onychina* has purplish cap and yellow gills.

Comments: Formerly known as *Tricholoma carneum* and *Clitocybe socealis*. The Pink Calocybe is uncommon, but one of our prettiest late-summer and fall lawn mushrooms.

97 Grayling
Cantharellula umbonata (Gmel. ex Fr.) Sing.
Tricholomataceae, Agaricales

Description: *Small, grayish, knobbed cap, off-white, forked gills, and off-white stalk; in moss.*
Cap: 1–2" (2.5–5 cm) wide; convex to flat and sunken in center, usually with small, distinct knob; dry, minutely hairy; grayish-brown to smoky-gray. Flesh white, bruising reddish.
Gills: *descending stalk*, close, narrow, thick, thick-edged, *regularly forked*; white to cream, bruising yellow or red.
Stalk: 1–3" (2.5–7.5 cm) long, ⅛–¼" (3–5 mm) thick; somewhat tough, off-white to grayish; stuffed.
Spores: 7–11 × 3–4 μ; spindle-shaped, smooth, colorless, amyloid. Spore print white.
Edibility: Good.
Season: August–November.
Habitat: Scattered to many, in haircap moss.
Range: Widespread in NE. North America.
Look-alikes: Small, gray *Agaricus** species in moss should be examined carefully. Gray *Lactarius** in moss have white latex.
Comments: Formerly known as *Cantharellus umbonatus*.

211 Imperial Cat
Catathelasma imperialis (Fr.) Sing.
Tricholomataceae, Agaricales

Description: *Large, brownish cap with buff gills descending long, double-ringed stalk.*
Cap: 5–16" (12.5–40 cm) wide; convex to flat; margin incurved at first; tacky to dry, breaking into small patches over center; dirty brownish. Flesh hard, thick, white. Odor and taste mealy.
Gills: descending stalk, close, broad; buff-yellow to olive-gray. Gill tissue divergent.
Stalk: 5–7" (12.5–18 cm) long, 2–3" (5–7.5 cm) thick, narrowing to pointed base; dry, covered to ring by yellowish to pinkish-brown sheath.

Veil: *partial veil double, leaving 2-layered, flaring ring on upper stalk:* upper layer membranous, persistent; lower layer slimy, adhering as sheath to stalk.
Spores: $11-14 \times 4-6 \mu$; long, cylindrical, smooth, colorless, amyloid. Spore print white.

Edibility: Edible.

Season: August–October.

Habitat: Single to scattered, under conifers in dense forests, especially in mountains.

Range: Rocky Mountains and Pacific NW.

Look-alikes: *C. ventricosa** is smaller, with short, stout stalk and dry cap. *C. singeri* has dingy yellow, glutinous cap.

Comments: Also known as *Armillaria imperialis*.

---

212 Swollen-stalked Cat

Catathelasma ventricosa (Pk.) Sing.
Tricholomataceae, Agaricales

Description: *Thick-fleshed, dirty white mushroom with gills descending short, double-ringed stalk.*
Cap: 3–6" (7.5–15 cm) wide; convex to broadly convex; dry, smooth, patchy with age; off-white to grayish. Flesh hard, thick, white. Taste unpleasant.
Gills: descending stalk, close to nearly distant, narrow to broad; whitish to buff. Gill tissue divergent.
Stalk: 2–5" (5–12.5 cm) long, 1–2" (2.5–5 cm) thick, narrowing, deep in soil; stout, dry, white to yellow-brown.
Veil: *partial veil double, leaving 2-layered, flaring ring:* upper layer hairy; lower membranous, persistent.
Spores: $9-12 \times 4-6 \mu$; elliptical, smooth, colorless, amyloid. Spore print white.

Edibility: Good.

Season: August–October.

Habitat: Scattered to numerous, under coniferous trees, especially spruce.

Range: N. North America, south to Colorado, N. California in mountains.

Look-alikes: *C. imperialis** is much larger, with sticky, dark cap when young. *C. macrospora* has broader spores.

Armillaria ponderosa* is spicy-fragrant and lacks double veil. *Amanita** species have single ring.

Comments: Once called *Armillaria ventricosa,* this is a good edible when cooked.

---

73 Rooting Redwood Collybia
Caulorhiza umbonata (Pk.) Lennox
Tricholomataceae, Agaricales

Description: *Brownish cap with knob, pale to buff gills, and long, fibrous, deeply rooting stalk; under redwood.*
Cap: 1–6″ (2.5–15 cm) wide, cone-shaped to convex to flat with knob; moist, smooth; reddish to orange-brown, fading to tannish.
Gills: notched, close, broad; off-white to pinkish-buff.
Stalk: 2–6″ (5–15 cm) long, ¼–¾″ (0.5–2 cm) thick, enlarging below; *deeply rooted up to twice its height;* brittle-tough, twisted-lined; buff tan.
Spores: 5–8 × 3–5 μ; elliptical, smooth, colorless, amyloid. Spore print white.
Season: October–January.
Habitat: Single to scattered, under coastal redwood.
Range: California.
Look-alikes: *Phaeocollybia** have brown spores.
Comments: Formerly known as *Collybia umbonata. C. hygrophoroides,* a similar eastern species, grows in spring in deciduous woods.

---

496 White Oysterette
Cheimonophyllum candidissimus (Berk. & Curt.) Sing.
Tricholomataceae, Agaricales

Description: *Small, white cap with white gills and stublike, lateral stalk; on dead wood.*
Cap: ⅟₁₆–¾″ (0.15–2 cm) wide; semicircular to kidney- or shell-shaped; margin radially lined, incurved at first; dry, minutely hairy, white.

Gills: descending stalk or attached to base; distant, broad, white.
Stalk (when present): stublike or lateral.
Spores: 5–6 × 4.5–5.5 μ; round, smooth, colorless. Spore print white.

Season: July–October.

Habitat: On decaying wood.

Range: Widely distributed in North America.

Look-alikes: *Panellus mitis* grows only on conifers, has partly slimy flesh and amyloid, sausage-shaped spores. *Crepidotus** species have ochre-brown spores.

Comments: Formerly known as *Pleurotus candidissimus*. These mushrooms often can be found in abundance and look like shining jewels on the dark wood.

279 Smoky-brown Clitocybe
Clitocybe avellaneialba Murr.
Tricholomataceae, Agaricales

Description: *Dark brown cap with blackish scales or fibers on center, whitish gills, brown stalk.*
Cap: 2–8″ (5–20 cm) wide; convex, becoming flat to somewhat sunken, sometimes vase-shaped with age; margin inrolled at first; usually knobbed; moist; smooth, fibrous, or scaly; blackish- to grayish-brown.
Gills: slightly descending stalk, close, narrow; white to cream-colored.
Stalk: 2–7″ (5–18 cm) long, ⅜–1¼″ (1–3 cm) thick, enlarging downward; smooth, brownish.
Spores: 8–11 × 4–5.5 μ; broadly spindle-shaped, smooth, colorless. Spore print white.

Season: October–November.

Habitat: On or near rotting alder wood.

Range: Pacific NW.

Look-alikes: *C. clavipes** has gray-brown cap and enlarged, spongy stalk base. *Clitocybula atrialba** grows on various trees, has scaly stalk and amyloid spores.

Comments: This mushroom can grow to be quite large. It can be common late into the fall, and sometimes grows in clusters.

320 **Fat-footed Clitocybe**
Clitocybe clavipes (Pers. ex Fr.) Kum.
Tricholomataceae, Agaricales

Description: *Gray-brown cap with white gills descending
stalk with enlarged, spongy base.*
Cap: 1–3" (2.5–7.5 cm) wide; convex
to flat to slightly sunken; moist,
smooth, grayish-brown. Flesh white,
thick at center, thinning outward.
Gills: deeply descending stalk, nearly
distant, narrow to broad; whitish.
Stalk: ¾–2⅜" (2–6 cm) long, ¼–⅜"
(0.5–1 cm) thick, enlarging toward
bulbous base; whitish to ash; spongy.
Spores: 5–8 × 3–5 μ; broadly
elliptical, smooth, colorless. Spore
print white.
Edibility: Edible with caution.
Season: July–November; December–February
in California.
Habitat: Single to several, on the ground under
conifers, especially white pine;
sometimes under deciduous trees.
Range: Widely distributed in North America.
Look-alikes: *C. nebularis** has crowded gills that only
slightly descend stalk and yellow spore
print. *C. gibba** has pinkish-tan cap.
Comments: When alcohol is consumed after this
mushroom, a headache and upper-body
flushing or transient rash may occur.

⊗ 241, 245 **Sweating Mushroom**
Clitocybe dealbata (Fr.) Kum. var. *sudorifica*
Pk.
Tricholomataceae, Agaricales

Description: *Small, smooth, dull white mushroom with
gills slightly descending stalk; in grass and
open woods.*
Cap: ⅜–1⅝" (1–4 cm) wide; convex to
flat, then sunken; margin incurved,
then wavy and upturned; dry, smooth,
white, then grayish to pinkish.
Gills: attached or slightly descending
stalk; close, narrow, white.
Stalk: 1–2" (2.5–5 cm) long, ¹⁄₁₆–¼"
(1.5–5 mm) thick; smooth, fibrous-

tough, white.
Spores: 5–6 × 3–4 μ; broadly elliptical, smooth, colorless. Spore print white.

Edibility: Poisonous.

Season: July–September; November–February in California.

Habitat: Scattered to numerous, on the ground in grass and in open woods.

Range: Widely distributed in North America.

Look-alikes: Several small, whitish *Clitocybes** are similar and not easily differentiated by field characteristics. *C. dilatata** is larger, grows in clusters in West. *Marasmius oreades** is buff, with distant, nearly free gills and rubbery stalk.

Comments: This little white mushroom often grows near Fairy Ring Mushrooms (*Marasmius oreades*) and is picked along with them by mistake. It contains the toxin muscarine, which causes transient, mild to severe poisoning, including perspiration, chills, and tunnel vision.

⊗ 238 Crowded White Clitocybe
Clitocybe dilatata Pers. ex Kar.
Tricholomataceae, Agaricales

Description: *Large, clustered, overlapping, fleshy, white mushroom with wavy, lobed to split margin.* Cap: 2–6″ (5–15 cm) wide; convex, becoming flat; margin incurved, becoming irregular and upturned; dry, smooth, whitish. Taste disagreeable. Gills: attached or slightly descending stalk, close, narrow; whitish to buff. Stalk: 2–5″ (5–12.5 cm) long, ¼–¾″ (0.5–2 cm) thick; dry, fibrous, white. Spores: 4.5–6.5 × 3–3.5 μ; elliptical, smooth, colorless, nonamyloid. Spore print white.

Edibility: Poisonous.

Season: May–November.

Habitat: Densely clustered along roads.

Range: Pacific NW.

Comments: Also known as *C. cerussata* var. *difformis*. This species is reported to cause muscarine poisoning.

262 Wood Clitocybe
Clitocybe ectypoides (Pk.) Sacc.
Tricholomataceae, Agaricales

Description: *Yellowish mushroom with depressed cap covered with small brown fibers; on wood.*
Cap: 1–2" (2.5–5 cm) wide; sunken in center to funnel-shaped; moist, watery grayish-yellow; with brown to black radial fibers and pointed scales.
Gills: descending stalk, almost distant, narrow, sometimes forked; yellowish.
Stalk: 1–2" (2.5–5 cm) long, ⅟₁₆–⅛" (1.5–3 mm) thick; grayish-yellow; solid.
Spores: 6.5–8 × 3.5–5 μ; elliptical, smooth, colorless, amyloid. Spore print white.
Season: July–September.
Habitat: On rotting logs, often of hemlock.
Range: E. Canada to North Carolina, west to Washington.
Look-alikes: *Gerronema chrysophylla** has golden yellow gills and yellow spore print.
Comments: Also known as *Pseudoarmillariella ectypoides.*

263 Funnel Clitocybe
Clitocybe gibba (Fr.) Kum.
Tricholomataceae, Agaricales

Description: *Sunken to funnel-shaped, red- to pink-tan cap with white gills descending stalk.*
Cap: 2–3" (5–7.5 cm) wide; convex to sunken, becoming funnel-shaped; dry, silky; reddish- to pinkish-tan.
Gills: descending stalk, close, narrow, whitish.
Stalk: 1⅝–3¼" (4–8 cm) long, ¼–⅜" (0.5–1 cm) thick, enlarging downward slightly; smooth, spongy, off-white.
Spores: 5–8 × 3–5 μ; elliptical, smooth, colorless. Spore print white.
Edibility: Edible.
Season: July–October; January–February in California.
Habitat: Under deciduous trees, especially oak.
Range: Quebec to Florida, to West Coast.

Look-alikes: *C. squamulosa* is brown, occurs under
conifers.

Comments: Also known as *C. infundibuliformis*. This
species is one of the first gilled ground
mushrooms to appear in summer. A
very large variety, *maxima,* also known
as *C. maxima,* is broadly knobbed to
flat, becomes funnel-shaped with age,
and has a stout stalk and strong odor.

264 Giant Clitocybe

Clitocybe gigantea (Sow. ex Fr.) Quél.
Tricholomataceae, Agaricales

Description: *Very large, white to buff cap with crowded,
white gills slightly descending short, stout,
whitish stalk.*
Cap: 4–18″ (10–46 cm) wide; broadly
convex to flat or funnel-shaped; margin
inrolled at first, spreading and
becoming furrowed; moist, smooth,
drying and becoming somewhat
cottony, soft to tough; white to buff.
Odor and taste mild to foul.
Gills: descending stalk, very crowded,
broad, sometimes forked; white to buff,
brownish when dried.
Stalk: 1–4″ (2.5–10 cm) long, 1–2″
(2.5–5 cm) thick; usually short and
stout, solid, dry, smooth; whitish.
Spores: 6–8 × 3–4.5 μ; elliptical,
smooth, colorless, amyloid. Spore print
white.

Edibility: Edible.

Season: August–October.

Habitat: Scattered to numerous or in fairy rings,
in open woods, waste places, and camps.

Range: Widely distributed in North America;
most common in Pacific NW.

Look-alikes: *C. gibba* var. *maxima* has pinkish-tan
cap and nonamyloid spores. *C. candida,*
more common in Northeast, is smaller.
C. robusta is smaller, has nonamyloid
spores and yellow spore print.

Comments: Also known as *Leucopaxillus giganteus*.
This large clitocybe is often found in
large arcs or fairy rings of 12–24
mushrooms.

267, 354 Cloudy Clitocybe
Clitocybe nebularis (Bat. ex Fr.) Kum.
Tricholomataceae, Agaricales

Description: *Large, grayish-brown cap with whitish
gills attached to long, thick, white stalk;
odor of skunk cabbage.*
Cap: 3–6″ (7.5–15 cm) wide; convex,
becoming flat to sunken; margin
incurved at first; dry, with radial fibers;
grayish. *Odor disagreeable.*
Gills: slightly descending stalk, close,
broad; white, becoming yellowish.
Stalk: 3–4″ (7.5–10 cm) long, 1–1⅝″
(2.5–4 cm) thick, with enlarged base;
whitish, with minute brownish fibers.
Spores: 6–7 × 3–4 μ; elliptical,
smooth, colorless. *Spore print pale yellow.*
Season: August–December; overwinters in
California.
Habitat: Scattered to several, in coniferous or
mixed woods.
Range: Rocky Mountains and West Coast.
Comments: Also known as *Lepista nebularis*, this is
sometimes parasitized by the Parasitic
Volvariella (*Volvariella surrecta*).

346 Blewit
Clitocybe nuda (Bull. ex Fr.) Big. & A.H.S.
Tricholomataceae, Agaricales

Description: *Large, violet cap, fading to tan, with violet
to buff gills and violet, often bulbous stalk;
in the open or in open woods.*
Cap: 2–6″ (5–15 cm) wide; convex to
flat or broadly sunken in center; margin
incurved, then wavy and upturned;
moist to dry, smooth; violet-gray or
tinged brownish. Odor fragrant.
Gills: notched, crowded, moderately
broad; violet to buff, discoloring.
Stalk: 1–3″ (2.5–7.5 cm) long, ⅜–1″
(1–2.5 cm) thick, usually with large,
bulbous base; solid, dry, scaly; violet to
gray or brown-tinged.
Spores: 6–8 × 4–5 μ; elliptical,
smooth to minutely roughened,
colorless. *Spore print pinkish-buff.*

Edibility: Choice.

Season: August—December; November—March in California.

Habitat: In open areas on compost, under blackberry stems, along paths and wood borders.

Range: Widely distributed in North America.

Look-alikes: *C. irina* lacks purplish coloration. *C. saeva* has buff cap and pale purplish stalk. *C. graveolens* lacks purplish coloration and has strongly unpleasant odor and taste. *C. tarda* is purplish but has much thinner, fibrous stalk. *Cortinarius alboviolaceus** has cobwebby veil when very young, gills that eventually turn rust-brown. *Entolomas** have angular, salmon-pink spores.

Comments: Also known as *Lepista nuda;* formerly known as *Tricholoma nudum* and, mistakenly, *T. personatum.* This strong-flavored mushroom is popular wherever it grows, and large quantities can usually be gathered late in the fall.

---

350 Anise-scented Clitocybe
Clitocybe odora (Bull. ex Fr.) Kum.
Tricholomataceae, Agaricales

Description: *Blue-green cap with whitish gills and whitish stalk; aniselike odor.*
Cap: 1–4" (2.5–10 cm) wide; convex to flat or somewhat sunken; margin incurved, becoming wavy and uplifted; moist, smooth; dingy green to bluish-green, sometimes blue or nearly white.
Gills: attached or slightly descending stalk, close, broad; whitish to buff or green-tinged.
Stalk: 1–3" (2.5–7.5 cm) long, ⅛–⅝" (0.3–1.5 cm) thick, sometimes enlarged above or below; whitish or tinged greenish; base often spongy or covered with white threads (mycelium).
Spores: 6–7 × 3–4 μ; elliptical, smooth, colorless. Spore print pinkish-cream.

Edibility: Edible.

Season: July—September; November—February

in California.

Habitat: Scattered to numerous, in oak woods.
Range: Widely distributed in North America.
Look-alikes: *C. fragrans** is smaller, whitish.
Comments: A West Coast variety, var. *pacifica*,
under conifers, is green all over.

---

237 **Fragrant Clitocybe**
Clitocybe suaveolens (Fr.) Kum.
Tricholomataceae, Agaricales

Description: *Light gray-brown to pinkish mushroom with
gills attached to or slightly descending
stalk; odor of anise.*
Cap: ⅜–2⅜" (1–6 cm) wide; convex,
becoming flat or depressed; margin
incurved, becoming uplifted; moist,
smooth; light gray-brown, tinted pink,
fading to pinkish. Odor of anise.
Gills: attached or slightly descending
stalk, close to nearly distant, broad;
pinkish- to ochre-buff.
Stalk: ¾–2⅜" (2–6 cm) long, ¹⁄₁₆–¼"
(1.5–5 mm) thick, sometimes
enlarging toward base; curved, silky-
shining, smooth, gray-brown to
pinkish; solid, becoming hollow.
Spores: 6.5–9 × 4–4.5 μ; elliptical,
smooth, colorless, nonamyloid. Spore
print pale pinkish-buff.
Season: July–December.
Habitat: On the ground under conifers.
Range: Colorado; Pacific NW. and California.
Look-alikes: *C. fragrans** is white.
Comments: This well-known western mushroom
bears a European name that has been
applied to several different mushrooms
with an anise odor.

---

352 **Black-and-white Clitocybula**
Clitocybula atrialba (Murr.) Sing.
Tricholomataceae, Agaricales

Description: *Blackish cap and scaly stalk with distant
whitish gills; on deciduous wood.*
Cap: 1–4" (2.5–10 cm) wide; convex,
expanding to flat, becoming sunken to

vase-shaped; margin decurved,
becoming raised; finely hairy to fibrous,
moist to dry, dark blackish-brown.
Gills: descending stalk, distant,
moderately broad, veined, sometimes
forked; grayish.
Stalk: 2–5″ (5–12.5 cm) long, ⅛–⅝″
(0.3–1.5 cm) thick, enlarged at base;
scaly and fibrous-streaked, brownish.
Spores: 9–11 × 7–9 μ and broadly
elliptical, or 6–10 μ and round;
smooth, colorless, amyloid. Spore print
white.

Season: May–November.

Habitat: Scattered to clustered, on deciduous
wood such as alder, maple, and oak.

Range: Pacific NW.

Look-alikes: *Clitocybe avellaneialba**, on rotting
alder, has brownish cap, smooth stalk.

Comments: Formerly called *Clitocybe atrialba,* this
species was transferred to *Clitocybula*
because of its amyloid spores.

---

22 Family Collybia
Clitocybula familia (Pk). Sing.
Tricholomataceae, Agaricales

Description: *Grayish-buff cap with incurved margin,*
crowded, ash gills, and pale gray stalk; in
large clusters on wood.
Cap: ⅜–1⅝″ (1–4 cm) wide; bell-
shaped, becoming convex, then flat;
margin incurved at first, spreading,
somewhat uplifted and torn with age;
moist, smooth; brownish-buff to cream-
buff. Flesh thin, fragile.
Gills: attached to nearly free, crowded,
narrow; whitish.
Stalk: 1⅝–3¼″ (4–8 cm) long, ⅟₁₆–⅛″
(1.5–3 mm) thick; smooth, grayish-
white, with white, flat hairs at base.
Spores: 3.5–4.5 μ; round, smooth,
colorless, amyloid. Spore print white.

Edibility: Edible.

Season: August–September.

Habitat: In large clusters on coniferous logs.

Range: Widely distributed in North America.

Look-alikes: *Mycena** species have straight cap

margin when young.

Comments: Formerly known as *Collybia familia*. Be certain that the mushroom identified as the Family Collybia is found on coniferous wood.

---

23 Clustered Collybia
Collybia acervata (Fr.) Kum.
Tricholomataceae, Agaricales

Description: *Reddish-brown, convex cap and stalk with off-white to pink-tinged gills; on wood.*
Cap: 1–2″ (2.5–5 cm) wide; convex to nearly flat; margin incurved at first; moist to dry, smooth; *reddish-brown,* drying to whitish-tan.
Gills: attached or nearly free, close, narrow; *whitish to pinkish-tinged.*
Stalk: 1¼–4″ (3–10 cm) long, ¹⁄₁₆–¼″ (1.5–5 mm) thick; dry, smooth, brittle; reddish-brown, base white-hairy; hollow.
Spores: 5.5–7 × 2.5–3 μ; elliptical, smooth, colorless. Spore print white.
Season: July–October; November in Texas.
Habitat: Clustered on coniferous logs and decayed wood.
Range: Northeast; Colorado and Texas.
Look-alikes: *Marasmius cohaerens** grows on the ground and has minute, spotlike hairs.
Comments: Because the gills of this species become pinkish-tinged as the mushroom ages, a spore print is necessary to identify it.

---

108 Little Brown Collybia
Collybia alkalivirens Sing.
Tricholomataceae, Agaricales

Description: *Small, dark brown mushroom with cap fading to buff, and dark brown gills.*
Cap: ⅜–1¼″ (1–3 cm) wide; convex, becoming flat or with broad knob; margin incurved at first, radially lined to furrowed; moist to dry, smooth; dark blackish- to reddish-brown, fading to brown or buff, often from disc outward. Cap and flesh greenish with KOH.

Gills: attached or nearly free, close to distant, narrow to broad; red-brown to dark brown or blackish.

Stalk: 1–3" (2.5–7.5 cm) long, $\frac{1}{32}$–$\frac{1}{8}$" (1–3 mm) thick; fibrous-brittle, smooth; dark reddish- to blackish-brown; greenish with KOH; hollow.

Spores: 5.4–8.6 × 2.2–5.4 μ; narrowly elliptical, smooth, colorless, nonamyloid; *greenish with KOH.* Spore print white (when obtainable).

Season: May–July in East; October–December in Northwest.

Habitat: Single to several or somewhat clustered, often near stumps, on decaying wood, soil, in moss, or on old fern hummocks.

Range: Widespread in N. North America.

Comments: This species is often common in eastern deciduous woods in the spring.

---

24 Tufted Collybia
Collybia confluens (Pers. ex Fr.) Kum.
Tricholomataceae, Agaricales

Description: *Brownish cap with whitish gills and densely white-hairy, buff-brown, rigid-brittle stalk; in dense clusters.*

Cap: 1–2" (2.5–5 cm) wide; convex to flat; margin incurved, becoming wavy; moist, smooth; reddish-brown, fading to grayish-pink or whitish.

Gills: *free or attached, very crowded,* narrow, whitish.

Stalk: 2–4" (5–10 cm) long, $\frac{1}{16}$–$\frac{1}{4}$" (1.5–5 mm) thick; compressed, brittle-tough; buff-brown, covered with dense, whitish hairs; base with cottony-white threads (mycelium); hollow.

Spores: 7–10 × 2–4 μ; narrowly elliptical, smooth, colorless. Spore print white to cream.

Edibility: Edible with caution.

Season: July–November.

Habitat: In clusters, on fallen leaves or needles.

Range: Quebec to North Carolina; also Colorado and California.

Look-alikes: *C. subnuda* has more widely spaced gills and bitter taste.

Comments: The Tufted Collybia seems to sit on a
bed of tangled white threads.

---

80 **Oak-loving Collybia**
Collybia dryophila (Bull. ex Fr.) Kum.
Tricholomataceae, Agaricales

Description: *Ochre cap with whitish gills and ochre,*
rigid-brittle stalk.
Cap: ⅜–2″ (1–5 cm) wide; convex to
flat; margin incurved, then upturned to
wavy; moist to dry, smooth; red- to
ochre-brown, fading to tan or yellow.
Gills: attached, crowded, narrow;
whitish to yellow-tinged.
Stalk: ⅜–3″ (1–7.5 cm) long, ¹⁄₁₆–¼″
(1.5–5 mm) thick, smooth, brittle;
reddish-brown to ochre or whitish,
with covering of white threads
(mycelium) at base; hollow.
Spores: 5–6 × 2–3 μ; elliptical,
smooth, colorless. Spore print white to
pale cream.
Edibility: Edible.
Season: May–November.
Habitat: Scattered to numerous, on wood and
leaves, in oak and pine woods.
Range: Widely distributed in North America.
Look-alikes: *C. butyracea* has cap with buttery feel,
pink-buff spore print, dextrinoid spores.
*Marasmius strictipes** has white stalk,
erect, dextrinoid cell rows in cap skin.
Comments: This is easy to spot when it is attacked
by the Collybia Jelly (*Christiansenia*
mycetophila), of which it is the sole host.
It probably should not be eaten when
the Jelly is growing on it.

---

106 **Violet Collybia**
Collybia iocephala (Berk. & Curt.) Sing.
Tricholomataceae, Agaricales

Description: *Violet cap and gills with whitish stalk.*
Cap: ⅜–1¼″ (1–3 cm) wide; convex to
flat; margin incurved at first, becoming
upturned and wavy; moist to dry,
wrinkled to lined; violet, fading;

turquoise with KOH. *Odor unpleasant.*
Gills: attached, distant, narrow, violet.
Stalk: 1–2" (2.5–5 cm) long, $\frac{1}{16}$–$\frac{1}{8}$"
(1.5–3 mm) thick, sometimes enlarged
below; minutely hairy, whitish.
Spores: 7–8.5 × 3–4 μ; elliptical,
smooth, colorless. Spore print white.

Season: August–October.
Habitat: On the ground on dead leaves.
Range: Massachusetts to Florida.
Comments: Also known as *Marasmius iocephalus.*

35 **Spotted Collybia**
   Collybia maculata (A. & S. ex Fr.) Kum.
   Tricholomataceae, Agaricales

Description: *Whitish cap, developing rusty spots, with*
white to buff gills and brittle stalk.
Cap: 2–6" (5–15 cm) wide; convex to
flat or with broad knob; margin
incurved at first, then wavy; dry,
smooth; white, becoming spotted and
streaked rust-red. Flesh thick, firm,
white. Taste bitter.
Gills: attached to nearly free, crowded,
narrow; whitish, discoloring with age.
Stalk: 2–6" (5–15 cm) long, $\frac{1}{4}$–$\frac{5}{8}$"
(0.5–1.5 cm) thick; dry, brittle-firm,
lined; whitish, almost rooting; hollow.
Spores: 5–6 × 4–5 μ; round to nearly
round, smooth, colorless; often
dextrinoid. *Spore print pinkish-buff.*
Edibility: Edible.
Season: July–November.
Habitat: Single to several, on decayed coniferous
wood or humus.
Range: Quebec to North Carolina.
Comments: Sometimes transferred to *Rhodocollybia*
because of its pinkish-buff spore print
and dextrinoid spores.   .

79 **Hairy-stalked Collybia**
   Collybia spongiosa (Berk. & Curt.) Sing.
   Tricholomataceae, Agaricales

Description: *Reddish-brown cap with whitish gills and*
stalk with dense reddish-brown hairs.

Cap: ⅜–1¼″ (1–3 cm) wide; convex to flat or sunken; margin incurved at first; moist to dry, smooth to wrinkled; reddish-brown, fading to tannish.
Gills: attached or notched at stalk, close to nearly distant, narrow to broad; whitish.
Stalk: 1–3″ (2.5–7.5 cm) long, ⅟32–⅟16″ (1–1.5 mm) thick, *with enlarged, spongy base; tough, reddish-brown, densely covered with coarse, reddish-brown to tawny hairs.* Flesh greenish with KOH.
Spores: 6–8.5 × 3–4 μ; elliptical, smooth, colorless. Spore print white.

Season: August–October.
Habitat: In woods on decaying oak leaves and twigs.
Range: Quebec to Florida, west to Michigan.
Look-alikes: *C. semihirtipes* lacks spongy stalk base. *C. subnuda* has stalk densely covered with small, white to grayish hairs.
Comments: Also known as *Marasmius spongiosus.*

---

11 Tuberous Collybia
Collybia tuberosa (Bull. ex Fr.) Kum.
Tricholomataceae, Agaricales

Description: *Small, white mushroom with small, brownish, tuberlike underground body; on ground or decaying mushrooms.*
Cap: 1¼–4″ (3–10 cm) wide; convex to flat or with broad knob; dry, smooth; whitish to buff.
Gills: attached, close, narrow; whitish.
Stalk: ⅜–1″ (1–2.5 cm) long, ⅟64–⅟32″ (0.5–1 mm) thick; dry, whitish or discoloring; *attached to reddish-brown, oval, tuberlike body.*
Spores: 4–6 × 2–3 μ; elliptical, smooth, colorless. Spore print white.

Season: August–November; overwinters in California.
Habitat: Numerous on decayed mushrooms.
Range: Widely distributed in North America.
Look-alikes: *C. cookei* has a small, round, yellowish-orange tuber. *C. cirrhata* has no tuber. *Strobilurus** species grow on cones. *Asterophora**, also on mushrooms, have

distant, poorly formed gills, lack tuber.
Comments: This very common fall species may
appear to be growing on the ground.

---

34 **Zoned-cap Collybia**
Crinipellis zonata (Pk.) Pat.
Tricholomataceae, Agaricales

Description: *Small cap with long, tawny hairs in zones,*
white gills, and hairy stalk; on wood.
Cap: ⅜–1″ (1–2.5 cm) wide; convex to
nearly flat, with small, cuplike
depression in center; dry; *densely covered*
with coarse, fibrous, tawny hairs set
radially, somewhat zoned, dextrinoid.
Gills: free to nearly free, close, narrow;
white.
Stalk: 1–2″ (2.5–5 cm) long, ⅟₃₂–⅟₁₆″
(1–1.5 mm) thick; densely tawny-
hairy, hollow.
Spores: 4–6 × 3–5 μ; elliptical,
smooth, colorless. Spore print white.
Season: August–September.
Habitat: On dead wood.
Range: Quebec to Florida; Indiana and Texas.
Look-alikes: *C. stipitaria* has minute nipple in center
of cuplike depression in cap; stalk
whitens as it dries. *C. campanella* grows
on twigs of dead arbor vitae.
Comments: Formerly known as *Collybia zonata*, the
Zoned-cap Collybia was transferred to
Crinipellis because of its dense covering
of coarse, fibrous, dextrinoid hairs.

---

54 **Golden-scruffy Collybia**
Cyptotrama asprata (Berk.) Redhead & Ginns.
Tricholomataceae, Agaricales

Description: *Small, yellow, scruffy-furrowed cap with*
whitish gills and yellow stalk; on wood.
Cap: ¼–¾″ (0.5–2 cm) wide; convex
to flat; margin radially lined to
furrowed; dry, wrinkled to furrowed,
scruffy-pitted with branlike pustules;
golden- to pale yellow.
Gills: attached or slightly descending
stalk, distant, broad; whitish.

Stalk: 1–2″ (2.5–5 cm) long, ⅟₃₂–⅛″
(1–3 mm) thick; solid, tough, cottony-
scaly with branlike pustules; golden- to
pale yellow.
Spores: 8–12 × 6–7 μ; broadly oval to
elliptical, smooth, colorless. Spore
print white.

Season: June–October; November in Texas.
Habitat: Scattered on hardwood twigs, oak logs.
Range: Quebec to Florida, west to Great Lakes
and Texas.
Comments: Formerly known as *Collybia lacunosa*
and *Xerula chrysopepla*.

---

63 Velvet Foot
Flammulina velutipes (Fr.) Kar.
Tricholomataceae, Agaricales

Description: *Sticky, tawny cap with whitish gills; stalk
pale yellowish above, dark velvet-brown
below; in clusters on wood.*
Cap: 1–2″ (2.5–5 cm) wide; convex to
flat or with broad knob; margin
incurved at first; slimy to tacky,
smooth; reddish-brown to reddish-
yellow or tawny.
Gills: attached, close to nearly distant,
broad; whitish to yellowish.
Stalk: 1–3″ (2.5–7.5 cm) long, ⅛–¼″
(3–5 mm) thick, narrowing downward
slightly; tough, yellowish to densely
velvety above, blackish-brown below.
Spores: 7–9 × 3–6 μ; elliptical,
smooth, colorless. Spore print white.
Edibility: Edible.
Season: October–May; July–August in
mountains.
Habitat: Clustered on deciduous logs,
particularly elm, aspen, poplar, and
willow.
Range: Widely distributed in North America.
Look-alikes: The deadly *Galerina autumnalis** has
brown spores and ring on stalk.
Comments: Formerly called *Collybia velutipes*. This
species is now cultivated and sold
commercially as Enotake, but the
commercial product looks very different
from the wild mushroom.

60   Golden-gilled Gerronema
Gerronema chrysophylla (Fr.) Sing.
Tricholomataceae, Agaricales

Description: *Scruffy-scaly, smoky-ochre cap with*
descending golden gills and smooth stalk; on
coniferous logs.
Cap: ¼–1⅝" (0.5–4 cm) wide; convex
to flat and sunken in center; margin
incurved at first; moist to dry, scruffy
to fibrous-scaly; ochre to smoky-ochre.
Gills: descending stalk, distant, broad;
golden- to apricot-yellow.
Stalk: 1–2" (2.5–5 cm) long, ¹⁄₁₆–⅛"
(1.5–3 mm) thick; somewhat curved,
smooth; yellow to smoky-ochre; stuffed
to hollow.
Spores: 8.5–15.5 × 4.5–7 μ; elliptical
to cylindrical, smooth, colorless. Spore
print yellow.
Season: July–August in East; August–October
in Rockies and Pacific NW.
Habitat: On coniferous logs.
Range: Quebec to Florida, west to Pacific NW.
and N. California.
Look-alikes: *G. luteicolor* has orange cap and stalk,
and orange to salmon-buff gills.
Comments: Also known as *Omphalina chrysophylla.*

---

433   Leaflike Oyster
Hohenbuehelia petaloides (Bull. ex Fr.) Schulz.
Tricholomataceae, Agaricales

Description: *Brownish, fan-shaped cap with grayish*
gills descending very short stalk.
Cap: 1–3" (2.5–7.5 cm) wide, 2–4"
(5–10 cm) long; fan- to spatula-shaped;
margin incurved, becoming lobed to
wavy; moist, *sometimes hoary-frosted at*
first, with gelatinous feel, smooth;
brownish or paler; base white-hairy.
Gills: descending stalk to base,
crowded, narrow; whitish to grayish,
edges fringed.
Stalk: very short, compressed, hairy.
Spores: 7–9 × 4.5–5 μ; elliptical,
smooth, colorless. Pleurocystidia huge,
abundant, with encrusted tips. Spore

print white.

Edibility: Edible.

Season: August in Colorado; fall–spring in California.

Habitat: On logs and stumps.

Range: Quebec south to North Carolina, west to Great Lakes States; reported from Colorado, California.

Look-alikes: Several oysterlike mushrooms grow on buried wood or wood debris.

Comments: Once called *Pleurotus petaloides,* this was transferred to *Hohenbuehelia* because of its gelatinous tissue layer and huge, thick-walled cystidia.

---

492 Elm Oyster
Hypsizygus tessulatus (Bull. ex Fr.) Sing.
Tricholomataceae, Agaricales

Description: *Large, whitish or tan cap, cracking with age, with whitish gills and off-center, whitish stalk; on wood.*
Cap: 2–6″ (5–15 cm) wide; convex to nearly flat and sunken; moist to dry, smooth to slightly hairy; becoming cracked with age and forming minute scales or broader patches; white, then buff-yellow. Flesh thick, firm, white.
Gills: attached, close to almost distant, broad; whitish, becoming cream.
Stalk: 2–4″ (5–10 cm) long, ⅜–¾″ (1–2 cm) thick, sometimes enlarged at base; off-center, stout, solid; dry, smooth, sometimes hairy, whitish.
Spores: 5–7 μ; round, smooth, colorless. Spore print white to buff.

Edibility: Edible.

Season: September–December.

Habitat: Single to several, on deciduous trees, especially elm.

Range: Across N. North America.

Look-alikes: *Pleurotus ostreatus** is stalkless or has gills descending short stalk. *P. elongatipes* grows on trees or logs, has long stalk that becomes hollow. *Panus strigosus* has broad cap and long, stout stalk covered with coarse hairs.

Comments: Formerly known as *Pleurotus ulmarius.*

This species should be looked for on the upper trunk of deciduous trees. It is edible but tough.

---

335 Common Laccaria
Laccaria laccata (Scop. ex Fr.) Cke.
Tricholomataceae, Agaricales

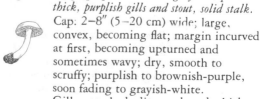

Description: *Pinkish-brown, sunken cap with thick, pale pink gills slightly descending pinkish stalk.*
Cap: ⅜–2″ (1–5 cm) wide; convex to flat, *usually with central depression;* margin upturned, wavy at times; moist to dry, smooth to scruffy; reddish- to pinkish-brown, fading to ochre-white.
Gills: attached or slightly descending stalk, distant, broad; pinkish-brown to pink.
Stalk: 1–3″ (2.5–7.5 cm) long, ⅛–¼″ (3–5 mm) thick; fibrous-tough, dry; pinkish to pinkish-brown.
Spores: 7–9 × 6–8 μ; round to broadly elliptical, spiny, colorless. Spore print white.
Edibility: Good, with caution.
Season: June–November; overwinters in California.
Habitat: Scattered to many, in poor soil, waste places, and sandy soil; also under pine.
Range: Widely distributed in North America.
Comments: The enormous variation probably means that this is a group of species.

254, 336 Purple-gilled Laccaria
Laccaria ochropurpurea (Berk.) Pk.
Tricholomataceae, Agaricales

Description: *Large, purplish-brown to grayish cap with thick, purplish gills and stout, solid stalk.*
Cap: 2–8″ (5–20 cm) wide; large, convex, becoming flat; margin incurved at first, becoming upturned and sometimes wavy; dry, smooth to scruffy; purplish to brownish-purple, soon fading to grayish-white.
Gills: attached, distant, broad, thick; purplish.

Stalk: 2–8″ (5–20 cm) long, ⅜–¾″ (1–2 cm) thick; often curved, stout, solid, dry, fibrous-rigid; grayish-white or with purplish or brownish tones.
Spores: 6–8 μ; round, spiny, colorless. Spore print white to pale violet.

Edibility: Good.
Season: July–November.
Habitat: On the ground in grassy areas and open oak woods.
Range: East of Great Plains.
Look-alikes: *Cortinarius alboviolaceus** and similar species have gills becoming rust-brown and cobwebby veil on young specimens. *L. amethystina* is smaller and violet overall.
Comments: This large, fleshy laccaria is often found in quantity in the fall and is good when mixed with other foods.

255 Sandy Laccaria
Laccaria trullisata (Ell.) Pk.
Tricholomataceae, Agaricales

Description: *Brownish, sunken cap with thick, purplish gills and sand-covered, pale brown stalk; in sand.*
Cap: 1–3″ (2.5–7.5 cm) wide; convex to nearly flat; margin incurved at first; moist, with minute radial fibers, somewhat scaly with age; dirty pink to dingy pale brown.
Gills: attached, distant, broad, thick; pinkish to purplish.
Stalk: 1–4″ (2.5–10 cm) long, ¼–¾″ (0.5–2 cm) thick, enlarged downward; curved, fibrous, covered with sand; dirty pink to dingy pale brown.
Spores: 16–22 × 6–9 μ; *large*, long-elliptical, *smooth*, colorless. Spore print white.
Edibility: Edible.
Season: August–October.
Habitat: Single to scattered, in shore areas.
Range: Atlantic Coast and Great Lakes region.
Comments: The Sandy Laccaria is often buried up to its cap in sand. It differs from other laccarias by having smooth spores.

423 Cockle-shell Lentinus
Lentinellus cochleatus (Pers. ex Fr.) Kar.
Tricholomataceae, Agaricales

Description: *Smooth, vase-shaped, red-ochre cap with toothed, pinkish-white gills descending grooved, solid stalk; in fused clusters.*
Cap: 1–2″ (2.5–5 cm) wide; flat with central depression to vase-shaped; margin incurved, becoming lobed to wavy; smooth, reddish-brown to grayish-buff.
Gills: descending stalk, close, broad; often torn, edges toothed; whitish, but tinged pinkish-cinnamon.
Stalk: 1–4″ (2.5–10 cm) long, ⅛–⅜″ (0.3–1 cm) thick, fused at base in dense clusters; central to off-center, deeply furrowed; reddish-brown or paler.
Spores: 4–5 × 3.5–4.5 μ; nearly round, ornamented with spines, amyloid, colorless. Spore print white.
Season: July–September.
Habitat: Clustered at base of deciduous stumps and on decayed or buried wood.
Range: E. North America to Great Lakes.
Look-alikes: *L. omphalodes** has furrowed but unfused, slender stalk. *Lentinus** species have central stalk and smooth, elliptical spores. *Panus** have smooth gill edges.
Comments: Formerly known as *Lentinus cochleatus.*

438 Stalked Lentinellus
Lentinellus omphalodes (Fr.) Kar.
Tricholomataceae, Agaricales

Description: *Buff to brown, smooth cap with toothed, pinkish-buff gills and thin, ridged stalk.*
Cap: 1–2″ (2.5–5 cm) wide; convex to nearly flat or sunken; moist, smooth; pinkish-brown, darkening to brown. Taste peppery.
Gills: attached, nearly distant, broad; edges toothed or torn; pinkish-brown.
Stalk: ¼–2″ (0.5–5 cm) long, ¹⁄₆₄–⅛″ (0.5–3 mm) thick, central to off-center, dry, furrowed; reddish-brown to brown.

Spores: 5—6.5 × 3.5—4.5 μ; short, elliptical, with amyloid spines. Spore print buff.

Season: August—November.

Habitat: On the ground or on deciduous or coniferous wood debris.

Range: Widely distributed in North America.

Look-alikes: *L. cochleatus** has fused stalks and is vase-shaped. *L. vulpinus* is robust, has densely hairy surface and fused stalks.

Comments: Although some mushrooms will show weathered gills that are torn, ragged, or appear to be toothed, in *Lentinellus* the toothing of the gill edge occurs because of uneven growth, not environmental conditions.

502 Bear Lentinus
Lentinellus ursinus (Fr.) Küh
Tricholomataceae, Agaricales

Description: *Stalkless, densely hairy, brownish cap with toothed, whitish gills and bitter taste; on dead wood.*

Cap: 1 4" (2.5—10 cm) wide; fan- to kidney-shaped; margin incurved; dry, base densely covered with stiff hairs; reddish-brown, paler toward margin. Taste acrid.

Gills: descending to base, close, broad; edges toothed and irregularly torn; whitish-pink to pinkish-brown.

Spores: 2.5—4.5 × 2—3.5 μ; nearly round, ornamented with amyloid spines, colorless. Spore print white.

Season: Summer and fall to early winter.

Habitat: Single to several on wood; conifers in West; hardwoods, often oak, in East.

Range: Widely distributed in North America.

Look-alikes: *L. vulpinus* has furrowed, stalklike base. *L. montanus* grows near western snowbanks in early spring. *Panellus** have smooth gills and oblong, smooth, amyloid spores. *Panus** and *Pleurotus** have smooth gills, nonamyloid spores.

Comments: Formerly known as *Lentinus ursinus*, this species differs from *Lentinus* by its rounded, amyloid-spined spores.

222 Scaly Lentinus
Lentinus lepideus (Fr. ex Fr.) Fr.
Tricholomataceae, Agaricales

Description: *White to buff cap with brownish scales, white gills with toothed edges, and ringed, tough, scaly stalk; on dead wood.*
Cap: 2–5″ (5–12.5 cm) wide; convex to nearly flat; margin incurved at first; tacky to dry, becoming scaly; whitish to buff. Flesh white, tough.
Gills: notched or slightly descending stalk, close to nearly distant, broad; edges toothed, often nearly even at first; whitish to buff, bruising brownish.
Stalk: 1–6″ (2.5–15 cm) long, 3/8–3/4″ (1–2 cm) thick, abruptly narrow at base; *central* (or nearly so), solid, scruffy-scaly below ring; whitish to brownish.
Veil: partial veil leaving a persistent, membranous, whitish ring on upper stalk.
Spores: 9–12 × 4–5 μ; almost cylindrical, smooth, colorless, nonamyloid. Spore print white.
Edibility: Edible.
Season: May–September; overwinters in California.
Habitat: Solitary to scattered, on decaying coniferous wood; sometimes on oak, fence posts, and railroad ties.
Range: Quebec to Florida, to West Coast.
Look-alikes: *L. ponderosus**, a very large western species, lacks veil.
Comments: Also called the "Train Wrecker" because it destroys railroad ties, causing derailments.

---

506 Large Lentinus
Lentinus ponderosus O.K.M.
Tricholomataceae, Agaricales

Description: *Large, fragrant, tough, dingy white to yellowish-brown, scaly mushroom with white spores; on coniferous wood.*
Cap: 4–20″ (10–50 cm) wide; convex to flat and somewhat sunken; margin

incurved; cottony, dry, smooth at first,
breaking into scales; dingy white,
darkening to yellowish, orange, or light
yellowish-brown. Flesh thick, hard,
whitish. Odor fragrant.
Gills: slightly descending stalk, close,
broad; toothed or torn on edge; whitish.
Stalk: 2–6″ (5–15 cm) long, 1–2″
(2.5–5 cm) thick, turnip-shaped, off-
center, stout, solid, hard, scaly;
whitish, then yellow to yellow-brown.
Spores: 8–12 × 4.5–5.5 μ; elliptical,
smooth, colorless, nonamyloid. Spore
print white.

Season: August–October.

Habitat: Single on coniferous wood; also
reportedly numerous on soil.

Range: W. United States.

Look-alikes: *Panus strigosus* has smooth-edged gills,
stiff hairs on cap and stalk, and grows
on deciduous wood.

Comments: This large, conspicuous, fragrant, scaly
mushroom is usually too tough to eat.

---

239 White Leucopax

Leucopaxillus albissimus (Pk.) Sing.
Tricholomataceae, Agaricales

Description: *Fleshy, white mushroom with crowded gills
and chalklike stalk set in dense, white mat.*
Cap: 1–4″ (2.5–10 cm) wide; convex
to flat; margin incurved at first, radially
lined at times; dry, smooth, white.
Odor fragrant; taste bitter.
Gills: attached or slightly descending
stalk, crowded, narrow; white.
Stalk: 2–3″ (5–7.5 cm) long, ¼–⅝″
(0.5–1.5 cm) thick, with basal bulb;
smooth, chalky; embedded in thick,
cottony, white threads (mycelium).
Spores: 5.5–8.5 × 4–6 μ; elliptical,
ornamented with amyloid warts,
colorless. Spore print white.

Season: August–October.

Habitat: Single, scattered to several, or
sometimes in fairy rings, in coniferous
woods, especially near redwood; near
spruce in East.

Range: Widespread in N. North America; in South in coastal mountains.

Look-alikes: *Melanoleuca albiflavida** and similar white *Clitocybe** lack dense white mycelium at stalk base.

Comments: This broadly defined species has many color varieties and is best identified by an examination of its spores. It is reportedly edible, but most varieties are unpalatable and hard to digest.

---

277 Bitter Leucopax

Leucopaxillus amarus (A. & S.) Küh.
Tricholomataceae, Agaricales

Description: *Fleshy, dry, reddish-brown cap and chalky stalk set in dense, white mat; taste bitter.*
Cap: 2–5″ (5–12.5 cm) wide; convex to nearly flat; margin incurved at first; dry, smooth; reddish-brown, pinkish toward margin. Odor and taste unpleasant, bitter.
Gills: attached, close, narrow; white.
Stalk: 2–3″ (5–7.5 cm) long, ¼–¾″ (0.5–2 cm) thick, with bulbous base; whitish to brown-tinged; embedded in thick, cottony-white mat (mycelium).
Spores: 4–6 × 3–5 μ; almost round, ornamented with amyloid warts, colorless. Spore print white.

Season: July–September.

Habitat: On the ground under conifers; also under live oak.

Range: W. North America, mostly near coast.

Comments: This fleshy, attractive, distinctive species is too bitter to eat. Also known as *L. gentianeus.*

---

266 Fried-chicken Mushroom

Lyophyllum decastes (Fr. ex Fr.) Sing.
Tricholomataceae, Agaricales

Description: *Grayish- to yellowish-brown cap with whitish gills and stalk.*
Cap: 1–5″ (2.5–12.5 cm) wide; convex to nearly flat; margin incurved at first, becoming upturned; moist, smooth;

gray to grayish-yellow or yellowish-brown.
Gills: attached or slightly descending stalk, close, moderately broad; whitish.
Stalk: 2–4″ (5–10 cm) long, ¼–¾″ (0.5–2 cm) thick; stout, smooth; whitish or somewhat discolored below.
Spores: 4–5.5 × 4–5 μ; nearly round, smooth, colorless; basidia with carminophilous granules. Spore print white.

Edibility: Good, with caution.

Season: June–October; overwinters in California.

Habitat: In dense clusters, on the ground in grassy areas, waste places, urban areas, and along roads.

Range: Widely distributed in North America.

Look-alikes: *L. loricatum* has blackish to brownish cap. *L. connatum* has whitish cap. *Clitocybe dilatata* * is whitish and usually misshapen. *Entoloma* * species have pinkish spore print and angular spores.

Comments: Also known as *Lyophyllum multiceps* and *Clitocybe multiceps*. When it is cooked, this abundant edible tastes like fried chicken.

---

90 **Cucumber-scented Mushroom**
Macrocystidia cucumis (Pers. ex Fr.) Heim
Tricholomataceae, Agaricales

Description: *Reddish-brown to buff, bell-shaped cap with attached white gills becoming pink, and red-brown stalk; odor of cucumber and fish.*
Cap: ⅜–2″ (1–5 cm) wide; bell-shaped, with small knob; margin straight; moist, smooth to silky; red-brown to ochre-tawny, fading to buff.
Gills: attached, close, broad; white, becoming pink.
Stalk: 1–2″ (2.5–5 cm) long, 1/16″ (1.5 mm) thick; round to slightly flattened, fibrous-brittle; dark red-brown; hollow.
Spores: 8–9 × 3–4 μ; elliptical, smooth, colorless to pale pink. *Cystidia huge,* bulbous below, narrowing above, on all surfaces. *Spore print pinkish.*

Season: May–December.

Habitat: Single to several, in grassy areas.
Range: Pacific NW.
Comments: This species was formerly included in the genus *Naucoria*. It has been placed in the large and diverse family Tricholomataceae until its relationships can be better understood.

---

488 White Marasmius
Marasmiellus albuscorticis (Secr.) Sing.
Tricholomataceae, Agaricales

Description: *Small, dry, white, pleated cap with white to pink-tinged gills and short, whitish stalk, black below; on twigs and stems.*
Cap: ⅜–1¼" (1–3 cm) wide; convex to flat; margin folded or wavy, becoming radially grooved; dry, minutely hairy; shiny white, spotted with age.
Gills: attached, distant, moderately broad; white to pinkish, with marked crossveins.
Stalk: ⅜–¾" (1–2 cm) long, ¹⁄₃₂–¹⁄₁₆"
(1–1.5 mm) thick, somewhat enlarged at base; central or off-center, solid; shiny white, blackish toward base.
Spores: $10–15 \times 5–6 \mu$; large, broad below, tapering to tip; smooth, colorless. Spore print white.
Season: July–October; winter in California.
Habitat: In rows, on twigs and stems of blackberries, raspberries, and other similar shrubs.
Range: Widely distributed in North America.
Comments: Also known as *Marasmius candidus* and *M. magnisporus*. It may be a complex of closely related species.

---

1 Black-footed Marasmius
Marasmiellus nigripes (Schw.) Sing.
Tricholomataceae, Agaricales

Description: *Small, downy, pleated white cap with whitish gills and tough black stalk, densely covered with white hairs; on twigs.*
Cap: ⅜–¾" (1–2 cm) wide; convex to flat; very thin, powdery, wrinkled,

chalk-white. *Flesh rubbery, reviving in moist weather.*
Gills: attached or slightly descending stalk, nearly distant, broad, *forked in part, with crossveins;* white, staining reddish.
Stalk: 1–2″ (2.5–5 cm) long, ⅟₃₂–⅟₁₆″ (1–1.5 mm) thick, narrowing downward; fibrous-tough, tubular; black, covered with minute white hairs; rootless, but surrounded underground by rootlike cords.
Spores: 8–9 μ; *triangular* to starlike, smooth, colorless. Spore print white.
Season: August–September.
Habitat: On twigs and leaves in mixed woods.
Range: New York to Michigan.
Comments: Formerly known as *Heliomyces nigripes,* this is a marasmiuslike mushroom with a gelatinous texture.

38 Fused Marasmius
Marasmius cohaerens (Pers. ex Fr.) Cke. & Quél.
Tricholomataceae, Agaricales

Description: *Brownish, velvety cap with off-white to brownish gills and tough, shiny, minutely dotted, brownish stalk.*
Cap: ⅜–1⅝″ (1–4 cm) wide; bell-shaped, becoming convex to flat or somewhat sunken in center; dry, dull, smooth to somewhat wrinkled, velvety; yellow-brown to amber or dark brown.
Gills: attached, distant, moderately broad; yellowish-white, becoming yellow-pink to faintly brownish.
Stalk: 1–3″ (2.5–7.5 cm) long, ⅟₃₂–⅛″ (1–3 mm) thick; round, straight to curved, pliant to horny, dry, smooth, shiny, minutely dotted; light yellowish- to dark brown; hollow.
Spores: 7–10 × 3–5 μ; spindle-shaped to elliptical, smooth, colorless. Setae present. Spore print white.
Season: Late spring into September.
Habitat: Clustered on leaves and humus of deciduous trees.

Range: Quebec to North Carolina, Great Lakes.

Look-alikes: *Collybia acervata** grows on decayed coniferous wood and lacks dots.

Comments: True to its name, this species has stalks that usually adhere to one another. Because of its changing gill color, a spore print should be made to verify the spore color. A variety, *lachnophyllus,* occurs on deciduous logs and twigs, has close, narrow gills, and abundant dots on the stalk.

17, 18 Fairy Ring Mushroom
Marasmius oreades (Bolt. ex Fr.) Fr.
Tricholomataceae, Agaricales

Description: *Dry, buff to tan, knobbed cap with off-white gills and felted, rubbery stalk; in grass.*
Cap: ⅜–1⅝" (1–4 cm) wide; bell-shaped with inrolled margin, becoming convex to broadly knobbed; margin upturned with age; dry, smooth, feltlike; pale yellowish-brown, fading to buff-white, sometimes grayish- to reddish-brown in part or overall.
Gills: attached or free, close to almost distant, broad; yellowish-white.
Stalk: ⅜–3" (1–7.5 cm) long, 1/16–¼" (1.5–5 mm) thick; straight, dry, rubbery, smooth or with long, twisted ribs, *minutely hairy or felted;* white to pale or orange-yellow, yellow- or light reddish-brown; solid to stuffed.
Spores: 7–10 × 4–6 μ; elliptical, smooth, colorless. Spore print white to buff.

Edibility: Good, with caution.

Season: May–September; year-round in California.

Habitat: In lawns and other grassy areas, often in fairy rings.

Range: Widely distributed in North America.

Look-alikes: The poisonous *Clitocybe dealbata** and allies have close, grayish gills somewhat descending stalk. *Inocybe umbratica* and allies have unpleasant odor and brown spores. *M. cystidiosus* grows in open

Comments: deciduous woods and has smooth stalk. Also known as the "Scotch Bonnet." This mushroom is an excellent edible, but it must be carefully examined to avoid being confused with a poisonous clitocybe or inocybe. It dries readily and becomes easily overlooked in grassy areas; then, when it rains, it revives and appears to have sprung up overnight. In early times, the rings were thought to be magical places where fairies danced in a circle on moonlit nights or gnomes buried their treasure. Fairy rings increase in size each year, as the mushroom's underground life-support system—a tangle of threads called the mycelium—grows outward, seeking nourishment. In the West, segments of some rings are 600 years old.

40 **Velvet-cap Marasmius**
Marasmius plicatulus Pk.
Tricholomataceae, Agaricales

Description: *Dry, velvety, rust-brown cap, becoming pleated, with whitish gills and tough, shiny, reddish-black stalk.*
Cap: ⅜–1¼" (1–3 cm) wide; bell-shaped to convex; radially pleated to furrowed with age; dry, velvety to hoary; dark burgundy to maroon or ochre-tawny toward margin.
Gills: attached or nearly free, broad; white to pinkish-white or buff.
Stalk: 2–5" (5–12.5 cm) long, 1/16–⅛" (1.5–3 mm) thick; fibrous-brittle, smooth, shiny; red-brown, paler at tip.
Spores: 11–15 × 5–6.5 μ; narrowly elliptical, smooth, colorless. Spore print white.
Season: September–October; November–February in California.
Habitat: Scattered to numerous, under conifers; also under oak.
Range: Washington and Idaho to California.
Comments: This species is one of many small, slender-stalked fungi that grow on leaf litter and decompose organic materials.

2 Pinwheel Marasmius
Marasmius rotula (Scop. ex Fr.) Fr.
Tricholomataceae, Agaricales

Description: _Small, dry, bell-shaped, pleated, white cap_
with sunken center, white gills, and tough,
shiny, black stalk.
Cap: 1/16–5/8" (0.15–1.5 cm) wide;
cushion-shaped, becoming convex and
broadly sunken with central cup;
radially pleated or furrowed; dry, dull,
smooth to velvety; light yellowish-
brown to yellowish-white or white.
Gills: attached either to stalk or collar,
distant, narrow to moderately broad;
yellowish-white.
Stalk: 5/8–3 3/8" (1.5–8.5 cm) long, 1/64–
1/32" (0.5–1 mm) thick; dry, fibrous to
wiry, smooth; yellowish-white,
becoming blackish-brown from base
upward; base often with rootlike, stiff
brownish hairs; hollow.
Spores: 6–10 × 3–5 μ; narrowly
elliptical, smooth, colorless. Spore
print white.
Season: May–October.
Habitat: In dense clusters, on decaying
deciduous wood.
Range: Widely distributed in North America.
Look-alikes: _M. capillaris_ typically grows on oak
leaves.
Comments: Like most marasmius species, this
revives with rain and seems to appear
overnight in the wet morning woods.

37 Garlic Marasmius
Marasmius scorodonius (Fr.) Fr.
Tricholomataceae, Agaricales

Description: _Small, dry, brownish cap with whitish gills_
and tough stalk; garlicky taste and odor.
Cap: 1/4–1 1/4" (0.5–3 cm) wide;
cushion-shaped, becoming convex to
nearly flat; margin incurved at first,
becoming wavy and upturned; dry,
smooth to minutely wrinkled; reddish-
to yellowish-brown, fading to pale
orange-yellow or pale yellow.

Gills: attached or nearly free, close to
almost distant, narrow, often forked;
yellowish-pink to white.
Stalk: ⅝–2⅜" (1.5–6 cm) long, ¹⁄₆₄–
⅛" (0.5–3 mm) thick, tapering below;
round to compressed, dry, shiny,
brittle to horny; yellowish-white to
dark brown.
Spores: 7–10 × 3–5 μ; elliptical,
smooth, colorless. Spore print white.

Edibility: Edible.

Season: July–October; overwinters in
California.

Habitat: Scattered to many, on twigs and debris.

Range: N. North America; California.

Look-alikes: *M. olidus* has yellowish-brown cap and
gills and grows on veins of oak leaves.
Other similar *Marasmius** species lack
garlicky odor. *Micromphale foetidum**
has foul odor and densely hairy stalk.

Comments: This garlicky species is edible and can
be used as a condiment to season
various dishes. It can be found on
living tree bark and dead wood, conifer
needles, and flattened grass.

67 Orange Pinwheel Marasmius
Marasmius siccus (Schw.) Fr
Tricholomataceae, Agaricales

Description: *Small, dry, bell-shaped, rust-orange,
pleated cap, often with sunken center, white
gills, and white to blackish-brown stalk.*
Cap: ⅛–1¼" (0.3–3 cm) wide;
cushion- to bell-shaped, becoming
convex, knobbed or sunken in center,
wrinkled to roughened, deeply pleated
to furrowed; minutely velvety; rust-
orange to orange. Flesh dextrinoid.
Gills: attached or free, distant,
moderately broad; white to pale yellow.
Stalk: 1–2½" (2.5–6.5 cm) long, ¹⁄₆₄–
¹⁄₃₂" (0.5–1 mm) thick; dry, brittle to
horny; smooth; white to yellowish,
brownish at base; hollow.
Spores: 16–21 × 3–4.5 μ; *large,*
spindle- to club-shaped, smooth,
colorless. Spore print white.

Season: July–October.
Habitat: Scattered to numerous, on deciduous
wood, leaves, and white pine needles.
Range: East Coast to Mississippi Valley.
Look-alikes: *M. pulcherripes* has pink to yellowish-
brown cap. *M. fulvoferrugineus* is more
rust-brown.
Comments: Related species differ mainly in color,
but all look like delicate umbrellas and
are among the first to appear after rain.

32 Orange-yellow Marasmius
Marasmius strictipes (Pk.) Sing.
Tricholomataceae, Agaricales

Description: *Moist, yellow to orange cap with yellowish
gills and off-white, rigid-brittle stalk.*
Cap: ¾–2½" (2–6.5 cm) wide; convex
to broadly knobbed; margin slightly
upturned with age; smooth, yellowish
to orange-yellow. Odor and taste
sometimes radishy. Flesh dextrinoid.
Gills: attached, close to crowded,
narrow; yellowish- to pinkish-buff.
Stalk: 1–3" (2.5–7.5 cm) long, ⅛–⅜"
(0.3–1 cm) thick; dry, minutely hairy
to fibrous; white to pale yellow; hollow.
Spores: 6–8 × 3–5 μ; elliptical,
smooth, colorless. Spore print white.
Season: July–September.
Habitat: Several to many, on fallen leaves.
Range: NE. North America to North Carolina,
west to Great Lakes.
Look-alikes: *Collybia butyracea* has buttery cap, lined
stalk, pinkish-buff spore print, and
grows under conifers. *C. dryophila** has
yellow to reddish-brown stalk. *M.
oreades** grows in grass.
Comments: Formerly called *Collybia strictipes*.

235 Yellowish-white Melanoleuca
Melanoleuca alboflavida (Pk.) Murr.
Tricholomataceae, Agaricales

Description: *Whitish to buff cap with white gills and
slender, straight, white stalk with twisted
lines of dark fibers.*

Cap: 2–4" (5–10 cm) wide; bell-shaped, expanding to flat; moist, smooth; yellowish-brown, becoming yellowish-buff to white.
Gills: attached, crowded, narrow, whitish.
Stalk: 1¼–6" (3–15 cm) long, ¼–⅜" (0.5–1 cm) thick; solid, brittle, fibrous-lined; whitish to buff.
Spores: 7–9 × 4–5.5 μ; oval, with amyloid ornamentation, colorless. Cystidia harpoonlike. Spore print white.

Edibility: Edible.
Season: June–September.
Habitat: Scattered to several, in deciduous woods.
Range: East of Great Plains.
Look-alikes: *Leucopaxillus** species have chalk-white stalk with conspicuous, matlike white mycelium. *Entolomas** have angular spores, salmon-pink spore print.
Comments: Formerly known as *Collybia albiflavida*. Because poisonous *Entoloma* species also have a fibrous-lined, straight stalk, a spore print should be taken.

359 Changeable Melanoleuca
Melanoleuca melaleuca (Pers. ex Fr.) Murr.
Tricholomataceae, Agaricales

Description: *Brownish cap with white gills and slender, straight, white stalk with twisted fibers.*
Cap: 1–3" (2.5–7.5 cm) wide; convex, expanding to flat or with low knob; moist, smoky-brown to dark brown, fading on drying.
Gills: notched, close, moderately broad; whitish.
Stalk: 1–3" (2.5–7.5 cm) long, ⅛–¾" (0.3–2 cm) thick; dark fibrous-lined, whitish; stuffed.
Spores: 7–8.5 × 5–5.5 μ; elliptical, with amyloid ornamentation, colorless. Spore print white.

Edibility: Edible.
Season: Summer–fall; fall–spring in California.
Habitat: Single to several, in lawns, waste

places, and along roads, under trees.
Range: Throughout North America.
Comments: This edible species was transferred to
Melanoleuca from the genus *Tricholoma*
because of spore characteristics.

---

28 Fetid Marasmius
Micromphale foetidum (Fr.) Sing.
Tricholomataceae, Agaricales

Description: *Brownish, sunken, pleated cap with
discolored gills and velvet-brown stalk; odor
strongly unpleasant; on wood.*
Cap: ⅜–1¼" (1–3 cm) wide; convex to
flat, with navel-like center; margin
incurved; dry, *radially pleated-folded;
reddish-brown,* drying tan. *Odor foul.*
Gills: *attached to collar,* distant,
moderately broad; yellowish, becoming
reddish-tinged.
Stalk: ¾–1¼" (2–3 cm) long, ¹⁄₃₂–¹⁄₁₆"
(1–1.5 mm) thick; brown, with velvety
hairs; base cottony; hollow.
Spores: 8.5–10 × 3.5–4 μ; elliptical,
smooth, colorless. Spore print white.
Season: July–September.
Habitat: On sticks, branches, and fallen wood.
Range: Quebec to North Carolina; Michigan.
Comments: Formerly known as *Marasmius foetidus.*
This species is usually found in clusters;
its foul odor makes it unmistakable.

---

68 Coral Spring Mycena
Mycena acicula (Schaeff. ex Fr.) Kum.
Tricholomataceae, Agaricales

Description: *Small, coral-red to yellow-tinged, bell-
shaped cap with orange to yellowish gills
and stalk; on debris.*
Cap: ⅛ –⅜" (0.3–1 cm) wide; bell-
shaped to convex, sometimes knobbed;
margin flaring or upturned with age;
smooth, with hoary bloom at first;
coral-red, then yellowish near margin,
fading to bright orange-yellow.
Gills: attached, close to almost distant,
moderately broad; pale orange to

whitish, with whitish edges.
Stalk: ⅜–2⅜" (1–6 cm) long, ½2" (1 mm) thick; orange- to lemon-yellow, densely white-hairy at first; with long, coarse, white hairs about base.
Spores: 9–11 × 2.5–5.4 μ; spindle-shaped, smooth, colorless. Spore print white.

Season: May–September.
Habitat: Singly or in groups, on leaves and twigs in wet woods.
Range: Quebec to North Carolina; also California.
Look-alikes: *M. strobilinoides** is scarlet to orange with scarlet gill edges. *M. amabilissima* is bright rose, fading to white, and grows on conifer needles. *M. adonis,* on conifer needles, is scarlet, fading to yellow. Small, brightly colored *Hygrophorus** species have a waxy feel.
Comments: This is one of the first gilled mushrooms to appear in the spring.

45 **Yellow-stalked Mycena**
Mycena epipterygia (Fr.) S.F.G.
Tricholomataceae, Agaricales

Description: *Small, slippery, dingy yellowish mushroom with long, thin stalk; under conifers.*
Cap: ⅜–¾" (1–2 cm) wide; egg-shaped, becoming convex with prominent knob; margin sometimes flaring with age; slimy-tacky, smooth, becoming radially furrowed; dark yellowish, tinged grayish-green at times, fading to white or flushed grayish-pink.
Gills: attached, almost distant, narrow; white to faintly yellow.
Stalk: 2–3" (5–7.5 cm) long, ½2–¹⁄₁₆" (1–1.5 mm) thick; slimy, with coarse hairs about base; chrome-yellow, fading to white.
Spores: 8–11 × 4–6 μ; oval, smooth, colorless, amyloid. Spore print white.
Season: Summer–fall; fall–winter in California.
Habitat: Scattered to numerous, in coniferous woods.

Range: Widely distributed in North America.
Look-alikes: *M. epipterygioides* has dark olive-gray cap. *M. viscosa* has strong odor and reddens with age.
Comments: The Yellow-stalked Mycena, including a bad-smelling variety that grows on coniferous wood, and its look-alikes are among the last gilled mushrooms to appear in the fall; often their season lasts into December.

6 Common Mycena
Mycena galericulata (Scop. ex Fr.) S.F.G.
Tricholomataceae, Agaricales

Description: *Brownish to grayish, bell-shaped cap with grayish gills and stalk; in clusters on decaying wood.*
Cap: 1–3" (2.5–7.5 cm) wide; bell-shaped, with broad knob; margin incurved at first, then spreading; moist, wrinkled to somewhat radially lined; brownish, fading to dirty tan or cinnamon-brown. Odor and taste sometimes strongly mealy.
Gills: attached, close to almost distant, moderately broad, with crossveins; whitish to grayish or pink-tinged.
Stalk: 2–5" (5–12.5 cm) long, $\frac{1}{16}$–$\frac{1}{4}$" (1.5–5 mm) thick; brittle, smooth or twisted; grayish-white above, darker gray to brown below; hollow.
Spores: 8–11 × 5.5–7 μ; elliptical, smooth, colorless, amyloid. Spore print white.
Season: June–October; November–February in California.
Habitat: In clusters, on rotten deciduous logs and stumps.
Range: Widely distributed in North America.
Look-alikes: *M. inclinata*, somewhat smaller, has stalk with white flecks at first, becoming rust-brown at base.
Comments: Its larger size and the incurved cap margin on young specimens distinguish this species from the many other grayish to brownish mycenas that grow on wood.

76 Bleeding Mycena
Mycena haematopus (Pers. ex Fr.) Kum.
Tricholomataceae, Agaricales

Description: *Small, reddish-brown, bell-shaped cap and brownish stalk, bleeding dark red when cut; clustered on decayed wood.*

Cap: ⅜–2" (1–5 cm) wide; egg-shaped, becoming bell-shaped to convex, then flat, knobbed, with upturned margin and scalloped edge; dry, with a bloom, becoming polished and radially lined; dark reddish-brown to reddish-gray toward margin. Flesh exudes thick, dark, blood-red juice when cut.

Gills: attached, close to almost distant, narrow to moderately broad; whitish, staining reddish-brown, with minutely cottony-white edges.

Stalk: 1–4" (2.5–10 cm) long, ⅟₃₂–⅛" (1–3 mm) thick; covered with dense coating of minute hairs, polished with age, with long, coarse hairs about base; whitish to reddish-tan, exuding dark blood-red juice when cut.

Spores: 8–11 × 5–7 μ; broadly elliptical, smooth, colorless, amyloid. Spore print white.

Season: June–September; may overwinter.

Habitat: Single to clustered, mostly on decaying wood.

Range: Widely distributed in North America.

Look-alikes: *M. sanguinolenta,* on the ground or in moss, has dark reddish-brown gill edges. *M. atkinsoniana,* on the ground, has yellowish gills with maroon edges.

Comments: This is one of the most frequently encountered mycenas.

64 Orange Mycena
Mycena leaiana (Berk.) Sacc.
Tricholomataceae, Agaricales

Description: *Sticky, shiny, orange mushroom with reddish-orange gill edges; clustered on wood.*

Cap: ⅜–2" (1–5 cm) wide; bell-shaped, becoming convex with sunken center; slimy-tacky, shiny, smooth to scruffy; reddish-orange, fading to orange-yellow.

Gills: attached, close to crowded, broad; ochre-pinkish, staining orange-yellow; edges brilliant reddish-orange.

Stalk: 1¼–2¾" (3–7 cm) long, ¹⁄₁₆–⅛" (1.5–3 mm) thick; slimy to tacky, fibrous-tough; orange to yellow; dense, long, coarse hairs about base.

Spores: 7–10 × 5–6 μ; elliptical, smooth, colorless, amyloid. Spore print white.

Season: June–September.

Habitat: Clustered on deciduous wood.

Range: Canada to North Carolina, west to Michigan.

Comments: This common mycena can be found throughout the summer, even when other mushrooms fail to appear.

---

52 Walnut Mycena
Mycena luteopallens (Pk.) Sacc.
Tricholomataceae, Agaricales

Description: *Small, yellow mushroom; on hickory nuts and walnuts.*

Cap: ⅜–⅝" (1–1.5 cm) wide; egg-shaped, becoming bell-shaped to flat or slightly knobbed; margin often wavy; translucent, hoary-downy, becoming smooth; yellow to orange, fading to whitish.

Gills: attached, almost distant, moderately broad, minutely hairy; yellow to pink-tinged, with pale edges.

Stalk: 2–4" (5–10 cm) long, ¹⁄₃₂–¹⁄₁₆" (1–1.5 mm) thick; yellowish, with long, coarse hairs about base.

Spores: 7–9 × 4–5.5 μ; elliptical, smooth to minutely roughened, colorless, amyloid. Spore print white.

Season: September–November.

Habitat: Scattered to several, on remains of hickory nuts and walnuts.

Range: New York to North Carolina, west to Michigan.

Comments: This beautiful yellow mycena can be easily recognized by its habitat.

98 Spotted Mycena
Mycena maculata Kar.
Tricholomataceae, Agaricales

Description: *Dark, bell-shaped cap with whitish gills developing reddish spots, and off-white to dirty reddish stalk.*
Cap: ¾–2" (2–5 cm) wide; bell-shaped to convex or nearly flat, with a distinct knob; margin flared or upturned with age; moist to tacky, smooth, radially lined to wrinkled; blackish-brown, then brown-gray and with reddish-brown spots, or watery-gray overall. Flesh grayish, bruising reddish-brown.
Gills: attached, close to nearly distant, narrow to moderately broad; whitish to pale grayish, becoming stained with reddish spots.
Stalk: ⅛–⅜" (0.3–1 cm) long, ¹⁄₁₆–¼" (1.5–5 mm) thick; brittle, smooth above; brownish, with dense white hairs near base; base staining reddish-brown; hollow.
Spores: 7–10 × 4–6 μ; elliptical, smooth, colorless, amyloid. Spore print white.

Season: August–October.
Habitat: Clustered on coniferous wood.
Range: N. North America south to Tennessee and California; common in Pacific NW.
Comments: This species, one of several brownish mycenas, can be readily identified by its reddish spots and stains.

94 Large Mycena
Mycena overholtzii A.H.S. & Sol.
Tricholomataceae, Agaricales

Description: *Grayish-brown cap, grayish gills, and hairy lower stalk; in clusters on coniferous wood, near melting snowbanks.*
Cap: 1–2" (2.5–5 cm) wide; convex to nearly flat with knob; moist, smooth, lined; grayish-brown to watery-gray.
Gills: attached or slightly descending stalk, almost distant, broad; off-white, staining gray.

Stalk: 2–4" (5–10 cm) long, ⅟₁₆–¼"
(1.5–5 mm) thick, enlarging
downward; dry, smooth, and pale
above; covered below with dense white
hairs; becoming reddish-brown.
Spores: 6–8 × 3.5–4 μ; elliptical,
smooth, colorless, amyloid. Spore print
white.

Season: May–August.

Habitat: Clustered on coniferous logs near
melting snowbanks.

Range: Mountainous areas of West.

Comments: Its appearance on spruce and fir near
melting snowbanks makes this brown
mycena distinctive.

---

104 **Pink Mycena**
Mycena pura (Pers. ex Fr.) Kum.
Tricholomataceae, Agaricales

Description: *Fleshy, rose to purplish mushroom with*
white to pinkish gills; on the ground.
Cap: 1–2" (2.5–5 cm) wide; egg-
shaped, becoming convex to flat,
sometimes with knob; margin straight
or slightly incurved; smooth, moist,
radially lined; variously colored: rose-
red, purplish, lilac-gray, or yellowish-
white with purplish tint at center. *Odor*
and taste somewhat radishlike.
Gills: attached, close to almost distant,
broad; whitish to gray or with pinkish
tones; edges whitish.
Stalk: 1–4" (2.5–10 cm) long, ⅟₁₆–¼"
(1.5–5 mm) thick, enlarging
downward; twisted-lined to smooth;
white to pink or purplish-tinged.
Spores: 5–9 × 3–4 μ; cylindrical to
elliptical, smooth, colorless, amyloid.
Spore print white.

Edibility: Edible with caution.

Season: July–September; November in Texas;
fall–early spring in California.

Habitat: Scattered to many, in woods.

Range: Canada to Florida, west across N.
North America to Pacific Coast; also
Texas, Colorado.

Comments: This species is atypical of its genus

because it grows singly on the ground, not clustered on wood. This mushroom has a radishlike taste, and although some people do eat it, it is reported to contain traces of muscarine, a toxin that affects the sympathetic nervous system.

70 Red-orange Mycena
Mycena strobilinoides Pk.
Tricholomataceae, Agaricales

Description: *Small, reddish-orange mushroom with bell-shaped cap; growing in mats under conifers.*
Cap: ⅜–¾" (1–2 cm) wide; conical, becoming bell-shaped; margin often scalloped and flaring with age; moist-tacky, smooth, radially lined; scarlet to red, fading to yellow.
Gills: attached, nearly distant, narrow; yellow to light pinkish-orange, with *scarlet edges.*
Stalk: 1¼–1⅝" (3–4 cm) long, ¹⁄₃₂–¹⁄₁₆" (1–1.5 mm) thick; covered with minute hairs, becoming smooth; orange-yellow, with long, coarse, orange fibers at base.
Spores: 7–9 × 4–5 μ; elliptical, smooth, colorless, amyloid. Spore print white.
Season: Late summer–fall.
Habitat: In conifer needles, especially pine.
Range: Mountainous areas of Pacific NW.; New England and Great Lakes area.
Look-alikes: Other reddish-orange *Mycena*s* have differently colored gill edges.
Comments: This species is often seen in great quantity in late fall.

102 Blue Mycena
Mycena subcaerulea (Pk.) Sacc.
Tricholomataceae, Agaricales

Description: *Small, bluish to brownish cap and stalk; gills whitish; on leaves or decayed wood.*
Cap: ¼–¾" (0.5–2 cm) wide; egg-shaped, becoming bell-shaped to broadly convex or flat; margin radially

lined; tacky, smooth, translucent; *pale blue to greenish-blue, becoming brownish, with margin usually remaining bluish.*
Gills: attached or nearly free, close to crowded, narrow to moderately broad; white to gray-tinged, with slightly fringed edges.
Stalk: 1–3″ (2.5–7.5 cm) long, ⅟₃₂–⅛″ (1–3 mm) thick; minutely hairy to smooth; bluish to greenish-blue, fading to grayish or grayish-brown.
Spores: 6–8 × 6–8 μ; round, smooth, colorless, amyloid. Spore print white.

Season: June–September.

Habitat: Single to scattered, in leaves, humus; on decayed wood of deciduous trees.

Range: East of Great Plains.

Look-alikes: *M. amicta,* in Northwest, grows on decaying conifers.

Comments: This can have 2 seasons, spring and fall. Note that this is a white-spored mushroom and does not bruise blue, unlike some hallucinogenic psilocybes that are similarly shaped.

---

27 Burn Site Mycena
Myxomphalia maura (Fr.) Hora
Tricholomataceae, Agaricales

Description: *Small, dark gray-brown, sunken cap with whitish gills and grayish brown, brittle stalk; in burned areas.*
Cap: ⅜–2″ (1–5 cm) wide; convex with incurved margin, becoming nearly flat with sunken center; slimy, smooth; dark gray-brown, then gray or white.
Gills: attached or slightly descending stalk, close, broad; white to grayish.
Stalk: 1–2″ (2.5–5 cm) long, ⅟₁₆–⅛″ (1.5–3 mm) thick; brittle, smooth, polished; grayish-brown, stuffed.
Spores: 4.5–6.5 × 3.5–4 μ; broadly elliptical to round, smooth, colorless, amyloid. Spore print white.

Season: May–June; September–October.

Habitat: On the ground on burn sites, under northern conifers.

Range: N. North America.

Comments: Formerly known as *Omphalina* and *Fayodia*, as well as *Mycena maura*. Its best field characteristic is its habitat.

---

10 Lichen Agaric
Omphalina ericetorum (Fr.) M. Lange
Tricholomataceae, Agaricales

Description: *Small, moist, brownish cap, fading to yellow, with cream-yellow gills descending brownish stalk; with lichens or in moss.*
Cap: ¼–1⅜" (0.5–3.5 cm) wide, but usually less than 1" (2.5 cm); flat to sunken in center or funnel-shaped; margin incurved at first, scalloped to wavy, radially lined; moist, smooth; brown, becoming yellowish to cream.
Stalk: ⅜–1¼" (1–3 cm) long, 1/32–⅛" (1–3 mm) thick, sometimes enlarged at base; pliant; smooth; brownish to pale yellow, with hairy base.
Spores: 7–9 × 4–6 μ; broadly elliptical, smooth, colorless. Spore print white.

Season: June–September; March–May and October–November on Pacific Coast.

Habitat: Scattered to several, in moss or soil or on conifer logs, but always associated with the lichen *Botrydina vulgaris*.

Range: Widespread in N. North America.

Comments: Also known as *Gerronema ericetorum*. This is reportedly the most common gilled mushroom in the Arctic. It occurs throughout the range of its lichen associate.

⊗ 310, 426, 483 **Jack O'Lantern**
Omphalotus olearius (DC. ex. Fr.) Sing.
Tricholomataceae, Agaricales

Description: *Orange to yellowish-orange mushroom with sharp-edged gills descending stalk; in clusters on wood or buried wood.*
Cap: 3–8" (7.5–20 cm) wide; convex to flat; usually circular at first, becoming sunken, with small central knob; margin incurved at first, then upturned

and wavy to lobed; dry, smooth, saffron-yellow.

Gills: descending stalk, close, narrow; yellow-orange.

Stalk: 3–8″ (7.5–20 cm) long, ⅜–⅝″ (1–1.5 cm) thick, narrowing at base; long, solid, smooth, dry, curved; saffron-yellow, darkening near base.

Spores: 3.5–5 μ; round or nearly round, smooth, colorless, nonamyloid. Spore print pale cream.

Edibility: Poisonous.

Season: July–November; November–March in California.

Habitat: Clustered at base of stumps and on buried roots of oak and other deciduous wood.

Range: E. North America and California.

Look-alikes: *Cantharellus cibarius** grows on ground, has wavy cap, blunt-edged, forked, gill-like folds with crossveins, and fragrant odor. *Hygrophoropsis aurantiaca**, on or near conifers, is smaller, with forked gills and typically orange-brown stalks.

Comments: Also called the "False Chanterelle," and often mistaken for the Chanterelle, this is a species complex. Some experts believe that the American forms may be distinct from the European form bearing the same name. If so, the eastern Jack is more properly called *O. illudens,* and the western form, *O. olivascens.* The eastern form has also been known as *Clitocybe illudens.* All forms are poisonous, typically causing gastric upset for a few hours to 2 days. When this species is gathered fresh and taken into a dark room, the gills give off an eerie green glow.

---

268   Rooted Oudemansiella
Oudemansiella radicata (Rel. ex Fr.) Sing.
Tricholomataceae, Agaricales

Description: *Sticky, smooth, gray-brown or paler cap with white gills and rigid-brittle, long-rooted stalk; on tree roots.*

Cap: 1–4" (2.5–10 cm) wide; convex to nearly flat or with broad knob; margin incurved at first; slimy to tacky or moist, smooth or somewhat wrinkled; dark brown to grayish- or smoky brown, sometimes nearly white.
Gills: attached, nearly distant, broad, thick; white.
Stalk: 2–8" (5–20 cm) long, 1/8–3/8" (0.3–1 cm) thick above ground; enlarging downward, with long, rootlike underground base tapering to point; dry, twisted-lined, rigid-brittle; whitish above, brownish below.
Spores: 12–16 × 9–11 μ; large, broadly elliptical, smooth, colorless. Spore print white.

Edibility: Edible.

Season: July–October; November in Texas.

Habitat: Single to several, on and about deciduous stumps and trees, especially beech.

Range: E. North America to Midwest.

Look-alikes: *O. longipes* has smaller, dry, velvet-brown cap. *Caulorhiza umbonata** and *Collybia hygrophoroides* have conical caps and amyloid spores.

Comments: Formerly known as *Collybia radicata*. This common summer-fall mushroom has a meaty, edible cap and a stalk that snaps on breaking.

---

498 **Late Fall Oyster**
Panellus serotinus (Fr.) Küh.
Tricholomataceae, Agaricales

Description: *Sticky, yellowish-green cap with yellowish gills and stublike stalk; on dead wood.*
Cap: 1–4" (2.5–10 cm) wide; fan- to kidney-shaped; margin incurved, becoming lobed to wavy; slimy when moist, smooth; yellowish-green, often with violet tones, to brownish-green.
Gills: attached, but appearing to descend stalk; close, narrow to broad; yellowish-white to orange-yellow.
Stalk: 1/4–3/4" (0.5–2 cm) long, 1/4–3/8" (0.5–1 cm) thick; stout, stublike,

lateral, hairy; yellow- to ochre-brown.
Spores: 4.5–5.5 × 1.5–2 μ; sausage-
shaped, smooth, colorless, amyloid.
Spore print yellowish.

Edibility: Edible.

Season: October–November, usually only after
first frost.

Habitat: Singly or in overlapping clusters, on
both coniferous and deciduous wood.

Range: Widely distributed in North America.

Comments: Also called the "Green Oyster," and
formerly known as *Pleurotus serotinus*.
The Late Fall Oyster is edible, but
requires long, slow cooking.

501 Luminescent Panellus
Panellus stipticus (Bull. ex Fr.) Kar.
Tricholomataceae, Agaricales

Description: *Small, tough, brownish cap with brownish*
gills and off-center, stublike, solid stalk;
clustered on dead wood.
Cap: ⅜–1¼" (1–3 cm) wide; kidney-
shaped, sunken at base; dry, hairy to
scruffy-scaly; brown to buff or tan.
Flesh thin, tough. *Taste sharply acrid.*
Gills: attached, but appearing to
descend stalk; close, narrow; pinkish-
brown.
Stalk: ¼–⅝" (0.5–1.5 cm) long, ⅛–
⅜" (0.3–1 cm) thick; very short, off
center, hairy; whitish-buff, often
compressed.
Spores: 3–6 × 2–3 μ; sausage-shaped,
smooth, colorless, amyloid. Spore print
white.

Season: May–December.

Habitat: Usually numerous, on logs and stumps.

Range: Widely distributed in North America.

Comments: Formerly known as *Panus stipticus,* this
species was transferred to *Panellus*
because of its amyloid spores. Since the
gills are often brownish, a spore print is
necessary to be certain this is not a
Crepidotus. Like the Jack O'Lantern, it
is luminescent, the gills giving off a
greenish light in darkness. It can
reportedly stop bleeding.

505 Ruddy Panus
Panus rudis Fr.
Tricholomataceae, Agaricales

Description: *Tough, densely hairy, tan, pink, or red-*
brown cap and off-center, stublike stalk.
Cap: 1–3″ (2.5–7.5 cm) wide; fan- or
kidney-shaped; margin incurved, lobed;
dry, velvety, coarsely hairy; pinkish-tan
to reddish-brown, or with violet tinge,
tan with age. Flesh thin, tough.
Gills: descending stalk, close, narrow;
whitish to tannish.
Stalk (when present): 3/8–3/4″ (1–2 cm)
long, 1/8–3/8″ (0.3–1 cm) thick; short,
off-center, stublike, densely hairy;
pinkish-brown to tan.
Spores: 5–6 × 2.5–3 μ; elliptical,
smooth, colorless. Spore print white.

Season: May–November.
Habitat: Several to clustered, on deciduous
wood.
Range: Widely distributed in North America.
Comments: This widespread species can be found
throughout the mushroom season.

500 Orange Mock Oyster
Phyllotopsis nidulans (Pers. ex Fr.) Sing.
Tricholomataceae, Agaricales

Description: *Stalkless, densely hairy, orange to buff cap*
with inrolled margin and orange-yellow
gills; odor bad.
Cap: 1–3″ (2.5–7.5 cm) wide;
semicircular to kidney-shaped; margin
inrolled at first; dry, coarsely hairy;
orange to orange-yellow to buff. Odor
and taste sharply disagreeable.
Gills: attached to hairy base, close to
distant, narrow to moderately broad;
orange-yellow to orange-buff.
Spores: 6–8 × 3–4 μ; sausage-shaped,
smooth, colorless to tinted. Spore print
pinkish.

Season: August in Colorado; September–
October in Northeast; November in
Texas; fall–winter in California.
Habitat: Single or in overlapping clusters, on

coniferous and deciduous wood.
Range: Widely distributed in North America.
Comments: Formerly known as *Pleurotus nidulans*
and *Claudopus nidulans*. This species is
not known to be poisonous, but it
tastes as bad as it smells.

---

493 Angel's Wings
Pleurocybella porrigens (Pers. ex Fr.) Sing.
Tricholomataceae, Agaricales

Description: *Broad, thin-fleshed, stalkless white cap*
with white gills; in clusters on wood.
Cap: 1–4″ (2.5–10 cm) wide; fan-
shaped to elongately ear-shaped, with
margin incurved; dry, smooth to
minutely hairy, white. Flesh thin,
white.
Gills: descending to stublike base,
crowded, narrow; white, then cream.
Spores: 6–7 × 5–6 μ; round, smooth,
colorless. Spore print white.
Edibility: Good.
Season: September–October.
Habitat: Scattered to many, on decaying
coniferous wood, particularly hemlock.
Range: N. North America south to North
Carolina, west to California.
Look-alikes: *Pleurotus ostreatus** is fleshier and grows
mostly on deciduous wood.
Comments: Formerly known as *Pleurotus porrigens*.
This species is often abundant.

---

495 Veiled Oyster
Pleurotus dryinus (Pers. ex Fr.) Kum.
Tricholomataceae, Agaricales

Description: *White, cottony-scaly cap with inrolled*
margin, evanescent partial veil, and white
gills; arising from short, stout, solid stalk.
Cap: 2–5″ (5–12.5 cm) wide; broadly
convex; margin inrolled at first, with
fragments of veil persisting; dry,
cottony-hairy, then somewhat scaly;
whitish. Odor aromatic to pungent.
Gills: descending stalk, almost distant
to close, narrow to broad; often with

crossveins on stalk; white, discoloring.
Stalk: 2–4″ (5–10 cm) long, ⅜–1¼″
(1–3 cm) thick; off-center, stout,
usually short, solid; whitish.
Veil: partial veil grayish; leaving
evanescent, membranous ring on upper
stalk, often fragments on cap margin.
Spores: 9–12 × 3.5–4 μ; elliptical,
smooth, colorless. Spore print white.

Edibility: Edible.
Season: July–October.
Habitat: Single to several, on deciduous trees.
Range: N. North America; Rockies.
Comments: Formerly called *Armillaria dryina* and
P. corticatus. The Veiled Oyster grows
on alder on the Pacific Coast, and on
hickory and maple in the East. It can
have a pleasing fragrance, and it is
edible, although somewhat tough.

484, 497 Oyster Mushroom
Pleurotus ostreatus Fr.
Tricholomataceae, Agaricales

Description. *Broad, fleshy, white, gray or brown cap
with broad, whitish or yellow-tinged gills
arising from attachment to wood or small,
hairy, stublike stalk; on wood.*
Cap: 2–8″ (5–20 cm) wide; oyster-
shaped, semicircular to elongated;
margin lobed to wavy at times; moist,
smooth; white to ash or brownish.
Flesh thick, white. Odor pleasant.
Gills: descending stalk, close to nearly
distant, narrow to broad, thick; white,
becoming yellowish.
Stalk (when present): ¼–⅜″ (0.5–1
cm) long, ¼–⅜″ (0.5–1 cm) thick; off-
center to lateral, short, stout, solid;
dry, white-hairy.
Spores: 7–9 × 3–3.5 μ; narrowly
elliptical, smooth, colorless. Spore
print white to pale lilac-gray.
Edibility: Choice.
Season: Year-round under favorable conditions.
Habitat: On many deciduous trees, especially
willow and aspen; rarely on pine and
hemlock; sometimes on buried stumps.

Range: Throughout North America.
Look-alikes: *Pleurocybella porrigens** is thin-fleshed, white, and grows on conifers. *Hypsizygus tessulatus** usually grows singly, has yellowing, cracking cap. *Lentinus** and *Lentinellus** species have toothed gill edges.
Comments: This species complex has many forms: in summer it is usually flat and whitish, but in fall and winter it is rounder and brownish. This excellent edible should be checked for white grubs. It is sometimes covered by the yellow, early stage of a slime mold. The Cornucopia Oyster (*P. sapidus*), also choice, is part of this complex. It has a lilac-gray spore print, a whitish cap, and usually an off-center stalk.

258 Amyloid Tricholoma
Porpoloma umbrosum (A.H.S. & Wall.) Sing.
Tricholomataceae, Agaricales

Description: *Fleshy, gray to grayish-brown mushroom, bruising reddish; under conifers.*
Cap: 2–4" (5–10 cm) wide; convex, becoming flat or somewhat sunken; margin incurved at first; dry, smooth, becoming cracked into small areas; gray, becoming grayish-brown. Flesh grayish, bruising reddish. Odor strong, resembling fresh cucumbers.
Gills: attached or slightly descending stalk, close, broad; grayish, bruising reddish.
Stalk: 1–2" (2.5–5 cm) long, ⅝–¾" (1.5–2 cm) thick, with narrowed base; gray to gray-brown, bruising reddish.
Spores: 7–9 × 3–4 μ; elliptical, smooth to faintly netted, colorless, *amyloid*. Spore print white.
Season: July–September.
Habitat: In grass under conifers, such as spruce and hemlock.
Range: Nova Scotia to North Carolina, west to Ohio.
Comments: First described as *Tricholoma umbrosum*. *Porpoloma* is a genus of mostly South

American species, and the Amyloid
Tricholoma may be the only North
American representative.

---

486 Black Jelly Oyster
Resupinatus applicatus (Bat. ex Fr.) S.F.G.
Tricholomataceae, Agaricales

Description: *Stalkless, minute, bluish-black, cuplike cap*
with grayish gills; on wood.
Cap: ⅛–¼″ (3–5 mm) wide; minute,
convex to cuplike; dry, hairy; grayish-
black, tinged blue. Flesh gelatinous.
Gills: arising from base, almost distant,
broad; white to grayish to blackish.
Spores: 4–6 μ; round, smooth,
colorless. Spore print white.
Season: June–November.
Habitat: On the underside of rotten logs.
Range: Quebec to Florida, west to Texas.
Comments: Formerly known as *Pleurotus applicatus.*
The Black Jelly Oyster is one of the
many mushrooms that must be searched
for; it is rarely found accidentally.

---

666 Netted Rhodotus
Rhodotus palmatus (Bull. ex Fr.) Maire
Tricholomataceae, Agaricales

Description: *Tough, ridged and pitted, pink to*
yellowish-orange cap with salmon gills and
off-center, short stalk; on wood.
Cap: 1–2″ (2.5–5 cm) wide; convex;
margin incurved; dry, netted with
ridges and pits; reddish to pinkish,
fading to orange-yellow. Flesh
gelatinous.
Gills: attached, close, moderately
broad, thin, with crossveins; becoming
pinkish.
Stalk: 1–2″ (2.5–5 cm) long, ⅛–¼″
(3–5 mm) thick; off-center, curved,
dry, fibrous-tough; reddish to pinkish.
Spores: 6–8 μ; round, warted, colorless
to tinted. Spore print pinkish.
Season: June–September.
Habitat: On decaying hardwood, often maple.

Range: E. Canada to Virginia and Michigan.
Comments: Formerly known as *Pleurotus
subpalmatus.* Its netted cap surface
makes this a most distinctive species.

59 Orange Moss Agaric
Rickenella fibula (Bull. ex Fr.) Raith.
Tricholomataceae, Agaricales

Description: *Small, orange cap with sunken center and
radially lined, inrolled margin; whitish
gills descending long, thin stalk; in moss.*
Cap: ⅛–⅝″ (0.3–1.5 cm) wide; convex
to flat, slightly to deeply sunken in
center; margin lined, inrolled to
straight; moist, ochre-orange to buff.
Gills: descending stalk, long, close to
almost distant, narrow to broad; often
with crossveins; white to buff-white.
Stalk: ⅜–2″ (1–5 cm) long, 1/32–1/16″
(1–1.5 mm) thick; fragile, finely hairy
at first; ochre-orange to buff; hollow.
Spores: 4–5 × 2–2.5 μ; elliptical,
smooth, colorless. Spore print white.
Season: June–November.
Habitat: In moss.
Range: Canada to Florida, west to Colorado.
Comments: This species looks like a pretty little
orange or yellow parasol in the bright
green moss. Known by many names,
including *Mycena, Omphalia,
Omphalina,* and *Gerronema,* it is now
placed in *Rickenella* because of its
nonamyloid spores and the size and
location of the cystidia.

243 Fringed Ripartites
Ripartites tricholoma (A. & S. ex Fr.) Kar.
Tricholomataceae, Agaricales

Description: *Silky, white cap with white gills becoming
pinkish-brown and yellowish-white stalk.*
Cap: 1–2″ (2.5–5 cm) wide; convex to
nearly flat, sunken in center; margin
incurved at first, radially lined with
long, coarse hairs; tacky to dry, silky-
soft, white.

Gills: attached, close, broad; white, becoming dull pinkish-brown.

Stalk: 1–2″ (2.5–5 cm) long, ⅛″ (3 mm) thick; fragile, somewhat minutely hairy; white to dingy brown; hollow.

Veil: partial veil evanescent.

Spores: 4–5 × 3.5–4 μ; oval to nearly round, warted, pale brown. Spore print pale brown.

Season: Late summer–fall.

Habitat: On humus or very decayed wood.

Range: Widely distributed in North America.

Comments: This species resembles a small clitocybe, but it has a veil and brownish, small, round, spiny spores.

7 Magnolia-cone Mushroom
Strobilurus conigenoides (Ell.) Sing.
Tricholomataceae, Agaricales

Description: *Small, whitish mushroom with white gills; on magnolia fruits.*

Cap: ¼–¾″ (0.5–2 cm) wide; convex to flat; margin incurved at first; dry, densely and minutely hairy; white.

Gills: attached, moderately crowded, broad; white.

Stalk: 1–2″ (2.5–5 cm) long, ¹⁄₃₂–¹⁄₁₆″ (1–1.5 mm) thick; brittle, dry, densely and minutely hairy; white.

Spores: 6–7 × 3–3.5 μ; elliptical, smooth, colorless. Spore print white.

Season: September–November.

Habitat: On magnolia fruits.

Range: Quebec to Florida, west to Texas.

Comments: Formerly known as *Collybia conigenoides*. This species is very common in the Southeast. Its range extends northward where forest magnolias grow.

8 Douglas-fir Collybia
Strobilurus trullisatus (Murr.) Lennox
Tricholomataceae, Agaricales

Description: *Small, pale, pinkish-white cap with pink-tinged to whitish gills and white to yellowish stalk; on Douglas-fir cones.*

Cap: ¼–¾″ (0.5–2 cm) wide; convex
to flat, becoming sunken; margin
incurved at first, flaring with age,
radially lined; dry, smooth to wrinkled,
densely and minutely hairy; white to
pinkish or sienna.
Gills: attached, moderately crowded,
broad; white to pale pink.
Stalk: 1–2″ (2.5–5 cm) long, ⅟₃₂–⅟₁₆″
(1–1.5 mm) thick; brittle, dry, densely
and minutely hairy; white above,
yellowish to cinnamon below; rootlike
extension covered with cottony, orange
threads (mycelium); stuffed to hollow.
Spores: 3–4.5 × 1.5–3 µ; elliptical,
smooth, colorless. Cystidia abundant on
cap surface, gills, and stalk. Spore print
white.

Season: Late August–November.
Habitat: On partially buried or old Douglas-fir
cones; rarely on pinecones.
Range: Idaho; West Coast.
Comments: Formerly known as *S. kemptonae* and
Collybia trullisata, this is the most
commonly collected species of
Strobilurus in North America.

---

503 Veiled Panus
Tectella patellaris (Fr.) Murr.
Tricholomataceae, Agaricales

Description: *Small, sticky to dry, brownish cap with*
incurved margin, veil covering buff-brown
gills, and stublike stalk; on twigs.
Cap: ⅜–⅝″ (1–1.5 cm) wide; small,
convex, hanging or cuplike; margin
strongly incurved; slimy to tacky,
becoming dry; fibrous, brown. Flesh
leathery-tough.
Gills: arising from stublike stalk or
point of attachment, distant, narrow,
buff-brown.
Stalk (when present): ⅟₃₂–⅛″ (1–3
mm) long, ⅟₆₄–⅟₁₆″ (0.5–1.5 mm)
thick; off-center, curved, dry, brown.
Veil: partial veil membranous, very
thin, pinkish-buff; leaving no ring.
Spores: 3–4 × 1–1.5 µ; sausage-

shaped, smooth, colorless, amyloid.
Spore print white.

Season: July–October.
Habitat: On logs and fallen branches of deciduous trees, especially alder, beech, birch, and willow.
Range: Widely distributed in North America.
Comments: Also known as *Panellus patellaris,* and formerly called *Panus operculatus.* This species is recognized by its small size and veil. Because the gills are colored, a spore print should be made to avoid mistaking this for a little brown-spored mushroom.

297 Veiled Trich
Tricholoma aurantium (Schaeff. ex Fr.) Rick.
Tricholomataceae, Agaricales

Description: *Sticky, reddish-orange, flattened-scaly cap and stalk with discoloring white gills and white zone at top of stalk.*
Cap: 1⅜–3½" (4–9 cm) wide; convex to nearly flat or with broad knob; margin incurved at first; slimy-tacky, with dense, flattened scales; reddish to tawny-ochre. Odor and taste of rancid meal.
Gills: notched, close, narrow; white, staining rust-brown.
Stalk: 2–3" (5–7.5 cm) long, ⅜–⅝" (1–1.5 cm) thick; solid, white, densely covered with reddish to tawny-ochre scales below ring zone.
Veil: partial veil small, cobwebby, leaving only a line on stalk.
Spores: 4–5.5 × 3–3.5 μ; elliptical, smooth, colorless. Spore print white.
Season: August–October; November–February in California.
Habitat: Scattered to numerous, in mixed woods of conifers, aspen, and tanoak-madrone.
Range: Widely distributed in North America.
Look-alikes: *Armillaria zelleri* * is usually larger and fleshier, has membranous ring.
Comments: Formerly known as *Armillaria aurantia.* While not toxic, this species is thoroughly unpalatable.

323 Canary Trich
Tricholoma flavovirens (Pers. ex Fr.) Lund.
Tricholomataceae, Agaricales

Description: *Sticky, yellow cap with brownish center, yellow gills, and yellowish-white stalk; under pine.*
Cap: 2–4″ (5–10 cm) wide; convex to nearly flat; slimy-tacky, slightly scaly about center; yellow to somewhat red-tinged.
Gills: notched to nearly free, close, broad; sulfur-yellow.
Stalk: 1–2″ (2.5–5 cm) long, ⅜–¾″ (1–2 cm) thick; stout, sometimes almost bulbous, solid; yellowish-white.
Spores: 6–8 × 3–5 μ; elliptical, smooth, colorless. Spore print white.

Edibility: Good.

Season: August–November; late fall–winter in California.

Habitat: Scattered to many, under pine in sandy areas.

Range: Widely distributed in North America.

Look-alikes: *T. leucophyllum* has white gills. *T. sulphureum* is offensively pungent. *T. sejunctum** has blackish radial fibers on cap and whitish gills.

Comments: Also known as the "Sandy Trich" and *T. equestre.* This is one of the last good edibles available during the season, and often appears as late as December in mild years.

---

281 Shingled Trich
Tricholoma imbricatum (Fr. ex Fr.) Kum.
Tricholomataceae, Agaricales

Description: *Dry, reddish-brown, fibrous-scaly cap with white gills, bruising reddish; solid stalk white above, reddish-brown below.*
Cap: 1–4″ (2.5–10 cm) wide; conical to convex and slightly knobbed; margin incurved when young, often wavy with age; dry, fibrous-scaly, *becoming coarsely scaly;* reddish-brown, paler toward margin.
Gills: notched, close, moderately broad;

whitish, becoming light yellowish-brown to dull brown.

Stalk: 2–4" (5–10 cm) long, ⅜–⅝" (1–1.5 cm) thick; solid, fibrous, mealy at top; white to buff, darkening to brownish below.

Spores: 5.5–6.5 × 3.5–4.5 μ; broadly elliptical, smooth, colorless. Spore print white.

Season: September–November; August in Colorado; November–March in California.

Habitat: Scattered to many, under conifers, especially pine.

Range: N. North America; California.

Look-alikes: *T. vaccinum** has brighter color and shaggy, cottony cap margin. *Inocybe** species have brown spore print.

Comments: The cap of the Shingled Trich becomes coarsely scaly with age.

⊗ 355 **Dirty Trich**
Tricholoma pardinum Quél.
Tricholomataceae, Agaricales

Description: *Dry, thick-fleshed, gray, scaly cap with white gills and stalk; under conifers.*
Cap: 2–6" (5–15 cm) wide; convex, broadly convex, or flat; dry, covered with whitish to grayish, dark fibrous scales. Odor and taste mealy.
Gills: notched, close, moderately broad, whitish.
Stalk: 2–6" (5–15 cm) long, ⅜–¾" (1–2 cm) thick, sometimes enlarged at base; dry, smooth, white.
Spores: 8–10 × 5.5–6.5 μ; elliptical, smooth, colorless. Spore print white.

Edibility: Poisonous.

Season: August in Colorado; September–October in Pacific NW.

Habitat: Several to many, under conifers, especially fir; mixed deciduous woods.

Range: Widespread in N. North America.

Look-alikes: Other dark-scaled *Tricholoma*s* are best differentiated microscopically.

Comments: Also called the "Poisonous Trich," this species causes severe gastric upset.

Some other, smaller, scaly species are reportedly edible, but should be avoided because of the difficulty in differentiating the many scaly, grayish tricholomas.

⊗ 296 **Red-brown Trich**
Tricholoma pessundatum (Fr.) Quél.
Tricholomataceae, Agaricales

Description: *Sticky, red-brown cap with red-staining white gills and white stalk; under pines.*
Cap: 2–6″ (5–15 cm) wide; convex, becoming flat; margin inrolled at first, becoming slightly upturned; sticky, smooth to slightly scaly; reddish-brown. Odor and taste mealy.
Gills: notched, close, moderately broad; white, staining reddish.
Stalk: 3–4″ (7.5–10 cm) long, ⅜–¾″ (1–2 cm) thick, sometimes narrowed below; stout, solid, smooth; white, becoming reddish-brown.
Spores: 4–6 × 2.5–3 μ; oval to elliptical, smooth, colorless. Spore print white.
Edibility: Poisonous.
Season: September–November; December–January in California.
Habitat: On the ground under conifers; also reported under live oak.
Range: W. North America; E. Canada.
Look-alikes: *T. populinum** grows under poplar.
Comments: This species is reportedly poisonous and should be avoided, along with all other red-brown tricholomas except the Poplar Trich.

---

257 **Poplar Trich**
Tricholoma populinum J. Lange
Tricholomataceae, Agaricales

Description: *Sticky, off-white to cinnamon-brown cap with white gills and stalk bruising reddish-brown; in sandy soil under poplars.*
Cap: 2–6″ (5–15 cm) wide; convex, becoming flat; margin inrolled,

becoming upturned and wavy; sticky, with radial fibers and water spots; reddish-brown, paler toward margin. Odor and taste mealy.

Gills: notched, close, narrow; white, staining reddish-brown.

Stalk: 1–3″ (2.5–7.5 cm) long, ⅜–1¼″ (1–3 cm) thick, sometimes enlarged below; white, staining reddish-brown.

Spores: 5–6 × 3.5–4 μ; oval, smooth, colorless. Spore print white.

Edibility: Edible.

Season: August–October.

Habitat: In dense clusters in sandy soil under cottonwoods and other poplars.

Range: Widespread in N. North America.

Comments: In the group of tricholomas with reddish-brown, slimy-tacky caps, the only well-known, safe edible is the Poplar Trich. It should be gathered only if found growing in large clusters under poplars, in grass or sandy soil.

---

353 Sticky Gray Trich
Tricholoma portentosum (Fr.) Quél.
Tricholomataceae, Agaricales

Description: *Slimy to sticky, smoky-gray cap with black radial fibers, white, yellow-tinted gills, and a stout, solid, fibrous stalk; under conifers.*
Cap: 2–5″ (5–12.5 cm) wide; convex to somewhat knobbed; margin incurved at first; slimy-sticky, with dark radial fibers; grayish, or with purplish or yellow tones. Odor and taste mealy.

Gills: notched, close, broad; whitish with grayish or yellow tones.

Stalk: 2–4″ (5–10 cm) long, ⅜–¾″ (1–2 cm) thick; stout, solid, fibrous, streaked; whitish or with yellow tones.

Spores: 5–6 × 3.5–5 μ; oval, smooth, colorless. Spore print white.

Edibility: Good, with caution.

Season: October; to February in California.

Habitat: Scattered to numerous, on sandy soil in mixed woods; also under pine and oak.

Range: Widespread in N. North America.

Look-alikes: *T. sejunctum** has yellowish cap with blackish fibers and bitter taste. The poisonous *T. pardinum** has cap without yellow tones.

Comments: This is an edible tricholoma, but it must be identified with care and a spore print taken so that a poisonous tricholoma or entoloma is not eaten.

358 Soapy Trich
Tricholoma saponaceum (Fr.) Kum.
Tricholomataceae, Agaricales

Description: *Lead-gray to olive-tinged cap with whitish gills and stalk, pinkish at base; odor soapy.*
Cap: 1⅝–3¼″ (4–8 cm) wide; convex to flat; margin incurved at first; dry to moist, smooth or becoming cracked into fine scales; lead-gray, often with olive or yellowish tones. Flesh white, staining pinkish. Odor and taste soapy to mealy.
Gills: attached or sometimes notched, almost distant, broad; white.
Stalk: 2–3¼″ (5–8 cm) long, ⅝–¾″ (1.5–2 cm) thick, widest at middle, narrowing or somewhat rooting below; stout, solid, minutely scaly at times; white, pinkish about base. Flesh becoming pinkish.
Spores: 5–6 × 3.5–4 μ; oval, smooth, colorless. Spore print white.

Season: July–November.

Habitat: Single to several, in deciduous woods.

Range: Widely distributed in North America.

Comments: This unpalatable mushroom can be distinguished by its soapy odor and the discoloration of the interior of the stalk.

319 Separating Trich
Tricholoma sejunctum (Sow. ex Fr.) Quél.
Tricholomataceae, Agaricales

Description: *Sticky to dry, yellowish cap with knob, streaked with blackish fibers, and with whitish gills and stalk.*
Cap: 2–3″ (5–7.5 cm) wide; convex to

broadly knobbed to flat; margin incurved at first; tacky to dry, fibrous; yellow, with conspicuous black radial fibers. Odor and taste mealy, somewhat bitter and unpleasant.

Gills: notched, close, broad; white, tinged yellowish with age.

Stalk: 2–4" (5–10 cm) long, ⅜–¾" (1–2 cm) thick; smooth to fibrous; white or tinged yellowish; stuffed.

Spores: 5–7 × 4–5.5 μ; oval, smooth, colorless. Spore print white.

Season: July–November.

Habitat: Scattered to numerous, in deciduous and coniferous woods, especially under pine, spruce, and hemlock; also oak.

Range: Widely distributed in North America.

Look-alikes: *T. flavovirens** has smooth cap and yellow gills. *T. leucophyllum* has smooth cap. *T. sulphureum* is sulfur-yellow overall and strongly malodorous.

Comments: Although most common in fall, this is one of few trichs to grow in summer.

---

270 Russet-scaly Trich
Tricholoma vaccinum (Pers. ex Fr.) Kum.
Tricholomataceae, Agaricales

Description: *Dry, reddish-brown, densely scaly cap with whitish or buff gills, bruising reddish, and hollow stalk with reddish-brown fibers; under conifers.*

Cap: 1–3" (2.5–7.5 cm) wide; conical to convex, becoming broadly convex to nearly flat, knobbed; margin incurved at first, cottony-fibrous; dry, fibrous to cracked-scaly, scales with downcurled tips; rust-brown, paler between fibrous scales.

Gills: notched, close, broad; whitish-buff, becoming reddish-cinnamon.

Stalk: 1–3" (2.5–7.5 cm) long, ⅜–⅝" (1–1.5 cm) thick, sometimes enlarged below; dry, sometimes with fibrous, downcurled scales; rust-brown or paler; usually hollow.

Veil: partial veil cobwebby, with light brown fibers; leaving remnants on cap

edge, but none on stalk.
Spores: 6–7.5 × 4–5 μ; oval-elliptical, smooth, colorless. Spore print white.

Season: July–December.
Habitat: Scattered to clustered, in coniferous woods, often under pine and spruce.
Range: Widespread in N. North America.
Look-alikes: *T. imbricatum** has dull brown cap, smooth at first, becoming coarsely scaly, and solid stalk.

Comments: This trich is not reported to be edible. Because it is densely fibrous-scaly, it could be mistaken for an inocybe. Since its gills discolor, a spore print should be made.

351 Fibril Trich
Tricholoma virgatum (Fr. ex Fr.) Kum.
Tricholomataceae, Agaricales

Description: *Dry, grayish cap with sharp knob, radially streaked with dark fibers, discoloring white gills, and smooth, white stalk; under conifers.*
Cap: 2–4" (5–10 cm) wide; conical to broadly conical with sharp knob; margin incurved at first; dry, radially streaked with fibers and scales; dark gray, sometimes with violet tones, paler toward margin. Taste acrid.
Gills: notched, close, broad; white.
Stalk: 3–5" (7.5–12.5 cm) long, ⅜–¾" (1–2 cm) thick; solid, smooth to slightly fibrous, white.
Spores: 6–7 × 5–6 μ; oval, smooth, colorless. Spore print white.
Season: August–October; fall and winter in California.
Habitat: Single to several, in coniferous and mixed woods.
Range: Widely distributed in North America.
Look-alikes: The poisonous *T. pardinum** is larger, stouter, and has convex cap. *T. terreum* has a flat, scaly cap. *T. saponaceum** has smooth cap.
Comments: This species is readily identified by its gray cap with blackish radial fibers, and its acrid taste.

302 Decorated Mop
Tricholomopsis decora (Fr.) Sing.
Tricholomataceae, Agaricales

Description: *Yellowish mushroom with blackish, fibrous scales over center; on coniferous wood.*
Cap: 1¼–2⅜" (3–6 cm) wide; convex, slightly sunken in center; moist, covered with minute, blackish, fibrous scales; yellowish-orange or with brown tones, darker in center.
Gills: attached, crowded, narrow; yellow.
Stalk: 1¼–2⅜" (3–6 cm) long, ⅛–⅜" (0.3–1 cm) thick; central to off-center, dotted with minute scales, yellow.
Spores: 6–7.5 × 4–5.5 μ; elliptical, smooth, colorless. Spore print white.

Season: Late June–October.
Habitat: Single to several, on coniferous logs.
Range: Throughout North America.
Look-alikes: *T. rutilans** has reddish scales. *T. sulfureoides** lacks colored fibers.
Comments: Formerly known as *Clitocybe decora,* this species was transferred to *Tricholomopsis* because of its wood habitat and its yellow gills with prominent cystidia.

---

265 Platterful Mushroom
Tricholomopsis platyphylla (Pers. ex Fr.) Sing.
Tricholomataceae, Agaricales

Description: *Brownish-gray cap with radial fibers, white gills, and stalk with white cords.*

Cap: 2–5" (5–12.5 cm) wide; bell-shaped, becoming convex to flat, often sunken in center; margin becoming wavy; streaked with dark radial fibers; grayish-brown or lighter.
Gills: attached, almost distant, broad, often splitting or eroded; white.
Stalk: 3–5" (7.5–12.5 cm) long, ⅜–¾" (1–2 cm) thick; stout, fibrous-lined with fibrous-tough rind; white, with thick, white runners (rhizomorphs) attached to base; solid, then hollow.
Spores: 7–9 × 5–7 μ; oval, smooth, colorless. Spore print white.

Edibility: Edible with caution.

Season: Late May—early October.

Habitat: Single to several, on and about deciduous logs, stumps, wood debris; also on buried wood.

Range: Quebec to Florida, west to Iowa; reported rarely along Pacific Coast.

Look-alikes: *Entoloma** species produce pink spore print. *Pluteus cervinus** has free gills and pink spores.

Comments: Formerly known as *Collybia platyphylla*, and also called *Oudemansiella platyphylla*. It is one of the first large, fleshy, gilled mushrooms to appear in the spring and is common throughout June and July. It is edible when very young and is generally considered safe, although a few instances of gastric upset have occurred.

---

327 Variegated Mop
Tricholomopsis rutilans (Schaeff. ex Fr.) Sing.
Tricholomataceae, Agaricales

Description: *Yellow cap and stalk with dense layer of reddish-fibrous scales and yellow gills; on coniferous wood.*
Cap: 2–4″ (5–10 cm) wide; convex, expanding to nearly flat; with margin inrolled at first; dry, yellowish, covered with dark reddish-hairy scales.
Gills: attached or notched, crowded, narrow; yellowish.
Stalk: 2–4″ (5–10 cm) long, 2–8″ (5–20 cm) thick; curved, yellowish, sometimes with reddish, fibrous scales; stuffed, becoming hollow.
Spores: 5–7 × 3–5 μ; broadly elliptical, smooth, colorless. Spore print white.

Season: May -June; August—November; November—February in California.

Habitat: Single to several, on coniferous wood, especially pine.

Range: Quebec to Florida, to West Coast.

Look-alikes: *T. decora** is yellowish-orange with blackish scales.

Comments: Formerly known as *Tricholoma rutilans*.

This species is more often found in cool weather. Its flavor is too unpleasant for it to be eaten.

303 Yellow Oyster Mop
Tricholomopsis sulfureoides (Pk.) Sing.
Tricholomataceae, Agaricales

Description: *Yellow mushroom with streaked cap; on coniferous wood.*
Cap: 1–3" (2.5–7.5 cm) wide; convex, expanding, sometimes with broad knob; fibrous to scaly, often streaked with age; yellow, drying bay-brown.
Gills: very slightly descending stalk or notched, close, broad; sulfur-yellow, drying bay-brown, with cottony-white edge.
Stalk: 1–3" (2.5–7.5 cm) long, $\frac{1}{4}$" (5 mm) thick; off-center, curved, fibrous; yellowish, drying bay-brown; stuffed, becoming hollow.
Spores: 5.5–6.5 × 4.5–5 μ; oval, smooth, colorless. Spore print white.
Season: Summer–fall.
Habitat: Single to several, on coniferous logs, especially eastern hemlock.
Range: Quebec to North Carolina, west to Great Lakes; Pacific NW.
Look-alikes: *T. decora** is yellowish-orange with blackish scales.
Comments: Formerly known as *Pleurotus sulfureoides.*

58 Fuzzy Foot
Xeromphalina campanella (Bat. ex Fr.) Küh. & Maire
Tricholomataceae, Agaricales

Description: *Small, orange-brown, sunken cap with off-white gills descending tough stalk with dense tuft of tawny hairs at base; clustered on wood.*
Cap: $\frac{1}{8}$–1" (0.3–2.5 cm) wide; convex to broadly convex with sunken center; margin incurved at first and radially lined; moist, smooth; ochre-tawny to cinnamon-brown.

Gills: descending stalk, nearly distant, narrow, with crossveins; yellowish to dull orange.

Stalk: 3/8–2" (1–5 cm) long, 1/64–1/8" (0.5–3 mm) thick, sometimes enlarging below with bulbous base; horny, yellow above, reddish-brown below; base covered with dense tuft of long, bright, tawny hairs.

Spores: 5–7 × 3–4 μ; elliptical, smooth, colorless, amyloid. Spore print light buff.

Season: May–November.

Habitat: Densely clustered on coniferous stumps and logs, especially well-decayed wood.

Range: Widely distributed in North America.

Comments: Formerly known as *Omphalina campanella*. There are at least 7 other species of *Xeromphalina* in North America, but this is probably the most abundant and widespread; several hundred mushrooms can cover one rotting hemlock stump.

STOMACH FUNGI
(Class Gasteromycetes)

This large group of fungi contains the puffballs, stalked puffballs, earthballs, earthstars, stinkhorns, and bird's-nest fungi, as well as the Hymenogastrales, an order of mushrooms that somewhat resemble deformed gilled mushrooms, boletes, or other fungi, but do not forcibly discharge their spores. Many of these mushrooms are common in summer and fall in urban parks, along roads and banks, in the wood-chip mulch placed about ornamental plantings, and especially in deserts and semiarid places, such as at high elevations, and along the seashore. Stomach fungi produce their spores within the fruiting body, rather than on an exposed hymenium. The interior of the fungus contains club-shaped structures, called basidia, that produce appendages (sterigmata) on which the spores develop. But stomach fungi do not forcibly discharge their spores; instead, they produce a spore mass, or gleba, from which spores are dispersed in a number of different ways. No mushrooms in this group produce a spore print. Although many species are edible, some are poisonous.

GILLED PUFFBALLS
(Order Hymenogastrales)

This artificial grouping of mushrooms includes many that resemble various gilled mushrooms and boletes, but whose basidia cannot forcibly eject their spores. Instead, the gills or tubes grow in a way that prevents the spores from being released or dispersed into the air. These mushrooms trap or enclose the spores, and dispersal is achieved through decay of the fruiting body or by the interference of animals. The spore mass (gleba) is typically fleshy and does not become powdery; the mushrooms do not produce a spore print. These are mushrooms of deserts, high mountains, and generally arid places, where there is an adaptive advantage in not exposing the developing basidia and immature spores to the excessive heat and evaporation of the environment. Family limits in this group are not well understood, so no families are described here; instead, 4 representative genera are included.

644 **Puffball Agaric**
Endoptychum agaricoides Czern.
Hymenogastraceae, Hymenogastrales

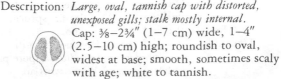

Description: *Large, oval, tannish cap with distorted, unexposed gills; stalk mostly internal.*
Cap: ⅜–2¾" (1–7 cm) wide, 1–4" (2.5–10 cm) high; roundish to oval, widest at base; smooth, sometimes scaly with age; white to tannish.

Gills: distorted, chambered; white, becoming brownish.

Stalk: internal, sometimes extending very slightly below; attached by cord to ground; whitish.

Spores: 6–9 × 5–7 μ; elliptical, smooth, pale brown.

Edibility: Edible with caution.

Season: May–October.

Habitat: In lawns, pastures, and gardens.

Range: Widely distributed in North America.

Look-alikes: *E. depressum* has blackish spore mass, longer stalk, grows in conifer woods.

Comments: Although this species resembles a gilled mushroom, it grows like a puffball. The gills retain the spores until they are dispersed by disintegration or by animals that eat them. It is edible when young, and tastes sweet and nutty.

662 American False Russula
Macowanites americanus Sing. & A.H.S.
Astrogastraceae, Hymenogastrales

Description: *Roundish, variably colored cap with deformed, ochre gills and short, white stalk.*
Cap: ⅜–2″ (1–5 cm) wide; roundish to somewhat convex at first, expanding to flat with downcurved margin; sticky at first, cracking into patches, smooth; yellow, pink, purple, red, blue, or olive, often mixed. Odor, taste mild.
Gills: attached or slightly descending stalk, irregular, convoluted, chambered; white, becoming ochre.
Stalk: ⅜–1¼″ (1–3 cm) long, ⅛–⅝″ (0.3–1.5 cm) thick; smooth, white.
Spores: 8.5–13.5 × 8–12 μ; short, elliptical to nearly round; with ridges and warts, amyloid, yellowish.

Edibility: Edible with caution.

Season: July–November.

Habitat: On the ground or buried in conifer duff, near fir, Douglas-fir, spruce, pine.

Range: Pacific NW. and Idaho.

Look-alikes: *Russula** species have normal gill development and produce a spore print. Other species of *Macowanites* are differently colored, strongly scented, bitter, or differ in spore size.

Comments: This species is common, especially in the mountains. It is reportedly edible, although little is known about it.

---

660 Western Rhizopogon
Rhizopogon occidentalis Zel. & Dod.
Rhizopogonaceae, Hymenogastrales

Description: *Whitish-yellow, irregularly shaped, potatolike, stalkless mushroom covered with a mesh of brownish strands.*
Mushroom: ⅜–2″ (1–5 cm) wide; somewhat pear-shaped; covered with small fibers and loose, cordlike, brownish strands (rhizomorphs); whitish, becoming lemon-yellow, bruising orange to reddish-brown. Spore mass pale yellow-orange, becoming brownish; minutely chambered. Odor of sourdough.
Spores: 5.5–7 × 2–3 μ; elliptical, smooth, cream.
Season: May–September.
Habitat: Single to several, under conifers, especially in sandy soil.
Range: Idaho, Pacific NW., and California.
Look-alikes: *R. luteolus* is smaller, yellowish-ochre, and has larger spores.
Comments: *Rhizopogon* is a large genus of more than 200 species; its members can be found throughout northern North America and south along the Atlantic, Pacific, and Gulf coasts. These species can be identified to genus when found covered with tangled rhizomorphs.

---

661 Stalked Yellow Trunc
Truncocolumella citrina Zel.
Rhizopogonaceae, Hymenogastrales

Description: *Irregularly rounded, yellowish mushroom with branching internal stalk.*
Mushroom: 1–3″ (2.5–7.5 cm) wide, ⅝–2″ (1.5–5 cm) high; depressed-roundish to kidney-shaped; shiny to

dull, cracking into patches; whitish, becoming lemon-yellow to ochre or olive-ochre. Spore mass brownish-yellow, becoming tawny-olive.
Stalk: internal, stout, stumplike at base, branching upward; yellowish.
Spores: 6–10 × 3.5–5 μ; elliptical, smooth, colorless, brownish in mass.

Season: August–November.

Habitat: On the ground or in duff under conifers, especially Douglas-fir.

Range: Pacific NW. and Idaho.

Comments: As this underground fungus develops, it bursts through the surface and is often seen partially exposed.

DESERT INKY CAP FUNGUS
(Order Podaxales)

This is an order with a single reported North American species, which somewhat resembles the Shaggy Mane (*Coprinus comatus*). Unlike the Shaggy Mane, however, species in this order lack gills; in cross-section, they reveal a spore mass (gleba) which becomes powdery with maturity. Spore dispersal is usually effected by disintegration of the outer covering (peridium) or by the actions of animals.

---

702 **Desert Inky Cap**
Podaxis pistillaris (L. ex Pers.) Fr.
Podaxaceae, Podaxales

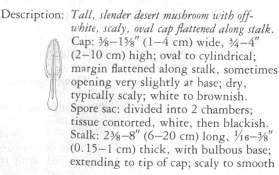

Description: *Tall, slender desert mushroom with off-white, scaly, oval cap flattened along stalk.*
Cap: ⅜–1⅝″ (1–4 cm) wide, ¾–4″ (2–10 cm) high; oval to cylindrical; margin flattened along stalk, sometimes opening very slightly at base; dry, typically scaly; white to brownish.
Spore sac: divided into 2 chambers; tissue contorted, white, then blackish.
Stalk: 2⅜–8″ (6–20 cm) long, 1⁄16–⅜″ (0.15–1 cm) thick, with bulbous base; extending to tip of cap; scaly to smooth

to somewhat twisted; whitish.

Spores: $10-15 \times 9-12$ μ; oval, smooth, yellow-brown; pore at tip.

Season: Spring; also summer and fall after rain.

Habitat: Single to clustered, in desert areas.

Range: Idaho; S. California.

Look-alikes: *Coprinus comatus** has gills and produces a spore print. *Phellorinia herculeana* has warted spores and outer covering that splits irregularly from above.

Comments: Also called the "False Shaggy Mane." This is a characteristic desert mushroom that survives its arid habitat by having very thick-walled spores, which are protected within the mushroom until they mature.

PLATED PUFFBALL FUNGI
(Order Gautieriales)

These are underground fungi that sometimes surface at maturity or are unearthed by burrowing animals. They are roundish to kidney-shaped, sometimes leafy to cauliflowerlike, and have a short, stalklike base and chambered interior. One family and genus have been described.

657 Plated Puffball
Gautieria morchelliformis Vitt.
Gautieriaceae, Gautieriales

Description: *Small, convoluted, with spore mass of short, radiating plates.*

Mushroom: ⅜–2″ (1–5 cm) wide; roundish to oblong; with thin outer cover soon disappearing. Spore mass white, becoming ochre-tawny to rust-brown; an arrangement of labyrinthine plates and broad cavities radiating from base. Odor often strong.

Stalk: composed of branching, rooting strands (rhizomorphs).

Spores: $12-24 \times 8-12.5$ μ; spindle-shaped, longitudinally lined, ochre.

Season: June–September.

Habitat: In clay soil under deciduous trees and shrubs; often in oak woods.

Range: New York to Ohio; also West Coast.

Comments: This species should be looked for in oak woods. Like many puffball-like fungi, it develops underground and emerges at maturity.

PUFFBALLS
(Order Lycoperdales)

At a distance, the Giant Puffball is easily mistaken for a soccer ball, and it can be many times larger. Most puffballs are much smaller, however, and range in size between that of a golf ball and a softball. They are very common from summer to late fall in lawns, urban parks, open woods, and in arid, inhospitable places like deserts and mountains. Most true puffballs are choice edibles when young.

Despite their diversity in size, structural features, surface, and color, they all have a spore mass (gleba) that is solid and white at first and becomes powdery as the spores mature. With age, the spore mass becomes yellowish to greenish-brown or purplish and, typically, is either dispersed through a hole that forms at the top (this is the "puff" of smoke one sees when a mature puffball is touched) or through the disintegration of the upper wall. The spores are usually small, roundish, and warted.

Puffballs are usually stalkless, although some have a pseudostalk composed of elongated tissue of the fruiting body. The covering (peridium) is usually 2-layered. The warted surface of some puffballs represents the remains of the outer cover; the inner covering surrounds the developing spore mass. In the earthstars, the outer covering splits into rays that curve back, producing a starlike pattern; the inner cover protects the spore mass.

Four families and 18 genera have been described; 3 families and 8 genera are included here.

646 Granular Puffball
Arachnion album Schw.
Arachniaceae, Lycoperdales

Description: *Small, white to buff puffball, disintegrating at maturity and containing a mass of grayish particles.*
Mushroom: ⅜–¾″ (1–2 cm) wide; roundish or with pointed base and small, tough, rootlike structure; smooth, dull white, becoming buff; covering single, thin, toughish, with no opening. Spore mass pure white, becoming grayish or brownish-olive; composed of separate minute chambers (peridioles) containing the spores. Spores: 4–5.5 × 3–4.5 μ; short, elliptical, smooth, brownish, with short stalk.
Season: June–September.
Habitat: On the ground in grass or sandy soil under trees; in flower beds.
Range: E. North America, especially Southeast.
Look-alikes: *Lycoperdon** species have double covering and powdery spore mass with capillitium at maturity.
Comments: As the puffball covering disintegrates, the granular, sandlike peridioles, too, fall apart and release the spores.

639 Arched Earthstar
Geastrum fornicatum (Huds.) Fr.
Geastraceae, Lycoperdales

Description: *Roundish ball with short stalk atop arched, starlike rays attached at tips to partially buried cup.*
Spore sac: ¾–1″ (2–2.5 cm) wide; roundish to compressed; with large, torn, mouthlike opening at top; velvety, brownish; attached below to short stalk. Spore mass white, firm,

becoming blackish-brown and powdery.
Rays: 1–2″ (2.5–5 cm) long; 4–5 in
number; brownish; upright, arching,
becoming curved back and down; tips
touching edge of thick mycelial disc.
Spores: 3.5–4.5 μ; round, warted,
dark brown.

Season: October–March.
Habitat: In leaves, twigs, and organic debris.
Range: Widely distributed in North America.
Look-alikes: *G. coronatum* is smaller, has mouthlike
opening outlined by circular groove.
Comments: This species is often found near stables
in the Southwest.

634 Beaked Earthstar
Geastrum pectinatum Pers.
Geastraceae, Lycoperdales

Description: *Roundish ball with beaked mouth; on long*
stalk atop arched, starlike rays.
Spore sac: ⅜–¾″ (1–2 cm) wide;
roundish to urn-shaped; with long,
beaked, deeply lined, mouthlike
opening at top; grayish-white to
purplish-brown; somewhat powdery,
underside often radially lined and
connected to long, slender stalk. Spore
mass white, firm, becoming brown and
powdery.
Rays: ⅝–¾″ (1.5–2 cm) long; 5–10 in
number; grayish-brown, upright,
arching, then curved back and down.
Spores: 4–6 μ; round, warted, brown.

Season: July–September.
Habitat: On the ground in open woods, near
cedar.
Range: E. North America west to Michigan.
Comments: This is one of the most distinctive
species of earthstar.

636 Rounded Earthstar
Geastrum saccatum Fr.
Geastraceae, Lycoperdales

Description: *Roundish sac encircled by starlike rays.*
Spore sac: ¼–¾″ (0.5–2 cm) wide;

roundish; with small, central, disclike
depression about mouthlike opening at
top; smooth, brownish or sometimes
nearly white. Spore mass white, firm,
becoming brown to purplish-brown and
powdery.
Rays: ⅝–¾" (1.5–2 cm) long; 5–7 in
number; ochre-brown or pinkish,
upright, then curved back and down.
Spores: 3.5–4.5 μ; rounded, warted,
brown.

Season: July–October.

Habitat: Several to many, in leaf litter under
trees.

Range: Widely distributed in North America.

Look-alikes: *G. fimbriatum* is smaller, and the pore
lacks distinct boundary line. *G. triplex**
has cuplike collar about spore sac.

Comments: This species is often found in large
quantity, usually in all stages of
development. When unopened, it
resembles a small puffball; when cut in
half, a rubbery outer skin and an
interior spore case are revealed.

635 Collared Earthstar
Geastrum triplex Jung.
Geastraceae, Lycoperdales

Description: *Roundish ball set in distinct, cuplike collar
and encircled by starlike rays.*
Spore sac: ¾–1¼" (2–3 cm) wide;
nearly round; with small, central,
disclike area about mouthlike opening
at top; smooth, gray- to reddish-brown;
set in bowl or cuplike collar of tissue.
Spore mass white, firm, becoming dark
brown and powdery.
Rays: ¾–1⅛" (2–4 cm) long; 4–8 in
number; brownish; coming to point at
top before opening; when open,
upright, slightly curving back and
under, cracking in places.
Spores: 3.5–4.5 μ; round, warted,
brown.

Season: August–October.

Habitat: Several to many, in leaves in open
deciduous woods.

Range: Widely distributed in North America.
Comments: The collar about the spore sac makes
this a distinctive species. Before
opening, the "button" is pointed rather
than rounded. One can often find the
remains of this and other earthstars in
winter and spring.

641 Tumbling Puffball
Bovista pila Berk. & Curt.
Lycoperdaceae, Lycoperdales

Description: *Round, becoming detached and rolling
about; with white outer skin flaking off to
reveal smooth, shiny, bronze spore sac.*
Mushroom: 1¼–3½" (3–9 cm) wide;
round or nearly so; attached to ground
by single small cord, which breaks at
maturity; outer skin pure white,
becoming dark pinkish and flaking off
in patches; inner skin smooth, shiny,
bronze, with cracks or pore opening at
top. Spore mass white, then brown.
Spores: 3.5–4.3 μ; round, smooth,
with short, colorless stalk; deep brown.
Season: June–October; overwinters.
Habitat: Single to scattered or many, in
pastures, around stables, and in open
woods.
Range: Widely distributed in North America.
Look-alikes: *B. plumbea* is smaller, bluish-gray with
age, and attached to ground by clump
of fibrous strands.
Comments: A common mushroom of grassy areas
and pastures, this species grows nearly
year-round; hardy ones overwinter and
tumble about, unattached to the
ground.

650 Sculptured Puffball
Calbovista subsculpta Morse
Lycoperdaceae, Lycoperdales

Description: *Large, white, nearly round, covered with
flattened to warted plates or scales; interior
white, becoming brownish.*
Mushroom: 3¼–6" (8–15 cm) wide;

2⅜–3½″ (6–9 cm) high; typically broader than tall, but nearly round at times; white, covered with low warts bearing brownish hairs at center, smooth near base. Spore mass white, becoming dark brown. Sterile base ¼–⅓ of mushroom; dull, white, firm. Spores: 3–5 μ; round, nearly smooth.

Edibility: Choice.
Season: April–August.
Habitat: Single to several, along roads and in open woods at high elevations.
Range: Rocky Mountains and Pacific Coast ranges.
Look-alikes: *Calvatia sculpta* has large, pyramidal warts. *C. subcretacea* is small, golf-ball-like.
Comments: This species of the high western mountains is edible only when the spore mass is white.

648 Western Giant Puffball
Calvatia booniana A.H.S.
Lycoperdaceae, Lycoperdales

Description: *Huge, white, flattened sphere with flattened, polygonal, buff warts; interior white, becoming green-brown and cracking with maturity.*
Mushroom: 8–24″ (20–60 cm) wide; 3–12″ (7.5–30 cm) high; sculpted, compressed, round, with a thick, rootlike attachment at base; white; warts broadening into shallow, buff plates. Spore mass white, becoming olive-brown, with no sterile base. Odor becoming unpleasant.
Spores: 4–6.5 × 3.5–5.5 μ; round to broadly elliptical, smooth to minutely ornamented; olive-brown.
Edibility: Choice.
Season: July–August.
Habitat: In semi-arid sagebrush areas.
Range: S. Idaho and adjacent states; reported in Southwest.
Comments: This species, also called "Boone's Puffball," is a good edible as long as its interior is white.

656 Skull-shaped Puffball
Calvatia craniformis (Schw.) Fr.
Lycoperdaceae, Lycoperdales

Description: *Skull-shaped, white, smooth, cracking into irregular areas and revealing white interior, which becomes yellow-green.*
Mushroom: 3¼–8" (8–20 cm) wide across top, 2⅜–8" (6–20 cm) high; top-shaped, with thick, stalklike base attached by fibrous strands; smooth, scaling off in thin plates, papery thin; white to tannish. Spore mass cottony and persistent, remaining intact; white, becoming greenish-yellow to yellow-brown.
Spores: 2.5–4.5 μ; round, nearly smooth, many with a short, blunt peg.
Edibility: Choice.
Season: August–October.
Habitat: On the ground in open woods, especially near oak.
Range: Widely distributed in North America.
Look-alikes: *C. bovista* has larger spores, is found in open areas. *C. cyathiformis** has purple spores, grows in grasslands and pastures. *C. rubroflava** bruises yellow.
Comments: This mushroom makes a choice edible, but only when immature, when it has a white spore mass. To be correctly identified to species, however, it must be mature, revealing its yellow-green or yellow-brown spore mass.

---

653 Purple-spored Puffball
Calvatia cyathiformis (Bosc) Morg.
Lycoperdaceae, Lycoperdales

Description: *Hemispherical, minutely cracked, tannish, with white interior becoming deep purple-brown, with large, persistent, dark violet, cuplike remains.*

Mushroom: 2¾–7" (7–18 cm) wide, 3½–8" (9–20 cm) high; hemispherical, smooth at first, becoming cracked into distinct areas; tannish. Spore mass white, becoming dark lilac to brownish-yellow to deep purple-brown.

Spores: 3.5—7.5 μ; round, with distinct spines.

Edibility: Choice.

Season: July—November; overwintering as dark violet cup.

Habitat: On the ground in grassy areas, parks, golf courses, and pastures.

Range: E. and C. United States.

Look-alikes: *C. fragilis* is smaller and has slight sterile base. *C. craniformis** has yellow-brown spore mass at maturity. *C. bovista* has dark olive-green spore mass at maturity.

Comments: This is a good edible when white within. The large, violet, cuplike base, which is left after the puffball disintegrates, often overwinters.

647 Giant Puffball
Calvatia gigantea (Bat. ex Pers.) Lloyd
Lycoperdaceae, Lycoperdales

Description: *Huge, white, smooth sphere, cracking irregularly at maturity; interior white, becoming yellowish-green to brownish.* Mushroom: 8—20″ (20—50 cm) wide, sometimes larger; round or nearly so, with rootlike attachment; smooth, kidlike, cracking irregularly; no pore formed at top; white. Spore mass white, becoming yellow-green to greenish-brown; *no sterile base evident.* Spores: 3.5—5.5 μ; round, smooth to minutely warted, greenish-brown.

Edibility: Choice.

Season: Late May to mid-July; August—October.

Habitat: Single or in arcs or fairy rings, in open woods, pastures, and urban areas.

Range: E. North America west to Ohio.

Look-alikes: A similar, if not identical, species occurs on the California coast.

Comments: Also known as *C. maxima* and *Lycoperdon giganteum.* This species is most common in parks and meadows in late August and September. When white within, it is an excellent edible mushroom that can be sautéed or batter-fried. An average-size specimen has been estimated to contain 7 trillion spores.

---

655 Orange-staining Puffball
Calvatia rubroflava (Cragin) Morg.
Lycoperdaceae, Lycoperdales

Description: *Flattened, white to pink-tinged sphere with very thin covering, turning rapidly bright yellow on bruising or cutting.*
Mushroom: ¾–4″ (2–10 cm) wide, ⅝–2″ (1.5–5 cm) high; almost round, with flat top, strongly pleated below, with more or less pointed base; centrally attached by 1 or more slender, rootlike strands; covering very thin, white to pink-tinged, rapidly turning bright yellow on bruising or cutting. Spore mass cottony and persistent (not powdery); white, becoming greenish-orange, bruising yellow; no sterile base. Spores: 3–5 μ; roundish, yellow-ochre.

Season: Late summer and fall.
Habitat: On the ground in gardens and lawns.
Range: New Jersey to Alabama, west to Ohio, Kansas, and Missouri.
Comments: While technically edible, this puffball is distinctly unpalatable. The odor of the mature mushroom is rank.

---

649 Spiny Puffball
Lycoperdon echinatum Pers.
Lycoperdaceae, Lycoperdales

Description: *Roundish, covered with clusters of long spines fused at tips; white, then brownish.*
Mushroom: 1–2″ (2.5–5 cm) wide, 1–1⅝″ (2.5–4 cm) high; roundish to flattened; covered with clusters of long spines, fused at tips, falling off and leaving marks on surface; white, becoming brownish. Spore mass white, becoming purplish-brown; sterile base small, white to grayish.
Spores: 4–6 μ; round, warted, purple-brown.

Season: August—November.
Habitat: Among leaves in woods.
Range: Quebec to Pennsylvania, west to Michigan.
Look-alikes: *L. pulcherrimum* has spines that do not darken with age and do not leave marks on surface. *L. perlatum** has separated spines and olive-brown spores. *L. marginatum* is similar, but fused spines break off in sheets.
Comments: This puffball is one of several possessing long, fused spines; its accurate identification to species requires seeing all stages of development.

652, 676 Gem-studded Puffball
Lycoperdon perlatum Pers.
Lycoperdaceae, Lycoperdales

Description: *White, round to turban-shaped mushroom with detachable, conical spines, developing pore at top; interior white, becoming greenish-brown.*
Mushroom: 1—2⅜″ (2.5—6 cm) wide, 1¼—3″ (3—7.5 cm) high; usually round to turban-shaped with somewhat elongated, stalklike base; covered with long and short spines, long ones readily breaking off and leaving distinct marks on surface; opening by pore at top; white, becoming buff. Spore mass white, becoming green-tinged ochre-brown. Sterile base sometimes elongated, stalklike, persistent.
Spores: 3.5—4.5 μ; round, minutely warted, olive-brownish.
Edibility: Choice.
Season: July—October.
Habitat: Single to scattered or clustered, in open woods, along roads, on ground in urban areas.
Range: Widely distributed in North America.
Comments: Formerly called *L. gemmatum.* This species is easy to recognize: any puffball that could be confused with it has purplish-brown spores. It is a choice edible if, when cut in half, it has undifferentiated white flesh.

664 Pear-shaped Puffball
Lycoperdon pyriforme Schaeff. ex Pers.
Lycoperdaceae, Lycoperdales

Description: *Pear-shaped, smoothish, yellow-brown*
mushroom with pore at top; on wood.
Mushroom: ⅝–1¾″ (1.5–4.5 cm)
wide, ¾–1¾″ (2–4.5 cm) high; pear-
shaped; developing pore at top; yellow-
brown, with minute warts and
granules. Spore mass white, becoming
greenish-yellow to deep olive-brown,
opening by pore at top; sterile base
small, attached by white strands.
Spores: 3–4.5 μ; round, smooth, deep
olive-brown.

Edibility: Choice.

Season: July–November.

Habitat: Scattered to densely clustered, on
wood, decaying logs, stumps, debris.

Range: Widely distributed in North America.

Comments: This is a very common and abundant
puffball. It grows in great masses on
decaying deciduous logs in the fall and,
if gathered when immature and pure
white, is a choice edible.

651 Tough Puffball
Mycenastrum corium (Guer.) Desv.
Lycoperdaceae, Lycoperdales

Description: *Large, roundish, with thick, whitish outer*
skin becoming flaky and patchlike; spore sac
thick, firm to corky, then deeply cracked.
Mushroom: 2–8″ (5–20 cm) wide;
round to pear-shaped; thick, feltlike,
whitish covering, cracking in patches,
exposing thick, corky-tough, brownish
spore sac. Spore sac becoming cracked
and split, sometimes curving back like
an earthstar; spore mass white,
becoming yellowish-green, eventually
purplish-brown.
Spores: 8–12 μ; round, warted, brown.

Season: July–October.

Habitat: Single to many, on soil and in pastures.

Range: Widely distributed in North America.

Comments: This species is common west of the

Mississippi, especially around stables
and in pastures. It becomes detached at
maturity and, rolling freely about,
often persists into the spring.

645 Western Lawn Puffbowl
Vascellum pratense (Pers.) Kreisel
Lycoperdaceae, Lycoperdales

Description: *White, turban-shaped lawn puffball,
leaving bowl-like base after disintegrating.*
Mushroom: ¾–1⅝" (2–4 cm) wide,
1–2" (2.5–5 cm) high; turban-shaped;
granular, white to tannish covering
wearing away and exposing shiny,
metallic brown spore sac, which
develops starlike lobes; upper part
disintegrates. Spore mass white, firm,
becoming olive-brown and powdery;
sterile base bowl-like, white, becoming
brownish.
Spores: 3–5.5 μ; round, smooth to
minutely ornamented, olive-brown.
Edibility: Edible.
Season: September–November.
Habitat: In lawns, golf courses, and pastures.
Range: Widely distributed in North America.
Comments: Also known as *V. depressum*. This is the
most common *Vascellum* species and
should be expected in lawns in autumn
along the Pacific Coast. Like other true
puffballs, it is edible when white
within.

BIRD'S-NEST FUNGI
(Order Nidulariales)

The bird's-nest fungi are so called
because they resemble cuplike birds'
nests with eggs inside. The "eggs" are
actually little seedlike cases, called
peridioles, that contain parts of the
spore mass, or gleba. The "nests" are
splash cups that disperse the eggs when
raindrops fall in. In most cases, the
nests begin as closed, cushionlike
structures with lids that open to permit

spore dispersal. Immature specimens or
empty nests may resemble some kinds
of cup fungi. The immature nests can
be easily distinguished by the presence
of the eggs, which contain basidia as
well as spores, and empty nests can be
distinguished microscopically from the
ascus-bearing cup fungi. One species,
the Cannon Fungus, is spherical rather
than nestlike, and has one peridiole.
Two families and 5 genera are included
here.

633   White-egg Bird's Nest
Crucibulum laeve (Huds.) Kamb.
Nidulariaceae, Nidulariales

Description: *Tawny, deeply cup-shaped, with smooth*
inner wall and white eggs attached by coiled
cords.
Mushroom: ¼–⅜″ (0.5–1 cm) high
and wide at rim of cup nest, narrowing
downward; almost round to short,
cylindrical; tawny-yellow, with velvety
outer surface and coarsely hairy lid;
inner surface smooth and shiny; mouth
smooth.
"Eggs": ¹⁄₁₆″ (1.5 mm) wide; thin,
uniform in shape; whitish; attached to
cup nest by long, thin cord.
Spores: $4-10 \times 4-6 \mu$; elliptical,
thick-walled, colorless, smooth.
Season: July–November.
Habitat: On dead wood and debris.
Range: Alaska to Mexico.
Look-alikes: *Cyathus** species have dark eggs.
*Nidula** and *Nidularia** have dark eggs
set in gel.
Comments: Also known as *C. vulgare.* Only this
bird's-nest fungus has white eggs.

632   Splash Cups
Cyathus striatus (Huds.) Willd. per Pers.
Nidulariaceae, Nidulariales

Description: *Broadly cup-shaped, narrowing below, with*
lined inner wall and dark eggs attached

by coiled cords.
Mushroom: ¼–⅝" (0.5–1.5 cm) wide and high; variable in shape, typically with slender base and upper ⅓ flaring outward; brownish to gray-buff to deep chocolate-brown; outer wall covered with shaggy or woolly hairs and faintly to strongly grooved; inner wall markedly grooved, not shaggy, shiny; lid white.
"Eggs": ¹⁄₃₂–¹⁄₁₆" (1–1.5 mm) wide; somewhat triangular, dark, with cord.
Spores: $15-20 \times 8-12 \mu$; elliptical, notched at 1 end, smooth.

Season: July–October.
Habitat: On dead wood, bark, twigs, and wood chips, in open woods.
Range: Widely distributed in North America.
Comments: *Cyathus* is the largest genus of bird's-nest fungi, and this species is the most distinctive and commonly encountered of its genus. The common name refers to the method of spore dispersal of these fungi: the sides of the cups are angled so that raindrops falling into the mushrooms dislodge the tiny spore sacs, or "eggs."

631 Common Gel Bird's Nest
Nidula candida Pk.
Nidulariaceae, Nidulariales

Description: *Deeply cup-shaped with smooth inner wall and light brown eggs embedded in a gel.*
Mushroom: ⅛–⅜" (0.3–1 cm) wide, ¼–¾" (0.5–2 cm) high; muglike, with vertical sides and wide, flaring mouth; gray to light brown, shaggy.
"Eggs:" ¹⁄₁₆–⅛" (1.5–3 mm) wide; large; gray to light brown; embedded in gel.
Spores: $6-10 \times 4-8 \mu$; round to elliptical, colorless.
Season: Late fall–early winter.
Habitat: In thickets, on old stems of blackberry and other *Rubus* species, and on rotting wood debris.
Range: Pacific NW. to C. California.

Look-alikes: *Crucibulum** and *Cyathus** species lack
gel, have eggs attached by cords.
Nidularia* is smaller, lacks lid.

Comments: The specific name *candida,* meaning
"shining white," is really a misnomer
for this fungus, which is scruffy and
light brown.

---

658   Pea-shaped Nidularia
Nidularia pulvinata (Schw.) Fr.
Nidulariaceae, Nidulariales

Description: *Round, tiny, fragile, not perfectly cup-*
shaped, without lid, with many dark brown
eggs embedded in gel.
Mushroom: $\frac{1}{16}$–$\frac{3}{8}$" (0.15–1 cm) wide;
round or nodular from pressure;
brownish, fading to dull gray-buff;
outer surface powdery, becoming
smoother; inner surface shiny, smooth,
brown.
"Eggs": $\frac{1}{32}$" (1 mm) wide; tiny,
irregularly shaped; dark gray-brown;
embedded in gel.
Spores: 6–10 × 4–7 μ; elliptical,
smooth, colorless, pointed at 1 end.

Season: August–October.

Habitat: On decaying wood and wood debris in
woods.

Range: Maine to Alabama and Louisiana, west
to Minnesota.

Look-alikes: *Nidula** species are cup-shaped and
have lid over nest until maturity.
Cyathus* and *Crucibulum** lack gel and
have cords attaching eggs.

Comments: Although widely distributed, this
species is infrequently collected.

---

551   Cannon Fungus
Sphaerobolus stellatus Tode ex Pers.
Sphaerobolaceae, Nidulariales

Description: *Tiny, orange-yellow sphere, developing*
starlike rays and containing spherical,
dark, slippery spore mass.
Mushroom: $\frac{1}{16}$" (1.5 mm) wide; almost
round; outer surface yellow-orange,

flaking and becoming whitish; opening
at top by 4–7 starlike rays. Spore ball
round, slippery, smooth, deep
chestnut-brown, divided into sections;
catapulted from mushroom in reaction
to moisture and light.
Spores: 7.5–10 × 3.5–5 μ; unevenly
oblong, smooth, colorless.

Season: May–September.

Habitat: On deciduous wood, twigs; also rotten
pine, packed sawdust, horse dung.

Range: Throughout E. United States, west to
Michigan, Ohio, Iowa, and Louisiana.

Comments: Also called the "Sphere Thrower."
Although easily overlooked in the field,
this is one of the most dramatic fungi:
it disperses its spores when, in reaction
to the presence of sufficient moisture
and light, the spore ball is catapulted
out of the mushroom. In a dark room,
turning on the light may trigger the
release of the spore ball, which can sail
through the air for several feet.

STINKHORNS
(Order Phallales)

The stinkhorns are fungi you usually
smell before you see. They begin as
egglike structures with an outer
covering (peridium) and a layered
interior that may be multicolored and is
partly gelatinous and odorless. The egg
opens by expansion from within, and a
single, stalklike extension or a number
of arms emerge. The spore mass (gleba)
is embedded in a green slime that
becomes strongly fetid as the
mushroom matures. This spore mass is
either atop the stalklike structure or
arms, or is deposited along the inner
surface of the arms. The odor attracts
insects; the spores, which are not
airborne, adhere to the bodies of the
insects and are dispersed by them.
Stinkhorns are very common in urban
areas, especially in parks and about
cultivated plantings, from summer to

fall. Six families and 22 genera have been described; 3 families and 7 genera are included here.

---

694 Columned Stinkhorn
Clathrus columnatus Bosc
Clathraceae, Phallales

Description: *Reddish, stalkless fungus with several columns fused at top, rising out of thick, whitish cup; odor fetid.*

egg

Mushroom: 1–2″ (2.5–5 cm) wide, 2–3″ (5–7.5 cm) high; composed of 2–5 stout, spongy, curved columns rising separately from base and fused at top into a flat roof; reddish; attached to ground by white, cordlike strand. Spore mass olive-brown, slimy, fetid; attached to underside of roof.
Spores: 3.7–4.8 × 1.8–2.4 μ; elliptical, smooth.

Season: October–March.

Habitat: In lawns and gardens and in sandy soil.

Range: North Carolina to Florida, west along Gulf Coast; also Mexico.

Look-alikes: *Pseudocolus schellenbergiae** has short stalk, brown cup at base, and green spore slime along inner sides of arms.

Comments: This species has been reported as far north as New York.

---

709 Lizard's Claw
Lysurus gardneri Berk.
Clathraceae, Phallales

Description: *Several short, pinkish, incurved or spreading arms atop a tall, white stalk; thick white cup about base; odor fetid.*
Head: ⅜–⅝″ (1–1.5 cm) wide, 1–2″ (2.5–5 cm) high; composed of 5–6 short, erect, pinkish arms, incurved and touching at tips (not fused). Spore mass olive-brown, slimy, fetid; enclosed within tips of arms.
Stalk: 3–4″ (7.5–10 cm) long, ⅜″ (1 cm) thick; white, hollow, with thick, white, saclike cup about base; attached

to ground by whitish cord.
Spores: 3–4 × 1–2 μ; elliptical,
smooth, olive-green.

Season: August–September.

Habitat: Single to several, in leaf litter or wood
debris; also in cultivated soil.

Range: Massachusetts to North Carolina, west
to Pennsylvania, Ohio; also California.

Look-alikes: *L. mokusin* has red arms, thinner stalk.
Simblum sphaerocephalum, in South, is
smaller, has round head with thick red
net.

Comments: Also called *Anthurus borealis* and
Lysurus borealis. The similar Small
Lizard's Claw (*L. mokusin*) is considered
a great delicacy by the Chinese.

---

695 Stinky Squid
Pseudocolus schellenbergiae Sum.
Clathraceae, Phallales

Description: *Three orange arms, arching outward and
fused at tips, arising from common short,
whitish stalk; thick, brownish cup about
base; odor fetid.*
Head: ¾–1¼″ (2–3 cm) wide, 1¼–
2⅜″ (3–6 cm) high; composed of 3–4
tapering orange arms rising from
common stalk, arching outward and
fused at tapering tips or free. Spore
mass greenish, slimy, *fetid;* distributed
along inner surfaces of arms.
Stalk: ¼–⅝″ (0.5–1.5 cm) long, ¼–
⅜″ (0.5–1 cm) thick; whitish, with
thick, brownish, saclike cup about base;
attached to ground by white cord.
Spores: 4.5–5.5 × 2–2.5 μ; elliptical-
oval, smooth, colorless.

Season: Summer–fall.

Habitat: Wood borders, parks, and gardens.

Range: Pennsylvania, New York; spreading.

Look-alikes: *Clathrus columnatus** has spore mass
only in area beneath fused tips of arms.

Comments: Also called *P. javanicus* and *P.
fusiformis.* First reported in North
America in Pittsburgh in 1915, this
fungus has spread widely across the
mid-Atlantic states.

701 Club-shaped Stinkhorn
Phallogaster saccatus Morg.
Hysterangiaceae, Phallales

Description: *Small, white to pinkish club containing greenish slime; odor fetid.*
Mushroom: ⅜–1¼" (1–3 cm) wide, 1–2" (2.5–5 cm) high; narrowing toward base; pear-shaped; smooth, whitish to pinkish or lilac-tinged; developing irregular depressions above that open eventually to release spore mass; lower part whitish within; attached to ground by maze of thick, whitish to pinkish strands. Spore mass greenish, slimy, becoming fetid; in distinct chambers.
Spores: 4–5 × 1.5–2 μ; nearly cylindrical, smooth, colorless to green-tinted.
Season: May–September.
Habitat: Single to numerous, on decayed wood.
Range: Widely distributed in North America.
Comments: This stinkhorn resembles a puffball when young, but a cross-section will reveal the green, chambered upper portion. The odor of the spore slime becomes fetid slowly.

707 Netted Stinkhorn
Dictyophora duplicata (Bosc) Fisch.
Phallaceae, Phallales

Description: *Greenish spore slime covering deeply pitted head on white stalk, encircled by white, skirtlike, netted veil; odor fetid.*
Head: 1¼–1⅝" (3–4 cm) wide, 2" (5 cm) high; deeply pitted; attached at tip to white, mouthlike circlet at tip of stalk; lower margin free, with white, fishnetlike veil, 3–5 cm long. Spore mass dark olive-green, slimy, fetid
egg
Stalk: 4–5" (10–12.5 cm) long, 1¼–1¾" (3–4.5 cm) thick; white, hollow; rising out of whitish to pinkish, thick, saclike cup; attached to ground by white cord.
Spores: 3.5–4.5 × 1–2 μ; elliptical, smooth, colorless.

Season: June–October.
Habitat: Single to many, on the ground about deciduous trees and stumps.
Range: Quebec to Florida and Alabama, west to Indiana.
Look-alikes: *Phallus impudicus* lacks skirtlike veil. *P. ravenelii** lacks veil and pitted head.
Comments: Also called *Phallus duplicatus*. This species can be grown indoors by placing the egglike immature mushroom on damp soil or wet paper towels and covering it with a glass bowl.

692 Elegant Stinkhorn
Mutinus elegans (Mont.) Fisch.
Phallaceae, Phallales

Description: *Greenish slime covering top part of long, tapered, pinkish, stalklike mushroom with whitish cup about base; odor fetid.*
Mushroom: ⅝–1″ (1.5–2.5 cm) wide, 4–7″ (10–18 cm) high; stalklike, thin, tapering, with small opening at tip; rosy-red above, pale to whitish below; hollow, rising out of flattened, saclike cup about base; attached to ground by white cord. Spore mass greenish, slimy, fetid; covering upper ⅓ of mushroom. Spores: 4–7 × 2–3 μ; elliptical, smooth.
Season: July–September.
Habitat: On wood debris, leaf litter, and rich soil, in woods and cultivated fields.
Range: Quebec to Florida, west to Great Lakes.
Look-alikes: *M. caninus* is less tapered, with clear separation of spore mass and stalk.
Comments: This stinkhorn can be found in the parks and gardens of many urban areas.

706 Ravenel's Stinkhorn
Phallus ravenelii Berk. & Curt.
Phallaceae, Phallales

Description: *Greenish spore slime covering granular head on white stalk, with white to pinkish cup about base; odor fetid.*
Head: ⅝–1⅝″ (1.5–4 cm) wide, 1¼–

egg

1¾" (3–4.5 cm) high; granular, attached at tip to white, mouthlike circlet at tip of stalk; lower margin free, with partly concealed, membranous veil. Spore mass olive-green, slimy, fetid; covering head.

Stalk: 4–6" (10–15 cm) high, ⅝–1¼" (1.5–3 cm) thick; yellowish to white, with white to pinkish, saclike cup about base; attached to ground by white cord.

Spores: 3–4.5 × 1–2 μ; elliptical, smooth, colorless.

Season: August–October.

Habitat: Single to clustered, on wood debris, rotten stumps, sawdust.

Range: Quebec to Florida, west to Ohio, Iowa.

Look-alikes: *P. impudicus* has deeply pitted head. *Dictyophora duplicata** has skirtlike, netted veil.

Comments: This eastern stinkhorn usually grows in clusters, and is common in urban areas.

FALSE PUFFBALLS
(Order Sclerodermatales)

Many of the mushrooms in this order bear a marked superficial resemblance to other species in the class Gasteromycetes. But unlike other stomach fungi, they have no well-defined fertile surface, or hymenium; instead, the microscopic spore-bearing structures (basidia) develop throughout the entire body of the mushroom. In addition, the spores, which are large and ornamented, are produced at different times during the growth of the mushroom, not simultaneously. The Sclerodermatales are either stalkless or develop a false stalk that is not different in tissue structure from the cap or head. They have a spore mass (gleba) that becomes powdery with maturity and they disperse their spores through an opening at the top or by disintegration of the spore-mass cover. These mushrooms are common in urban

parks, along the seashore, and in
woods; most grow from summer to fall,
but a few overwinter. Some fungi in
this order are known to be poisonous,
and it is important to distinguish the
toxic species from their edible look-
alikes. Four families and 9 genera have
been described; 2 families and 4 genera
are included here.

638 Barometer Earthstar
Astraeus hygrometricus (Pers.) Morg.
Astraeaceae, Sclerodermatales

Description: *Roughened, roundish ball encircled by*
cracked, starlike rays which close in dry
weather.
Spore sac: ⅜–1¼" (1–3 cm) wide;
nearly round, with irregular mouthlike
opening at top; felty-rough; whitish,
becoming gray to brown. Spore mass
white, then brownish and powdery.
Rays: 1–2" (2.5–5 cm) long; 6–12 in
number, formed from smooth, grayish-
brown skin covering spore sac, splitting
into starlike rays and opening in wet
weather; upright at first, curving back
and becoming distinctly cracked;
closing in dry weather.
Spores: 7–11 μ; round, warted, brown.
Season: September–November.
Habitat: In groups in sandy soil.
Range: Widely distributed in North America.
Look-alikes: *A. pteridis* is western, much larger, and
has rays cracking in quiltlike pattern.
Comments: This mushroom is often found near
conifers, and at first may look more like
a spider than a mushroom. Geasters of
the order Lycoperdales do not open and
close in response to the weather.

637 Saltshaker Earthstar
Myriostoma coliforme (Pers.) Corda
Astraeaceae, Sclerodermatales

Description: *Roundish ball with several holes, set on*
short stalks atop spreading, starlike rays.

Spore sac: ⅝–2″ (1.5–5 cm) wide; rounded to compressed, with several small, porelike openings about upper surface; supported by several short stalks; rough, grayish-brown. Spore mass white, becoming brown and powdery.

Rays: ⅜–1¼″ (1–3 cm) long; 5–7 in number, formed from smooth, brownish skin covering spore sac, splitting into starlike rays and curving back.

Spores: 4–6 μ; round, warted, brown.

Season: July–October.

Habitat: In sandy soil in open areas, near deciduous trees.

Range: Widely distributed in North America.

Comments: This rare but unmistakable earthstar develops several pore openings for spore release instead of just one.

---

698 Dye-maker's False Puffball
Pisolithus tinctorius (Pers.) Cok. & Couch
Sclerodermataceae, Sclerodermatales

Description: *Large, powdery-brown, broad-topped, club-shaped fungus.*

Mushroom: 2–4″ (5–10 cm) wide, 2–8″ (5–20 cm) high; roundish to pear-shaped, becoming elongated and club-shaped, narrowing below to stout, stalklike base; ochre-brown and shiny at first, becoming powdery above.

Spore mass: upper half a mass of small, whitish to yellow or brown, egglike spore sacs embedded in blackish jelly; slowly maturing, turning to brownish powder, and disintegrating through weathering; lower part sterile, fibrous, attached to ground by numerous cordlike, persisting strands.

Spores: 7–12 μ; round, spiny, cinnamon-brown.

Season: July–October.

Habitat: In sandy soil in open woods, old pastures, lawns, and areas with pines.

Range: Widely distributed in North America.

Comments: The immature stage is reportedly eaten

in parts of Europe, while the whole fresh or dried mushroom is used in making rich golden-brown to black dyes. It is known to be mycorrhizal with pines, and greatly aids in forestry projects in poor or sandy soils.

⊗ 654 **Pigskin Poison Puffball**
Scleroderma citrinum Pers.
Sclerodermataceae, Sclerodermatales

Description: *Warted, thick-skinned, yellow-brown ball, opening by pore at top.*
Mushroom: 1–4″ (2.5–10 cm) wide, ¾–1⅛″ (2–4 cm) high; roundish to somewhat flattened; covered with rough warts, rindlike; yellow-brown. Rind flesh white, opening by pore at top. Spore mass whitish, soon becoming marbled and purplish-black to black. Spores: 8–12 μ; round, ornamented, blackish-brown.
Edibility: Poisonous.
Season: July–November.
Habitat: Single to many, on the ground, on wood debris, near trees, in woods and sandy soil.
Range: Widely distributed in North America.
Comments: Also known as *S. aurantium.* This mushroom is the most common species of *Scleroderma.* In parts of its range, it is sometimes parasitized by the Parasitic Bolete (*Boletus parasiticus*). It is often thought to be some kind of puffball— hence the common name—but its thick, rindlike skin and marbled, purplish-black spore mass distinguish it. Although sometimes used in place of truffles, it is known to cause nausea and vomiting.

---

642 **Earthstar Scleroderma**
Scleroderma geaster Fr.
Sclerodermataceae, Sclerodermatales

Description: *Dark brownish, powdery ball encircled by large, thick, spreading, starlike rays.*

Spore mass: 1–2″ (2.5–5 cm) wide; roundish; firm at first, brown to purple-brown; becoming powdery and blackish-brown.

Rays: 2–3″ (5–7.5 cm) long; 4–8 in number, formed from thick, rough, tan skin covering spore ball; splitting into irregular starlike rays, blackish and variously cracked, opening upright and expanding with age.

Spores: 5–10 μ; round, warted, purple-brown.

Season: August–November.

Habitat: In sandy soil, clay banks, lawns, and rich soil.

Range: Widely distributed in North America.

Comments: This has an unusually thick skin and no spore sac. It is common in sandy areas, and the flattened, blackened rays are often found through the winter.

STALKED PUFFBALLS
(Order Tulostomatales)

The Tulostomatales, or stalked puffballs, grow in deserts and dry soil. Like other fungi in the class Gasteromycetes, these mushrooms have a spore mass (gleba) that becomes powdery with maturity and that is dispersed through an opening at the top of the mushroom. Some genera in this order resemble the "true" puffballs, but the fruiting body that rests on the ground has a long, rigid, underground stalk. Others in this group are very different indeed. They have several distinct kinds of tissue, including a gelatinous covering, a pigmented layer, a hard covering (peridium) over the spore sac, and a cheeselike spore mass. Unlike the Sclerodermatales, the Tulostomatales have stalks composed of tissue different from that of the fruiting body. And unlike other Gasteromycetes, these have basidia that mature over a period of time, not simultaneously.

700 **Stalked Puffball-in-aspic**
Calostoma cinnabarina Desv.
Calostomataceae, Tulostomatales

Description: *Thick, clear jelly covering short, red spore sac, with raised, red, crosslike mouth at top; stalk short, stout, rooting.*
Mushroom: ⅜–⅝″ (1–1.5 cm) wide, 1¼–2″ (3–5 cm) high; layered: outer skin thick, gelatinous, transparent; inner skin thin, bright red, breaking into red, seedlike pieces. Spore sac cover (peridium) dusted with fine, cinnabar-red powder, wearing away. Spore sac light, clear yellow, with crosslike opening at top. Entire mushroom grows in gelatinous pulp. Spores: 14–20 × 6.3–8.5 μ; oblong-elliptical; distinctly pitted, cream-yellow.
Stalk: ⅝–1¼″ (1.5–3 cm) long, ¼–⅝″ (0.5–1.5 cm) thick; rooting, short, bulky, cinnabar-red.

Season: April–May; September–October; overwinters.

Habitat: On the ground in woods

Range: Massachusetts to Florida, west to Ohio and Texas.

Look-alikes: *C. lutescens* is long-stalked and yellowish. *C. ravenelii* is yellowish, with no gelatinous layer, and does not break into pieces.

Comments: This striking, brightly colored mushroom is readily recognizable. Fairly common in eastern woods.

---

731 **Desert Stalked Puffball**
Battarrea phalloides (Dicks.) Pers.
Tulostomataceae, Tulostomatales

Description: *Tall, shaggy-woody stalk with woody cup at base and small, compressed to oval spore sac on top.*
Spore sac: 1–2″ (2.5–5 cm) wide, 1–1⅝″ (2.5–4 cm) high; oval to compressed, splitting horizontally about center; brownish. Spore mass sticky, brown

Stalk: 6–16″ (15–40 cm) long, ¼–⅝″ (0.5–1.5 cm) thick; slender, ragged-scaly, brownish; woody, with woody cup about base.
Spores: 5.5–8.5 × 4.5–6.5 μ; almost round, warted, brown.

Season: Year-round, occurring after rains.

Habitat: Single to several, in desert areas and sandy soil along coast.

Range: Washington to California; reported in Rocky Mountains.

Comments: This is one of a fair number of variously shaped desert puffballs. Other genera have developed different mechanisms for spore dispersal, but all have adapted to arid conditions and can survive the heat and desiccation that would destroy large, fleshy gilled mushrooms.

730 Buried-stalk Puffball

Tulostoma simulans Lloyd
Tulostomataceae, Tulostomatales

Description: *Small puffball with raised, rounded, mouthlike pore at top; stalk usually buried.* Spore sac: ⅜–⅝″ (1–1.5 cm) wide; roundish to acorn-shaped, with small, tubelike mouth projecting at top; sand-covered, dark reddish-brown. Spore mass whitish, becoming brown.
Stalk: ⅝–1¼″ (1.5–3 cm) long, ⅛″ (3 mm) thick; scaly, fibrous rust-brown; often entirely buried.
Spores: 4–6 μ; round, warted, yellowish.

Season: April–December.

Habitat: Several to many, in sandy soil.

Range: Widely distributed in North America.

Comments: Although distinctive, *Tulostoma* species are difficult to identify in the field because distinguishing characteristics are in part microscopic. Members of the genus are found at sea level in sandy areas, but also in interior deserts and arid mountains up to 9000 feet. Except for the long, fibrous stalk, they usually resemble simple puffballs.

SLIME MOLDS
(Subdivision Myxomycotina)

Slime molds are the Dr. Jekylls and Mr. Hydes of the plant world. The transformations they go through from the time they first appear to their disintegration—often within 24 hours —are so complete that one may know a species at a particular stage of development and not recognize it at another. Slime molds are slimy and moldlike only when they first emerge; they soon change color, shape, and texture as they develop. A slimy, white mass, for example, produces a group of tiny, stalked, glossy yellow balls, which after spore dispersal resemble fuzzy cones. There are several hundred slime molds in North America, although any given area may yield no more than 50 species. They can be found everywhere there is moisture, preferring wet, decayed logs and leaves, but also colonizing lawn grass, living plant stems, and even old discarded mattresses and cotton clothing. They are especially common in spring and fall, or during times of high precipitation and moderate temperatures.

Slime molds differ from other fungi by beginning life as protoplasm, with the amoebalike ability to move and ingest nutrients. In some species, this movement can be the conspicuous flowing of a gelatinous substance called plasmodium, which is usually encased in a transparent sheath except at its margin. The slime mold propels itself through "protoplasmic streaming": a series of expansions and contractions that continues until fruiting occurs. The plasmodium may be extensive and easy to observe, or hidden in bark, wood, or leaves.

The slime molds are classified into 2 major groups on the basis of whether their spores are produced externally or in some kind of spore case. In the genus *Ceratiomyxa*, the plasmodium produces stalks bearing naked spores; all other slime molds have spore cases, which grow out of a horny, spongy, or limy base (hypothallus). These spore cases may appear in one of 3 basic forms. The first is a plasmodiocarp, a stalkless, raised network that follows the outline of the plasmodium. Another form is a sporangium, which is usually stalked and either solitary or in small, dense clusters, with cylindrical to round, single to multiheaded, smooth to fuzzy spore-bearing cases. The third kind is an aethalium, which is stalkless and resembles a cushion or lump of dough, becoming crusty and splitting as the spores within reach maturity.

Those slime molds with spore cases are divided into orders on the basis of spore color, the presence or absence of capillitia—threadlike tubes mixed with the spores in the spore cases—and the presence of lime crystals or granules. The order Liceales has pale to brightly colored spores and lacks a true capillitium. The order Trichiales has pale to brightly colored spores and abundant capillitia. The order Physarales has dark spores and lime deposits in the outer wall of the spore case, the capillitium, or both. The order Stemonitales has dark spores but no lime deposits in the wall of the spore case or the capillitium.

740 Coral Slime
Ceratiomyxa fruticulosa (Müll.) Mac.
Ceratiomyxaceae, Ceratiomyxales

Description: *Translucent mass, producing spread of minute, translucent, white to yellowish, erect "icicles."*
Mushroom: $\frac{1}{64}$–$\frac{1}{32}$" (0.5–1 mm) wide, $\frac{1}{32}$–$\frac{3}{8}$" (0.1–1 cm) high; irregular; clustered, erect, branched or simple "icicles"; translucent, whitish or yellowish.
Spores: 10–13 × 6–7 μ; round to elliptical, smooth, translucent; white in mass.

Season: June–October.

Habitat: On rotten wood; sometimes on leaves or litter.

Range: Widely distributed in North America.

Comments: This is one of the most widely distributed slime molds, and one of the most beautiful, resembling a thick tangle of tiny coral or icicles. There are 2 common forms: var. *flexuosa* (branched) and var. *poroides* (porelike), which may only be responses to environmental conditions, not true varieties. Except for a minute Florida species, the Morel-like Slime (*C. morchella*), which somewhat resembles a morel in shape, this is the only species of *Ceratiomyxa* in North America; it differs from all other slime molds because it develops its spores externally on individual stalks. It can sometimes cover an area of 3–5 feet.

559 Scrambled-egg Slime
Fuligo septica (L.) Wigg.
Physacaceae, Physarales

Description: *Yellowish or white mass, producing a large, fluffy, soft, yellowish, irregular form, becoming crusty and blackish within.*
Mushroom: 1–8" (2.5–20 cm) long and nearly as wide, $\frac{3}{8}$–$1\frac{1}{4}$" (1–3 cm) thick; cushion-shaped to irregular; slimy, white to yellow or ochre. Flesh

whitish to yellowish, darkening with age, becoming black, bleeding when bruised.

Spores: 6–9 μ; round, minutely spiny; pinkish-brown; dull black in mass. Capillitium transparent, with white, yellow, or reddish, spindle-shaped nodes.

Season: May–October.

Habitat: On rotten wood, litter, or compost at first, migrating to live plants and soil.

Range: Widely distributed in North America.

Comments: This common and conspicuous slime mold is usually seen in cool weather, in spring and fall. It can be found on stumps, logs, and leaf litter compost, from which the plasmodium migrates to the stems and leaves of living plants. It is variable in size, shape, and color; with age it appears crusty and cakelike when cut. Occasionally, it is attacked by a purplish cup fungus, the Violet Nectria (*Nectria violacea*). A similar species, on rotting straw and manure, the Ashy Fuligo (*F. cinerea*), is whitish and arises from a base mass (plasmodium) that becomes ashy in color.

554 Insect-egg Mass Slime
Leocarpus fragilis (Dicks.) Rost.
Physacaceae, Physarales

Description: *Orange-yellow mass, producing cluster of shiny, yellow to brown, egglike cases.* Mushroom: 1/64–1/16" (0.5–1.5 mm) wide, 1/16–1/8" (1.5–3 mm) high; oval to nearly round, growing out of constricted base; smooth, shiny, brittle; pale yellow or ochre to reddish-brown. Flesh core membranous, transparent, surrounded by limy layer and brittle outer surface.

Stalk: very short, fragile, yellowish-white.

Spores: 11–16 μ; round, coarsely warted, brownish; black in mass. Capillitium slender, transparent, tubular, with limy, white nodes.

Season: June—October.

Habitat: Several to clustered, on dead leaves, rotten wood, or living plants.

Range: Widely distributed in North America.

Comments: The 3 layers of flesh and the capillitium, which consists of limy and limeless tubes, make this species distinctive.

561, 562 Many-headed Slime
Physarum polycephalum Schw.
Physacaceae, Physarales

Description: *Many-branched, veined, fan-shaped, sheetlike, yellowish mass, expanding and contracting, producing many grapelike heads on unbranched stalk.*
Mushroom: $1/16-1/8''$ (1.5—3 mm) wide, $1/32-1/4''$ (1—5 mm) high; heads roundish, brain-shaped to clustered; granular-limy, black to greenish. Stalk yellow.
Spores: $9-11\ \mu$; round, minutely spiny, violet-brown; black in mass. Capillitium dense, long, slender, translucent-yellowish, with yellow to white, spindle-shaped or irregular nodes.

Season: June—October.

Habitat: On dead wood and fleshy fungi.

Range: Widely distributed in North America.

Comments: Also called the "Grape Cluster Slime." This is most conspicuous in its beginning stage, as a spreading, yellow to greenish-yellow, branching network. A very common related species, the Ashy Physarum (*P. cinereum*), created a stir in Texas when people thought it was a blob from outer space. When the fire department hosed it down, it split apart, and the moisture provided it with the means to spread and grow substantially larger. People demanded that the governor call in the National Guard, but a scientist dispelled the fear by identifying the foreign substance as a common slime mold.

665 Wolf's-milk Slime
Lycogala epidendrum (L.) Fr.
Reticulariaceae, Liceales

Description: *Reddish mass, producing small, pinkish-gray to brownish, rounded cushions that exude a pinkish paste.*
Mushroom: ⅛–⅝" (0.3–1.5 cm) wide and high; round to compressed, cushion-shaped; warted to roughened; pinkish-gray to yellowish-brown or greenish-black; opening at top. Flesh pinkish, pastelike, becoming ochre and powdery as spores mature.
Spores: 6–7.5 μ; round, netted, lavender to pale ochre; pinkish-gray to pale ochre in mass. Pseudocapillitium long, branching, flattened, tubular, with transverse folds and wrinkles.

Season: June–November.

Habitat: Scattered to clustered, on dead wood, especially large logs.

Range: Widely distributed in North America.

Look-alikes: *L. flavofuscum* can grow on living trees, is usually larger and more cushion-shaped, and has hard, thick, brittle outer wall.

Comments: Formerly called *Lycoperdon epidendrum.* This puffball-like slime is frequently encountered. It is also called the "Toothpaste Slime" because the substance that oozes out of the immature broken cushion shapes resembles pink toothpaste.

751 Red Raspberry Slime
Tubifera ferruginosa (Batsch.) Gmel.
Reticulariaceae, Liceales

Description: *Transparent mass, becoming whitish to red, then brown, producing red, raspberrylike forms, turning purplish, then glossy brown and cigarlike.*
Mushroom: less than ¹⁄₆₄" (0.5 mm) wide, ⅛–¼" (3–5 mm) high, up to 6" (15 cm) or more in length; cylindrical to oval, usually crowded together and compressed; thin, translucent, with

iridescent outer wall; reddish to
reddish-brown to purple-brown;
opening at top; base spongy, off-white
to brownish.
Spores: 6–8 μ; round, netted, brown;
dark brown in mass.

Season: June–November.
Habitat: On dead wood, leaves, or litter.
Range: Widely distributed in North America.
Comments: When several of these mushrooms are
compressed together and reddish in
color, they look like raspberries, but
they soon turn purplish to brown and
then resemble a bunch of miniature
cigars.

732 Japanese-lantern Slime
Dictydium cancellatum (Batsch.) Mac.
Cribrariaceae, Liceales

Description: *Purplish-black mass, producing minute,*
round, basketlike heads on long, thin stalks;
on dead wood.
Mushroom: ¹⁄₆₄″ (0.5 mm) wide, ¹⁄₃₂–
¼″ (1–5 mm) high; round or nearly
round, basketlike; reddish-brown to
brownish-purple; netted, crossveins
thin except at top, with dark granules
on all parts of net.
Stalk: long, thin, brownish.
Spores: 5–7 μ; round, nearly smooth,
with dark granules, pale red; reddish to
purplish in mass.

Season: July–September.
Habitat: On dead wood.
Range: Widely distributed in North America.
Look-alikes: *D. mirabile,* reported in California,
grows on decayed coniferous wood and
is sparingly netted below, more coarsely
netted over top of basket. Species of
Cribraria have shorter, thinner threads
about basketlike head, connecting with
thickened nodes.
Comments: This is a very common and widespread
slime mold. Although it is very small,
it is nonetheless rather easily identified.

555 Carnival Candy Slime
Arcyria denudata (L.) Wett.
Trichiaceae, Trichiales

Description: *Glistening whitish mass, producing pinkish-reddish, cottony, cylindrical tops on tiny stalks; on dead wood.*
Mushroom: ¹⁄₆₄–¹⁄₃₂″ (0.5–1 mm) wide, ¹⁄₁₆–¹⁄₄″ (1.5–5 mm) high; oval to cylindrical, crowded; cottony-thready, reddish, becoming dingy ochre.
Stalk: dark or reddish, broadening into cuplike container above; rising from small sheetlike base, ¹⁄₆₄–¹⁄₁₆″ (0.5–1.5 mm) long.
Spores: 6–8 μ; round, with a few scattered warts, transparent; red to reddish-brown in mass. Capillitium erect, elastic; bright red, fading to brown or yellow; with distant spokes or partially spiral rings.
Season: June–November.
Habitat: On dead wood, mostly deciduous.
Range: Widely distributed in North America.
Look-alikes: *A. incarnata* is brighter red, with shorter stalk and broader, shallower cup.
Comments: Carnival Candy Slime is usually found fruiting in very dense clusters.

553, 557 Yellow-fuzz Cone Slime
Hemitrichia clavata (Pers.) Rost.
Trichiaceae, Trichiales

Description: *Whitish mass, producing thick stalk and glossy yellowish-orange top, which opens and leaves a fuzzy cone; on dead wood.*
Mushroom: ¹⁄₆₄–¹⁄₃₂″ (0.5–1 mm) wide, ¹⁄₃₂–¹⁄₁₆″ (1–1.5 mm) high; club- to gondola-shaped, stalked, crowded; shiny, olive-yellow to orange; opening above and leaving a goblet-shaped cup filled with yellowish, fuzzy threads.
Stalk: short, yellowish above, reddish-brown below; rising from sheetlike, reddish-brown base.
Spores: 7–9 μ; round or nearly round, coarsely warted to netted, pale yellow;

yellow in mass. Capillitium elastic, yellow, minutely velvety; closely wound with spirals.

Season: June–November.
Habitat: In dense clusters on dead wood.
Range: Throughout N. North America.
Look-alikes: *H. stipitata,* a more southern species that is often common in the Northeast, grows scattered and has slender, nearly cylindrical stem, much shallower cup with large, out-turned lobes, and yellow plasmodium that turns orange-red before fruiting.
Comments: As with many other slime molds, the transformations of this species from beginning to end of the life cycle are dramatic. Often, it is possible to recognize the Yellow-fuzz in one stage of development, but not in another.

560 Pretzel Slime
Hemitrichia serpula (Scop.) Rost.
Trichiaceae, Trichiales

Description: *Whitish to yellowish mass, producing a large, pretzel-like network of thickened, yellow strands.*
Mushroom: 1–2″ (2.5–5 cm) wide or wider; slender, branchlike, interwoven; thin, transparent, bright yellow to tawny; develops on pretzel-like network.
Spores: 11–16 μ; round, coarsely netted, pale yellow; golden-yellow in mass. Capillitium tangled, long, yellow; ringed or in spiny spirals.
Season: June–August.
Habitat: On dead wood, leaves, and plant litter.
Range: Widely distributed in North America.
Comments: This is one of the easiest slime molds to identify because of its pretzel-like shape. Unlike most slime molds, it doesn't develop into round balls or stalked rods or cones, but simply grows larger, with a fruiting body producing spores along the netted branches.

556   **Multigoblet Slime**
Metatrichia vesparium (Batsch.) Nan.-Brem.
Trichiaceae, Trichiales

Description: *Dark, reddish-black mass, producing many-headed stalk that opens, leaving empty "goblets"; on wood.*
Mushroom: $\frac{1}{64}$" (0.5 mm) wide, $\frac{1}{16}$–$\frac{1}{8}$" (1.5–3 mm) high; a cluster of oval heads, $\frac{1}{32}$–$\frac{1}{16}$" (1–1.5 mm) high, $\frac{1}{64}$" (0.5 mm) wide, rising up from broad, brownish-red, sheetlike base; shiny, metallic, dark red to maroon or blackish; dome-shaped lid opening, leaving empty "goblet."
Stalk: $\frac{1}{32}$–$\frac{1}{16}$" (1–1.5 mm) high; single to fused, fairly thick, solid, brick-red.
Spores: 8–12 μ; round, minutely warted, reddish-orange; brownish-red in mass. Capillitium long, folded over and coiled about, reddish; with spines at tips, and ornamented with spiny, spiral bands.
Season: July–October.
Habitat: On rotten wood.
Range: Widely distributed in North America.
Comments: This is the only North American species of *Metatrichia;* it is distinguished by its many-headed, shiny, dark, metallic lids.

594, 596   **Tapioca Slime**
Brefeldia maxima (Fr.) Rost.
Stemonitaceae, Stemonitales

Description: *White mass, producing large, whitish, tapiocalike form; becoming cushion-shaped and pinkish, then crusty and brown to black.*
Mushroom: 1–12" (2.5–30 cm) wide, $\frac{1}{4}$–$\frac{5}{8}$" (0.5–1.5 cm) high; large, cushion-shaped; borne on widespread, silvery, shiny, horny, sheetlike base; nipplelike, becoming smooth, breaking apart; pinkish to purplish-black.
Spores: 9–12 μ; round, warted, yellow-brown; brownish-black to purplish-

black in mass. Capillitium dark, netted, with multicellular, bladderlike sacs at nodes.

Season: July–October.
Habitat: On dead wood and litter.
Range: Throughout N. North America.
Comments: This is one of the largest slime molds and can be recognized by its size alone, as well as by its silvery, shiny base and black, crusty, cakelike cushion.

552 White-footed Slime
Diachea leucopodia (Bull.) Rost.
Stemonitaceae, Stemonitales

Description: *Glossy white mass, producing cylindrical, white tops, turning pink, then blue, on thick, white stalks.*
Mushroom: 1/64" (0.5 mm) wide, 1/32–1/16" (1–1.5 mm) high; heads cylindrical to elliptical; metallic-blue, iridescent-purple, or bronze.
Stalk: limy, brittle, stout, snow-white; rising from white, veiny, limy, sheetlike base.
Spores: 8–11 μ; round, minutely roughened; dull lilac to nearly black in mass. Capillitium branching, brownish.
Season: June–October.
Habitat: On fallen sticks and leaves; migrating onto living plants.
Range: Widely distributed in North America.
Comments: This highly distinctive, easily identifiable slime mold has been reported to cause damage in strawberry beds and sweet potato fields.

550, 736 Chocolate Tube Slime
Stemonitis splendens Rost.
Stemonitaceae, Stemonitales

Description: *White mass, producing clusters of brownish, cylindrical tops on thin, short, shiny, black stalks.*
Mushroom: 1/32–1/16" (1–1.5 mm) wide, 1/4–3/4" (0.5–2 cm) high; slender,

cylindrical, densely clustered; purplish-brown to black, rising from silvery to purplish, sheetlike base.

Stalk: ⅛–¼" (3–5 mm) long; shiny, black.

Spores: 7.5–9 μ; round, faintly warted, violet-brown; purplish-black in mass. Capillitium branching, forming irregular surface net.

Season: May–October.

Habitat: On dead wood and leaves.

Range: Widely distributed in North America.

Look-alikes: *S. axifera* is bright rust-brown, with smaller, smooth spores. *S. fusca* has stalk usually more than ¼ total height, and warted-netted spores.

Comments: This common, abundant species resembles delicate, densely packed, miniature pipe cleaners, giving rise to the alternate common name, the "Pipe-cleaner Slime."

Part III
Appendices

SPORE PRINT COLOR CHART

This chart will help to identify gilled
mushrooms to the genus level by means
of the color of spore prints and such
field characteristics as type of gill
attachment, the presence or absence of
veils, and habitat. (See the Introduction
for information on gill attachment, how
veils develop, and how to make a spore
print.) In the chart, a typical
mushroom is described for each genus.
The information applies particularly to
those species that are included in this
guide. Remember that young fungi
may differ significantly from older
specimens and that in each genus some
species may be atypical. Thus, although
most mushrooms in a genus may
usually grow on the ground, a few
species may grow on wood. Similarly,
although most mushrooms in a genus
may have free gills, in a few species
they may be slightly attached. Many of
these exceptions are noted in the section
labeled "Distinctive Features."

Spore Print First, take a spore print and determine
Color: to which of the following color groups
it belongs:

White to Cream or Yellow
Pink to Salmon, Brownish-pink, or
Reddish
Ochre- to Rust- or Chocolate-brown

Purple-brown to Purple-black, Smoky-black, or Black
Gray-green to Olive

Gill Attachment (GA): Next, determine whether the mushroom's gills are attached to the stalk (A), descending it (D), or free from it (F). In some genera, the caps are stalkless (S) and the gills radiate from the point where the mushroom is attached to wood or the ground.

Veils (V): See if the mushroom you have found has a partial veil (P) covering the gills, or remnants of one (usually indicated by a ring or ring zone around the stalk or a fringe of material about the cap margin). Check for signs of a universal veil (U), often indicated by a cup at the stalk base or patches about the stalk base and on the cap surface.

Habitat (H): Note the habitat of the mushroom. Most grow on the ground (G) or on wood (W), but a few are found on dung (D), cones (C), or on other mushrooms (M).

Distinctive Features: For each genus, key field characteristics of a typical adult mushroom are listed. These may include cap, gill, or stalk features, as well as details on size, odor, or habitat. Prominent exceptions are included as well.

Using the Chart: Within the correct spore print color group, find the genera that correspond to the characteristics of your mushroom. For each genus, distinctive field characteristics are listed. These should lead you to the genera in which your mushroom is likely to belong. Finally, turn to the page numbers indicated for full descriptions of the species within each genus.
The text entries, color photographs, and line drawings should help you to narrow your identification down to species. Be sure to read the text sections

on look-alikes to rule out possible misidentification. Finally, in some cases microscopic examination is essential for accurate identification. *Unless you have a positive identification—based on both field and microscopic characteristics—do not eat any mushroom that could possibly be confused with a poisonous species.*

Abbreviations: The following abbreviations are used in the chart:

Gill Attachment:		
	A	Gills attached
	D	Gills descending stalk
	F	Gills free from stalk
	S	Stalkless and shelflike; gills radiate from point of attachment

Veils:		
	–	No veils
	P	Partial veil or veil remnants present
	P±	Partial veil or veil remnants present in some species
	U	Universal veil or veil remnants present
	U±	Universal veil or veil remnants present in some species
	U&P	Universal and partial veils or veil remnants present
	U/P	Universal or partial veils or veil remnants present

Habitat:		
	C	Cones
	D	Dung
	G	Ground
	M	Mushrooms
	W	Wood

Genera	GA	V	H	Distinctive Features

WHITE TO CREAM OR YELLOW SPORE PRINT

Genera	GA	V	H	Distinctive Features
Strobilurus p. 797	A	–	C	Small; on fallen cones and magnolia fruits
Baeospora p. 738	A	–	C/W	Small; on fallen cones or decaying wood
Calocybe p. 740	A	–	G	Cap pinkish to reddish; often in grassy areas
Caulorhiza p. 743	A	–	G	Stalk long and rooting; under redwoods
Collybia p. 753	A	–	G	Margin incurved when young; gills crowded, in some species free; stalk brittle-tough; some on wood, a few on rotting mushrooms
Melanoleuca p. 776	A	–	G	Gills crowded; stalk fibrous-lined, stiff
Porpoloma p. 794	A	–	G	Gills notched at stalk; often bruising reddish
Crinipellis p. 758	A	–	G/W	Small; cap and stalk brown-hairy; gills sometimes free
Marasmiellus p. 770	A	–	G/W	Small; stalk rubbery; typically on sticks, berry canes; revives
Marasmius p. 771	A	–	G/W	Small; gills sometimes free; stalk rubbery, often very thin; revives

Genera	GA	V	H	Distinctive Features
Micromphale p. 778	A	–	G/W	Small; flesh gelatinous; stalk tough; odor foul; revives
Oudemansiella p. 788	A	–	G/W	Stalk long, rooting, brittle
Asterophora p. 738	A	–	M	Small; cap or gills becoming powdery; gills often malformed; on rotting mushrooms
Cyptotrama p. 758	A	–	W	Small; cap and stalk golden and granular; gills white, distant
Flammulina p. 759	A	–	W	Stalk velvety; in clusters
Hypsizygus p. 761	A	–	W	Cap surface often cracking; stalk off-center, often long
Tricholomopsis p. 807	A	–	W	Often yellowish; stalk off-center in some species
Armillaria p. 731	A	P	G	Often fragrant; stalk scaly-sheathed and typically with ring at top
Cystoderma p. 510	A	P	G	Cap granular; stalk sometimes with ring; sheathing veil when young
Tricholoma p. 799	A	P±	G	Gills notched at stalk; stalk fleshy, typically ringless
Dissoderma p. 512	A	U	G	Base tuberous; sheathing veil granular-warty

Genera	GA	V	H	Distinctive Features
Squamanita p. 524	A	U&P	G	Cap scaly; base tuberous
Clitocybe p. 744	A/D	–	G	Cap pale; stalk typically fleshy; some on wood
Laccaria p. 762	A/D	–	G	Cap finely scaly; gills violet to pinkish, thick, waxy
Lactarius p. 680	A/D	–	G	Cap often zoned; gills shatter when dry; latex present
Leucopaxillus p. 767	A/D	–	G	Large; stalk fleshy; embedded in mycelial mat; sometimes on well-decayed wood
Lyophyllum p. 768	A/D	–	G	Often densely clustered and along roadsides; stalk fleshy
Russula p. 697	A/D	–	G	Margin often radially lined or furrowed; stalk chalklike; gills shatter
Myxomphalia p. 786	A/D	–	G	Small; on burn sites
Mycena p. 778	A/D	–	G/W	Small, fragile; cap bell-shaped; some in dense clusters on wood
Clitocybula p. 751	A/D	–	W	Small; typically clustered
Xeromphalina p. 809	A/D	–	W	Small, toughish; stalk base with tufted tawny hairs; often clustered

Genera	GA	V	H	Distinctive Features
Armillariella p. 736	A/D	P±	W	Cap with erect hairs; black rhizomorphs usually present; clustered; sometimes on ground on buried wood
Lentinus p. 766	A/D	P±	W	Gill edges toothed; stalk typically central
Hygrophorus p. 654	A/D	U±; P±	G/W	Often small and colorful; gills waxy, wedge-shaped; veils typically slimy, usually absent
Cantharellula p. 741	D	–	G	Cap grayish, with knob; gills white to cream, blunt, regularly forked; in haircap moss
Rickenella p. 796	D	–	G	Small; fragile; colorful; in moss
Gerronema p. 760	D	–	G/W	Small; brightly colored; spore print white, cream, yellow, orange, or salmon; some with lichen; in forests
Hohenbuehelia p. 760	D	–	G/W	Flesh partly gelatinous; stalk off-center to lateral
Hygrophoropsis p. 669	D	–	G/W	Margin incurved when young; gills orange to pink and irregularly forked
Omphalina p. 787	D	–	G/W	Small; often gray to brownish; spore print whitish; often in exposed areas; in moss, often with lichen

Genera	GA	V	H	Distinctive Features
Omphalotus p. 787	D	–	W	Large, orange; gills close, sharp-edged, luminescent (when fresh) in the dark; clustered
Catathelasma p. 741	D	P	G	Large; veil double; under conifers
Cheimonophyllum p. 743	D/S	–	W	Small; stalk (when present) stubby and lateral
Lentinellus p. 764	D/S	–	W	Gill edges toothed; stalk (when present) off-center or lateral; sometimes on ground
Panellus p. 789	D/S	–	W	Fleshy-tough; stalk (when present) lateral and stubby
Panus p. 791	D/S	–	W	Tough-leathery; stalk (when present) off-center
Pleurocybella p. 792	D/S	–	W	Whitish; flesh thin; stalk (when present) lateral
Resupinatus p. 795	D/S	–	W	Small; blackish; flesh gelatinous
Tectella p. 798	D/S	P	W	Small, tough; stalk (when present) off-center and stubby; veil conspicuous when young
Pleurotus p. 792	D/S	P±	W	Often stalkless; spore print often tinted lilac-gray
Lepiota p. 513	F	P	G	Cap typically scaly, often with knob in center; stalk often with movable ring

Genera	GA	V	H	Distinctive Features
Amanita p. 526	F	U; P±	G	Cap often with veil patches; gills sometimes somewhat attached; stalk usually with ring; cup or remnants typically about stalk base
Limacella p. 554	F	U; P±	G	Cap and veil slimy; stalk sometimes with ring

PINK TO SALMON, BROWNISH-PINK, OR REDDISH SPORE PRINT

Genera	GA	V	H	Distinctive Features
Clitocybe p. 744	A	–	G	Gills notched at stalk; spore print pinkish-buff
Entoloma p. 641	A	–	G	Cap sometimes pointed, silky; some on decayed wood
Macrocystidia p. 769	A	–	G	Small; odor of cucumber and raw fish
Pouzarella p. 648	A	–	G	Cap and stalk often coarsely scaly; stalk base with stiff hairs
Rhodotus p. 795	A	–	W	Cap ridged, pitted; stalk short, off-center
Rhodocybe p. 648	A/D	–	G	Gills often crowded and sometimes deeply descending stalk; some taste bitter
Clitopilus p. 641	D	–	G/W	Whitish; cap often with bloom; margin incurved; stalk sometimes off-center

Genera	GA	V	H	Distinctive Features
Pluteus *p. 673*	F	–	W	Flesh soft, watery; gills attached in some; often on ground on buried wood
Melanophyllum *p. 523*	F	P	G	Gills reddish when young, sometimes somewhat attached; spore print dries cinnamon-purple
Volvariella *p. 677*	F	U	G/W	Stalk base with saclike cup; 1 species on mushrooms
Phyllotopsis *p. 791*	S	–	W	Cap orange, hairy; odor foul

OCHRE- TO RUST- OR CHOCOLATE-BROWN SPORE PRINT

Genera	GA	V	H	Distinctive Features
Bolbitius *p. 559*	A	–	G	Small; fragile; cap slimy, lined; gills cinnamon; sometimes on wood debris or dung
Asterophora *p. 738*	A	–	M	Small; cap or gills becoming powdery; gills often incompletely formed; on rotting mushrooms
Phaeolepiota *p. 523*	A	P	G	Large; cap golden-granular; sheathing, granular veil and ring
Phaeomarasmius *p. 711*	A	P	W	Small; cap scaly, powdery

Genera	GA	V	H	Distinctive Features
Pholiota p. 712	A	P	W	Cap typically scaly, often slimy; stalk usually with ring; some on ground, may be on buried wood
Alnicola p. 611	A	P±	G	Stalk without ring; near alders
Conocybe p. 559	A	P±	G	Small; fragile; gills cinnamon; stalk sometimes with ring; some on wood
Hebeloma p. 624	A	P±	G	Slimy cap; ring on some; radishy odor
Inocybe p. 626	A	P±	G	Cap fibrous, often with knob; gills typically grayish to gray-brown; stalk without ring; odor typically spermatic
Agrocybe p. 556	A	P±	G/W	Cap hairless; surface often cracking; ring often present
Galerina p. 620	A	P±	G/W	Small; those in moss have thin, bell-shaped caps and fragile stalks; those on wood have fleshy, convex caps; stalk sometimes with ring
Gymnopilus p. 623	A	P±	W	Orange; stalk sometimes with ring; spore print orange-tinted
Simocybe p. 638	A	P±	W	Small; cap velvety; stalk without ring and sometimes off-center

Genera	GA	V	H	Distinctive Features
Cortinarius p. 611	A	U±; P	G	Cap, stalk, and young gills often colorful; stalk sometimes with ringlike fibrous zone; partial veil cobwebby
Rozites p. 635	A	U&P	G	Cap frosted at center; stalk with central ring and no cup
Ripartites p. 796	A/D	P±	G/W	Small; margin bristly to woolly; and incurved when young
Tubaria p. 638	A/D	P±	G/W	Small; fragile; cap often finely flecked or fringed; gills usually pale brown
Phaeocollybia p. 634	A/F	–	G	Cap slimy, often pointed; stalk deeply rooting
Phylloporus p. 672	D	–	G	Boletelike, but with gills; gill layer separable from cap flesh; spore print yellow-brown to olive
Paxillus p. 670	D/S		G/W	Margin incurved; gill layer separable from cap flesh; stalk (when present) often off-center
Agaricus p. 500	F	P	G	Gills often pinkish when young; stalk with ring; lower veil surface sometimes with patches; spore print purple-brown in some

Genera	GA	V	H	Distinctive Features
Crepidotus p. 636	S	–	W	Small to medium

PURPLE-BROWN TO PURPLE-BLACK, SMOKY-BLACK, OR BLACK SPORE PRINT

Genera	GA	V	H	Distinctive Features
Naematoloma p. 708	A	P	G/W	Stalk ringless; often clustered
Stropharia p. 725	A	P	G/W	Often large; stalk with ring; rhizomorphs often present; 1 species on dung
Panaeolus p. 602	A	P±	D	Gills mottled; stalk often stiff
Psathyrella p. 604	A	U±; P±	G/W	Margin often with veil remnants; stalk whitish, hollow, fragile, rarely with ring; cup absent; 1 species on mushrooms; spore print reddish in a few
Psilocybe p. 719	A/D	P±	G/W /D	Small; cap sticky to moist; stalk with ring in 2 species; some species bruising blue
Chroogomphus p. 650	D	P	G	Flesh pink to orange; gills distant; stalk ringless; spores smoky-black to smoky-olive; under conifers
Gomphidius p. 652	D	P±	G	Flesh white; cap and veil slimy; gills distant; stalk ringless, base often yellow; under conifers

Genera	GA	V	H	Distinctive Features
Coprinus p. 596	F/A	U±; P±	G/W /D	Cap elongate when young; gills dissolve; cup absent

GRAY-GREEN TO OLIVE SPORE PRINT

Phylloporus p. 672	D	–	G	Boletelike, but with gills; gill layer separable from cap flesh; spore print yellow-brown to olive
Chlorophyllum p. 509	F	P	G	Large; gills white, becoming slowly gray-green; stalk with usually movable ring; typically in grassy areas

MUSHROOM POISONING

No simple test can determine whether a mushroom is edible or poisonous. The only way to be certain is to know exactly what species you have found. Only experience can teach you to recognize characteristics that differentiate edible from poisonous species.

There are some rules of thumb that can be helpful: 1) do not eat any *Amanita* species and be especially careful in identifying *Amanita* look-alikes or any other white mushroom; 2) avoid little brown mushrooms ("LBM's") and large brownish mushrooms, especially those with pinkish, brownish, purple-brown, or blackish gills; 3) avoid false morels. Only a few mushrooms cause life-threatening illness. Many others cause mild to severe poisoning.

Hallucinogenic mushrooms containing psilocybin cause other and sometimes unpredictable reactions. Some mushroom toxins affect the central nervous system, others the peripheral nervous system, but most cause mild to severe gastrointestinal upset. Some people react adversely to species that are harmless for most, and some react adversely to species they have eaten before without ill effects. Typical symptoms include nausea, diarrhea, cramps, and vomiting.

Before you eat any wild mushroom, be sure to read the appendix on cooking and eating wild fungi.

Here we describe known mushroom toxins and list the species in this guide that contain them, the symptoms they produce, and how much time elapses before the symptoms occur. If you suspect you have mushroom poisoning, seek medical attention immediately and bring along whole, uncooked samples of the mushrooms you have eaten.

Amanitin:
: *Amanita* species, including. *A. phalloides* and the *A. virosa* complex; *Galerina* species, including *G. autumnalis* and *G. venenata;* *Lepiota* species, including *L. josserandii; Conocybe filaris.* Symptoms occur 6–24 hours after eating, typically in 10–14 hours and rarely 48 hours later. They include abdominal pains, nausea, vomiting, and diarrhea, lasting for 1 or more days. A short remission takes place, then pain recurs, with liver dysfunction, jaundice, renal failure, convulsions, coma, and often death. With sustained medical assistance, recovery can take place in 1–2 weeks.

Monomethylhydrazine (MMH):
: *Gyromitra* species, including *G. esculenta.* Symptoms occur 6–12 hours after ingestion, or, rarely, 2 hours later. They include a bloated feeling, nausea, vomiting, watery or bloody diarrhea, abdominal pains, muscle cramps, faintness, loss of coordination, and, in severe cases, convulsions, coma, and death. With medical attention, recovery can occur within hours.

Orellanin:
: *Cortinarius* species, including *C. gentilis.* Symptoms occur 3–14 days after ingestion, and consist of acute or chronic renal failure, which can result in death. A kidney transplant is sometimes required, and recovery can take as long as 6 months.

Muscarine:
: *Clitocybe* species, including *C. dealbata* and *C. dilatata;* most *Inocybe* species. Symptoms include profuse perspiration,

salivation, tears, blurred vision, abdominal cramps, watery diarrhea, constriction of the pupils, a fall in blood pressure, and slowing of the pulse. Although symptoms usually subside in 6–24 hours, severe cases may require hospitalization, and death has been reported in people with preexisting illnesses.

Ibotenic acid and muscimol: *Amanita* species, including *A. muscaria* and *A. pantherina.* Symptoms occur 30 minutes to 2 hours after ingestion. They include dizziness, lack of coordination, delusions, staggering, delirium, muscular cramps and spasms, hyperactivity, and deep sleep. Recovery usually takes place within 4–24 hours; some cases require hospitalization.

Coprine: *Coprinus atramentarius; Clitocybe clavipes.* Symptoms can occur as long as 5 days after eating the mushrooms. They are precipitated by the ingestion of alcohol, usually about 30 minutes after the alcohol is taken. Symptoms include flushing of the face and neck, distension of neck veins, swelling and tingling of hands, a metallic taste in the mouth, palpitations, and hypotension; nausea, vomiting, and sweating may then occur. Recovery usually occurs within 2–4 hours.

Psilocybin and psilocin: *Psilocybe* species, including *P. baeocystis, P. caerulipes, P. coprophila, P. cubensis, P. cyanescens, P. pelliculosa, P. semilanceata, P. stunzii; Conocybe smithii; Gymnopilus spectabilis; Panaeolus subbalteatus.* Symptoms occur within 30–60 minutes, rarely as long as 3 hours later. They include mood shifts, which can range from the pleasant to the apprehensive; symptoms may often include unmotivated laughter, hilarity, compulsive movements, muscular weakness, drowsiness, visions, then sleep. Recovery usually takes place within 6 hours. The victim should be assured that the symptoms will pass.

Immune injury: *Paxillus involutus.* Symptoms occur 1–3 hours or more after ingestion. They result from a gradually acquired sensitivity to the species, and include the destruction of red blood cells, vomiting, diarrhea, and cardiovascular irregularity. They usually disappear in 2–4 days, but can last much longer in severe cases, and may require hospitalization.

Diverse, mostly unknown toxins: *Chlorophyllum molybdites; Entoloma sinuatum, E. strictius, E. vernum; Omphalatus olearius; Tricholoma pardinum.* Symptoms occur 30 minutes to 3 hours after ingestion. They include serious, severe nausea, vomiting, diarrhea, and abdominal pain. Recovery can take several hours or days, depending on the species, the amount eaten, and the health of the victim. Hospitalization is sometimes required. *Agaricus californicus, A. hondensis, A. meleagris, A. xanthodermus; Amanita brunnescens; Boletus subvelutipes; Gomphus floccosus; Hebeloma crustuliniforme; Lactarius representaneous, L. rufus, L. scrobiculatus, L. torminosus, L. uvidus, L. vinaceorufescens; Lepiota cepaestipes, L. clypeolaris, L. cristata, L. lutea; Naematoloma fasciculare; Phaeolepiota aurea; Pholiota squarrosa; Ramaria formosa, R. gelatinosa; Russula emetica; Scleroderma citrinum; Tricholoma pessundatum.* Symptoms usually occur within 30 minutes to 3 hours of ingestion. They include nausea, vomiting, diarrhea, and abdominal pain. Recovery can take from 1 hour to 2 days, depending on the species, the amount eaten, and the general health of the victim.

Edible mushrooms known to cause occasional adverse reactions: *Armillariella mellea, A. tabescens; Calvatia gigantea; Laetiporus sulphureus; Lepiota naucina, L. rachodes; Morchella elata; Suillus luteus; Tricholompsis platyphylla;* and others. Symptoms occur within 2 hours. They include nausea, vomiting, and diarrhea. Recovery usually takes place within a few hours.

COOKING AND EATING WILD MUSHROOMS

Before you eat a wild mushroom, be absolutely sure your identification is correct and that the mushroom is a safe edible. The first time you eat any species, take only a small portion and do not drink any liquor. If you experience no side effects, try a slightly larger portion the next time. Don't eat a large quantity, no matter how often you have eaten a particular mushroom; mushrooms, in general, are indigestible.

Unless otherwise advised, cook all wild mushrooms. Some people may experience upset after eating even cooked fungi. Others inexplicably develop a reaction to a species they have eaten for years. Some people are allergic to even the common cultivated mushroom. Don't serve wild fungi to children, the elderly, or the sick; these individuals are most susceptible to poisoning. And don't force anyone who is unwilling to share your enthusiasm for wild foods. Be sure to read the appendix on mushroom poisoning and learn to recognize its symptoms. If you think you have been poisoned, seek medical aid immediately.

Collect only firm, fresh mushrooms for the table. Cut them in half and check for insects and worms. Do not collect more mushrooms than you can use, and always be sure to bring some home intact to confirm your identifications. Remember to keep all species separate, either by enclosing them in paper bags or by wrapping them in wax paper. Once at home, refrigerate your mushrooms and use them as soon as possible, since most fungi are perishable. Use inky caps immediately or parboil them to keep them from liquefying. Always set aside a few of the uncooked mushrooms; if you become ill, their toxins can be identified.

Almost all mushrooms are best sautéed in butter or oil. Most require seasoning with salt and pepper, or with onions, garlic, shallots, or scallions. Lemon juice heightens the taste of many mushroom dishes. Sour cream is helpful when you want to rescue a dish that is too spicy or harsh. Use cream as a base for mushroom soups and sauces. Taste whatever you cook before you serve it to others and avoid alcohol or drink it in moderation.

Mushrooms can be preserved in a number of ways. Most dry well and can be stored for quite a long time. To dry mushrooms, slice them very thin and place them in a food dryer or in a very cool oven. To rehydrate, soak for 1–3 hours in warm cream or wine. In some, such as morels and a few boletes, the flavor improves with age; other mushrooms may become leathery with drying. If this should occur, make a mushroom powder of the dried mushrooms, and use the powder with salt, pepper, and garlic flavor as a seasoning.

Most mushrooms can be frozen for later use. Clean, then blanch or parboil the mushrooms, then plunge them in ice water, drain, and pack in freezer-safe containers. Frozen mushrooms should not be kept for longer than 1 year. Almost all mushrooms can be made into a rich paste called *duxelles*. For every half pound of mushrooms, use 2 tablespoons of butter and 1 tablespoon of oil. Begin by chopping the mushrooms as fine as possible (a food processor can be used); put the mushrooms in a clean dishcloth or in cheesecloth and wring out as much moisture as possible. Save the liquid for soups and sauces. Add minced scallions, shallots, or garlic, and salt and pepper to taste. Sauté the mixture in the oil and butter over medium-high heat, stirring frequently until all the moisture has evaporated. *Duxelles* can be frozen and used in almost any dish. Many mushrooms add a new dimension to sauces. Chop mushrooms very fine; cook them first in butter or oil, or blanch or parboil them, and then add to a sauce or gravy, and cook all ingredients together slowly.

Many cookbooks give specific instructions for preparing different mushrooms. Included here are general guidelines for various groups of edible fungi.

Morels: Clean morels in the field, cutting off the bottom of the stalk. At home, slice them in half, then soak them in water to get rid of insects and debris. Morels are best dried, then rehydrated, preferably in cream, and then cooked. Sauté them and serve with eggs or veal. The cream in which you rehydrate them can be heated and used as a sauce or a soup.

Truffles: North American truffles are too little known to be considered edible for most. Imported truffles are grated over pasta or into omelettes to release their pungent odor.

Tree-Ears: Slice these jelly fungi, combine them with other vegetables, and sauté; the fungi add a distinctive crunchy texture and color contrast. Tree-Ears dry easily and store well.

Chanterelles: Chanterelles are among the most delicious edibles. Clean them in the field; at home wash them thoroughly, then slice and sauté. If too much moisture is released as they cook, increase the heat and stir continuously until the liquid evaporates. Chanterelles are spicy and need little seasoning. They can be dried, but sometimes become leathery when rehydrated; it is better to freeze them after they are cooked. Leathery, dried chanterelles work well when powdered.

Black Trumpets: These mushrooms are too aromatic to be eaten by themselves, but they make a fine condiment. They are also good in patés or egg dishes. They dry and store very well.

Corals: These large, fleshy fungi are good sautéed and served with mixed vegetables. The Flat-topped Coral is one of the sweetest mushrooms; sauté it briefly, then add it to fruit compotes.

Tooth Fungi: The large, fleshy tooth fungi that grow on trees can be cleaned, sliced, and then sautéed in garlic butter. Any bitterness will be removed by parboiling them first. The meaty and flavorful Sweet Tooth is also good when sautéed. Although many people dislike the harsh flavor of the Scaly Tooth, young specimens can be eaten if carefully cleaned, parboiled, and then sautéed slowly, and seasoned with garlic or spices.

Cauliflower Mushrooms: Slice these firm, attractive edibles, then sauté, season, and serve. One cauliflower mushroom is often large enough to make an entire vegetable dish.

Polypores: Although most polypores are too tough to eat or are good only when young, a few are choice: the Chicken Mushroom (provided it is not growing on eucalyptus), the Hen of the Woods, and the Umbrella Polypore are excellent. Collect fresh mushrooms, clean them well, cut in small pieces, then sauté very slowly in butter in a covered skillet. The Hen of the Woods is also good pickled or in soups. The

Chicken Mushroom makes a fine poultry substitute. The Beefsteak Polypore is one of the few wild mushrooms that can be eaten raw. Slice it into thin strips and add it to salads. It has a tart flavor, which can be accentuated with lemon juice.

Boletes: Slice boletes in half; if worms and insects are present, discard the stalks. If the stalks are scaly, peel them. If the tubes are not firm and compact, cut them away from the cap. If the caps have a thick slime layer, it should be removed because it may cause stomach upset. Sautéed caps are the best part of these fungi; the stalks can be used, but they should be cooked longer. Most boletes dry well and store easily; many taste even better rehydrated than fresh.

Gilled Mushrooms: Clean mushroom caps with a damp cloth or brush; do not wash them because they soak up water and become soggy. Most species can be sautéed. Dry gilled mushrooms for future use, or parboil or sauté them and then freeze. Don't freeze mushrooms before cooking them; they lose their flavor and texture. Edible *Agaricus* species such as the Meadow Mushroom can be used whenever recipes call for cooked mushrooms.

The Honey Mushroom is often parboiled before it is sautéed. Like almost all species, this one is especially good when it is young.

Use only the caps of the Fairy Ring Mushroom, and be sure to cook them until they are almost dry.

The aborted stage of the Aborted Entoloma is delicious sautéed and served with black butter and capers. Cook edible *Lactarius* species slowly, sautéeing them in a covered skillet. Use inky caps in soups, especially those that are cream-based. The Shaggy Mane can be sliced lengthwise, then cooked and served like asparagus.

The Fragrant Armillaria and White Matsutake are so fragrant and spicy that only small amounts of them need be used. Slice them very thin, sauté, and then add them to rice dishes. Their caps can be broiled, and are especially good if they have first been marinated in soy sauce.

Oyster mushrooms are used as a substitute for clams or oysters; they are excellent broiled or in casseroles, as are Blewits and parasols. Blewits with too strong a flavor should be blanched or parboiled first, then baked. Parasols may be broiled after the tough stalks have been removed.

Puffballs: Eat only puffballs that are young and have a pure white interior. Clean puffballs, then slice them and cook slowly; season to taste and serve. They may also be dipped in batter and deep fried. Some cooks use the large species as an eggplant substitute; first slice them thin, then bread and sauté.

GLOSSARY

Agaric A mushroom bearing gills on the undersurface of its cap.

Amyloid Turning blue-black in Melzer's reagent; used in reference to the spore walls or the surface ornamentation of the spores.

Ascus The saclike cell in which spores are produced by mushrooms of the subdivision Ascomycotina (plural, asci).

Basidium The club-shaped cell in which spores are produced by mushrooms of the subdivision Basidiomycotina (plural, basidia).

Bolete A fleshy mushroom bearing a tubelike layer on the undersurface of its cap and belonging to the family Boletaceae.

Bruising Changing color when handled.

Button The immature stage of a mushroom; typically used in reference to amanitas, which have a universal veil.

Cap The top or head of a mushroom.

Capillitium Sterile threads mixed with the spores in the spore cases of some puffballs and slime molds (plural, capillitia).

Carminophilous Becoming dark-dotted in a solution of

acetocarmine; used in reference to microscopic basidia.

Chlamydospore An asexual spore developing from unspecialized hyphae.

Clamp A bridgelike structure attached to two adjacent hyphae over a cell cross-wall in some basidiomycetes.

Cluster A group of mushrooms rising together from the same spot, typically touching and often attached at the base.

Conidiospore An asexual spore developing from specialized hyphae.

Coniferous Cone-bearing.

Conk A large, woody, hoof-shaped polypore growing on wood.

Cortina A cobwebby tissue that covers the immature gills in *Cortinarius* and some related genera.

Cup The saclike tissue at the stalk base of some volvariellas and amanitas, left by the universal veil after it has ruptured.

Cuticle The surface cell layer of a mushroom.

Cyanophilous Turning blue in a solution of cotton-blue; used in reference to a spore wall.

Cystidium A sterile cell on the cap, gills, or stalk of many basidiomycetes, especially gilled mushrooms.

Deciduous Shedding leaves annually, and leafless for part of the year.

Deliquesce To dissolve; used in reference to the gills of inky caps.

Depressed Sunken.

Descending Running down the stalk; used in reference to gills or pores.

Dextrinoid   Turning reddish-brown in Melzer's reagent; used in reference to the spore walls or their surface ornamentation.

Disc   The central portion of the cap.

Egg   The immature button stage of amanitas and stinkhorns; one of the spore sacs in a bird's-nest fungus.

Evanescent   Disappearing quickly.

Fairy ring   An arc or circle of gilled mushrooms or puffballs, arising from a mycelium that expands outward from a central point.

Fertile surface   The spore-bearing surface, or hymenium, of a mushroom.

Flesh   The interior tissue of a mushroom.

Free   Not attached to the stalk; used in reference to gills.

Fruiting body   The reproductive portion of a fungus; typically appearing above ground.

Fungus   An organism, traditionally included in the plant kingdom, that lacks chlorophyll and possesses spores.

Gill   One of the radial, bladelike plates that bear spores, located on the undersurface of the cap of many mushrooms.

Gleba   The spore mass of a gasteromycete.

Hymenium   The spore-bearing surface (plural, hymenia).

Hypha   One of the filamentous threads that make up the body of a fungus (plural, hyphae).

Incurved   Rolled or bent inward.

KOH   A 3–10 percent solution of potassium hydroxide in water; used to test color reactions.

Lateral   Attached at the edge; used in reference to stalks that are thus attached to caps.

Latex   A clear, milky, or colored liquid that exudes from cut surfaces, especially in the genus *Lactarius*.

Margin   The edge of a mushroom cap.

Melzer's reagent   A solution of 20 cc of water, 1.5 gm of potassium iodide, 0.5 gm of iodine, and 20 gm of chloral hydrate; used to test color reactions in spores or certain mushroom tissues.

Micron   One one-thousandth of a millimeter; the unit used to measure microscopic features.

Mushroom   The fruiting body of a fungus.

Mycelium   The vegetative portion of a fungus (plural, mycelia).

Mycology   The scientific study of fungi.

Mycorrhiza   A symbiotic association between a fungus and the root ends of a flowering plant (plural, mycorrhizae).

Ornamented   Decorated with warts, ridges, wrinkles, or a netlike pattern; used in reference to spore surfaces.

Parasitic   Living in or on another animal or plant and deriving food from it.

Partial veil   A tissue that covers and protects the immature gills or tubes of some gilled mushrooms and boletes.

Peridiole   A small, egglike structure that contains the spores; found in some gasteromycetes.

Peridium   The outer layer of the fruiting body of a gasteromycete.

Perithecium   A minute flask-shaped vessel containing

the asci in a pyrenomycete (plural, perithecia).

Pore The mouth or opening of the tube in boletes and polypores.

Revive To resume an earlier shape and function when exposed to water.

Rhizomorph A cordlike strand of mycelium at the base of a mushroom.

Ring The remnants left by a partial veil after it has ruptured; located on the stalk.

Ring zone The remnants left on a stalk by a thin or cobwebby partial veil after it has ruptured.

Saprophyte An organism that takes nourishment from dead organic matter.

Scale A torn piece of the cap or stalk surface.

Seta A bristle-shaped sterile cell, usually colored, on the surface of some polypores and a few agarics (plural, setae).

Shelflike Stalkless; typically growing from wood.

Solid Filled with dense flesh.

Spore The reproductive unit in a fungus.

Spore mass The portion of a gasteromycete containing the spores.

Spore print The pattern made by the spores as they are discharged from gills or tubes.

Stalk The portion of a mushroom that supports the cap and elevates it sufficiently for adequate spore dispersal.

Sterigma A short stalk, usually 1 of a group of 4, arising from the top of the basidium of a basidiomycete and from which the spores emerge (plural, sterigmata).

Stroma   A cushionlike structure in which or on which the fruiting bodies are produced in a pyrenomycete (plural, stromata).

Stuffed   Filled with loose flesh.

Symbiont   An organism that lives in a mutually beneficial relationship, or symbiosis, with another organism.

Toadstool   Popular term for a poisonous mushroom.

Tube   A hollow cylinder that contains the basidia where spores are produced in a bolete or polypore.

Universal veil   A tissue that encloses the entire immature stage of some gilled mushrooms and boletes.

Veil   A tissue that covers and protects the immature stage of some gilled mushrooms and boletes; called a universal veil when it encloses the entire immature mushroom and a partial veil when it covers only the gills or tubes.

Volva   A saclike cup or tissue surrounding the stalk base, left after the universal veil has ruptured.

Zoned   With distinct bands.

AUTHORS OF FUNGI

The following abbreviations are used in this book to refer to the names of authors of fungi.

Afz.	Afzelius	DeCh.	De Chambre
A.H.S.	A. H. Smith	DeNot.	De Notaris
A. & S.	Albertini and	Desm.	Desmazieres
	Schweinitz	Desv.	Desvaux
Atk.	Atkinson	Dicks.	Dickson
Bal.	Ballen	Dit.	Ditmar
Ban.	Banning	Dod.	Dodge
Bank.	Banker	Ear.	Earle
Bat.	Bataille	Ehr.	Ehrhart
Beard.	Beardslee	Ell.	Ellis
Berk.	Berkeley	Ev.	Everhart
Ber.	Berthet	Fisch.	Fischer
Bert.	Bertillon	Fkl.	Fuckel
Big.	Bigelow	Fr.	Fries
Boid.	Boidin	M. Gees.	Maas Geesteranus
Boif.	Boiffard	Gen.	Genervier
Bolt.	Bolton	Gil.	Gilbert
Bon.	Bonorden	Gilb.	Gilbertson
Bond.	Bondartsev	Gill.	Gillet
Boud.	Boudier	Gmel.	Gmelin
Br.	Broome	Grev.	Greville
Bref.	Brefeld	Guer.	Guersent
Bres.	Bresadola	Hark.	Harkness
Britz.	Britzelmayr	Henn.	Hennings
Brond.	Brondeau	Hes.	Hesler
Bull.	Bulliard	Hoeh.	Hoehnel
Burd.	Burdsall	Holmsk.	Holmskjold
Burl.	Burlingham	Hon.	Honey
Carp.	Carpenter	Hook.	Hooker
Ces.	Cesati	Horn.	Hornemann
Cke.	M. C. Cooke	Hot.	Hotson
Cok.	Coker	Hub.	Hubbard
Cor.	Corner	Huds.	Hudson
Cunn.	Cunningham	Jacq.	Jacquin
Curt.	M. A. Curtis	Jung.	Junghuhn
Czern.	Czerniaier	Kalchb.	Kalchbrenner
DC.	De Candolle	Kamb.	Kamby

Kan.	Kanouse	Rog.	Rogers
Kar.	Karsten	Rost.	Rostafinski
Kauff.	Kauffman	Russ.	Russell
Kl.	Klotzch	Ryv.	Ryvarden
Kon.	Konrad	Sacc.	Saccardo
Kotl.	Kotlaba	Sant.	Santi
Kromb.	Krombholz	Schaeff.	Jacob Schaeffer
Küh.	Kühner	J. Schaeff.	Julius Schaeffer
Kum.	Kummer	Schrad.	Schrader
L.	Linnaeus	Schroet.	Schroeter
Lamb.	Lambotte	Schulz.	Schulzer
Let.	Letellier	Schum.	Schumacher
Lév.	Léveille	Schw.	Schweinitz
Ley.	Leysser	Scop.	Scopoli
Lindbl.	Lindblad	Secr.	Secretan
Litsch.	Litschauer	S.F.G.	S. F. Gray
Lor.	Lorinser	Sing.	Singer
Lund.	Lundell	Sol.	Solheim
Mac.	Macbride	Som.	Sommerfelt
Mar.	Martin	Sow.	Sowerby
Mass.	Massee	Stde.	Staude
Matt.	Mattuschka	Steud.	Steudl
Maub.	Maublanc	Sum.	Sumstine
Mazz.	Mazzer	Sunh.	Sunhede
Melz.	Melzer	Tal.	Talbot
Mér.	Mérat	Thax.	Thaxter
Mét.	Métrod	Torr.	Torrey
Mit.	Mitchell	Tr.	Tracy
Mont.	Montagne	Tul.	Tulasne
Morg.	Morgan	Tyl.	Tylutki
Müll.	Müller	Under.	Underwood
Murr.	Murrill	Vitt.	Vittadini
Nan.-	Nannenga-	Viv.	Viviani
Brem.	Bremekamp	Wall.	Wallroth
Nannf.	Nannfeldt	Wat.	Watling
Neu.	Neuhoff	Weinm.	Weinmann
Nyl.	Nylander	Wett.	Wettstein
O.K.M.	O. K. Miller, Jr.	Willd.	Willdenow
Opat.	Opatowski	Wigg.	Wiggers
Parm.	Parmasto	Wkfld.	Wakefield
Pat.	Patouillard	Wulf.	Wulfen
Pers.	Persoon	Zel.	Zeller
Pet.	Petersen		
Pil.	Pilát		
Pk.	Peck		
Pom.	Pomerleau		
Pouz.	Pouzar		
Quél.	Quélet		
Rait.	Raitvir		
Raith.	Raithelhuber		
Rde.	Reade		
Rel.	Relhan		
Ret.	Retzius		
Rick.	Ricken		
Rio.	Riousset		

PICTURE CREDITS

The numbers in parentheses are plate numbers. Some photographers have pictures under agency names as well as their own. Agency names appear in boldface. Photographers hold copyrights to their works.

Catherine Ardrey (14, 15, 20, 23, 40, 161, 165, 167, 168, 176, 178, 189, 194 right, 197, 199, 212, 217, 228 right, 232, 271, 284, 289, 301, 305, 318, 319, 323, 330, 338, 341, 347, 353, 358, 398, 414, 415, 470, 612, 625, 719)

Arthur S. Bailie (123, 164, 174, 206, 231, 326, 331, 511, 514, 718)

Harley Barnhart (7, 54, 101, 140, 160, 170, 180, 202, 244, 250, 293, 300, 337, 356, 376, 403, 409, 411, 426, 432, 471, 542, 572, 598, 607, 615, 623, 633, 685, 687, 689, 752)

Arne Benson (696)

Michael Beug (24, 74, 77, 83, 91, 95, 131, 186, 201, 211, 218, 221, 227 left, 230, 237, 252, 287, 296, 297, 299, 302, 306, 346, 354, 360, 386, 388, 393, 401, 420, 431, 433, 438, 444, 448, 463, 519, 549, 604, 626, 631, 650, 661, 708, 722, 737, 739, 755)

H. E. Bigelow (32, 38, 59, 105, 117, 193 left and right, 224, 245, 255, 279, 307, 512, 525, 646)

Edmond R. Boynton (96, 266, 364, 717)

Sonja Bullaty (518)

Charles F. Coffill (664)

Bruce Coleman Ltd.
Adrian Davies (563) John Markham

(632) R. K. Murton (321) John Shaw (736) R. Tidman (143) Peter Ward (541)

Cornell Laboratory of Ornithology
C. B. Moore (34, 61, 236, 295, 329, 361, 487, 528, 601, 649, 658)

Frances V. Davis (111, 325, 332, 355, 363, 422, 440, 459, 461, 652, 713)
E. R. Degginger (45, 132, 173, 240, 339, 344, 475, 485, 515, 548, 555, 667)
Jim Driscole (79)
John Durkota (113, 371)
William E. Ferguson (599)
David Fisher (497)
Steve Fox (474, 596)
Richard Homola (569, 570, 608)
J. Q. Jacobs (226 right)
Emily Johnson (4, 25, 41, 49, 52, 57, 126, 127, 136, 156, 228 left, 248, 261, 274, 286, 290, 292, 294, 309, 327, 343, 362, 381, 382, 383, 384, 412, 435, 445, 494, 509, 568, 605, 619, 654, 695, 700, 741, 742, 743)
Peter Katsaros (29, 68, 99, 110, 190, 223, 235, 314, 315, 400, 413, 423, 429, 500, 510, 550, 553, 556, 557, 558, 635, 663, 692, 751)
David P. Lewis (43, 278)
Al Leyenberger (561)
Gary H. Lincoff (5, 30, 51, 71, 135, 153, 169, 183, 203, 227 right, 253, 254, 313, 370, 374, 379, 473, 477, 507, 577, 585, 620, 637, 660, 670, 672, 674, 675, 676, 678, 680, 707, 723)
John R. MacGregor (1, 2, 48, 55, 56, 72, 391, 439, 447, 617, 653, 686, 728)
Tom Martin (552, 616, 740)
John Menge (450, 624)

National Audubon Society Collection/Photo Researchers, Inc.
John Bova (97) Ken Brate (328) Peter Katsaros (6, 419, 476, 482, 490, 603, 733) Robert Lee (430) Howard A.

Miller, Sr. (67) Charlie Ott (516, 738, 746) Noble Proctor (656) Louis Quitt (268) L. West (342)

Chie Nishio (710)
Robert T. Orr (3, 37, 103, 138, 157, 317, 333, 340, 392, 502, 504, 532, 659, 684, 730)
Robert Peabody (520)
A. Petersen (441)
Mary Plant (121)
Samuel S. Ristich (18, 36, 137, 159, 397, 425, 442, 489, 517, 521, 530, 533, 534, 535, 537, 538, 544, 546, 559, 571, 573, 574, 575, 576, 578, 579, 580, 581, 582, 583, 584, 586, 588, 589, 590, 591, 592, 593, 594, 595, 642, 669, 673, 677, 683, 691, 753)
Kit Scates (8, 10, 11, 12, 16, 17, 19, 21, 22, 27, 31, 33, 35, 53, 58, 60, 62, 63, 66, 69, 70, 73, 75, 78, 81, 82, 85, 88, 89, 90, 100, 104, 106, 115, 118, 119, 122, 128, 133, 141, 145, 149, 152, 154, 155, 162, 166, 171, 177, 182, 184, 185, 188, 191, 192, 195 left and right, 196, 204, 208, 209, 210, 213, 222, 225, 233, 238, 239, 241, 242, 243, 246, 247, 251, 256, 257, 258, 259, 260, 262, 265, 267, 273, 275, 277, 280, 281, 282, 283, 285, 288, 298, 304, 311, 316, 334, 335, 336, 351, 359, 366, 369, 372, 387, 394, 396, 402, 427, 434, 437, 449, 451, 454, 455, 456, 460, 462, 464, 465, 466, 467, 469, 479, 480, 488, 493, 496, 499, 501, 505, 506, 508, 522, 524, 526, 527, 529, 531, 539, 540, 543, 560, 567, 602, 609, 610, 611, 621, 622, 627, 628, 640, 655, 671, 679, 704, 709, 714, 715, 720, 724, 745, 747, 748, 749)
Milton J. Schwartz (109, 562, 630, 634, 688, 727, 732, 754)
Alex Shigo (597)
William A. Smith (163, 179, 198, 207, 214, 350, 357, 410, 472, 492, 551, 587, 701, 705, 721)

INDEX

Numbers in boldface type refer to color plates. Numbers in italics refer to pages. Circles preceding common names make it easy for you to keep a record of the mushrooms you have seen.

A

Acetabula ancilis, 331

Agaric
- ○ Fly, **143, 144, 680**, *539*
- ○ Lichen, **10**, *787*
- ○ Orange Moss, **59**, *796*
- ○ Puffball, **644**, *811*
- ○ Yellow-orange Fly, **137, 677**, *540*

Agaricaceae, *500*

Agaricales, *499*

Agarics and Boletes, *499*

Agaricus
- ○ Abruptly-bulbous, **122**, *500*
- ○ Bleeding, **160**, *505*
- ○ California, **151**, *504*
- ○ Eastern Flat-topped, *508*
- ○ Felt-ringed, **158**, *506*
- ○ Red-gilled, **159**, *523*
- ○ Spring, **157**, *503*
- ○ Western Flat-topped, **152**, *507*
- ○ Wine-colored, **161**, *508*
- ○ Yellow-foot, **155**, *509*

Agaricus
abruptibulbus, **122**, *500*
abruptibulbus-silvicola
complex, *502*

albolutescens, 501, 503, 509
arvensis, **156**, *501*
augustus, **150**, *502*
bisporus, *500, 505*
bitorquis, **157**, *503*
brunnescens, 505
californicus, **151**, *504*
campestris, **153, 154**, *505*
echinatum, 523
haemorrhoidarius, **160**, *505*
hondensis, **158**, *506*
meleagris, **152**, *507*
placomyces, 508
rodmani, 503
silvaticus, 506
silvicola, 501
subrutilescens, **161**, *508*
xanthodermus, **155**, *509*

Agaricus and Lepiota
Family, *500*

Agrocybe
- ○ Hard, **207**, *557*
- ○ Hemispheric, **47**, *557*
- ○ Maple, **223**, *556*
- ○ Spring, **225**, *558*

Agrocybe
acericola, **223**, *556*
dura, **207**, *557*
pediades, **47**, *557*
praecox, **225**, *558*

Albatrellus
avellaneus, 445
caeruleoporus, 457, 441
confluens, 444
cristatus, 454, 442
ellisii, 420, 456, 442
flettii, 462, 443
hirtus, 445
ovinus, 460, 444
peckianus, 442, 443
pescaprae, 464, 445
sublividus, 445
subrubescens, 444
sylvestris, 442, 443, 445

Aleuria aurantia, 603, 349

Aleurodiscus
amorphus, 417
oakesii, 571, 417

Alnicola
melinoides, 30, 611
scolecina, 611

Amanita
○ Booted, 127, 532
○ Citron, 125, 531
○ Cleft-foot, 126, 134, 147, 527
○ Coker's, 166, 531
○ Gemmed, 128, 129, 537
○ Gray-veil, 149, 544
○ Hated, 114, 548
○ Jonquil, 537
○ Orange Spring, 148, 550
○ Powder-cap, 119, 120, 533
○ Rag-veil, 164, 675, 530
○ Russulalike, 537
○ Salmon, 117, 553
○ Smith's, 165, 546
○ Stout stalked, 146, 547
○ Strangulated, 133, 538
○ Volvate, 116, 552
○ Western Woodland, 167, 546

Amanita
aestivalis, 528
aspera, 145, 526

bisporigera, 551
brunnescens, 126, 134, 147, 527
brunnescens var. *pallida,* 528, 531
caesarea, 142, 681, 528
calyptrata, 529
calyptroderma, 138, 671, 529
cinereopannosa, 164, 675, 530
citrina, 125, 531
cokeri, 166, 531
cothurnata, 127, 532
farinosa, 119, 120, 533
flavoconia, 136, 139, 534
flavorubescens, 135, 678, 535
frostiana, 535
fulva, 115, 674, 536
gemmata, 128, 129, 537
inaurata, 133, 538
junquillea, 537
muscaria var. *flavivolvata,* 539
muscaria var. *formosa,* 137, 677, 539
muscaria var. *muscaria,* 143, 144, 680, 538
ocreata, 529, 551
pachycolea, 112, 549
pantherina, 130, 131, 679, 541
parcivolvata, 140, 542
peckiana, 552
phalloides, 113, 673, 543
porphyria, 149, 544
rubescens, 132, 545
russuloides, 537
silvicola, 167, 546
smithiana, 165, 546
spissa, 146, 547
spreta, 114, 548
umbonata, 528
umbrinolutea, 112, 549
vaginata var. *alba,* 550
vaginata var. *livida,* 550
vaginata var. *vaginata,* 112, 549

velosa, 148, *550*
velutipes, 541
verna, 551, 553
virosa, 123, 124, 672, *551*
volvata, 116, *552*
wellsii, 117, *553*

Amanitaceae, *525*

Amanita Family, *525*

○ **Amanita Mold,** 699, *372*

Amanitopsis
parcivolvata, 542
strangulata, 538
vaginata, 549
vaginata var. *fulva, 536*
velosa, 550
volvata, 553

Amylocystis lapponicus, 492

Anellaria separata, 603

○ **Angel's Wings,** 493, *792*

Anthopeziza floccosa, 343

Anthracobia melaloma, 569, *352*

Anthurus borealis, 833

○ **Antlers, Carbon,** 738, *375*

Aphyllophorales, *386*

Apiocrea chrysosperma, 372

Arachniaceae, *817*

Arachnion album, 646, *817*

Arcyria
denudata, 555, *850*
incarnata, 850

Armillaria
○ Fetid, 210, *735*
○ Fragrant, 209, *732*
○ Pungent, *734*
○ Scaly Yellow, 205, *734*
○ Shaggy-stalked, 196, *731*

Armillaria
albolanaripes, 196, *731*

aurantia, 799
caligata, 209, *732*
dryina, 793
imperialis, 742
luteovirens, 735
mellea, 736
ponderosa, 213, *733*
straminea var. *americana,*
205, *734*
umbonata, 525
ventricosa, 743
viscidipes, 734
zelleri, 210, *735*

Armillariella
mellea, 182, 206, *736*
tabescens, 272, 313, *737*

Ascobolaceae, *326, 346*

Ascobolus carbonarius, 570, *346*

Ascocoryne sarcoides, 616, *362*

Ascomycetes, *323*

Ascomycotina, *323*

Asterophora
lycoperdoides, 12, *738*
parasitica, 738

Astraeaceae, *837*

Astraeus
hygrometricus, 638, *837*
pteridis, 837

Astrogastraceae, *812*

Aurantioporus croceus, 465

Auricularia
auricula, 617, *380*
polytricha, 380

Auriculariaceae, *380*

Auriculariales, *380*

Auriscalpiaceae, *427*

Auriscalpium vulgare, 373, *426*

B

Baeospora
○ Conifer-cone, 9, *738*
○ Lavender, 105, *739*

Baeospora
myosura, 9, *738*
myriadophylla, 105, *739*

Basidiomycetes, *377*

Basidiomycotina, *377*

Battarrea phalloides, 731, *841*

Bird's Nest
○ Common Gel, 631, *829*
○ White-egg, 633, *828*

Bird's-nest Fungi, *827*

Bisporella citrina, 568, *362*

Bjerkandera
adusta, 582, *445*
fumosa, 446

○ **Bladder Stalks, 724,** *411*

○ **Blewit, 346,** *749*

○ **Blue Legs, Stuntz's, 84, 95,** *725*

○ **Blusher, 132,** *545*
○ Yellow, 135, 678, *535*

Bolbitiaceae, *555*

○ **Bolbitius, Yellow, 51,** *559*

Bolbitius Family, *555*

Bolbitius vitellinus, 51, *559*

Boletaceae, *562, 649, 668*

Bolete
○ Admirable, 396, *569*
○ Ash-tree, 452, *564*
○ Bay, 395, *565*
○ Bitter, 382, *593*
○ Black Velvet, 377, *590*
○ Bluing, 402, *576*
○ Burnt-orange, 397, *591*

○ Chestnut, 384, *575*
○ Chrome-footed, 410, *591*
○ Dark, 378, *595*
○ Frost's, 408, *568*
○ Gilled, 324, 326, *672*
○ Graceful, 385, *594*
○ King, 405, *568*
○ Lilac-brown, 380, *592*
○ Ornate-stalked, 418, *570*
○ Parasitic, 370, *571*
○ Peppery, 398, *571*
○ Pitch Pine, *589*
○ Powdery Sulfur, 417, *579*
○ Red-cracked, 413, *567*
○ Red-mouth, 409, *572*
○ Rosy Larch, 404, *574*
○ Russell's, 403, *563*
○ Shaggy-stalked, 407, *563*
○ Spotted, 383, *565*
○ Two-colored, 412, *566*
○ Violet-gray, 381, *594*
○ Yellow-cracked, 399, *572*
○ Zeller's, 414, *573*

Bolete Family, *562*

Boletellus
betula, 407, *563*
chrysenteroides, 567
russellii, 403, *563*

Boletes, Agarics and, *499*

Boletinellus merulioides, 452, *564*

Boletinus, 586
porosus, 565

Boletopsidaceae, *440*

Boletopsis
○ Gray, *447*
○ White-black, *447*

Boletopsis, 440
grisea, 447
leucomelaena, 447
subsquamosa, 458, *446*

Boletus
affinis var. *maculosus,* 383, *565*

badius, 395, 565
bicolor, 412, 566
chrysenteron, 413, 567
edulis, 405, 568
frostii, 408, 568
griseus, 570
luridus, 573
mirabilis, 396, 569
ornatipes, 418, 570
parasiticus, 370, 571
piperatus, 398, 571
projectellus, 570
retipes, 570
sensibilis, 567
subtomentosus, 399, 572
subvelutipes, 409, 572
truncatus, 567
variipes, 568
zelleri, 414, 573

Bondarzewia, 440
berkeleyi, 476, 477, 447
montana, 463, 448

Bondarzewiaceae, 440

Bovista
pila, 641, 820
plumbea, 820

Brain, Red Tree, 565, 423

Brefeldia maxima, 594, 596, 852

Brick Tops, 36, 710

Bulgaria inquinans, 610, 363

C

Calbovista subsculpta, 650, 820

Calocera
cornea, 381
viscosa, 735, 380

Calocybe, Pink, 280, 740

Calocybe
carnea, 280, 740
onychina, 740

persicolor, 740

Caloporus dichrous, 590, 449

Caloscypha fulgens, 602, 349

Calostoma
cinnabarina, 700, 841
lutescens, 841
ravenelii, 841

Calostomataceae, 841

Calvatia
booniana, 648, 821
bovista, 822, 823
craniformis, 656, 822
cyathiformis, 653, 822
fragilis, 823
gigantea, 647, 823
maxima, 823
rubroflava, 655, 824
sculpta, 821
subcretacea, 821

Calycella citrina, 363

Camarophyllus pratensis, 665

Cantharellaceae, 387

Cantharellula umbonata, 97, 741

Cantharellus
cibarius, 427, 387
cinnabarinus, 421, 388
ignicolor, 307, 389
infundibuliformis, 393
lateritius, 428, 390
luteocomus, 391, 394
lutescens, 394
minor, 57, 391
morgani, 670
subalbidus, 432, 392
tubaeformis, 441, 392
umbonatus, 741
xanthopus, 435, 393

Cap
◯ Death, 113, 673, 543
◯ Liberty, 85, 723

○ Mica, 42, 600
○ Powder, 12, 738

○ **Carbon Balls**, 668, 374

○ **Carbon Cushion**, 669, 375

Cat
○ Imperial, 211, 741
○ Swollen-stalked, 212, 742

Catathelasma
imperialis, 211, 741
macrospora, 742
singeri, 742
ventricosa, 212, 742

Cauliflower Mushroom
○ Eastern, 756, 411
○ Rooting, 755, 412

Caulorhiza
hygrophoroides, 743
umbonata, 73, 743

Cenagium, 417

Cenangium furfuraceum, 366

Ceratiomyxa, 844
fruticulosa, 740, 845
fruticulosa var. *flexuosa*, 845
fruticulosa var. *poroides*, 845
morchella, 845

Ceratiomyxaceae, 845

Cerrena unicolor, 540, 450

○ Chanterelle, 308, 427, 387
○ Cinnabar-red, 421, 388
○ Clustered Blue, 444, 397
○ False, 311, 669
○ Flame-colored, 307, 389
○ Fragrant, 425, 395
○ Scaly Vase, 430, 431, 396
○ Small, 57, 391
○ Smooth, 428, 390
○ Trumpet, 441, 392
○ White, 432, 392
○ Yellow-footed, 435, 393

Chanterelle Family, 387

Cheilymenia, 353

Cheimonophyllum candidissimus, 496, 743

Chlorociboria aeruginascens, 598, 361

Chlorophyllum molybdites, 169, 170, 509

Chondrostereum purpureum, 539, 418

Christiansenia mycetophila, 564, 418

Chromocrea gelatinosa, 371

Chroogomphus
○ Brownish, 220, 650
○ Wine-cap, 219, 651
○ Woolly, 221, 650

Chroogomphus
leptocystis, 651
pseudovinicolor, 651
rutilus, 220, 650
tomentosus, 221, 650
vinicolor, 219, 651

○ Ciboria, Common Wood, 361

Ciboria peckiana, 361

Clathraceae, 832

Clathrus columnatus, 694, 832

Claudopus nidulans, 792

Clavaria
amethystina, 402
botrytis, 407
cristata, 403
formosa, 408
fusiformis, 359, 400
kunzei, 410
ligula, 403
mucida, 406
pistillaris, 404
purpurea, 737, 400
pyxidata, 401
stricta, 410
truncata, 405

vermicularis, **739,** *400*
zollingeri, 402

Clavariaceae, *398*

Clavariadelphus
borealis, 404
ligula, **689,** *403*
mucronatus, 404
pistillaris, **688,** *403*
sachalinensis, 403
truncatus, **439,** *404*

Clavicipitaceae, *368*

Clavicorona
avellanea, 401
pyxidata, **744,** *401*

Clavulina
amethystina, **752,** *402*
cinerea, 402, 403
cristata, **743,** *402*
rugosa, 403

Clavulinopsis
aurantiocinnabarina, 399
fusiformis, **734,** *399*
helvola, 399
laeticolor, 399
pulchra, 399

Climacodon septentrionale,
520, *427*

Clitocybe
○ Anise-scented, **350,** *750*
○ Cloudy, **267, 354,** *749*
○ Crowded White, **238,** *746*
○ Fat-footed, **320,** *745*
○ Fragrant, **237,** *751*
○ Funnel, **263,** *747*
○ Giant, **264,** *748*
○ Smoky-brown, **279,** *744*
○ Wood, **262,** *747*

Clitocybe
atrialba, 752
avellaneialba, **279,** *744*
candida, 748
cerussata var. *difformis, 746*
clavipes, **320,** *745*
dealbata var. *sudorifica,*
241, 245, *745*

decora, 807
dilatata, **238,** *746*
ectypoides, **262,** *747*
gibba, **263,** *747*
gibba var. *maxima, 748*
gigantea, **264,** *748*
graveolens, 750
illudens, 788
infundibuliformis, 748
irina, 750
maxima, 748
monadelpha, 737
morgani, 670
multiceps, 769
nebularis, **267, 354,** *749*
nuda, **346,** *749*
odora, **350,** *750*
odora var. *pacifica, 751*
robusta, 748
saeva, 750
socealis, 740
squamulosa, 748
suaveolens, **237,** *751*
tarda, 750

○ Clitocybula, Black-and-
white, **352,** *751*

Clitocybula
atrialba, **352,** *751*
familia, **22,** *752*

Clitopilus
abortivus, 642
noveboracensis, 649
orcellus, 641
prunulus, **242,** *641*

Club
○ Green-headed Jelly, **727,**
365
○ Ochre Jelly, **726,** *364*
○ Water, **729,** *366*

○ Coccora, **138, 671,** *529*
○ Green-capped, *529*

Collybia
○ Clustered, **23,** *753*
○ Douglas-fir, **8,** *797*
○ Family, **22,** *752*
○ Golden-scruffy, **54,** *758*

○ Hairy-stalked, 79, 756
○ Little Brown, 108, 753
○ Oak-loving, 80, 755
○ Rooting Redwood, 73, 743
○ Spotted, 35, 756
○ Tuberous, 11, 757
○ Tufted, 24, 754
○ Violet, 106, 755
○ Zoned-cap, 34, 758

Collybia
acervata, 23, 753
albiflavida, 777
alkalivirens, 108, 753
butyracea, 755, 776
cirrhata, 757
confluens, 24, 754
conigena, 739
conigenoides, 7, 797
cookei, 757
dryophila, 80, 755
familia, 753
hygrophoroides, 789
iocephala, 106, 755
lacunosa, 759
maculata, 35, 756
myriadophylla, 740
platyphylla, 808
radicata, 789
semihirtipes, 757
spongiosa, 79, 756
strictipes, 776
subnuda, 754, 757
trullisata, 798
tuberosa, 11, 757
umbonata, 743
velutipes, 759
zonata, 758

Coltricia, 438, 440, 472
cinnamomea, 445, 450
montagnei var. *greenei*, 450, 452
montagnei var. *montagnei*, 449, 451
perennis, 451, 452

Coniophora
cerebella, 415

puteana, 592, 415

Coniophoraceae, 414

○ Conk, Artist's, 518, 460

Conocybe
○ Bog, 74, 561
○ Deadly, 81, 559

Conocybe
crispa, 560
filaris, 81, 559
lactea, 5, 560
smithii, 74, 561
tenera, 75, 561

Coprinaceae, 596

Coprinus
○ Non-inky, 4, 598
○ Orange-mat, 41, 602
○ Woolly-stalked, 705, 599

Coprinus
atramentarius, 19, 596
comatus, 704, 597
disseminatus, 4, 598
insignis, 597
lagopides, 599
lagopus, 705, 599
laniger, 602
micaceus, 42, 600
plicatilis, 3, 600
quadrifidus, 703, 601
radians, 41, 602

Coral
○ Clustered, 749, 407
○ Cotton-base, 748, 405
○ Crested, 743, 402
○ Crown-tipped, 744, 401
○ Flat-topped, 439, 404
○ Gray, 403
○ Jellied False, 741, 385
○ Jelly-based, 745, 408
○ Light Red, 750, 406
○ Pestle-shaped, 688, 403
○ Purple Club, 737, 400
○ Spindle-shaped Yellow, 734, 399
○ Straight-branched, 733, 409

) Strap-shaped, **689**, *403*
) Violet-branched, **752**, *402*
) White, **742**, *410*
) White False, *386*
) White Green-algae, **690**, *406*
) White Worm, **739**, *400*
) Yellow-tipped, **746**, *408*

Coral and Pore Fungi and Allies, *386*

Coral Fungus Family, *398*

Cordyceps
) Beetle, **691**, *369*
) Goldenthread, **683**, *370*
) Headlike, **728**, *368*
) Trooping, **687**, *369*

Cordyceps
canadensis, 368
capitata, **728**, *368*
melolonthae, **691**, *369*
militaris, **687**, *369*
ophioglossoides, **683**, *370*

Coriolus versicolor, 489

) **Corpse Finder, 278,** *625*

Cort
) Blood-red, **333**, *618*
) Bracelet, **331**, *612*
) Bulbous, **347**, *616*
) Cinnabar, **332**, *613*
) Deadly, **298**, *615*
) Poznan, *615*
) Pungent, **341**, *619*
) Red-gilled, **300**, *618*
) Ringed, *613*
) Saffron-colored, **299**, *614*
) Silvery-violet, **342**, *611*
) Slimy-banded, **276**, *614*
) Variable, **305**, *617*
) Violet, **340**, *620*
) Viscid Violet, **344**, *617*

Corticiaceae, *416*

Corticium
bombycinum, **576**, *419*
caeruleum, 420

Cortinariaceae, *610*

Cortinarius
alboviolaceus, **342**, *611*
armillatus, **331**, *612*
camphoratus, 619
cinnabarinus, **332**, *613*
cinnamomeus, 615
collinitus, **276**, *614*
croceofolius, **299**, *614*
fragrans, 619
gentilis, **298**, *615*
glaucopus, **347**, *616*
haematochelis, 613
heliotropicus, 617
iodeoides, 617
iodes, **344**, *617*
multiformis, **305**, *617*
obliquus, 612
orellanus, 615
pseudoarquatus, 616
pyriodorus, 619
sanguineus, **333**, *618*
semisanguineus, **300**, *618*
traganus, **341**, *619*
vanduzerensis, 614
violaceus, **340**, *620*

Cortinarius Family, *610*

Cotylidia diaphana, **436**, *498*

Craterellus
cantharellus, 391
cinereus, 395
confluens, 395
cornucopioides, 395
fallax, **443**, *394*
odoratus, **425**, *395*

Creopus gelatinosa, 371

Crep
○ Flat, **494**, *636*
○ Jelly, **504**, *637*

Crepidotaceae, *636*

Crepidotus
applanatus, **494**, *636*
fulvotomentosus, 638
malachius, 637

mollis, **504,** 637

Crepidotus Family, 636

Cribraria, 849

Cribrariaceae, 849

Crinipellis
campanella, 758
stipitaria, 758
zonata, 34, 758

○ Crown, Pink, **640,** *346*

Crucibulum
laeve, **633,** 828
vulgare, 828

Crust
○ Brown-toothed, **577,** *438*
○ Buff, **576,** *419*
○ Reddish-brown, **578,** *439*

Crust Fungus Family, *416*

Cryptochaete rufa, 423

Cryptoporus volvatus, **528,** 452

○ Cudonia, Yellow, **725,** *363*

Cudonia
circinans, 364
lutea, **725,** 363
monticola, 364

Cup
○ Acorn, *363*
○ Black Rubber, **611,** *342*
○ Bladder, **601,** *348*
○ Blueberry, **624,** *360*
○ Blue-staining, **602,** *349*
○ Brown-haired White, **629,** *351*
○ Burn Site Ochre, **569,** *352*
○ Burn Site Shield, **570,** *346*
○ Common Brown, **625,** *347*
○ Crustlike, **619,** *331*
○ Elf, **626,** *353*
○ Eyelash, **604,** *353*
○ Hairy Black, **612,** *341*
○ Hairy Rubber, **614,** *339*

○ Long-stalked Gray, **623,** *335*
○ Pink Burn, **608,** *354*
○ Pixie, **627,** *350*
○ Recurved, **620,** *347*
○ Ribbed-stalked, **622,** *332*
○ Scarlet, **605,** *343*
○ Scurfy Alder, **575,** *365*
○ Shaggy Scarlet, **607,** *343*
○ Stalked Hairy Fairy, **630,** *361*
○ Stalked Scarlet, **606,** *344*
○ Veined, **621,** *330*

Cup Fungi and Allies, *325*

Cups
○ Splash, **632,** 828
○ Yellow Fairy, **568,** *362*

Cyathus striatus, **632,** 828

Cyclomyces greenei, 452

Cyptotrama asprata, 54, 758

Cystoderma
○ Common Conifer, **191,** *511*
○ Pungent, **194,** *510*
○ Tuberous, **195,** *512*

Cystoderma
amianthinum var.
amianthinum, 511
amianthinum var.
rugusoreticulatum, **194,** *510*
echinatum, 523
fallax, **191,** *511*
granosum, 512
paradoxum, 512

D

Dacrymyces palmatus, **567,** *381*

Dacrymycetaceae, *380*

Dacrymycetales, *380*

Daedalea
confragosa, 454

quercina, 467, 453
unicolor, 450

Daedaleopsis confragosa,
481, 454

Daldinia concentrica, 668,
374

Dasyscyphus virgineus, 630,
361

◯ Death Cap, 113, 673,
543

Dendrothele candida, 417

Dentinum
repandum, 455, 428
umbilicatum, 428

Dermatiaceae, 361

Dermocybe
cinnabarina, 613
croceofolia, 615
sanguinea, 618
semisanguinea, 619

Desert Inky Cap Fungus,
814

◯ Destroying Angel, 123,
124, 672, 551

Diachea leucopodia, 552,
853

Dictydium
cancellatum, 732, 849
mirabile, 849

Dictyophora duplicata, 707,
834

◯ Disc, Hophornbeam,
571, 417

Disc Fungi, 325

Discina
perlata, 618, 331, 396
venosa, 330

Disciotis venosa, 621, 330

Discomycetes, 325

Dissoderma paradoxum, 195,
512

Dry Rot Family, 414

Dunce Cap
◯ Brown, 75, 561
◯ White, 5, 560

E

Earthstar
◯ Arched, 639, 817
◯ Barometer, 638, 837
◯ Beaked, 634, 818
◯ Collared, 635, 819
◯ Rounded, 636, 818
◯ Saltshaker, 637, 837

Earth Tongue(s), 356
◯ Irregular, 686, 359
◯ Orange, 685, 358
◯ Velvet, 682, 357

Echinodontiaceae, 429

Echinodontium tinctorium,
532, 429

Elaphomycetaceae, 355

Encoelia furfuracea, 575,
365

Endoptychum
agaricoides, 644, 811
depressum, 812

Entoloma
◯ Aborted, 253, 670, 641
◯ Blue-toothed, 101, 644
◯ Early Spring, 282, 646
◯ Hairy-stalked, 100, 648
◯ Midnight-blue, 369, 642
◯ Salmon Unicorn, 49, 644
◯ Straight-stalked, 274, 646
◯ Violet, 368, 647
◯ Yellow Unicorn, 50, 643

Entoloma
abortivum, 253, 670, 641
cuspidatum, 643
lividum, 645

madidum, 369, *642*
murraii, 50, *643*
porphyrophaeum, 648
salmoneum, 49, *644*
serrulatum, 101, *644*
sinuatum, 275, 367, *645*
strictius, 274, *646*
vernum, 282, *646*
violaceum, 368, *647*

Entoloma Family, *640*

Entolomataceae, *640*

Exidia
alba, 383
glandulosa, 563, *382*
nucleata, 383
recisa, 383

Exposed Hymenium
Fungi, *379*

F

○ Fairy Fan, Velvet, 693, *359*

○ False Chanterelle, 311, *669*

False Coral
○ Jellied, 741, *385*
○ White, *386*

False Morel
○ Conifer, 714, *336*
○ Gabled, 716, *335*
○ Saddle-shaped, 719, 720, *339*
○ Snowbank, 715, 717, *338*
○ Thick-stalked, 718, *337*

False Puffball(s), *836*
○ Dye-maker's, 698, *838*

○ **False Russula, American,** 662, *812*

○ **False Truffle, Fuzzy,** 643, *350*

○ **False Turkey-tail,** 536, *497*

Favolus
alveolaris, 508, *455*
canadensis, 455
europaeus, 455

Fayodia maura, 787

Fiber Head
○ Black-nipple, 316, *629*
○ Blushing, 14, *632*
○ Caesar's, 314, *627*
○ Green-foot, 269, *627*
○ Lilac, 13, *631*
○ Pungent, 317, *632*
○ Scaly, 183, *633*
○ Straw-colored, 318, *628*
○ Torn, 315, *630*
○ White, 16, *629*
○ White-disc, 15, *626*
○ Woolly, 271, *630*

○ **Fingers, Dead Man's,** 697, *376*

Fistulina, 440
hepatica, 513, *455*
pallida, 457

Fistulinaceae, *440*

Flammula penetrans, 623

Flammulina velutipes, 63, *759*

Flask Fungi, *367*

Flasks, Ostiole, *367*

Fomes
annosus, 466
connatus, 473
fomentarius, 527, *457*
igniarius, 476
laricis, 459
officinalis, 459
pini var. *abietis,* 474
pinicola, 459
rimosus, 477
subroseus, 458

Fomitopsis
cajanderi, 524, *458*
officinalis, 529, *458*

pinicola, 531, 459
rosea, 458

Fuligo, Ashy, 846

Fuligo
cinerea, 846
septica, 559, 845

Fungi
Bird's-nest, 827
Coral, 386
Cup and Allies, 325
Exposed Hymenium, 379
Flask, 367
Jelly, 379
Plated Puffball, 815
Pore, 386
Stomach, 810

Fungus
Cannon, 551, 830
Fluted-stalked, 696, 332
Indian Paint, 532, 429
Silver Leaf, 539, 418
Stalked Cauliflower, 753, 345

Fungus Family
Coral, 398
Crust, 416
Parchment, 494
Tooth, 426

Fuscoboletinus
ochraceoroseus, 404, 574
paluster, 575, 583

Fuzzy Foot, 58, 809

G

Galerina
Deadly, 228, 620
Deadly Lawn, 39, 622
Sphagnum-bog, 43, 621

Galerina
autumnalis, 228, 620
marginata, 621, 623, 716
tibicystis, 43, 621
venenata, 39, 622

Galiella rufa, 614, 339

Ganoderma, 440
applanatum, 518, 460
curtisii, 461, 462
lucidum, 514, 460
oregonense, 461, 462
tsugae, 515, 461

Ganodermataceae, 440

Gasteromycetes, 810, 836

Gautieriaceae, 817

Gautieriales, 815

Gautieria morchelliformis, 657, 817

Geastraceae, 817

Geastrum
coronatum, 818
fimbriatum, 819
fornicatum, 639, 817
pectinatum, 634, 818
saccatum, 636, 818
triplex, 635, 819

Geoglossaceae, 357

Geoglossum, 357, 358

Geopora
cooperi, 643, 350
harknessi, 350

Geopyxis
carbonaria, 627, 350
cupularis, 354
vulcanalis, 351, 354

Gerronema, Golden-gilled, 60, 760

Gerronema
chrysophylla, 60, 760
ericetorum, 787
fibula, 796
luteicolor, 760

Gill
Common Split, 487, 493
Crimped, 472, 493

Gilled Puffballs, 811

Globifomes graveolens, **544,** *462*

Gloeophyllum
sepiarium, **516,** *463*
trabeum, *463*

○ Goat's Foot, **464,** *445*

○ Goblet, Rose, **628,** *340*

Gomphidiaceae, *649*

Gomphidius
○ Clustered, **216,** *652*
○ Rosy, **217,** *653*
○ Slimy, **218,** *652*

Gomphidius
glutinosus, **218,** *652*
largus, *652*
oregonensis, **216,** *652*
rutilus, *650*
subroseus, **217,** *653*
tomentosus, *651*
vinicolor, *651*

Gomphidius Family, *649*

○ **Gomphus, Pig's Ear,**
440, *396*

Gomphus
bonari, *397*
clavatus, **440,** *396*
floccosus, **430, 431,** *396*
kauffmanii, *397*

○ Grayling, **97,** *741*

Grifola frondosa, **474, 475,**
463

○ Grisette, **112,** *549*
○ Tawny, **115, 674,** *536*

Guepiniopsis alpinus, **599,**
382

Gym
○ Big Laughing, **214,** *623*
○ Little, **301,** *623*

Gymnopilus
penetrans, **301,** *623*
sapineus, *623*

spectabilis, **214,** *623*

○ Gypsy, **201,** *635*

Gyrodon merulioides, *565*

Gyromitra, *331*
ambigua, *339*
brunnea, **716,** *335*
caroliniana, *336*
esculenta, **714,** *336*
fastigiata, **718,** *337*
gigas, **715, 717,** *338*
infula, **719, 720,** *339*
korfii, *338*

○ Gyroporus, Red, **411,**
576

Gyroporus
castaneus, **384,** *575*
cyanescens, **402,** *576*
purpurinus, **411,** *576*

H

Haematostereum
gausapatum, *495*
sanguinolentum, **541,** *494*

Hapalopilus nidulans, **510,**
464

Hebeloma
crustuliniforme, **259,** *624*
syriense, **278,** *625*

Heliomyces nigripes, *771*

Helotiales, *356*

Helotium citrinum, *363*

Helvella
○ Fluted Black, **721,** *334*
○ Fluted White, **722,** *333*
○ Smooth-stalked, **723,** *334*

Helvella
acetabulum, **622,** *332*
crispa, **722,** *333*
elastica, **723,** *334*
esculenta, *337*
griseoalba, *333*
infula, *339*

lacunosa, 721, *334*
macropus, 623, *335*
stevensii, *334*
underwoodii, *336*

Helvellaceae, *326, 331*

Hemitrichia
clavata, 553, 557, *850*
serpula, 560, *851*
stipitata, *851*

Hen of the Woods, 474, 475, *463*

Hericium
abietis, *430*
coralloides, 548, *429*
erinaceus, 547, *430*
ramosum, 549, *431*

Heterobasidion annosum, 519, *465*

Hirschioporus pargamenus, *490*

Hohenbuehelia petaloides, *433, 760*

Humaria hemisphaerica, 629, *351*

Humidicutis marginata, *663*

Hyaloscyphaceae, *361*

Hydnaceae, *426*

Hydnellum
aurantiacum, 479, *431*
caeruleum, 469, *432*
diabolus, *433*
peckii, 470, *433*
pineticola, *434*
spongiosipes, 446, *434*
suaveolens, *433*

Hydnochaete olivaceum, 577, *438*

Hydnoporia fuscescens, *438*

Hydnum, *436*
auriscalpium, *427*
chryscomum, *424*

imbricatum, 466, *434*
repandum, *428*
scabrosum, *435*
septentrionale, *428*

Hygrocybe
cantharellus, 657
coccinea, 658
conica, 659
flavescens, 660
marginata, 663
miniata, 664
psittacina, 666

Hygrophoraceae, *654*

Hygrophoropsis, Fragrant, *434*, 669

Hygrophoropsis
aurantiaca, 311, 669
olida, *434*, 669

Hygrophorus
acutoconicus, 659
agathosmus, 360, *654*
amarus, 667
bakerensis, 273, *655*
calophyllus, 656
camarophyllus, 365, *656*
cantharellus, 56, *656*
chlorophanus, 48, *660*
chrysodon, 260, *657*
coccineus, 65, *658*
conicus, 71, 72, *658*
eburneus, 236, *659*
erubescens, *667*
flavescens, 48, *660*
flavodiscus, 261, *660*
fuligineus, *661*
gliocyclus, *661*
hypothejus, 304, *661*
inocybiformis, 366, *662*
marginatus, 53, *662*
miniatus, 69, *663*
olivaceoalbus, 199, *664*
pratensis, 309, *664*
psittacinus, 103, *665*
pudorinus, 246, *666*
puniceus, 658
purpurescens, *667*
ruber, *663*

russula, 345, 667
speciosus, 325, 667
speciosus var. *kauffmanii*,
668

Hygrophorus Family, 654

Hymenochaetaceae, 437

Hymenochaete
badio-ferruginea, **578**, 439
tabacina, 439

Hymenochaete Family,
437

Hymenogastraceae, 811

Hymenogastrales, 811

Hymenomycetes, 379

Hymenoscyphus fructigenus,
363

○ Hypocrea, Yellow
Cushion, 663, 371

Hypocrea
citrina, 371
gelatinosa, 663, 371
sulphurea, 371

Hypocreaceae, 371

Hypomyces
○ Golden, 400, 371
○ Yellow-green, 451, 373

Hypomyces
chrysospermus, 400, 371
hyalinus, 699, 372
lactifluorum, 429, 373
luteovirens, 451, 373

○ Hypoxylon, Red
Cushion, 667, 374

Hypoxylon
argillaceum, 375
fragiforme, 667, 374
fuscum, 375

Hypsizygus tessulatus, 492,
761

Hysterangiaceae, 834

I

Inky
○ Alcohol, 19, 596
○ Japanese Umbrella, 3, 600

Inky Cap
○ Desert, 702, 814
○ Scaly, 703, 601

Inky Cap Family, 596

Inocybe
albodisca, 15, 626
caesariata, 314, 627
calamistrata, **269**, 627
fastigiata, 318, 628
fastigiella, 628
fuscodisca, 316, 629
geophylla, 16, 629
geophylla var. *lilacina*, 632
hirsuta var. *maxima*, 628
lacera, 315, 630
lanuginosa, **271**, 630
lilacina, 13, 631
pudica, 14, 632
sororia, 317, 632
terrigena, 183, 633
umbratica, 626, 772
violacea, 632

Inonotus, 438, 440
dryadeus, 534, 466
obliquus, 597, 467
radiatus, 475

Irpex
cinnamomeus, 438
lacteus, 467
mollis, 488

Ischnoderma
benzoinum, 468
resinosum, 525, 468

J

○ Jack O'Lantern, 310,
426, 483, 787

Jelly
○ Apricot, 422, 383
○ Collybia, 564, 418
○ Orange, 567, 381

Jelly Club
⊃ Green-headed, 727, *365*
⊃ Ochre, 726, *364*

⊃ **Jelly Cone, Golden, 599,** *382*

Jelly Drops
⊃ Black, 610, *363*
⊃ Purple, 616, *362*

Jelly Fungi, *379*

⊃ Jelly Leaf, 754, *384*
⊃ Yellow, *384*

Jelly Roll
⊃ Amber, *383*
⊃ Black, 563, *382*
⊃ Granular, *383*
⊃ Pale, *383*

⊃ Jelly Tooth, 459, *383*

K

⊃ Knot, Sweet, 544, *462*

Kuehneromyces mutabilis,
716

⊃ Kurotake, 458, *446*

L

Laccaria
⊃ Common, 335, *762*
⊃ Purple-gilled, 254, 336, *762*
⊃ Sandy, 255, *763*

Laccaria
amethystina, 740, *763*
laccata, 335, *762*
ochropurpurea, 254, 336, *762*
trullisata, 255, *763*

⊃ **Lactarius, Slimy,** 364, *688*

Lactarius, 449
aquifluus, 256, *680*
atroviridis, 693

camphoratus, 295, *681*
chrysorheus, 696
controversus, 247, *681*
corrugis, 293, *682*
croceus, 696
deceptivus, 248, *683*
delíciosus, 312, *683*
fallax, 685, 687
fragilis, 681
fumosus, 687
gerardii, 362, *684*
gerardii var. *fagicola,* 685
gerardii var. *subrubescens,* 685
hygrophoroides, 290, *685*
indigo, 337, *686*
kauffmanii, 688
lignyotus, 363, *686*
luteolus, 250, *687*
maculatus, 695
mucidus, 364, *688*
paradoxus, 349, *688*
peckii, 292, *689*
piperatus, 240, *690*
piperatus var. *glaucescens,* 690
pseudomucidus, 688
pubescens, 694
representaneus, 284, *690*
rufulus, 681, *692*
rufus, 289, *691*
scrobiculatus, 285, *692*
sordidus, 361, *693*
speciosus, 691
subpurpureus, 244, *693*
thyinos, 684
torminosus, 251, *694*
torminosus var. *nordmanensis,* 694
uvidus, 338, *695*
vellereus, 683, 690, 698
vinaceorufescens, 288, *695*
volemus, 291, *696*

Laetiporus
sulphureus, 478, 511, *468*
sulphureus var. *semialbinus,* 469

Laricifomes officinalis, 459

Laxitextum bicolor, 573, 420

○ Lead Poisoner, 275, 367, 645

Leccinum
atrostipitatum, 577
aurantiacum, 388, 577
chromapes, 592
insigne, 389, 578
scabrum, 390, 578

Lentaria byssiseda, 748, 405

○ Lentinellus, Stalked, 438, 764

Lentinellus
cochleatus, 423, 764
montanus, 765
omphalodes, 438, 764
ursinus, 502, 765
vulpinus, 765

Lentinus
○ Bear, 502, 765
○ Cockle-shell, 423, 764
○ Large, 506, 766
○ Scaly, 222, 766

Lentinus
cochleatus, 764
lepideus, 222, 766
ponderosus, 506, 766
ursinus, 765

Lenzites
betulina, 535, 469
saepiaria, 463

Leocarpus fragilis, 554, 846

Leotia
atrovirens, 364, 365
lubrica, 726, 364
viscosa, 727, 365

Leotiaceae, 362

Lepidella, 532, 547

Lepiota
○ Black-disc, 178, 514
○ Deadly, 162, 517
○ Green-spored, 169, 170, 509
○ Lemon-yellow, 180, 518
○ Malodorous, 177, 517
○ Onion-stalked, 179, 515
○ Reddening, 173, 174, 513
○ Red-tinged, 175, 522
○ Shaggy-stalked, 176, 516
○ Sharp-scaled, 163, 513
○ Smooth, 121, 519

Lepiota
acutesquamosa, 163, 513
americana, 173, 174, 513
atrodisca, 178, 514
cepaestipes, 179, 515
clypeolaria, 176, 516
cristata, 177, 517
echinatum, 523
flammeatincta, 514
glioderma, 554
helveola, 518
josserandii, 162, 517
lutea, 180, 518
morgani, 510
naucina, 121, 519
procera, 172, 520
rachodes, 171, 521
rubrotincta, 175, 522
solidipes, 555

Lepiota Family, Agaricus and, 500

Lepista
nebularis, 749
nuda, 750

Leptonia, 641
serrulata, 645

Leptopodia elastica, 334

Leucoagaricus
naucinus, 520
rubrotinctus, 523

Leucocoprinus
birnbaumii, 519
cepaestipes, 516

Leucopax
○ Bitter, 277, 768

○ White, **239**, 767

Leucopaxillus
albissimus, **239**, 767
amarus, **277**, 768
gentianeus, 768
giganteus, 748

Liceales, *844*

Limacella
○ Ringed, **168**, 554
○ Slimy-veil, **141**, 554

Limacella
glioderma, **141**, 554
lenticularis, 555
solidipes, **168**, 554

○ **Ling Chih**, **514**, *460*

○ **Lizard's Claw**, **709**, *832*
○ Small, *833*

Lopharia cinerascens, **591**, *495*

Lycogala
epidendrum, **665**, 848
flavofuscum, 848

Lycoperdaceae, *820*

Lycoperdales, *816*

Lycoperdon
echinatum, **649**, 824
epidendrum, 848
gemmatum, 825
giganteum, 823
marginatum, 825
perlatum, **652**, **676**, 825
pulcherrimum, 825
pyriforme, **664**, 826

Lyophyllum
connatum, 769
decastes, **266**, 768
loricatum, 769
multiceps, 769

Lysurus
borealis, 833
gardneri, **709**, *832*
mokusin, 833

M

Macowanites americanus,
662, *812*

Macrocystidia cucumis, **90**,
769

Macrolepiota
procera, 521
rachodes, 522
rhacodes, 522

Macroscyphus macropus, 335

Macrotyphula fistulosa, 404

Marasmiellus
albuscorticis, **488**, 770
nigripes, **1**, 770

Marasmius
○ Black-footed, **1**, 770
○ Fetid, **28**, 778
○ Fused, **38**, 771
○ Garlic, **37**, 770
○ Orange Pinwheel, **67**, 775
○ Orange-yellow, **32**, 776
○ Pinwheel, **2**, 774
○ Velvet-cap, **40**, 773
○ White, **488**, 770

Marasmius
candidus, 770
capillaris, 774
cohaerens, **38**, 771
cohaerens var. *lachnophyllus,*
771
cystidiosus, 772
foetidus, 778
fulvoferrugineus, 776
iocephalus, 756
magnisporus, 770
olidus, 775
oreades, **17**, **18**, 772
plicatulus, **40**, 773
pulcherripes, 776
rotula, **2**, 774
scorodonius, **37**, 774
siccus, **67**, 775
spongiosus, 757
strictipes, **32**, 776

○ **Matsutake, White,** 213, *733*

Melanoleuca
○ Changeable, 359, *777*
○ Yellowish-white, 235, *776*

Melanoleuca
alboflavida, 235, *776*
melaleuca, 359, *777*

Melanophyllum echinatum, 159, *523*

Meripilus giganteus, 471, *470*

Merulius
○ Coral-pink, 523, *421*
○ Trembling, 586, *421*

Merulius
ambiguus, 422
corium, 422
incarnatus, 523, *421*
lacrimans, 416
tremellosus, 586, *421*

Metatrichia vesparium, 556, *852*

Microglossum rufum, 685, *358*

Micromphale foetidum, 28, *778*

Microstoma floccosa, 607, *343*

Milky
○ Aromatic, 295, *681*
○ Buff Fishy, 250, *687*
○ Burnt-sugar, 256, *680*
○ Chocolate, 363, *686*
○ Common Violet-latex, 338, *695*
○ Corrugated-cap, 293, *682*
○ Deceptive, 248, *683*
○ Dirty, 361, *693*
○ Gerard's, 362, *684*

○ Hygrophorus, 290, *685*
○ Indigo, 337, *686*
○ Northern Bearded, 284, *690*
○ Orange-latex, 312, *683*
○ Peck's, 292, *689*
○ Peppery, 240, *690*
○ Pink-fringed, 251, *694*
○ Red-hot, 289, *691*
○ Silvery-blue, 349, *688*
○ Spotted-stalked, 285, *692*
○ Variegated, 244, *693*
○ Voluminous-latex, 291, *696*
○ Willow, 247, *681*
○ Yellow-latex, 288, *695*

Mitrula
irregularis, 359
paludosa, **684,** *358*
phalloides, 359

○ **Mo-Ehr,** 380

Monilinia vaccinii-corymbosi, 624, *360*

○ **Moose Antlers,** 615, *345*

Mop
○ Decorated, 302, *807*
○ Variegated, 327, *808*
○ Yellow Oyster, 303, *809*

Morchella
angusticeps, 326
bispora, 329
conica, 326
crassipes, 328
deliciosa, 327
elata, 713, *326*
esculenta, 710, *327*
semilibera, 711, *328*

Morchellaceae, *326*

Morel
○ Black, 713, *326*
○ Conical, *326*
○ Conifer False, 714, *336*
○ Gabled False, 716, *335*
○ Half-free, 711, *328*

○ Peck's, *326*
○ Saddle-shaped False, 719, 720, *339*
○ Snowbank False, 715, 717, *338*
○ Thick-footed, *328*
○ Thick-stalked False, 718, *337*
○ White, *327*
○ Yellow, 710, *327*

Mucronoporaceae, 438, 440

Mucronoporus, 472

Multiclavula mucida, 690, *406*

Mushroom
○ American Caesar's, 142, 681, *528*
○ Brown Alder, 30, *611*
○ Chicken, 478, 511, *468*
○ Cucumber-scented, 90, *769*
○ Eastern Cauliflower, 756, *411*
○ Fairy Ring, 17, 18, *772*
○ False Caesar's, 140, *542*
○ Fawn, 231, 232, *675*
○ Fried-chicken, 266, *768*
○ Honey, 182, 206, *736*
○ Horse, 156, *501*
○ Lawn Mower's, 26, *606*
○ Lobster, 294, 429, *373*
○ Magnolia-cone, 7, *797*
○ Meadow, 153, 154, *505*
○ Oyster, 484, 497, *793*
○ Parrot, 103, *665*
○ Platterful, 265, *807*
○ Ringless Honey, 272, 313, *737*
○ Rooting Cauliflower, 755, *412*
○ Straw, *677*
○ Sweating, 241, 245, *745*
○ Sweetbread, 242, *641*

Mutinus
caninus, 835
elegans, 692, *835*

Mycena
○ Bleeding, 76, *781*
○ Blue, 102, *785*
○ Burn Site, 27, *786*
○ Common, 6, *780*
○ Coral Spring, 68, *778*
○ Large, 94, *783*
○ Orange, 64, *781*
○ Pink, 104, *784*
○ Red-orange, 70, *785*
○ Spotted, 98, *783*
○ Walnut, 52, *782*
○ Yellow-stalked, 45, *779*

Mycena
acicula, 68, *778*
adonis, 779
amabilissima, 779
amicta, 786
atkinsoniana, 781
epipterygia, 45, *779*
epipterygioides, 780
fibula, 796
galericulata, 6, *780*
haematopus, 76, *781*
inclinata, 780
leaiana, 64, *781*
lilacifolia, 740
luteopallens, 52, *782*
maculata, 98, *783*
maura, 787
overholtzii, 94, *783*
pura, 104, *784*
sanguinolenta, 781
strobilinoides, 70, *785*
subcaerulea, 102, *785*
viscosa, 780

Mycenastrum corium, 651, *826*

Mycoacia fragilissima, 424

Mycorrhaphium adustum, 542, *435*

Myriostoma coliforme, 637, *837*

Myxomphalia maura, 27, *786*

Myxomycotina, *843*

N

Naematoloma
○ Dispersed, 77, 709
○ Smoky-gilled, 62, 708

Naematoloma
capnoides, 62, 708
dispersum, 77, 709
fasciculare, 61, 709
squamosa, 724
sublateritium, 36, 710
udum, 709

Naucoria, 556, 611
centuncula, 638
cucumis, 770
semiorbicularis, 558

○ **Nectria, Violet**, 846

Nectria violacea, 846

Neolecta irregularis, 686, 359

Neournula
nordmanensis, 340
pouchetii, 628, 340

Nidula candida, 631, 829

○ **Nidularia, Pea-shaped,**
658, 830

Nidulariaceae, 828

Nidulariales, 827

Nidularia pulvinata, 658, 830

Nolanea, 641
murraii, 643
nodospora, 648
salmonea, 644
strictior, 646
verna, 647

Nyctalis asterophora, 738

O

○ **Old Man of the Woods,**
379, 580

Omphalia fibula, 796

Omphalina
campanella, 810
chrysophylla, 760
ericetorum, 10, 787
fibula, 796
maura, 787

Omphalotus
illudens, 788
olearius, 310, 426, 483, 787
olivascens, 788

Onnia, 438, 440
circinata, 452, 471
tomentosa, 448, 471

○ **Orange Peel**, 603, 349

Osteina obducta, 461, 472

Ostiole Flasks, 367

Ostropales, 366

Otidea
alutacea, 352
leporina, **600**, 351
onotica, 352
smithii, 352

○ **Oudemansiella, Rooted,**
268, 788

Oudemansiella
longipes, 789
platyphylla, 808
radicata, **268**, 788

Oxydontia fragilissima, 424

Oxyporus populinus, 521, 472

Oyster
○ Black Jelly, 486, 795
○ Cornucopia, 794
○ Elm, 492, 761
○ Late Fall, 498, 789
○ Leaflike, 433, 760
○ Orange Mock, 500, 791
○ Veiled, 495, 792

○ **Oysterette, White**, 496, 743

P

Pachyella clypeata, 348

Panaeolus
- ○ Bell-cap, 87, *602*
- ○ Girdled, 91, *604*
- ○ Semi-ovate, 20, *603*

Panaeolus
acuminatus, 604
campanulatus, 87, 602
phalaenarum, 603
retirugis, 603
semiovatus, 20, 603
separatus, 603
subbalteatus, 91, 604

○ Panellus, Luminescent, 501, 790

Panellus
mitis, 744
patellaris, 799
serotinus, 498, 789
stipticus, 501, 790

○ Panther, 130, 131, 679, 541

Panus
- ○ Ruddy, 505, *791*
- ○ Veiled, 503, *798*

Panus
operculatus, 799
rudis, 505, 791
stipticus, 790
strigosus, 761, 767

○ Parasol, 172, *520*
○ Shaggy, 171, *521*

Parchment
- ○ Bleeding Conifer, 541, 494
- ○ Bristly, 591, 495
- ○ Ceramic, 572, 498
- ○ Conifer, 583, 422
- ○ Crowded, 545, 496
- ○ Hairy, 537, 496
- ○ Silky, 574, 497
- ○ Two-toned, 573, 420

Parchment Fungus Family, 494

○ Patches, Yellow, 136, 139, *534*

○ Pax, Velvet-footed, 322, *670*

Paxillaceae, *649, 668*

Paxillus
- ○ Poison, 287, *671*
- ○ Stalkless, 499, *672*

Paxillus
atrotomentosus, 322, 670
corrugatus, 672
involutus, 287, 671
panuoides, 499, 672
vernalis, 671

Paxillus Family, 668

Paxina acetabulum, 333

Peckiella viridis, 373

Peniophora
gigantea, 583, 422
rufa, 565, 423

Peziza
acetabula, 333
badia, 347
badio-confusa, 625, 347
cerea, 348
macropus, 335
pendulus, 485
perlata, 331
repanda, 620, 347
sylvestris, 348
varia, 348
venosa, 330
vesciculosa, 601, 348

Pezizaceae, *326, 346*

Pezizales, *325*

Phaeocollybia
- ○ Kit's, 89, *634*
- ○ Pretty, 88, *634*

Phaeocollybia
californica, 635
fallax, 88, 634
scatesiae, 89, 634

Phaeohelotium flavum, 362

Phaeolepiota aurea, 192, 523

Phaeolus schweinitzii, 480, 473

Phaeomarasmius confragosus, 639
erinaceellus, 190, 711

Phaeophlebia strigoso-zonata, 425

Phallaceae, 834

Phallales, 831

Phallogaster saccatus, 701, 834

Phallus
duplicatus, 835
impudicus, 835, 836
ravenelii, 706, 835

Phanerochaete chrysorhiza, 558, 423

Phellinus, 438, 440
chrysoloma, 579, 474
everhartii, 476, 477
gilvus, 517, 475
igniarius, 530, 475
laevigatus, 476
pini, 474
pini var. *abietis*, 474
pomaceus, 476
rimosus, 476
robustus, 476, 477

Phellodon
atratus, 436
niger, 468, 436
niger var. *alboniger*, 436
tomentosus, 436

Phellorinia herculeana, 815

Phlebia
○ Radiating, 587, 424
○ Zoned, 580, 425

Phlebia
gigantea, 423
radiata, 587, 424
strigoso-zonata, 425

Phlogiotis helvelloides, 422, 383

Pholiota
○ Bitter, 66, 712
○ Burnt-ground, 189, 713
○ Changing, 227, 715
○ Destructive, 208, 714
○ Golden, 186, 712
○ Golden False, 192, 523
○ Ground, 181, 718
○ Powder-scale, 190, 711
○ Scaly, 187, 716
○ Sharp-scaly, 184, 188, 717
○ Yellow, 185, 715

Pholiota, 556
astragalina, 66, 712
aurea, 524
aurivella, 186, 712
autumnalis, 621
carbonaria, 189, 713
confragosa, 639
destruens, 208, 714
erinaceella, 711
flammans, 185, 715
fulvozonata, 714
hiemalis, 713
highlandensis, 714
limonella, 713
mutabilis, 227, 715
praecox, 558
squarrosa, 187, 716
squarrosoides, 184, 188, 717
subcaerulea, 726
terrestris, 181, 718
terrigena, 634
veris, 716
vernalis, 716

Pholiotina filaris, 560

Phylloporus, 668
rhodoxanthus, 324, 326, 672

Phyllotopsis nidulans, 500, 791

Physacaceae, 845

Physalacria inflata, 724, 411

Physarales, 844

○ **Physarum, Ashy,** 847

*Physarum
cinereum,* 847
polycephalum, 561, 562, 847

○ **Pig's Ears,** 618, *331*

Piptoporus betulinus, 477

Pisolithus tinctorius, 698, 838

Plated Puffball Fungi, *815*

Plectania melastoma, 609, *341*

Pleurocybella porrigens, 493, *192,* 794

*Pleurotus
applicatus,* 795
candidissimus, 744
corticatus, 793
dryinus, 495, 792
elongatipes, 761
nidulans, 792
ostreatus, 484, 497, 793
petaloides, 761
porrigens, 792
sapidus, 794
serotinus, 790
subpalmatus, 796
sulfureoides, 303, 809
ulmarius, 761

Plicaturopsis crispa, 472, 493

Pluteaceae, 673

Pluteus
○ Black-edged, 233, 674
○ Golden Granular, 230, 675
○ Pleated, 234, 676
○ Yellow, 229, 673

*Pluteus
admirabilis,* 229, 673
atricapillus, 676
atromarginatus, 233, 674
aurantiorugosus, 230, 675
californicus, 638
caloceps, 675
cervinus, 231, 232, 675
cervinus var. *alba,* 676
coccineus, 675
flavofuligineus, 674
leoninus, 674
longistriatus, 234, 676
magnus, 676
seticeps, 677
umbrosus, 674

Pluteus Family, *673*

Podaxaceae, *814*

Podaxales, *814*

Podaxis pistillaris, 702, 814

○ **Poison Pie,** 259, 624

Polyozellus multiplex, 444, 397

Polyporaceae, *438, 439*

Polypore
○ Beefsteak, 513, *455*
○ Berkeley's, 476, 477, *447*
○ Birch, 491, *477*
○ Birch Crust, *476*
○ Bitter Iodine, 465, *480*
○ Black-footed, 543, *478*
○ Black-staining, 471, *470*
○ Blue Cheese, 522, *490*
○ Blue-pored, 457, *441*
○ Bondarzew's, 463, *448*
○ Bone, 461, *472*
○ Cinnabar-red, 512, *486*
○ Clinker, 597, *467*

- ◯ Conifer-base, 519, *423, 465*
- ◯ Cracked-cap, 533, *476*
- ◯ Crested, 454, *442*
- ◯ Dye, 480, *473*
- ◯ Elegant, 453, *483*
- ◯ Flecked-flesh, 530, *475*
- ◯ Flett's, 462, *443*
- ◯ Fruit Tree, *476*
- ◯ Gelatinous-pored, 590, *449*
- ◯ Golden Spreading, 579, *474*
- ◯ Green's, 450, *452*
- ◯ Hexagonal-pored, 508, *455*
- ◯ Larch, 529, *458*
- ◯ Little Nest, 485, *485*
- ◯ Milk-white Toothed, 584, *467*
- ◯ Montagne's, 449, *451*
- ◯ Mossy Maple, 521, *472*
- ◯ Mossy Maze, 540, *450*
- ◯ Multicolor Gill, 535, *469*
- ◯ Mustard-yellow, 517, *475*
- ◯ Ochre-orange Hoof, *474*
- ◯ Orange Sponge, 747, *486*
- ◯ Pale Beefsteak, 374, *456*
- ◯ Pendulous-disc, 489, *484*
- ◯ Red-belted, 531, *459*
- ◯ Resinous, 525, *468*
- ◯ Resinous Conifer, *468*
- ◯ Rooting, 419, 447, *480*
- ◯ Rosy, 524, *458*
- ◯ Scaly Yellow, 420, 456, *442*
- ◯ Sheep, 460, *444*
- ◯ Shiny Cinnamon, 445, *450*
- ◯ Smoky, 582, *445*
- ◯ Split-pore, 585, *487*
- ◯ Spongy Toothed, 546, *488*
- ◯ Spring, 509, *478*
- ◯ Staining Cheese, 589, *492*
- ◯ Tender Nesting, 510, *464*
- ◯ Thick-maze Oak, 467, *453*
- ◯ Thin-maze Flat, 481, *454*
- ◯ Tinder, 527, *457*
- ◯ Umbrella, 424, 473, *482*
- ◯ Veiled, 528, *452*
- ◯ Violet Toothed, 538, *490*
- ◯ Warted Oak, 534, *466*
- ◯ White Cheese, 490, *491*
- ◯ White Spongy, 526, *488*
- ◯ Winter, 375, *479*
- ◯ Woolly Velvet, 448, *471*
- ◯ Yellow-red Gill, 516, *463*

Polypore Family, *439*

Polyporus, 472
adustus, 446
albellus, 491
alboluteus, 486
arcularius, 509, 478
badius, 543, 478
berkeleyi, 448
betulinus, 477
brumalis, 375, 479
caeruleoporus, 441
caesius, 491
cinnabarinus, 487
cinnamomeus, 451
conchifer, 485
cristatus, 442
dichrous, 449
dryadeus, 466
elegans, 483
ellisii, 443
fagicola, 482
flettii, 444
fragilis, 492
frondosus, 464
giganteus, 471
gilvus, 475
graveolens, 463
griseus, 447
hirtus, 465, 480
illudens, 464
lentus, 482
leucospongius, 489
mcmurphyi, 482
melanopus, 479, 481, 483
montagnei, 452
montanus, 449
mori, 455
nidulans, 465
osseus, 472
ovinus, 445
pargamenus, 490

pescaprae, 445
picipes, 479
pocula, 485
radicatus, 419, 447, 480
resinosus, 468
schweinitzii, 474
squamosus, 507, 481
sulphureus, 469
tuberaster, 481
tulipiferae, 468
umbellatus, 424, 473, 482
varius, 453, 483
versicolor, 489
volvatus, 453

Pore Fungi, 386

○ Poria, Orange, 581, 484

Poria
mutans, 484
obliqua, 467
spissa, 581, 484
taxicola, 484
versipora, 487

Porodisculus pendulus, 489, 484

Poronidulus conchifer, 485, 485

Porphyrellus, 594

Porpoloma umbrosum, 258, 794

Pouzarella
babingtonii, 648
dysthales, 648
nodospora, 100, 648
strigosissima, 648

○ Prince, 150, 502

Psathyrella
○ Clustered, 93, 607
○ Common, 25, 604
○ Corrugated-cap, 92, 608
○ Parasitic, 21, 606
○ Red-gilled, 107, 605
○ Ringed, 96, 608
○ Velvety, 99, 609

○ Wrinkled-cap, 609

Psathyrella, 599
candolleana, 25, 604
castaneifolia, 607
conissans, 107, 605
delineata, 609
epimyces, 21, 606
foenisecii, 26, 606
hydrophila, 93, 607
longistriata, 96, 608
rugocephala, 92, 608
spadicea, 605
sublateritia, 605
velutina, 99, 609

Pseudoarmillariella
ectypoides, 747

Pseudocolus
fusiformis, 833
javanicus, 833
schellenbergiae, 695, 833

Pseudocoprinus, 599

Pseudofistulina, 440
radicata, 374, 456

Pseudohydnum gelatinosum, 459, 383

Pseudoplectania nigrella, 612, 341

Psilocybe
○ Blue-foot, 33, 719
○ Bluing, 31, 721
○ Common Large, 226, 721
○ Conifer, 44, 723
○ Dung-loving, 82, 720
○ Mountain Moss, 83, 722
○ Potent, 86, 719
○ Scaly-stalked, 55, 724

Psilocybe
baeocystis, 86, 719
caerulipes, 33, 719
coprophila, 82, 720
cubensis, 226, 721
cyanescens, 31, 721
merdaria, 720
montana, 83, 722

pelliculosa, 44, *723*
semilanceata, **85**, *723*
silvatica, *723*, *724*
squamosa, 55, *724*
strictipes, *719*, *722*
stuntzii, 84, 95, *725*
thrausta, *724*

Pterula, 376

Ptychoverpa bohemica, *329*

Puffball(s), 816
Buried-stalk, 730, *842*
Desert Stalked, 731, *841*
Dye-maker's False, 698, *838*
False, *836*
Gem-studded, 652, 676, *825*
Giant, 647, *823*
Gilled, *811*
Granular, 646, *817*
Orange-staining, 655, *824*
Pear-shaped, 664, *826*
Pigskin Poison, 654, *839*
Plated, 657, *815*
Purple-spored, 653, *822*
Sculptured, 650, *820*
Skull-shaped, 656, *822*
Spiny, 649, *824*
Stalked, *840*
Tough, 651, *826*
Tumbling, 641, *820*
Western Giant, 648, *821*

Puffball-in-aspic, Stalked, 700, *841*

Puffbowl, Western Lawn, 645, *827*

Pulcherricium caeruleum, 595, *420*

Pulveroboletus ravenelii, 417, *579*

Punctulariaceae, *425*

Punctularia strigoso-zonata, 580, *425*

Pustularia rosea, *354*

Pycnoporellus alboluteus, 747, *486*

Pycnoporus
cinnabarinus, 512, *486*
sanguineus, *487*

Pyrenomycetes, 367

Pyronema omphalodes, *352*

Pyronemataceae, *326*, *349*

R

Rabbit Ears, Yellow, 600, *351*

Ramaria
araiospora, 750, *406*
araiospora var. *rubella*, *407*
aurea, *408*
botrytis, 749, *407*
formosa, 746, *408*
gelatinosa var. *gelatinosa*, *409*
gelatinosa var. *oregonensis*, 745, *408*
strasseri, *407*
stricta, 733, *409*
stricta var. *concolor*, *410*
stricta var. *stricta*, *410*
subbotrytis, *407*

Ramariopsis kunzei, 742, *410*

Resupinatus applicatus, 486, *795*

Reticulariaceae, *848*

Rhizina
inflata, *332*
undulata, 619, *331*

Rhizopogon, Western, 660, *813*

Rhizopogon
luteolus, *813*
occidentalis, 660, *813*

Rhizopogonaceae, *813*

Rhodocollybia, 756

○ **Rhodocybe, Cracked-cap, 283,** *648*

Rhodocybe mundula, **283,** *648*

Rhodophyllus, 640

○ **Rhodotus, Netted, 666,** *795*

Rhodotus palmatus, **666,** *795*

Rickenella fibula, **59,** *796*

○ **Ripartites, Fringed, 243,** *796*

Ripartites tricholoma, **243,** *796*

Rot
○ Dry, **593,** *415*
○ Wet, **592,** *415*

Rozites caperata, **201,** *635*

Russula
○ Almond-scented, **306,** *704*
○ American False, **662,** *812*
○ Blackish-red, **329,** *703*
○ Emetic, **328,** *701*
○ Firm, **286,** *699*
○ Fragile, **343,** *702*
○ Graying Yellow, **321,** *698*
○ Green Quilt, **356,** *700*
○ Purple-bloom, **339,** *705*
○ Red-and-black, **249,** *701*
○ Rosy, **334,** *705*
○ Shellfish-scented, **330,** 707
○ Short-stalked White, **252,** *698*
○ Tacky Green, **357,** *697*
○ Variable, **348,** *706*

Russula, 449
aeruginea, **357,** *697*
atropurpurea, 704
brevipes, **252,** *698*
cascadensis, 698
claroflava, **321,** *698*

compacta, **286,** *699*
crustosa, **356,** *700*
cyanoxantha var. *variata,* 707
decolorans, 700
delica, 698
densifolia, 701
dissimulans, **249,** *701*
emetica, **328,** *701*
flava, 699
fragilis, **343,** *702*
fragrantissima, 704
krombholzii, **329,** *703*
laurocerasi, **306,** *704*
lutea, 699
mariae, **339,** *705*
nigricans, 701
ochroleuca, 699
olivacea, 697
rosacea, **334,** *705*
silvicola, 702
subfoetens, 705
variata, **348,** *706*
ventricosipes, 705
virescens, 700
xerampelina, **330,** *707*

Russulaceae, *499, 679*

Russula Family, *679*

Russulales, *449*

S

○ **Saddle, Dryad's, 507,** *481*

Sarcoscypha
coccinea, **605,** *343*
floccosa, 343
occidentalis, **606,** *344*

Sarcoscyphaceae, *326, 343*

Sarcosoma
globosa, 342
latahensis, **611,** *342*
mexicana, 342

Sarcosomataceae, *326, 339*

Sarcosphaera
coronaria, 347

crassa, 640, *346*
eximia, 347

Scaber Stalk
○ Aspen, 389, *578*
○ Common, 390, *578*
○ Red-capped, 388, *577*

Schizophyllaceae, *492*

Schizophyllum commune,
487, *493*

Schizophyllum Family,
492

○ Scleroderma, Earthstar,
642, *839*

*Scleroderma
aurantium,* 654, *839*
citrinum, 654, *839*
geaster, 642, *839*

Sclerodermataceae, *838*

Sclerodermatales, *836*

Sclerotinia, 360

Sclerotiniaceae, *360*

Scutellinia scutellata, 604,
353

*Serpula
himantioides, 416*
lacrimans, 593, *415*

○ Shaggy Mane, 704, *597*

○ Shelf, Hemlock Varnish,
515, *461*

*Simblum sphaerocephalum,
833*

○ Simocybe, American, 29,
638

Simocybe centunculus, 29,
638

Slime
○ Carnival Candy, 555, *850*
○ Chocolate Tube, 550, 736,
853

○ Coral, 740, *845*
○ Insect-egg, 554, *846*
○ Japanese-lantern, 732, *849*
○ Many-headed, 561, 562,
847
○ Morel-like, *845*
○ Multigoblet, 556, *852*
○ Pretzel, 560, *851*
○ Red Raspberry, 751, *848*
○ Scrambled-egg, 559, *845*
○ Tapioca, 594, 596, *852*
○ White-footed, 552, *853*
○ Wolf's-milk, 665, *848*
○ Yellow-fuzz Cone, 553,
557, *850*

Slime Molds, *843*

○ Slippery Jack, 401, *586*

○ Slippery Jill, 371, *589*

*Sparassis
crispa,* 756, *411*
herbstii, 412
radicata, 755, *412*

*Spathularia
flavida, 360*
spathulata, 360
velutipes, 693, *359*

Spathulariopsis velutipes, 360

Sphaeriales, *367*

Sphaerobolaceae, *830*

Sphaerobolus stellatus, 551,
830

Spongipellis pachyodon, 546,
488

Spongiporus leucospongia,
526, *488*

Spragueola irregularis, 359

○ Spread, Velvet Blue,
595, *420*

○ Squamanita, Knobbed,
193, *524*

Squamanita
odorata, 525
paradoxum, 512
umbonata, 193, 524

○ Squid, Stinky, 695, 833

○ Stain, Green, 598, 361

Stalked Puffballs, 840

Steccherinum, 436
ochraceum, **588**, 437
septentrionale, 428

Stemonitaceae, 852

Stemonitales, 844

Stemonitis
axifera, 854
fusca, 854
splendens, 550, 736, 853

Stereaceae, 494

○ Stereum, Stalked, 436, 498

Stereum
cinerascens, 495
complicatum, 545, 496
diaphanum, 499
fasciatum, 497
frustulosum, 498
fuscum, 421
hirsutum, 537, 496
lobatum, 497
ostrea, 536, 497
purpureum, 418
rameale, 496
rufa, 423
sanguinolentum, 495
sericeum, 498
striatum, 574, 497
strigoso-zonata, 425
versicolor, 497

Stictidaceae, 366

Stinkhorn(s), 831
○ Club-shaped, 701, 834
○ Columned, 694, 832
○ Elegant, 692, 835

○ Netted, 707, 834
○ Ravenel's, 706, 835

Stomach Fungi, 810

Strobilomyces
confusus, 581
floccopus, 379, 580

Strobilurus
conigenoides, 7, 797
kemptonae, 798
trullisatus, 8, 797

Stropharia
○ Garland, 215, 727
○ Green, 198, 725
○ Hard's, 202, 728
○ Lacerated, 200, 728
○ Questionable, 197, 726
○ Round, 46, 730
○ Wine-cap, 203, 204, 729

Stropharia
aeruginosa, 198, 725
ambigua, 197, 726
coronilla, 215, 727
hardii, 202, 728
hornemannii, 200, 728
riparia, 727
rugosoannulata, 203, 204, 729
semiglobata, 46, 730
squamosa, 724

Strophariaceae, 708

Stropharia Family, 708

Suillus
○ Blue-staining, 394, 582
○ Chicken-fat, 416, 581
○ Dotted-stalk, 376, 584
○ Hollow-stalked Larch, 393, 583
○ Larch, 406, 584
○ Painted, 391, 587
○ Pungent, 386, 588
○ Short-stalked, 387, 582
○ Tomentose, 415, 590
○ Western Painted, 392, 585
○ White, 372, 587

Suillus
acidus, 588
albidipes, 584
americanus, 416, 581
brevipes, 387, 582
caerulescens, 394, 582
cavipes, 393, 583
cothurnatus, 586, 589
granulatus, 376, 584
grevillei, 406, 584
lakei, 392, 585
luteus, 401, 586
pictus, 391, 587
pinorigidus, 589
placidus, 372, 587
ponderosus, 583
proximus, 585
pseudobrevipes, 582
punctatipes, 584
pungens, 386, 588
sibiricus, 581
subluteus, 371, 589
tomentosus, 415, 590

○ **Swamp Beacon, 684,** *358*

T

Tarzetta
cupularis, 626, 353
rosea, **608,** *354*

Tectella patellaris, **503,** *798*

Thelephora
griseozonata, 413
intybacea, 413
terrestris, 437, 413
vialis, 442, 413

Thelephoraceae, 437

Thelephore, Vase, 442,
413

Thimble-cap
○ Smooth, **708,** *329*
○ Wrinkled, 712, *329*

Tooth
○ Bearded, 547, *430*
○ Bear's Head, **548,** *429*

○ Black, **468,** *436*
○ Bluish, 469, *432*
○ Comb, 549, *431*
○ Jelly, 459, *383*
○ Kidney-shaped, 542, *435*
○ Northern, 520, *427*
○ Ochre Spreading, 588, *437*
○ Orange Rough-cap, 479, *431*
○ Pinecone, 373, *426*
○ Red-juice, 470, *433*
○ Scaly, 466, *434*
○ Spongy-footed, 446, *434*
○ Spreading Yellow, 558, *423*
○ Sweet, 455, *428*

○ **Tooth Fungus Family,** *426*

Trametes
hirsuta, 489
velutina, 489
versicolor, **482,** 489

○ **Tree-Ear, 617,** *380*

Tremella
foliacea, 754, *384*
frondosa, 384
lutescens, 385
mesenterica, 566, *385*
mycetophila, 419

Tremellaceae, *382*

Tremellales, 379, *382*

Tremellodendron
candidum, 386
pallidum, 741, *385*
schweinitzii, 386

Trich
○ Canary, 323, *800*
○ Dirty, 355, *801*
◔ Fibril, 351, *806*
○ Poplar, 257, *802*
○ Red-brown, 296, *802*
○ Russet-scaly, 270, *805*
○ Separating, 319, *804*
○ Shingled, 281, *800*
○ Soapy, 358, *804*

○ Sticky Gray, 353, *803*
○ Veiled, 297, *799*

Trichaptum
abietinus, 490
biformis, **538,** *490*

Trichiaceae, 850

Trichiales, *844*

Trichoglossum, *358*

Trichoglossum
hirsutum, **682,** *357*
velutipes, 357

○ **Tricholoma, Amyloid,**
258, *794*

Tricholoma
aurantium, **297,** *799*
caligatum, 733
carneum, 740
equestre, 800
flavovirens, **323,** *800*
imbricatum, **281,** *800*
leucophyllum, 800, 805
melaleuca, 778
nudum, 750
pardinum, **355,** *801*
personatum, 750
pessundatum, **296,** *802*
ponderosum, 734
populinum, **257,** *802*
portentosum, **353,** *803*
russula, 667
rutilans, 808
saponaceum, **358,** *804*
sejunctum, **319,** *804*
sulphureum, 800, 805
terreum, 806
umbrosum, 794
vaccinum, **270,** *805*
virgatum, **351,** *806*
zelleri, 735

Tricholoma Family, *731*

Tricholomataceae, *731*

Tricholomopsis
decora, **302,** *807*
platyphylla, **265,** *807*

rutilans, **327,** *808*
sulfureoides, **303,** *809*

Trogia crispa, 493

Truffle(s), *354*
○ Fuzzy False, **643,** *350*
○ Oregon White, **659,** *355*

○ **Trumpet, Black, 443,**
394

○ **Trunc, Stalked Yellow,**
661, *813*

Truncocolumella citrina,
661, *813*

Tubaria
○ Fringed, **78,** *639*
○ Ringed, **224,** *638*

Tubaria
confragosa, **224,** *638*
furfuracea, **78,** *639*
pellucida, 640

Tuberaceae, *355*

Tuberales, *354*

Tuber gibbosum, **659,** *355*

Tubifera ferruginosa, **751,**
848

○ **Tuft, Sulfur, 61,** *709*

Tulostoma simulans, **730,**
842

Tulostomataceae, *731*

Tulostomatales, *840*

Tuning Fork
○ Clublike, *381*
○ Yellow, **735,** *380*

○ **Turkey-tail, 482,** *489*
○ False, **536,** *497*

Tylopilus
alboater, **377,** *590*
ballouii, **397,** *591*
chromapes, **410,** *591*
eximius, **380,** *592*
felleus, **382,** *593*

gracilis, **385**, *594*
indecisus, *593*
plumbeoviolaceus, **381**, *594*
porphyrosporus, *595*
pseudoscaber, **378**, *595*
rubrobrunneus, *592*

Tyromyces
caesius, **522**, *490*
chioneus, *490*, *491*
fragilis, **589**, *492*
lacteus, *491*
mollis, *492*
tephroleucus, *491*

U

Underwoodia columnaris,
696, *332*

Urn
○ Devil's, **613**, *342*
○ Jellylike Black, **609**, *341*

Urnula craterium, **613**, *342*

Ustulina deusta, **669**, *375*

V

Vascellum
depressum, *827*
pratense, **645**, *827*

○ Vase, Common Fiber,
437, *413*

○ Veil, Western Yellow,
145, *526*

○ Velvet Foot, **63**, *759*

Verpa
bohemica, **712**, *329*
conica, **708**, *329*

Vibrissea truncorum, **729**,
366

Volvariella
○ Parasitic, **111**, *679*
○ Smooth, **118**, *678*
○ Tiny, **110**, *678*
○ Tree, **109**, *677*

Volvariella
bombycina, **109**, *677*
pusilla, **110**, *678*
speciosa, **118**, *678*
surrecta, **111**, *679*
volvacea, *677*

W

Waxy Cap
○ Chanterelle, **56**, *656*
○ Dusky, **365**, *656*
○ Fading Scarlet, **69**, *663*
○ Golden, **48**, *660*
○ Golden-spotted, **260**, *657*
○ Gray Almond, **360**, *654*
○ Inocybelike, **366**, *662*
○ Larch, **325**, *667*
○ Late Fall, **304**, *661*
○ Orange-gilled, **53**, *662*
○ Russulalike, **345**, *667*
○ Salmon, **309**, *664*
○ Scarlet, **65**, *658*
○ Slimy-sheathed, **199**, *664*
○ Tawny Almond, **273**, *655*
○ Turpentine, **246**, *666*
○ White, **236**, *659*
○ Yellow-centered, **261**, *660*

○ Witches' Butter, **566**, *385*
○ Yellow, *385*

○ Witch's Hat, **71**, **72**, *658*

Wynnea
americana, **615**, *345*
sparassoides, **753**, *345*

X

Xeromphalina campanella,
58, *809*

Xerula chrysopepla, *759*

Xylaria
hypoxylon, **738**, *375*
polymorpha, **697**, *376*

Xylariaceae, *374*

Xylobolus frustulatus, **572**,
498

STAFF

Prepared and produced by
Chanticleer Press, Inc.

Publisher: Paul Steiner
Editor-in-Chief: Gudrun Buettner
Executive Editor: Susan Costello
Managing Editor: Jane Opper
Project Editor: Mary Beth Brewer
Associate Editor: Ann Whitman
Art Director: Carol Nehring
Art Associate: Ayn Svoboda
Production: Helga Lose
Picture Library: Edward Douglas,
Dana Pomfret
Silhouettes: Paul Singer
Drawings: Bunji Tagawa

Design: Massimo Vignelli

THE AUDUBON SOCIETY FIELD GUIDE SERIES

Also available in this unique, all-color, all-photographic format:

Birds (*Eastern Region*)

Birds (*Western Region*)

Butterflies

Fishes, Whales, and Dolphins

Fossils

Insects and Spiders

Mammals

Reptiles and Amphibians

Rocks and Minerals

Seashells

Seashore Creatures

Trees (*Eastern Region*)

Trees (*Western Region*)

Wildflowers (*Eastern Region*)

Wildflowers (*Western Region*)